Ecological Studies, Vol. 146

Analysis and Synthesis

Ecological Studies

Volumes published since 1994 are listed at the end of this book.

Springer

Berlin
Heidelberg
New York
Barcelona
Hong Kong
London
Milan
Paris
Singapore
Tokyo

S. Porembski W. Barthlott (Eds.)

Inselbergs

Biotic Diversity of Isolated Rock Outcrops in Tropical and Temperate Regions

With 157 Figures, 15 in Color, and 47 Tables

Springer

Prof. Dr. Stefan Porembski
Universität Rostock
Institut für Biodiversitätsforschung
Allgemeine und Spezielle Botanik
Wismarsche Str. 8
18051 Rostock
Germany

Prof. Dr. Wilhelm Barthlott
Universität Bonn
Botanisches Institut
Meckenheimer Allee 170
53115 Bonn
Germany

Cover illustration: Inselberg near Camponario, Brazil.
Photograph taken by Stefan Porembski

ISSN 0070-8356
ISBN 978-3-642-64120-6 Springer-Verlag Berlin Heidelberg New York

Library of Congress Cataloging-in-Publication Data.

Inselbergs : biotic diversity of isolated rock outcrops in tropical and temperate regions /
S. Porembski, W. Barthlott (eds.)
p. cm. – (Ecological studies, ISSN 0070-8356 ; v. 146)
Includes bibliographical references.

ISBN-13: 978-3-642-64120-6 e-ISBN-13: 978-3-642-59773-2
DOI: 10.1007/978-3-642-59773-2

1. Rock plants–Ecophysiology. 2. Inselbergs. 3. Plant communities. I. Porembski, S. (Stefan), 1960- II. Barthlott, Wilhelm. III. Series
QK938.R6 I57 2000
581.7'53–dc21 00-055698

Springer-Verlag Berlin Heidelberg New York
a member of BertelsmannSpringer Science+Business Media GmbH

Softcover reprint of the hardcover 1st edition 2000

Cover design: *design & production* GmbH, Heidelberg
Typesetting: Bader • Damm • Kröner, Heidelberg

SPIN 10575065 31/3130 YK – 5 4 3 2 1 0 – Printed on acid free paper

Preface

Biodiversity is a highly specific quality of our planet and its enormous extent was recognized only very recently. Today, the term biodiversity is in the focus of discussions within political, cultural and biological contexts. Despite its ubiquity in public awareness, remarkably little is known about the essential components of biodiversity, e.g., the number of species and the mechanisms which control the species richness of ecosystems. Estimates of global species numbers range between 5 and 80 millions with no exact data available up to now. There is no doubt, however, that species richness increases from the polar towards the equatorial regions, with the tropical rainforests forming the most speciose ecosystem on earth. Human-caused destruction of tropical rainforests has reached dramatic dimensions during the past decades, resulting in the conversion of previously contiguous forests into a patchwork consisting of pastures, plantations, and various stages of secondary forests. Only rough estimates are possible about the accompanying loss of species, since detailed knowledge about extinction rates does not exist for most groups of organisms. That reliable data on different aspects of biodiversity are largely lacking is widely accepted as a serious deficit, and this circumstance has been lamented by both biologists and politicians. Consequently, the CONVENTION ON BIOLOGICAL DIVERSITY (CBD, Rio de Janeiro 1992) emphasized the responsibility of mankind to protect and to use earths ecosystems in a sustainable way. A major precondition to reach this target is a more comprehensive understanding of ecological properties that set apart the species-rich tropical biomes from their less diverse counterparts at higher latitudes.

The task of identifying significant factors controlling the spatially highly variable distribution of species diversity is considerably simplified by concentrating on certain groups of organisms or distinct ecosystems. In terrestrial biota vascular plants form the most relevant (e.g., in terms of biomass production, contribution to ecosystem structure) and relatively easy to investigate entity which is comparatively well known and thus provides a highly appropriate group for the purpose of biodiversity studies.

Inselbergs (from German Insel = island and Berg = mountain) represent a very distinct ecosystem and form excellent model systems for addressing central questions of biodiversity research (e.g., minimum size of populations, consequences of fragmentation). These frequently huge, millions of years-old monoliths bear a unique and somewhat bizarre vegetation and are ecologically clearly circumscribed. Radiating into temperate regions, they form in particular in the tropics a geologically almost uniform system which is highly suited for comparative analyses throughout all continents. In being remarkable landscape elements, inselbergs play an important role in various cultures for religious motives. They are thus often sanctuaries or forbidden places like the Mont Niénokoué which is feared as a montagne fétiche. During our first attempt to climb this monolithic outcrop, situated in dense primary rainforest in the southwest of the Côte d'Ivoire, our local guides refused to accompany us for fear of a punishment by their gods.

During his journey along the Orinoco river, the great Alexander von Humboldt (1819) was fascinated by the blackish coloration of the inselbergs in this region. He collected rock samples which he presented to the British Museum in London. According to the chemist Children, the blackish coloration was due to a cover of manganese and iron oxide. This, however, proved to be wrong, as it was noticed later that cryptogams (i.e., cyanobacteria and lichens) are responsible for the characteristic dark color of inselberg rocks.

Historically, it is hardly understandable that 200 years had to pass after the observations of A. v. Humboldt before the extraordinary biology of the ecosystem inselberg could be presented within this monographic treatment. One can only speculate about the lack in attractiveness for biologists of the seemingly bare and hardly accessible monoliths. How could we otherwise understand that Charles Darwin did not mention them in his journal on the occasion of his visit in spring 1832 to Rio de Janeiro, which is surrounded by a most scenic inselberg landscape? He could not imagine that the inselbergs situated in the wet forests of the Mata Atlântica, like the Pão do Açúcar, harbor endemics whose next relatives occur in semiarid and arid regions far away. Forming islands surrounded by rainforest, they bear dry-adapted cacti of the genus *Coleocephalocereus*, which is in full adaptive radiation like Darwin's finches and giant tortoises of the Galápagos archipelago.

Similarly, subsequent generations of biologists have largely ignored tropical inselbergs. As a result, during the first decades of the 20th century only a few studies (e.g., Engler 1910; Mildbraed 1922) were occupied with this topic. For geologists and geomorphologists, however, inselbergs attracted continuous scientific interest over a long period (see reviews in Bremer and Jennings 1978). It is therefore not by chance that the term inselberg was coined by the geologist Wilhelm Bornhardt (1900) during his work in East Africa. In contrast to the situation in the tropics in the USA and in Australia, granitic outcrops have long been studied extensively in an ecosystemary context (see surveys by Quarterman et al. 1993; Shure 1999; Hopper et al. 1997).

In 1973 the coeditor of this volume (W.B.) had a first opportunity to study inselbergs in Brazil and subsequently (1976) in the Côte d'Ivoire within a long-term research project. It was fascination from the first moment – which increased in the following years when visiting inselbergs in Madagascar and Brazil in the 1980s. However, only in 1990 in cooperation with the senior editor was a thorough research program established and we became fully aware of the scientific potential of inselbergs as a model for comparative biodiversity analyses on a broad geographic scale, which allows addressing general questions of biodiversity research in an interdisciplinary manner. It was a unique and exciting opportunity to analyze a globally more or less uniform azonal ecosystem – which, due to its low agricultural potential, frequently comprises in many regions the last unaltered habitats not dominated by humans (Porembski et al. 1998). In total, some 600 inselbergs in Africa (Bénin, Burkina Faso, Congo, Côte d'Ivoire, Equatorial Guinea, Guinea, Malawi, Rwanda, Zimbabwe), on Madagascar and the Seychelles, in Australia and South America (Bolivia, Brazil, French Guiana, Venezuela) were investigated by our inselberg working group. This monographic treatment is aimed as a summary of our knowledge about Inselbergs: Biotic Diversity of Isolated Rock Outcrops in Tropical and Temperate Regions. Apart from the recently published report on the Granite Outcrops Symposium held at the University of Western Australia in 1996 (Withers and Hopper 1997), it forms the first comprehensive account of the various abiotic and biotic characteristics of granitic and gneissic inselbergs that covers a nearly global scale.

In order to provide our readers with as much information about inselbergs as possible this work is accompanied by colour plates which are aimed to provide vivid examples of characteristic plants and habitats on inselbergs throughout the world. The colour plates can be found behind the Preface.

Detailed information on flora and vegetation of the inselbergs studied in the framework of this book is available under the following internet address: http://www.uni-rostock.de/fakult/manafak/biologie/abt/botanik/inselbergs/indexeng.html

Editors of books like this obviously owe a great debt to a large number of individuals and institutions. We acknowledge the funding of our research project (Vegetation of inselbergs: structure, diversity and ecogeographical differentiation of a tropical plant community) by the

Deutsche Forschungsgemeinschaft (DFG) from 1990 to 1997. This project was embedded within an interdisciplinary program (Mechanisms of maintaining tropical diversity) initiated and chaired by K.E. Linsenmair (Würzburg), which became a focal point of German biodiversity research in the tropics. This program has been of outstanding importance for various biological disciplines and has contributed considerably to the capacity building among younger German scientists. Without the initiative and competence of its chairman, his ability to encourage colleagues, and to persuade financial supporters this program would never have been so successful. We should thus like to express our sincere thanks to Prof. Dr. Karl Eduard Linsenmair.

Second, we thank our authors and coauthors (including to some extend our own postgraduates within our inselberg project) for their contributions and making their data available for publication.

The editors specifically thank many colleagues who provided invaluable assistance, in particular: Laurent Aké Assi (Abidjan), His Excellency Erik Becker-Becker (Ambassador of Venezuela, Bonn), Robert Faden (Washington), Paul Goetghebeur (Gent), Otto Huber (Caracas), Jean Lejoly (Bruxelles), Sylvia Phillips (Kew), Hildemar Scholz (Berlin-Dahlem), Brice Sinsin (Cotonou), Dieter Supthut (Zürich), Norbert Wilbert (Bonn), and the German embassies in various countries.

We should like to extend our indebtedness to the staff of the Springer-Verlag and are particularly grateful for the patience of Dr. Andrea Schlitzberger.

Finally, a personal acknowledgement should be added to a book that appears exactly at the 100th anniversary of the term inselberg. Our intensified research, which started in 1976 in the Côte d'Ivoire, developed into fruitful cooperation which continues until today. Despite being in full agreement with the Convention on Biological Diversity, this is no longer self-evident. We thus dedicate this book to the Baulé, Sénoufo, Lobi, and all other ethnic groups of the République de la Côte d'Ivoire. We were frequently surprised by their extraordinary knowledge about the species richness of their country and we are persuaded that they will conserve and use sustainably the inherited profusive biodiversity of the Côte d'Ivoire.

Bonn and Rostock
September 2000

Stefan Porembski
Wilhelm Barthlott

References

Barthlott W, Winiger M (eds) (1998) Biodiversity. A challenge for development research and policy. Springer, Berlin Heidelberg New York

Bornhardt W (1900) Zur Oberflächengestaltung und Geologie Deutsch-Ostafrikas. Reimer, Berlin

Bremer H, Jennings J (eds) (1978) Inselbergs/Inselberge. Z Geomorphol, N F Supplem 31

Engler A (1910) Die Pflanzenwelt Afrikas. Engelmann, Leipzig

Hopper SD, Brown AP, Marchant NG (1997) Plants of Western Australian granite outcrops. In: Withers PC, Hopper SD (eds) Granite outcrops symposium. J R Soc West Aust 80:141–158

Humboldt A v (1819) Relation historique du Voyage aux regions équinoxiales du Nouveau Continent, ait en 1799–1804 par A. de Humboldt et A. Bonpland, vol II. Maze, Paris

Mildbraed J (1922) Wissenschaftliche Ergebnisse der zweiten deutschen Zentral-Afrika-Expedition 1910–1911. Band II: Botanik. Klinkhardt & Biermann, Leipzig

Porembski S, Martinelli G, Ohlemüller R, Barthlott W (1998) Diversity and ecology of saxicolous vegetation mats on inselbergs in the Brazilian Atlantic rainforest. Divers Distrib 4:107–119

Quarterman E, Burbanck MP, Shure DJ (1993) Rock outcrop communities: limestone, sandstone, and granite. In: Martin WH, Boyce SG, Echternacht AC (eds) Biodiversity of the southeastern United States. Upland terrestrial communities. Wiley, New York, pp 35–86

Shure DJ (1999) Granite outcrops of the southeastern United States. In: Anderson RC, Fralish JS, Baskin JM (eds) Savannas, barrens, and rock outcrop plant communities of North America. Cambridge University Press, Cambridge, pp 99–118

Withers PC, Hopper SD (eds) (1997) Granite outcrops symposium. J R Soc West Aust 80

Contents

1	**Why Study Inselbergs?**	1
	W. Barthlott, S. Porembski	
1.1	Inselbergs – Model Ecosystems for Biodiversity Studies	1
1.2	Perspectives of Inselberg Research	3
1.3	Linking Inselberg Research to Other Ecosystems	5
References		6
2	**Inselbergs: Geomorphology and Geoecology**	7
	H. Bremer, H. Sander	
2.1	Introduction	7
2.2	Diversity of Forms and Terminology	7
2.3	Genesis of Inselbergs	8
2.4	Ayers Rock	18
2.5	Microforms and Substrata – Examples from the Côte d'Ivoire	20
2.6	Discussion	32
References		34
3	**Xeric Islands: Environmental Conditions on Inselbergs**	37
	J. Szarzynski	
3.1	Introduction	37
3.2	The Study of Microclimate on Inselbergs in the Côte d'Ivoire	38
3.3	Conclusions	46
References		47

4 **Islands on Islands: Habitats on Inselbergs** 49
S. Porembski, U. Becker, R. Seine

4.1 Introduction 49
4.2 Habitat Types on Inselbergs 51
4.3 Floristic Affinities to Other Rock Outcrop Ecosystems . . 65
References 66

5 **Algae, Fungi, and Lichens on Inselbergs** 69
B. Büdel, U. Becker, G. Follmann, K. Sterflinger

5.1 Introduction 69
5.2 Algae 72
5.3 Fungi 75
5.4 Lichens 76
5.5 Habitat Separation 79
5.6 Ecophysiology 83
5.7 Conclusion 87
References 88

6 **Bryophytes** 91
J.-P. Frahm

6.1 Introduction 91
6.2 Floristic Diversity 92
6.3 Habitats 97
6.4 Structural Adaptations 99
6.5 Dispersal 100
6.6 Distribution 101
6.7 Comparison with Flowering Plants 101
References 102

7 **Vascular Plants on Inselbergs: Systematic Overview** . . . 103
W. Barthlott, S. Porembski

7.1 Introduction 103
7.2 Data Availability 104
7.3 Systematic Distribution and Species Numbers 104

7.4 Comparison with Vascular Epiphytes 114
References 115

8 Vascular Plants on Inselbergs: Vegetative and Reproductive Strategies 117
N. Biedinger, S. Porembski, W. Barthlott

8.1 Introduction 117
8.2 Adaptational Aspects of Plant Life- and Growth-Forms on Inselbergs 117
8.3 Reproductive Strategies 133
8.4 Plant Functional Groups and Redundancy in the Vegetation of Inselbergs 138
References 140

9 Ecophysiology of Vascular Plants on Inselbergs 143
M. Kluge, J. Brulfert

9.1 Introduction 143
9.2 Environmental Factors as Stressors 144
9.3 Adaptation of Inselberg Plants to Scarcity of Water 145
9.4 Adaptation of High Irradiance and Temperatures 157
9.5 Adaptation to Deficiency of Mineral Nutrients: Carnivorous Plants 160
9.6 Modes of Photosynthesis in Plants of Mt. Angavokely (Central High Plateau of Madagascar) 161
9.7 Conclusions 168
References 170

10 Variations on One Theme: Regional Floristics of Inselberg Vegetation
Authors according to region

10.1 West African Inselberg Vegetation 177
S. Porembski

10.1.1 Geography and Geology 177
10.1.2 Climate 178
10.1.3 Vegetation 179

10.1.4 Inselberg Vegetation 180
10.1.5 Life-Forms and Ecological Adaptations 196
10.1.6 Systematic and Ecological Affinities 199
10.1.7 Phytogeography and Endemism 203
References 208

10.2 East and Southeast Africa 213
R. Seine, U. Becker

10.2.1 Geography and Geology 213
10.2.2 Climate 213
10.2.3 Vegetation 214
10.2.4 Inselberg Vegetation 216
10.2.5 Life-Forms 228
10.2.6 Vegetative Adaptations 228
10.2.7 Reproductive Adaptations 230
10.2.8 Systematic and Ecological Characteristics 230
10.2.9 Biogeography, Endemism, and Speciation 232
References 234

10.3 The Arid Scenario: Inselbergs in the Namib Desert Are Rich Oases in a Poor Matrix (Namibia and South Africa) 237
N. Jürgens, A. Burke

10.3.1 Geography and Geology 237
10.3.2 Geomorphological Features 238
10.3.3 Soils 241
10.3.4 Climate 242
10.3.5 Flora, Vegetation, and Life-Forms 243
10.3.6 Human Impact and Conservation 255
10.3.7 Research Needs 256
References 257

10.4 Vegetation of Malagasy Inselbergs 259
E. Fischer, I. Theisen

10.4.1 Introduction 259
10.4.2 Geography and Geology 259

10.4.3 Climate 261
10.4.4 Vegetation 261
10.4.5 Inselberg Vegetation 263
10.4.6 Systematic Characteristics 268
10.4.7 Life-Forms and Adaptations 269
10.4.8 Biogeography and Endemism 271
10.4.9 Considerations upon Diversity 274
References 274

10.5 Seychelles 277
N. Biedinger, K. Fleischmann

10.5.1 Introduction 277
10.5.2 Geography and Geology 277
10.5.3 Climate 278
10.5.4 Vegetation of the Granitic Islands 279
10.5.5 Inselberg Vegetation 280
10.5.6 Life-Forms and Adaptations 283
10.5.7 Systematic and Ecological Characteristics 285
10.5.8 Biogeography and Endemism 286
10.5.9 Conservation 287
References 288

10.6 Flora and Vegetation of Inselbergs of Venezuelan Guayana 291
A. Gröger

10.6.1 Geography and Geology 291
10.6.2 Climate 291
10.6.3 Vegetation of Venezuela 293
10.6.4 Inselberg Vegetation 295
10.6.5 Life-Forms and Adaptations 299
10.6.6 Systematic Characteristics 302
10.6.7 Biogeography and Endemism 304
10.6.8 Appendix 309
References 312

10.7	**The Guianas (Guyana, Suriname, French Guiana)** U. P. D. Raghoenandan	315
10.7.1	Introduction	315
10.7.2	Geography and Geology	315
10.7.3	Climate	318
10.7.4	Vegetation of the Guianas	319
10.7.5	Inselberg Vegetation	320
10.7.6	Life-Forms and Adaptations	327
10.7.7	Systematic and Ecological Characteristics	328
10.7.8	Biogeography and Endemism	334
10.7.9	Conclusion	335
References		336
10.8	**Southeast Brazil** H. D. Safford, G. Martinelli	339
10.8.1	Introduction	339
10.8.2	Geography	340
10.8.3	Geology and Palaeoclimatology	343
10.8.4	Modern Climate	345
10.8.5	Vegetation of Southeastern Brazil	347
10.8.6	Inselberg Vegetation of Southeastern Brazil	352
10.8.7	Life-Forms and Adaptations	358
10.8.8	Systematics	362
10.8.9	Biogeography and Endemism	380
10.8.10	Summary	384
10.8.11	Postscript	385
References		385
10.9	**Floristics of Australian Granitoid Inselberg Vegetation** S. D. Hopper	391
10.9.1	Introduction	391
10.9.2	Geography and Geology	392
10.9.3	Flora, Climate, and Vegetation	392
10.9.4	Inselberg Vegetation – Life-Forms, Adaptations, Systematics	394
10.9.5	Biogeography, Endemism, and Speciation	400

10.9.6 Biodiversity of Australian Inselbergs 404
References 404

10.10 Flora and Vegetation of Granite Outcrops in the Southeastern United States 409
R. Wyatt, J. R. Allison

10.10.1 Geography and Geology 409
10.10.2 Climate 412
10.10.3 Zonal Vegetation 415
10.10.4 Inselberg Vegetation 416
10.10.5 Life-Forms and Adaptations 419
10.10.6 Systematic and Ecological Characteristics 421
10.10.7 Biogeography and Endemism/Speciation 423
10.10.8 Appendix 428
References 430

11 Phytogeography 435
R. Seine, S. Porembski, U. Becker

11.1 Introduction 435
11.2 Systematic Composition of Inselberg Florulae 435
11.3 General Tendencies in Distribution Patterns 437
11.4 Endemism 441
11.5 Influence of Regional Diversity 441
11.6 Influence of Paleogeography and Paleoclimate 443
11.7 Speciation 444
11.8 Exotics and Weeds 445
References 448

12 Factors Controlling Species Richness of Inselbergs 451
S. Porembski, R. Seine, W. Barthlott

12.1 Introduction 451
12.2 Species Richness of Inselbergs: the Relationship Between Local and Regional Diversity 453
12.3 Species Richness, Size, and Isolation of Inselbergs 457
12.4 Species Richness: Diversity Patterns on Inselbergs 462

12.5 Species Richness of Inselbergs: Coexistence of Equilibrium and Nonequilibrium Communities 464
12.6 Conclusions 475
References 478

13 The Fauna of Inselbergs 483
M. A. Mares, R. H. Seine

13.1 Introduction 483
13.2 How Inselbergs Influence Faunal Development 484
13.3 Appendix: Animal Species Closely Associated with Inselbergs 487
References 490

14 Human Dimensions and Conservation 493
R. Seine

14.1 Introduction 493
14.2 Cultural Aspects 493
14.3 Human Impact 499
14.4 Genetic Resources 500
14.5 Proposals for Conservation and Sustainable Use 501
14.6 Appendix 502
References 505

15 Biodiversity of Terrestrial Habitat Islands – the Inselberg Evidence 507
S. Porembski

15.1 Introduction 507
15.2 Inselbergs – an Old and Environmentally Harsh Ecosystem 507
15.3 The Importance of Cryptogams on Inselbergs 509
15.4 Adaptive Traits Towards Life on Inselbergs 509
15.5 Phytogeographical Aspects 511
15.6 Species Richness of Inselberg Plant Communities 511
References 513

Subject Index 515

Species Index 519

Contributors

W. Barthlott

Universität Bonn, Botanisches Institut, Meckenheimer Allee 170, 53115 Bonn, Germany

U. Becker

Universität zu Köln, Botanisches Institut, Lehrstuhl I, Gyrhofstr. 15, 50931 Köln, Germany

N. Biedinger

Universität Rostock, Institut für Biodiversitätsforschung, Allgemeine & Spezielle Botanik, Wismarsche Str. 8, 18051 Rostock, Germany

H. Bremer

Geographisches Institut der Universität zu Köln, Albertus-Magnus Platz, 50923 Köln, Germany

J. Brulfert

Lab. Biochim. Fonc. Membranes Veget., CNRS, F-91198 Gif-sur Yvette, France

B. Büdel

Universität Kaiserslautern, Fachbereich Biologie, Allgemeine Botanik, Geb. 13/2, Postfach 3049, 67653 Kaiserslautern, Germany

A. Burke

Enviroscience, P.O. Box 90230, Windhoek, Namibia

E. Fischer

Institut für Biologie der Universität Koblenz, Rheinau 1, 56075 Koblenz, Germany

K. Fleischmann

ETH, Geobotanisches Institut, Zürichbergstr. 38, 8044 Zürich, Switzerland

J.-P. Frahm

Universität Bonn, Botanisches Institut, Meckenheimer Allee 170, 53115 Bonn, Germany

A. Gröger

Botanischer Garten München, Menzinger Str. 65, 80638 München, Germany

S.D. Hopper

Kings Park and Botanic Garden, West Perth, Western Australia 6005, Australia

N. Jürgens

Universität zu Köln, Botanisches Institut, Lehrstuhl I, Gyrhofstr. 15, 50931 Köln, Germany

M. Kluge

Technische Hochschule, Institut für Botanik, FB Biologie, Schnittspahnstr. 3–5, 64287 Darmstadt, Germany

M.A. Mares

University of Oklahoma, Oklahoma Museum of Natural History and Department of Zoology, Norman, Oklahoma 73019-0606, USA

G. Martinelli

Instituto de Pesquisas Jardim Botânico do Rio de Janeiro, Rua Pacheco Leâo, 915, CEP 22460-030, Rio de Janeiro, Brazil

S. Porembski

Universität Rostock, Institut für Biodiversitätsforschung, Allgemeine & Spezielle Botanik, Wismarsche Str. 8, 18051 Rostock, Germany

U.P.D. Raghoenandan

University of Suriname, National Herbarium, P.O. Box 9212, Paramaribo, Suriname

H.D. Safford

University of California, Division of Environmental Studies and Section of Evolution and Ecology, Davis, California 95616, USA

H. Sander

Geographisches Institut der Universität zu Köln, Albertus-Magnus Platz, 50923 Köln, Germany

R. Seine

Universität Bonn, Botanisches Institut, Meckenheimer Allee 170, 53115 Bonn, Germany; present address: European Astronaut Centre, Porz-Wahnheide, Linder Höhe, 51147 Köln, Germany

R. H. Seine

Universität Bonn, Zoologisches Institut, Poppelsdorfer Schloß, 53115 Bonn, Germany

K. Sterflinger

Carl von Ossietzky Universität Oldenburg, AG Geomikrobiologie, ICBM, Carl von Ossietzky Str. 9–11, Postfach 2503, 26111 Oldenburg, Germany

J. Szarzynski

Universität Mannheim, Geographisches Institut, L9, 1–2, 68131 Mannheim, Germany

I. Theisen

Universität Bonn, Botanisches Institut, Meckenheimer Allee 170, 53115 Bonn, Germany

R. Wyatt

University of Georgia, Institute of Ecology, Athens, Georgia 30602, USA

Plate 1a–f. South America. **a** Steep-sided dome-shaped inselbergs in the Brazilian state of Espirito Santo. (Photograph S. Porembski). **b** Small granite outcrop harboring typical Cactaceae and Bromeliaceae in southern Bahia, Brazil. (Photograph W. Barthlott). **c** Cyanobacteria crust on inselberg in southern Venezuela. (Photograph A. Gröger). **d** Mats formed by Bromeliaceae and the narrow endemic Amaryllidaceae *Worsleya rayneri* near Petropolis, Brazil. (Photograph W. Barthlott). **e** Monodominant mat formed by *Pitcairnia geyskesii*, Nouragues (French Guiana). (Photograph S. Porembski). **f** Species-rich mats are frequent on Brazilian inselbergs, with genera like *Coleocephalocereus, Orthophytum, Vellozia,* and *Selaginella* being particularly characteristic (Minas Gerais, Brazil). (Photograph S. Porembski)

Plate 2a–f. USA. **a** During spring, rock pools on inselbergs in the southeastern USA bear a dense vegetation cover of annuals. The reddish coloration is due to the Crassulaceae *Diamorpha smallii*. **b** The rare endemic quillwort *Isoetes tegetiformans* colonizes rock pools on only a few inselbergs in the southeastern USA. **c** Tree establishment is only possible in larger vegetation islands with deeper soils. *Pinus taeda* is a characteristic element of this formation on rock outcrops in Georgia. **d** Several poikilohydric species of *Selaginella* occur as mat formers on inselbergs in the southeastern USA. **e** The minute annual *Diamorpha smallii* frequently uses moss cushions as growth sites on inselbergs. **f** Over shallow soils on flat rocks *Diamorpha smallii* forms dense stands that are widely visible because of their striking reddish coloration. (Photographs **a–f** S. Porembski)

Plate 3a–f. West Africa. **a** Mt. Niangbo covers an area of 7 km² and is the largest inselberg in the Côte d'Ivoire. More than 200 species of vascular plants occur on this outcrop. **b** Isolated boulders on inselbergs are differentiated from the surrounding rocks by their coverage consisting of green algae lichens. The brownish coloration of the surrounding rock is due to cyanobacteria and cyanobacterial lichens. **c** Succulents are comparatively rare on inselbergs in the Upper Guinea Region. *Euphorbia unispina* preferentially occurs in *Afrotrilepis pilosa* mats. **d** On inselbergs situated in the rainforests of Gabon and Equatorial Guinea *Euphorbia etestui* is a characteristic colonizer of *Afrotrilepis pilosa* mats. **e** The ephemeral flush vegetation comprises numerous annuals and is particularly rich in carnivorous plants, e.g., *Utricularia* spp. **f** *Burmannia madagascariensis* and *Drosera indica* are likewise typical elements of ephemeral flush vegetation. (Photographs **a–f** S. Porembski)

Plate 4a–f. West Africa. **a** The island-like attributes of inselbergs are particularly well developed when surrounded by perhumid rainforest with which almost no species are in common. **b** Due to the sparse vegetation cover most rainfall is lost by runoff to the surroundings. Drainage channels are testimony to the erosional power of flowing water over very long time spans. **c** The poikilohydrous Cyperaceae *Afrotrilepis pilosa* is the most important mat-forming species on West African inselbergs. During the dry season this poikilochlorophyllous species looses all photosynthetic pigments. **d** In the vicinity of villages, enhanced fire frequencies have caused morphological responses of *Afrotrilepis pilosa*. Old individuals of *Afrotrilepis* are relatively fire-resistant due to the possession of a protective sheath of adventitious roots. **e** Solutional processes have led to the erosion of rock, which resulted in the development of oriçangas. They contain a shallow substrate and provide growth sites for short-lived species like *Cyanotis lanata*. **f** In historical times, humans used level rocks on inselbergs for grinding cereals. Today, these grinding holes are no longer used and are colonized by aquatics, e.g., the Scrophulariaceae *Dopatrium longidens*. (Photographs **a–f** S. Porembski)

Plate 5a–f. Madagascar and East Africa. **a** The poikilohydric Velloziaceae are typical elements on east African inselbergs. *Xerophyta splendens* may attain tree size on Mt. Mulanje (Malawi). (Photograph S. Porembski). **b** *Xerophyta dasylirioides* (*left*) and *Pachypodium densiflorum* (*center*) are frequent colonizers of *Coleochloa setifera* mats on Madagascan inselbergs. (Photograph W. Barthlott). **c** The neotenic Cactaceae *Rhipsalis baccifera* var. *horrida* is strictly limited to rock outcrops in Madagascar. In the **foreground** the ascelpiad *Stapelianthus decaryii* is to be seen. (Photograph W. Barthlott). **d** The resurrection plant *Myrothamnus moschata* is endemic to Madagascan outcrops. In the dry state the species survives long periods of drought. (Photograph W. Barthlott). **e** Succulents (e.g., *Pachypodium densiflorum*) and poikilohydric vascular plants, such as *Coleochloa setifera*, are richly represented on Madagascan inselbergs. The inselberg in the background has been infested by the neophytic *Furcraea gigantea*. (Photograph W. Barthlott). **f** On East African inselbergs *Coleochloa setifera* is the most important mat-forming species. The desiccation-tolerant *Myrothamnus flabellifolia* frequently occurs in these mats. (Photograph S. Porembski)

Plate 6a–f. Seychelles. **a** Rock outcrops are a common view on the granitic islands of the Seychelles where they are often surrounded by rainforest. **b** *Dracaena reflexa* and *Pandanus* sp. forming a vegetation island in a deep cleft. **c** In particular, *Pandanus* trees form characteristic elements of the Seychellan inselberg vegetation. **d** The endemic rhizomatous Cyperaceae *Lophoschoenus hornei* forms dense stands in exposed localities on rock outcrops. **e** The pitcher plant *Nepenthes pervillei* is the only carnivorous species on Seychellan inselbergs. **f** *Sarcostemma viminale* belongs to the few succulents which occur on Seychellan inselbergs. (Photographs **a–f** S. Porembski)

Plate 7a–f. Australia. **a** The geologically oldest Australian inselbergs occur in Western Australia (e.g., Boulder Rock) and form part of the Yilgarn Craton. The granites are between 2700 and 2600 Ma old. **b** Wave Rock near Hyden (Western Australia). Striking are the blackish Tintenstrich formations which result from a cover of the rock surface by cyanobacteria. **c** The monocotyledonous genus *Borya* (rainy season aspect) is the most important representative of vascular resurrection plants on Australian inselbergs. **d** During the dry season *Borya* loses most of the water content and survives for several months in a desiccated state. **e** The tiny Centrolepidaceae *Aphelia brizula* develops during the rainy season over shallow soils in ephemeral flush vegetation on Western Australian inselbergs. **f** Gnammas are seasonally water-filled rock pools that comprise a highly specialized flora including *Isoetes, Glossostigma*, and *Myriophyllum*. (Photographs **a–f** S. Porembski)

Plate 8a–f. South America. **a, b** Brazilian inselbergs are characterized by a high degree of spatial species turnover between outcrops situated in close distance. On the left (**a**) *Encholirium* sp. is the dominant mat-former whereas *Tillandsia kurt-horstii* and *Buiningia brevifolia* dominate on an inselberg nearby (**b**). **c** Several species of the genus *Vellozia* occur on inselbergs in southeastern Brazil where they attain a height of sometimes more than 3 m. **d** The pseudostems of Veloziaceae (here *Vellozia candida* in Minas Gerais, Brazil) provide growing sites for highly specialized orchids (e.g., *Pseudolaelia vellozicola*) which are restricted to this particular substrate. **e** In particular southeastern Brazilian inselbergs are characterized by large numbers of succulents and xerophytes, e.g., *Buiningia*, *Orthophytum* and *Anthurium*. **f** In southeastern Brazil even tiny rock outcrops are colonized by long-lived columnar cacti (*Coleocephalocereus* spec., *Pilosocereus glaucescens*). (Photographs **a–f** W. Barthlott)

Plate 9a–f. Monocotyledonous poikilohydric rosette trees. **a** *Borya sphaerocephala* (Boryaceae), Western Australia. (Photograph S. Porembski). **b** *Xerophyta splendens* (Velloziaceae), Malawi. (Photograph S. Porembski). **c** *Microdracoides squamosus* (Cyperaceae), Guinea. (Photograph W. Barthlott). **d** *Bulbostylis leucostachya* (Cyperaceae), Venezuela. (Photograph W. Barthlott). **e** *Afrotrilepis pilosa* (Cyperaceae), West Africa. (Photograph S. Porembski). **f** *Afrotrilepis pilosa* (Cyperaceae), with the orchid *Polystachya microbambusa*. (Photograph W. Barthlott)

1 Why Study Inselbergs?

W. Barthlott and S. Porembski

1.1 Inselbergs – Model Ecosystems for Biodiversity Studies

The study of islands has provided fundamental insights for our understanding of ecological and evolutionary processes that affect the biodiversity of ecosystems. Experimental studies and observations on islands have a long tradition, with Darwin (1859) and Wallace (1881) advancing the ideas of evolution and speciation. More recently, the island biogeography theory of MacArthur and Wilson (1967) provided an important boost for ecology which became increasingly important in view of the rapidly increasing fragmentation of habitats all over the world. Current investigations on islands take different aspects of biodiversity research into consideration, such as species-area relationships and patterns of species richness (for survey see Adsersen 1995; Vitousek et al. 1996).

In addition to oceanic islands there exist terrestrial habitats which are ecologically isolated from the surrounding area and which either form fragments (i.e., surrounded by a mosaic of habitats at least partly tolerable by fragment species) or islands (i.e., surrounded by an inhospitable matrix). Comparative research on their diversity used different types of habitat isolates, such as caves, mountains, and even individual plants. One of the most advanced projects devoted to understand the consequences of fragmentation of formerly contiguous ecosystems is the Biological Dynamics of Forest Fragments Project which is located to the north of Manaus (Amazonas, Brazil). This long-term study measures the relationship between tropical forest remnant size and their species diversity (survey in Bierregaard Jr. and Stouffer 1997).

Granitic and gneissic inselbergs (from German Insel = island and Berg = mountain, a term coined by Bornhardt 1900) form naturally occurring terrestrial islands or fragments depending on the habitat requirements of the species considered. Since granite is the main component of the Earth's continents, granite outcrops are landforms which have developed under a wide range of climates (Campbell 1997; Myers 1997). In both temperate

Ecological Studies, Vol. 146
S. Porembski and W. Barthlott (eds.) Inselbergs

and tropical regions granite outcrops attracted the interest of early naturalists. For example, in the southeastern USA and in Australia the study of the biotic and physical characteristics of plant communities on inselbergs has a long tradition and has provided detailed insights into, e.g., composition and structural and functional aspects (see the contributions of Wyatt and Allison, Chap. 10.10, and Hopper, Chap. 10.9, this Vol.). However, despite their widespread occurrence in many tropical regions, inselbergs were largely ignored as subjects of ecosystem research. Only rarely could tropical inselbergs attract the interest of early naturalists. For example, Alexander von Humboldt (1819), while traveling along the Orinoco in southern Venezuela, provided us with first descriptive accounts of the blackish crust on the rocks of inselbergs.

It is only since a relatively short time that the global importance of inselbergs for several aspects of biodiversity research has been recognized. In being scattered like islands throughout a matrix of, e.g., forest or savanna, inselbergs offer excellent research opportunities for addressing different topics of biodiversity research and conservation management. The complexity of most ecosystems is an important obstacle to their understanding. In contrast to this, the comparatively low structural richness of inselbergs makes quantitative and qualitative assessments of their inventory easier. However, despite their relatively limited structural richness they comprise a broad array of life-forms. Because of the particular growth conditions on inselbergs, both nonvascular and vascular plants are represented by specialized species adapted to drought stress, heat, and high irradiation (see Büdel et al., Chap. 5, and Biedinger et al., Chap. 8, this Vol.). In this context, it is important to note that the harsh environmental conditions on outcrops have been the driving force in the evolution of poikilohydric vascular plants on inselbergs (i.e., the complete desiccation tolerance of vascular plants, see Kluge and Brulfert, Chap. 9, this Vol.), whereas this ecophysiological trait plays at best a minor role in other ecosystems.

Inselbergs occur in a broad range of sizes and in highly varying degrees of isolation throughout all major biomes and allow for the rapid experimental testing of hypotheses. Unlike most other ecosystems, inselbergs maintain their typical attributes irrespective of geographic location, thus enabling broad-scale comparisons between very different regions (i.e., different in diversity, vegetation type, etc.). In forming original habitat islands, inselbergs offer the opportunity to compare fragmentation effects, which are a usual precondition here, with the consequences of fragmentation in formerly contiguous ecosystems. Moreover, one has to note that in many regions inselbergs belong to the least-disturbed ecosystems and are therefore of special interest. This is in strong contrast to most oceanic islands, which were severely affected by human influences already long ago. However, inselbergs too are not free from signs of human impact, as is demonstrated by the invasion of weeds on rock outcrops located near settlements. Similarly to oceanic islands, inselbergs form useful models for analyzing the dynamics and consequences of plant invasions because of their relatively low complexity, which is an important prerequisite for a regular monitoring.

In contrast to a number of comprehensive accounts on the geology and geomorphology of inselbergs (e.g., Bremer and Jennings 1978; Thomas 1994), no such contributions are available on the biology of this ecosystem to date despite a considerable number of individual publications. This disregard is even more remarkable because other rock outcrop formations, e.g., cliffs (Larson 1990; Matthes-Sears et al. 1997; Larson et al. 1999) and tepuis (i.e., sandstone table mountains), have attracted much scientific interest over the past years. A symposium on granite outcrops held at the University of Western Australia (Perth) in 1996 brought together a considerable number of researchers concerned with a broad spectrum of different aspects of inselberg research. This meeting formed an important milestone toward this first comprehensive treatment of inselberg studies.

1.2 Perspectives of Inselberg Research

Even though the number of studies concerned with inselbergs has increased over the past years, large gaps still exist in regard to our understanding of basic attributes of this ecosystem. Apart from a lack of knowledge about the basic constituents of the vegetation of inselbergs in certain regions (e.g., India), there exists an even larger deficit concerning questions of ecosystem structure and function. Moreover, even in more traditional fields of botany, like morphology and anatomy, essential functional characteristics of the sometimes bizarre growth forms (e.g., caulescent rosette trees) are barely known. In addition, important physiological properties (e.g., poikilohydry of vascular plants) of many inselberg specialists are only poorly studied.

Obviously, a positive correlation exists between outcrop size and species diversity (Porembski et al., Chap. 4, this Vol.) which apparently is a function of increasing habitat diversity. For a number of species it is known that there is a certain threshold of inselberg size below which they do not occur. An analysis of the determinants of minimum habitat sizes might offer opportunities to test hypotheses concerning the development of diversity in fragmented landscapes. Inselbergs thus offer a great potential for studying issues which today are of global concern and which might affect conservation biology.

There is still much controversy concerning the relationship between the species richness of an ecosystem and its response to disturbing effects. The examination of inselberg habitats over longer time scales might therefore provide valuable information. Experimental permanent plots which included habitats of varying diversity were established on Ivorian inselbergs in 1990. Preliminary results indicate a considerable degree of variation (i.e., local extinctions, immigrations) between individual habitats with climatic disturbances acting as a highly significant factor which controls habitat-specific turnover rates (Porembski and Barthlott 1997). However, only limited data are available in regard to differences between plant strategies (e.g., reproductive effort, population size), which might explain the long-term existence of species-rich communities in habitat fragments. Likewise, nothing is known about the dynamics of metapopulations and the number and size of subpopulations needed for long-term survival. Investigations devoted to the examination of the genetic diversity of spatially isolated inselberg populations could become very useful concerning the deciphering of the relationship among degree of isolation, population size, disturbance, and exchange rates between disjunct populations.

Granitic and gneissic inselbergs consist of a largely uniform substrate (at least from the viewpoint of plants). Large differences in location, however, provide good opportunities for comparative studies along ecological gradients. Moreover, inselbergs provide excellent opportunities to evaluate the relationship between local and regional diversity, which is still not understood (Caley and Schluter 1997). A comparison between selected inselbergs of nearly identical size surrounded by a similar matrix (e.g., rainforest) is expected to demonstrate the nature of this relationship. First results already indicate that inselberg species richness is positively influenced by a high regional diversity, as could be shown for mat communities on rock outcrops situated in Brazilian Atlantic rainforest (Porembski et al. 1998; Seine et al., Chap. 11, this Vol.). Additionally, several other aspects concerning the diversity of species communities in habitat fragments could be addressed. For example, one could ask whether habitats which are rich in plant species are more speciose on other trophic levels as well when compared to less species-rich habitats. Up to now these questions simply cannot be answered, as is the case with far more complex problems, e.g., on the level of ecosystem functions (e.g., energy and nutrient flow). Because of their simplicity in ecosystem structure, inselbergs provide good opportunities for documenting the relationship between diversity and ecosystem function. This interaction could be tested experimentally by the removal or addition of dominant species in order to evaluate the consequences in regard to ecosystem function (e.g., resistance to alien species).

Data obtained from the hitherto established permanent plots on inselbergs in the Côte d'Ivoire have shown that their vegetation is characterized by a rapid response to environmental fluctuations. This fact and the widespread occurrence of inselbergs offer interesting possibilities for the assessment of the relationship between inselberg diversity and global environmental change. In being geologically very stable and constituting old landscape elements, inselbergs experienced in the past dramatic fluctuations of the surrounding vegetation as a consequence of climatic oscillations. In comprising a broad spectrum of plant growth forms with very different ecological affinities, future changes in seasonality, temperature, and precipitation could result in a relatively rapid shift in the composition of the inselberg vegetation. In many regions, therefore, a regular monitoring of inselbergs in different regions could contribute to the recognition of possibly changing climatic conditions.

1.3 Linking Inselberg Research to Other Ecosystems

Oceanic islands and their terrestrial counterparts have proved to be very useful in studies of biodiversity. The understanding of inselbergs with respect to population and community ecology is, however, still in its infancy. Despite this fact, certain general tendencies are already visible concerning significant factors influencing the species richness of inselberg plant communities. Future research on inselbergs would benefit considerably from a comparative analysis of other ecosystems that are characterized by similar properties and which likewise underlie extreme environmental conditions. Of particular interest in this respect are other geographically widespread azonal ecosystems which are well delimited against their surroundings, like cliffs and ferricretes. Since there is great similarity in important growth conditions, it would also be rewarding to evaluate the interactions between species richness and disturbance by comparing the vegetation of inselbergs with epiphytic plant communities in tropical forests.

Another important aim of future inselberg research is to understand the reasons for their susceptibility towards invasive weeds and, on the other hand, to identify those inselberg specialists which might develop into weeds when given the opportunity to reach habitats outside inselbergs via roads or forest clearings. In this case, a close link exists with oceanic island research, which is particularly aware of the danger of invading alien species.

Acknowledgements. We thank the Deutsche Forschungsgemeinschaft for supporting our research on inselbergs. Many colleagues have contributed considerably in providing us with information on different aspects of this ecosystem. We should like to thank them all.

References

Adsersen H (1995) Research on islands: classic, recent, and prospective approaches. In: Vitousek PM, Loope LL, Adsersen H (eds) Islands. Biological diversity and ecosystem function. Ecological Studies 115. Springer, Berlin Heidelberg New York, pp 7–21

Bierregaard RO Jr, Stouffer PC (1997) Understory birds and dynamic habitat mosaics in Amazonian rainforests. In: Laurance WF, Bierregaard RO Jr (eds) Tropical forest remnants. Ecology, management, and conservation of fragmented communities. University of Chicago Press, Chicago, pp 138–155

Bornhardt W (1900) Zur Oberflächengestaltung und Geologie Deutsch-Ostafrikas. Reimer, Berlin

Bremer H, Jennings J (eds) (1978) Inselbergs/Inselberge. Z Geomorphol, N F, Supplem 31

Caley MJ, Schluter D (1997) The relationship between local and regional diversity. Ecology 78:70–80

Campbell EM (1997) Granite landforms. In: Withers PC, Hopper SD (eds) Granite outcrops symposium. J R Soc West Aust 80:101–112

Darwin C (1859) On the origin of species by means of natural selection. John Murray, London

Humboldt A v (1819) Relation historique du Voyage aux regions équinoxiales du Nouveau Continent, ait en 1799–1804 par A. de Humboldt et A. Bonpland, vol. II Maze, Paris

Larson DW (1990) Effects of disturbance on old-growth *Thuja occidentalis* at cliff edges. Can J Bot 68:1147–1155

Larson DW, Matthes U, Kelly PE (1999) Cliff ecology: pattern and process in cliff ecosystems. Cambridge University Press, Cambridge

MacArthur RH, Wilson EO (1967) The theory of island biogeography. Princeton University Press, Princeton

Matthes-Sears U, Gerrath JA, Larson DW (1997) Abundance, biomass, and productivity of endolithic and epilithic lower plants on the temperate-zone cliffs of the Niagara Escarpment, Canada. Int J Plant Sci 158:451–460

Myers JS (1997) Geology of granite. In: Withers PC, Hopper SD (eds) Granite outcrops symposium. J R Soc West Aust 80:87–100

Porembski S, Martinelli G, Ohlemüller R, Barthlott W (1998) Diversity and ecology of saxicolous vegetation mats on inselbergs in the Brazilian Atlantic rainforest. Divers Distrib 4:107–119

Porembski S, Barthlott W (1997) Seasonal dynamics of plant diversity on inselbergs in the Ivory Coast (West Africa). Bot Acta 110:466–472

Thomas MF (1994) Geomorphology in the tropics. A study of weathering and denudation in low latitudes. Wiley, New York

Vitousek PM, Loope LL, Adsersen H, D'Antonio CM (1996) Island ecosystems: do they represent "natural experiments" in biological diversity and ecosystem function? In: Mooney HA, Cushman JH, Medina E, Sala OE, Schulze E-D (eds) Functional roles of biodiversity: a global perspective. Wiley, Chichester, pp 245–259

Wallace, AR (1881) Island life. Harper and Brothers, New York

2 Inselbergs: Geomorphology and Geoecology

H. BREMER and H. SANDER

2.1 Introduction

Inselbergs are isolated rises above a plain which consist of hard bedrock. If they have a soil cover, then this is very sparse. They vary in height depending on their development, and they take on different forms, as far as both ground plan and cross section are concerned, according to their genesis and lithology. Microforms caused by weathering can have formed on the rock surfaces.

When describing inselberg formation, one must take into account weathering and erosion, which are above all regulated by soaking. This, in turn, is controlled by climate, soil cover, the lithology, the water balance, and the vegetation. In addition, the production of organic acids is an important factor for weathering.

Inselbergs are clearly defined landscape elements, which allows observation of the interplay of various physiogeographic factors. This geoecological viewpoint is more comprehensive than merely stating that inselbergs are hard or massive rock. On the other hand, inselbergs are independent systems, which can be better understood by taking into account geomorphological processes and their evolution. In *Phytogeomorphology* Howard and Mitchell (1985) stress the interdependence of land forms and plants. Whilst the former grew over a long period of time and therefore can show a trend in their development, the latter have conformed to today's conditions. "Thus, in combination, the two form a powerful tool for the survey, management, and planning of our environment."

2.2 Diversity of Forms and Terminology

Bornhardt (1900) described inselbergs or inselberg landscapes as isolated rises which he had recognized as being an independent type of relief in

Ecological Studies, Vol. 146
S. Porembski and W. Barthlott (eds.) Inselbergs

East Africa. They were discussed in more detail in the first half of the 20th century. The discussion was reported in English (Thomas 1978), German (Hövermann 1978), and French (Birot 1978) in a supplementary issue of the *Zeitschrift für Geomorphologie*. Thomas (1994) contains a comprehensive recent discussion, particularly of English literature. Bornhardt and Monadnock are also used as synonyms for inselberg, although these are often restricted to rock forms which are called Härtlinge in German. Furthermore, in German geomorphology one distinguishes, regarding escarpments: outlier (Auslieger) and mounted (Aufsitzer) inselbergs; in areas of divides there are Fernlinge (residual forms are not restricted to this). Depending on their form, inselbergs with bulging convex, steep, often little-structured slopes are called Glockenberge or Zuckerhüte (depending on the type locality in Rio de Janeiro), in English, domes or whalebacks (particularly those which are long). Tafelberge in the sedimentary rock correspond to buttes and mesas in English. Blockinselberge (castle koppjes - in South Africa - or castellated hills) are either totally, mainly, or partially covered by blocks (Wollsäcke in the granite or Kernblöcke - corestones). Smaller forms can be made up of blocks only, although the transition to tors is gradual. Very small rises (up to around 5 m), which are still connected to the bedrock, are called Klippen (tor in English) or Büßersteine. Flatter, but elongated forms are called Schildinselberge (shield inselbergs). Büdel (1977) uses the term Grundblöcke (basal blocks) for individual blocks which swim in the soil cover in isolation or are exposed; for small hillocks he uses the term Grundhöcker (basal knob).

2.3 Genesis of Inselbergs

If a rock outcrop is exposed, the rain water can trickle off more quickly, it is less soaked than the area that lies around it, weathering and erosion are greatly delayed and are nearly brought to a complete halt. However, weathering and erosion continue to take their toll on the surrounding soil, so that these areas are lowered (Fig. 2.1). The rock outcrop develops - seen relatively - into an inselberg. These are self-amplifying processes. This opposition has been described as diverging weathering and denudation (Bremer 1971). A prerequisite for these processes is deep weathering, as described by Büdel (1957) in the model of double planation (Fig. 2.3).

Inselbergs are such prominent land forms that they cannot be explained in terms of subordinate differences in rock. They are, moreover, only to be understood within the frame work of general relief development. Figure 2.2 gives a very schematic representation of the best-known models,

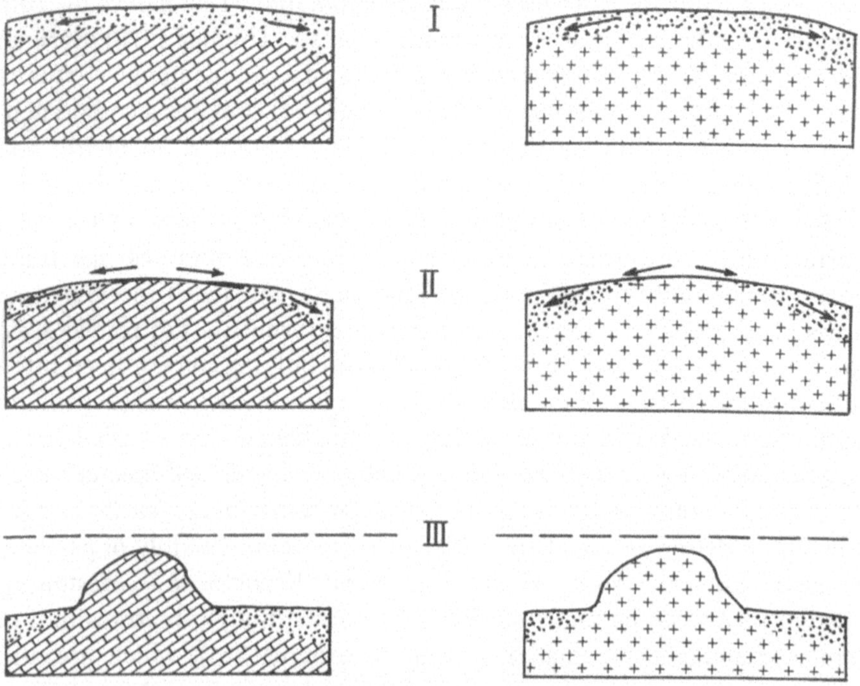

Fig. 2.1I–III. Inselbergs grow out of a plain, seen relatively, when moisture (*arrows*) stays in the soil (*dotted*) for a long time and accumulates by running off the rock. Because of this, the rock along the lower boundary of the soil undergoes further weathering. The loosened material can be eroded from the surface, whereby this is then lowered in the vicinity of the inselberg. Very intensive tropical weathering can take place independently of the lithology, as can be shown, for example, by looking at Ayers Rock (*left*, steep arkose) or inselbergs in eastern Sri Lanka (*right*, granite or massive gneiss). Adaptation to the rock structures occurs when the weathering is less intensive. **I** Area with very intensive weathering (oxisol) and little relief. **II** Shield inselberg with surrounding, very intensive weathering. **III** Dome (Glockenberg) on top of an area with oxisol/ultisol

whereby the classic models of Davis, Penck, or King only presuppose a humid, not necessarily a tropical, climate. The main difference to Büdel's model (1957, 1978) is the fact that river deepening must first take place and that the magnitude of the erosion depends on the slope. However, a circumdenudation, which could lead to the formation of an inselberg, can hardly be explained in this way. Jessen's model (1936) is similar to Büdel's, but also to the other two as well. Only in Büdel's model can inselbergs be explained without recourse to divides and without any particular rock hardness, as can be observed in numerous examples in the terrain. Such cases are frequent enough, so that an explanation of inselbergs in terms of

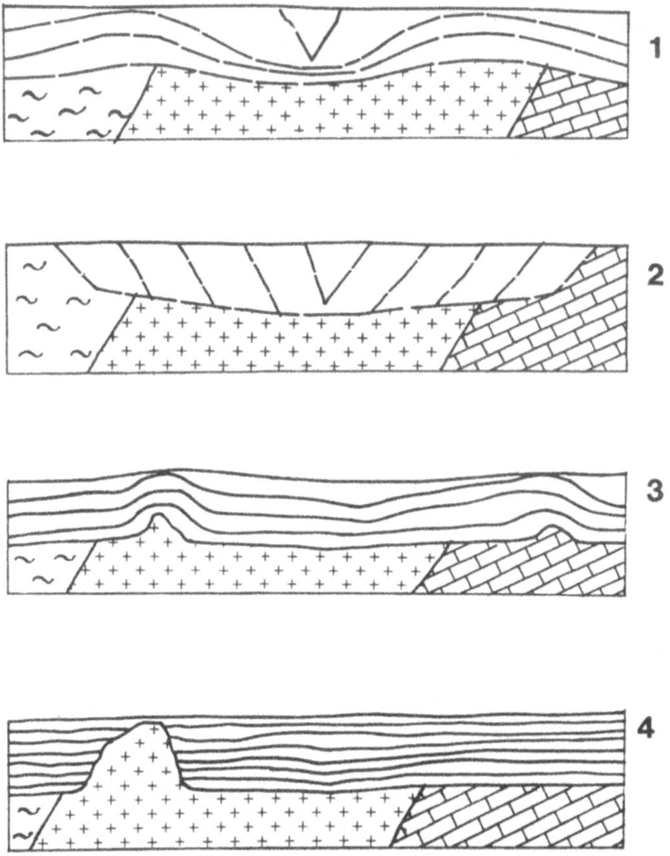

Fig. 2.2 1–4. Various ideas concerning the development of reliefs, extremely simplified. **1** Davis assumed that rising took place rapidly. Steep v-valleys develop into wide box valleys; at the same time, the areas containing water divides are lowered until the peneplain arises as the final stage. **2** In contrast to this, the slopes retreat in parallel according to Penck and King. The pediment or the pediplain comes to a halt in front of the footslope. The genesis of inselbergs is hardly to be explained in terms of these classic representations, unless one takes into consideration particularly resistant rock segments. **3** According to Jessen (1936), the lowering is two-dimensional and occurs in very wide swales. The rivers hardly precede the two-dimensional erosion. The areas of water divides become smaller and smaller and even inselbergs are reduced over the course of time. **4** Büdel (1957, 1978) in particular stresses intensive, deep weathering as a prerequisite for the genesis of a plain which caps rock of varying resistance. River and plain are lowered at the same time. Inselbergs grow out in relation to the rest of the plain, indeed in those places where there is little initial, self-amplifying, soaking. They are hardly reduced in size afterwards due to the diverging weathering (Bremer 1971, 1981)

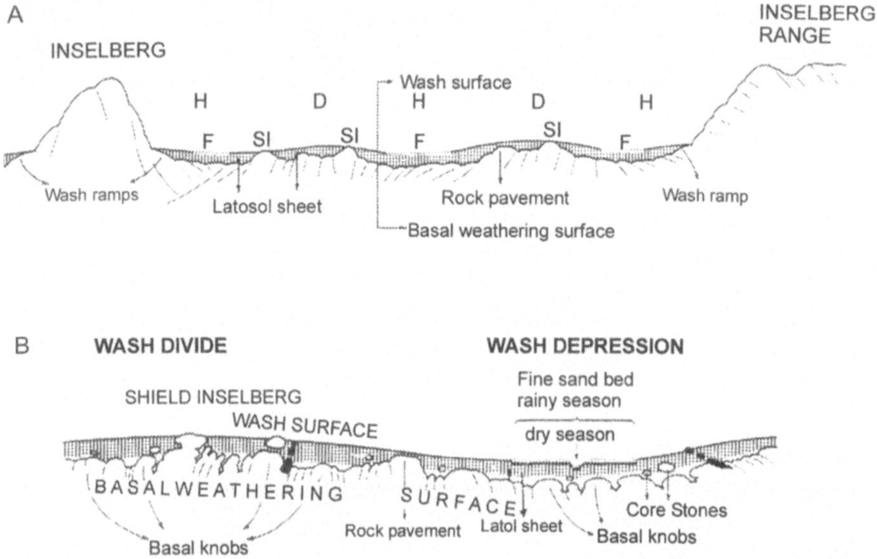

Fig. 2.3.A Typical N-S cross section through the Tamilnad Plain (E side of the Deccans) as an example of an etchplain. *M* Spülmulde (wash depression); *S* Spülscheide (wash divides); *SI* Schildinselberg (shield inselbergs); *F* von Feinsand erfülltes Flußbett (fine sand in the rainy season riverbed). **B** Wash divide and wash depression, detail of **A.** (Büdel 1977: p. 97, Büdel 1982: p. 126)

hard rock compartments is not generally applicable and therefore insufficient, although it seems evident in some cases. However, there are no inselbergs in loose rock or slates, since the weathering is so even here that no diverging takes place. In contrast, granite and massive rock such as coarse-grained migmatites or gneisses are preferred (Fig. 2.4).

2.3.1 Forms of Inselbergs

Inselbergs can vary greatly in size, from 10 m to several hundred meters in diameter; from 1–2 m (shield inselbergs) to several hundred meters in height. The ratio of diameter to height can be 1:1, or 10:1, or more. A large, structured rise above a plain is called an inselgebirge. Size and circumference of inselbergs are not very meaningful in morphogenetic terms, unless one can determine the level of the lowering of the surrounding area on the basis of their height. It is the forms that are more interesting for morphogenesis.

Depending on the intensity of the weathering, domes or castle koppjes can form (Fig. 2.4). If the weathering is very even, then planational lowering

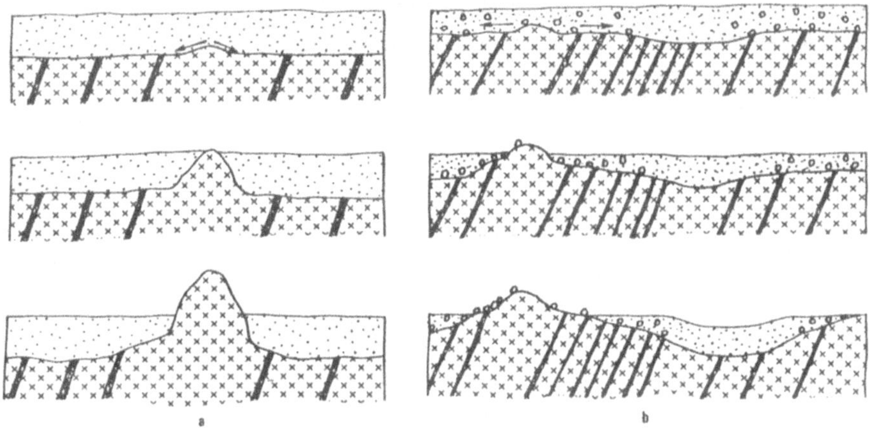

Fig. 2.4a,b. Inselbergs evolve from a cover of deep and intensive weathering. The first stages in an unevenness of the weathering front may be due to rock compactness, e.g., different distances of joints, position of a later evolving escarpment or a divide (**a**). Once soil water movement is diverted, a feedback process takes place. If the weathering cover is less uniform (**b**) blocks may be left out from the decay, and might cover the inselberg after exposure. This can be due to a more semihumid climate, to faster downwasting or to a special lithology. Other factors being equal, climate is the decisive component. As the water can also intrude vertically through joints in the granite or migmatites, inselbergs form particularly easily in these rocks, often with bulging convex slopes

takes place without the formation of inselbergs. The number and distribution of inselbergs is therefore dependent on the general genesis of the relief. It is quite possible that parts of the inselberg can be covered by soil and vegetation. This can have evolved afterwards, or at the same time as exposure, because the soil was not completely washed away in places, or was formed anew. Soils can be very stable even on very steep slopes, because they are highly porous, which is why the water can escape very quickly. The transition from totally weathered soil to hard rock takes places within a few centimeters. Späth (1981) speaks of internal divergence.

The initial exposure of a rock surface can occur by chance. However, if one attempts to deduce the different soaking which takes place, there are usually indications that there are areas with less moisture. Thus, an area with little jointing, i.e., massive rock, is less soaked. The same applies to hard rocks, whose mineral balance, containing, for example, a large proportion of quartz, can lead to delayed weathering. These are thus preferred locations for inselbergs. However, the composition of the rock is not solely responsible for the formation of inselbergs, since on the one hand the ground plan of the inselberg is seldom identical to the distribution of the rock (already mentioned in Bornhardt 1900). On the other hand, patterns of joints are repeated in similar forms without inselbergs emerging. Inselbergs are particularly striking in granite and coarse-grained migmatites, since they tend to be round forms and to have blocks. Often enough, lithological conditions are also postulated which cannot be verified given the deep weathering in the environment of the inselbergs. Taking into account the morphological situation (Bremer 1989) usually yields a more comprehensive explanation. When escarpments form, outlier and mounted inselbergs can arise (Fig. 2.5). Inselbergs on divides or on secondary divides can be explained due to a differentiation of the soaking when the rivers are lowered. If this occurs rapidly, inselbergs can emerge right next to rivers. These in particular are counterexamples to the theory that inselbergs arise due to slope retreat. (see above, Fig. 2.2; cf. also Thomas 1994).

The slow development of inselbergs means that these landforms can be very old (cf. also Twidale 1978). It can take an estimated 2–6 Ma to lower a plain by 100 m (Bremer 1982a). An inselberg that is 3–400 m high may thus be 10–20 Ma old. This alone proves that they are highly stable, which is confirmed by the microforms of weathering (see below). It is thus no wonder that inselbergs can be widespread in other climates as paleoforms of a humid climate (Büdel 1978), or that in fact they were subject to different climates without being destroyed. In the case of Ayers Rock this can be proved by means of the microforms (see below). Above all, one cannot explain inselbergs on the basis of the present climate of the location. Recently, they are most widespread in seasonal tropical climates, although they probably arose in a perhumid climate. The mechanism involved in double planation with intensive weathering speaks in favour of this. Since the limit of the recent deep oxisol formation is 1650–2000 mm annual rainfall (Späth 1981; Bremer 1994), the inselbergs were at least first formed in a perhumid climate. Furthermore, the bulging convex forms in the upper part also speak in favor of this. Very high inselbergs usually occur in isolation, i.e., they may have been first formed by accident during a period of otherwise very intensive weathering. Numerous inselbergs in a particular area, on the other hand, point to somewhat restricted weathering. The tallest inselbergs then indicate the highest level of a lowered plain, since the smaller ones were successively exposed. These ideas were developed on the basis of the general differences in the relief and the frequency of the inselbergs in the humid southwest and the seasonal climate of the east and north of Sri Lanka.

The lower slopes and footslopes may have developed in a different climate than was prevalent when the inselberg was first formed (Fig. 2.6). This would explain the different types of slope. In the nature of things one also has to take into account local soaking as well as the climate. Thus, at

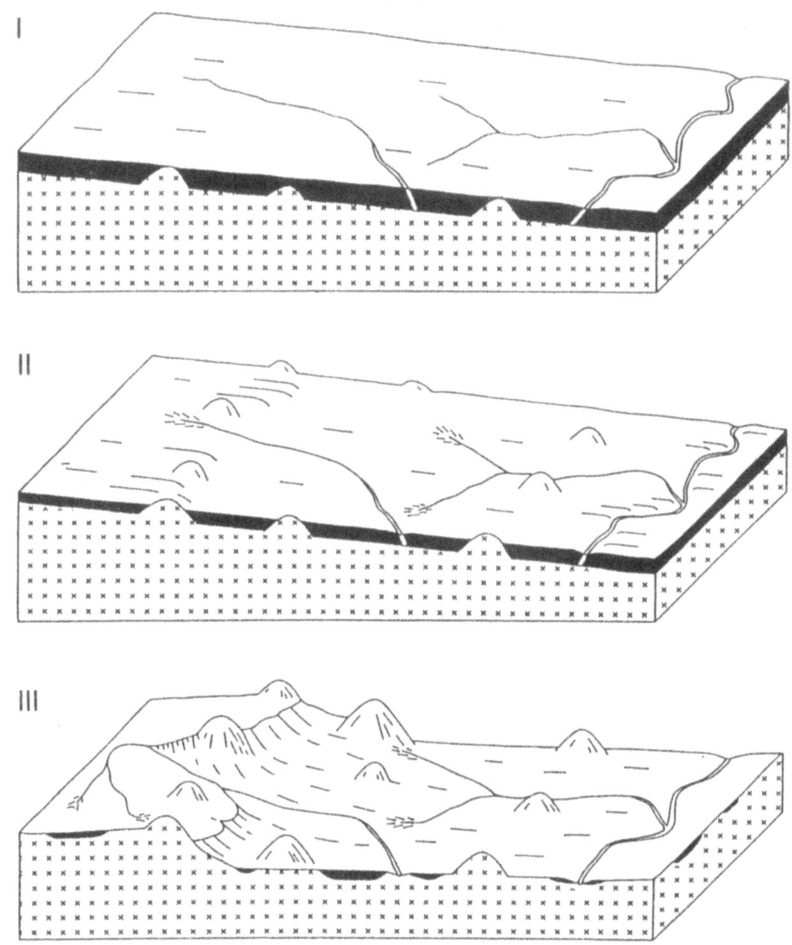

Fig. 2.5I–III. Development of escarpment and inselbergs near Kabba, SW Nigeria, simplified after Bremer (1971). During and after uplift planational lowering is more active near the base level (Niger River) where soil moisture and therefore weathering and denudation are somewhat higher (stage **I**). Once a rock surface is exposed, inselbergs and slight rises develop (stage **II**). As the network of rivers is inherited, the rises have many wide openings. Rivers also attack the rise from the rear. The proximal one may have been completely planated, (stage **III**) and may only be deduced from a cluster of inselbergs. The distal rise evolves into an escarpment with inselbergs in front and on top and with wide indentations (triangular reentrant after Büdel 1977). The original weathering cover has been largely removed. The rivers today are slightly incised and accompanied by some alluvium. This is most likely due to climatic change. (Bremer 1993: p. 192)

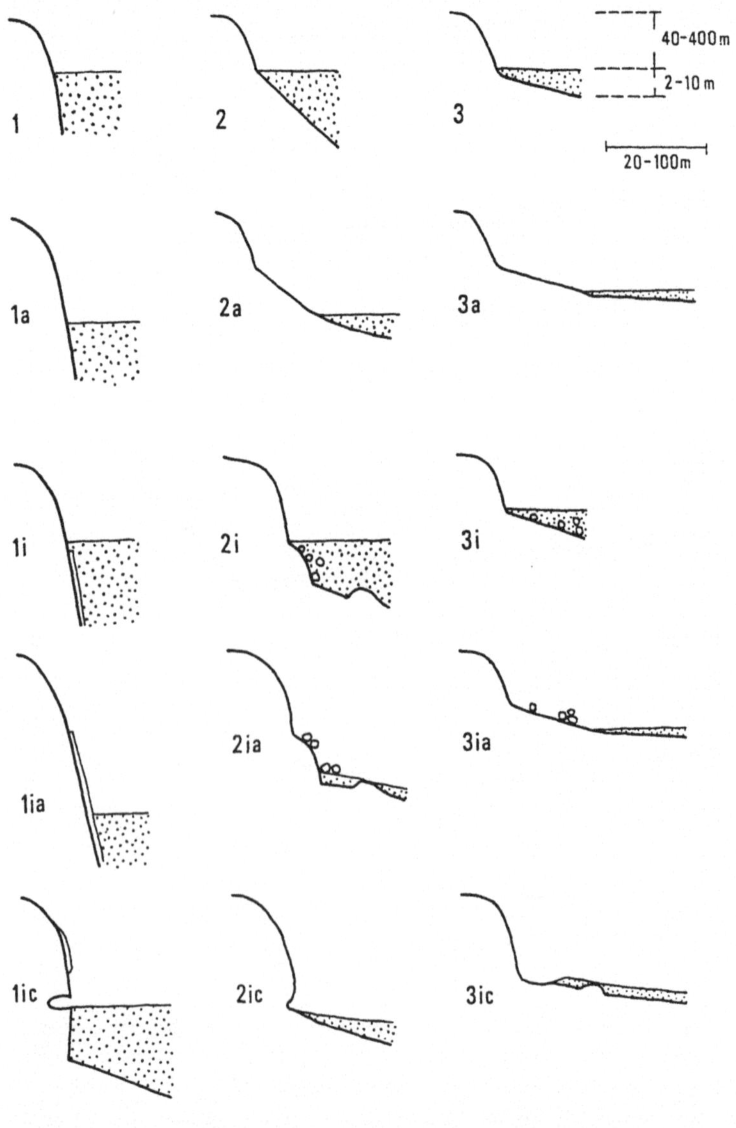

Fig. 2.6 1–3. The development of slopes on inselbergs. Weathering and exposure are regulated by soaking, which in turn is dependent on climate, rock, and local water supply. The more intense the soaking, the more intensive the weathering, the steeper the slope under the weathering cover (**1–3**) and the flatter the lower slope after exposure (**1a–3a**). Exfoliation sheets in particular may be situated in the weathering cover (**1i** and **1ia**). This is very probable for blocks and basal knobs. They are exposed shortly afterwards (**2ia**). Blocks are formed especially when the weathering is a little less intensive (**3i** and **3ia**). Depressions probably arise when the planational lowering stops. Whether hollows (**1ic**, Sri Lanka, India) arose in a very humid (paleo) climate, flares (**2ic**, Australia) and moats (**3ic**, West Africa) in a less humid environment is still unclear. Schematic representation. The measurements given are only orders of magnitude

Ayers Rock there is a flat slope on a spur even with individual blocks. In the embayment of the little Uluru valley a cavetto has developed. In between, often straight slopes with a steep lower slope have been documented. In individual cases large blocks lie on the pediment, which were probably removed by large flakes and which fell down. As with other inselbergs, there are no large, regular talus slopes which would have been supplied by the slope here. Inselbergs can, however, have been sedimentated, as can be seen in brown coal mines exposed in central Germany. It is seldom that several blocks on top of one another can be made out on the pediment or on the lower slope. In fact, often blocks are on the bulging convex slopes and are then better described as basal blocks. Similarly, several exfoliation sheets lying on top of one another are seldom. On the Glashouse Mountains, basalt plugs in east Queensland, an individual bulging convex sheet which caps the basalt structures has developed in several places. Apparently it has developed in the course of weathering. Exfoliation sheets, thus, do not seem to be very frequent and do not seem to be phenomena which occur repeatedly. A reduction in the size of the inselbergs on account of the sheets is limited. Decay of inselbergs as postulated by Brook (1978) is unlikely.

2.3.2 Microforms of Weathering

Microforms due to weathering can be found fairly frequently on inselbergs. As a rule, they developed after exposure of the inselberg, in so far as they are not forms of the footslope. Both, however, indicate stability of the inselberg's slope (Bremer 1982b).

Flat basins of several decameters up to 1–2 m diameter and 1–2 dm depth are found most frequently. They are called oriçangas (northern South America), gnammas (Australia), solution pans (English), Opferkessel (especially forms in the German low mountain ranges, which can be deeper and less wide), cuvettes (French; this term is also used to describe large depressions such as small basins). Further forms of weathering are large Rinnen (runnels) of up to several meters in width and 1–2 m depth. Their dimensions are larger than karren in limestone. On the other hand, they cannot be regarded as valleys, as they have a round bottom, even on steep slopes.

Hollows and deepenings can be found in various forms and sizes. Honeycomb weathering of only a few centimeters diameter is as seldom as tafoni, which are over 10 m, even several decameters high at Ayers Rock, and several meters deep. They have irregular peaks and belts. Smooth ground plans and walls are characteristic of hollows which developed at the level of the plain, as often a flat slope follows on the downward slope. They can also be several meters deep and high and often have a low opening. In India (Ajanta, in the northwestern Deccan in basalt, however on the valley slope) and in Sri Lanka (for example Dambulla), famous religious sites and places of cultural interest are to be found inside them. Flares are more common in Australia, otherwise they are seldom.

The surface of the rock often has small flakes. These flakes have dimensions ranging in length and breadth from a few centimeters to 1–2 dm; they are several millimeters to 2 cm thick. In addition, the surface of the inselberg is often rough, which can be explained by the granular disintegration of individual grains. The question of whether flaking and granular disintegration are purely mechanical or aided by chemical processes such as hydratation and salt wedging remains unanswered. Likewise, it is still unclear how polygonal or mural weathering arises. This term describes small, irregular runnels, which occur in polygonal structures or in triangles. They are seldom fully evolved and are usually to be found on small areas or blocks. The release of pressure is seen by many authors to be the cause of large sheets. However, since there is nearly always only one sheet and this is also only on one side of the inselberg and apparently only disintegrates very slowly, one has to ask why the sheet follows the form of the inselberg and this although the inclination of the slope changes, probably according to the very great variation in pressure. It is possible that this is due to weathering in the soil parallel to the rock surface (Bremer 1981). It is only seldom that one can observe fresh flakes breaking away. Screes or talus slopes with a grading or sorting of the debris such as is the case in the Alps cannot be found. Block fields on the lower slope almost always show that the blocks are very stable. These are probably for the most part basal blocks which remained behind when the fine-textured soil material was carried off. These blocks are subject to the flaking of small particles and granular disintegration just as is the case with the rock surface of the inselberg. They can be split by radical cracks. All in all, the blocks break up very slowly, if at all.

Bulging convex slopes, even halfdomes can occur on escarpments just as can microforms of weathering. This shows the same genesis in principle through planational lowering with diverging weathering. Describing halfdomes as inselbergs (Hövermann 1978), however, ignores the main criterion, namely that they rise above an etchplain. Conspicuous individual mountains in a hilly area can have been inherited from an old plain, but remains of this should then be found.

2.4 Ayers Rock

This inselberg in central Australia is unique because several hypotheses can be derived, or rather confirmed. Consisting of an arkose, it shows that inselbergs are not restricted to crystalline rock. The neighboring Olga Mountains show similar manifestations in a fixed conglomerate. The bulging convex forms in thin layers in sedimentary rock are particularly striking. From a distance one could also take them to be granite inselbergs. Secondly, it can be proved that these inselbergs were exposed to very different climates after their genesis, without the entire forms being lowered or changed. Inselbergs can, thus, be very stable over several million years. Thirdly, it can be proved by means of micromorphological tests, using Ayers Rock as an example, that the red color arose due to iron compounds and in part also due to silica, which forms a thin outer layer on top. The substances did not originate inside the rock, but must have been blown up onto it and were then changed on the surface.

Microforms of weathering on Ayers Rock are interlocked in such a way that age sequencing can be carried out (Bremer 1965). By comparing them with similar forms, such as is described in detail by Wilhelmy (1981), it was possible to derive climatic phases for each form. These were also exhibited in the weathering and formation of the relief in the surrounding plains. The old age of the inselberg was confirmed through drilling on their surface, which reached vales and graben in which tertiary lignite was found. These studies, put forward in 1965, could be confirmed in their entirety during later journeys (1975, 1988). Small caves, especially in the eastern part, are an addition. These are similar to forms in Sri Lanka and probably arose in a perhumid climate. Twidale and Bourne (1975) are right in saying that these hollows developed at the level of the plain, which, however, is probably not true of the tafoni.

During the first tests in 1961, 12 small rock samples were taken from various morphologic positions, along with 5 samples from the pediment. They were supplemented by 4 further rock samples in 1975. They are for the most part samples of the youngest flakes, which cover all the forms. Since they are fixed and there is hardly any debris, it was decided early on that the flakes are not recent. This finding was confirmed in full by an analysis of the thin section.

The unaltered rock consists of quartz and feldspar, is thus an arkose, which is grey in color. The red color of Ayers Rock is thus due to a coating which covers all forms apart from the innermost part of the tafoni, the black stripes in the large runnels and the base of the oriçanga at the top of the plain. There is also generally surprisingly little weathering close to the surface of the arkose, even feldspar has hardly been attacked. The coating consists of a rind of iron-rich gel or a clay-iron complex, which is up to 0.2 mm thick. Since there are only individual mica or other minerals in the arkose which could supply the iron, this must have been supplied externally. This is confirmed by the fact that hairline cracks in the flakes have been filled from the outside. These fissures are up to 0.2 mm thick and usually run subparallel to the surface. They are 5–10 mm in length. Their proportions are thus at least one dimension smaller than the flakes. The fissures can have spread to the outside. Sometimes, especially in the case of violet flakes which occur on the lower slope and on the walls of tafoni, but also in the case of a flake from the edge of an oriçanga, a sequence of the filling could be made out: the darker, usually violet clay-iron filling has cracked again and has been filled with salt or gypsum; the opening is closed off by means of a thinner and lighter clay-iron crust. All this suggests that the fissures are not very active. However, it is true that small particles can flake off now and again. This had previously been deduced from the rough surface. In contrast to the original supposition, these are not only grains which flake off, but also arkose particles. The hairline cracks generally run through grain boundaries.

The coating thus shows different stages. The older clay-iron complexes may be due to a process of crystallizing out through aging. The substances which are easily soluble, such as salts and gypsum, apparently remain only in very sheltered locations. The weathering rind with its red or brightred color and various thicknesses consists of an iron gel which has crystallized relatively little. This substratum can also be found on and in the algae which have grown vertically in tubes and which form the black stripes in the runnels.

At least as far as the outer rind is concerned, the material was imported. This is conceivable given the not infrequent dust storms of central Australia. Since this coating covers the whole of the mountain, there not only has to be fairly regular distribution, it also has to have been fixed. Apparently the humidity and the infrequent dew is sufficient to effect the fixation as well as for it to be able to be transformed into gel, and even to set off crystallization. In the case of old, very thick rinds, growth in layers can be discerned, whereby even the finest quartz grit (up to 0.1 mm) can be caked in. This applies in particular to samples from the tafoni and from the bottom of a thick sheet, where rock soaking can cause a somewhat moist surrounding. A sample from a belt in a tafoni even shows typical kaolinite rolls. Over and above the very first stages of weathering, new formations of minerals are also possible. This is, however, an exception, in which weathering is perhaps assisted by interstitial water and organic acid from bats.

The samples from the pediment show a sandy and gritty clay. Old clay-iron or iron-silica combinations make cementation possible. Grains from a sample taken at 40 cm point to clay films (B_t). Kunkar forms at a distance of 150 m from the foot of the mountain show two phases due to iron infiltration. The samples confirm that today only very little erosion takes place on Ayers Rock, although since the mountain arose (in the Upper Cretaceous/Palaeogene?) this has never been very much anyway.

These findings from Ayers Rock are particularly interesting for ecological questions in so far as they show that material that has been blown up can certainly remain there for a long time in order for it to be chemically transformed. Thus, one can also deduce that material that has been blown up can be available for plants.

2.5 Microforms and Substrata – Examples from the Côte d'Ivoire

2.5.1 Inselbergs Examined

The following detailed tests were carried out on inselbergs in the Côte d'Ivoire. Table 2.1 shows that perhumid and seasonal climatic zones in the country were included. However, this is of only limited importance for the development of the substrata mentioned below, since

1. the inselbergs have a microclimate which is clearly different from that of the surrounding area (cf. Szarzynski, Chap. 3, this Vol.; Bakker 1957), and
2. for at least parts of the Comoé National Park a humid paleoclimate is probable.

The investigation comprises very different inselberg forms: from the shield inselbergs of the Comoé National Park, to the partially exposed outcrops of the Taï National Park, to the whaleback and the complicated step-like formation of the Mt. Niangbo. In view of the parent rocks, however, the differences are minimal. In all the cases we are dealing with acid coarse-grained, unstratified rock (see Fig. 2.7).

Table 2.1. Classification of the inselbergs examined according to landscape zones (after Eldin 1971)

Dry period (months)	Annual rainfall (mm)	Inselberg examined	Height of the surrounding area (m NN)	Height of the inselberg above the area (m)	Lithology	Form
> 8	1100–1700					
7–8	1100–1700	Shield inselberg in the Comoé National Park	200–280	0.5–5	Granite	Shield
		Niangbo		300	Granite[a]	Regular, step-like change between flat areas and slopes
5–6	1100–1600	Sénéma	300	310	Granite	Whaleback
4–5	1200–1800	Mafa	60	140	Granite	With 3 peaks, close to the form of the whaleback
3–4	1600–2500	Stations inselberg in the Taï National Park	170	25	Migmatite	Partly exposed surface of an outcrop
2–3	1900–2500	Mt. Niénokoué (Taï National Park)	140	250	Granite	With 2 peaks, close to the form of the whaleback

[a] Partially foliated.

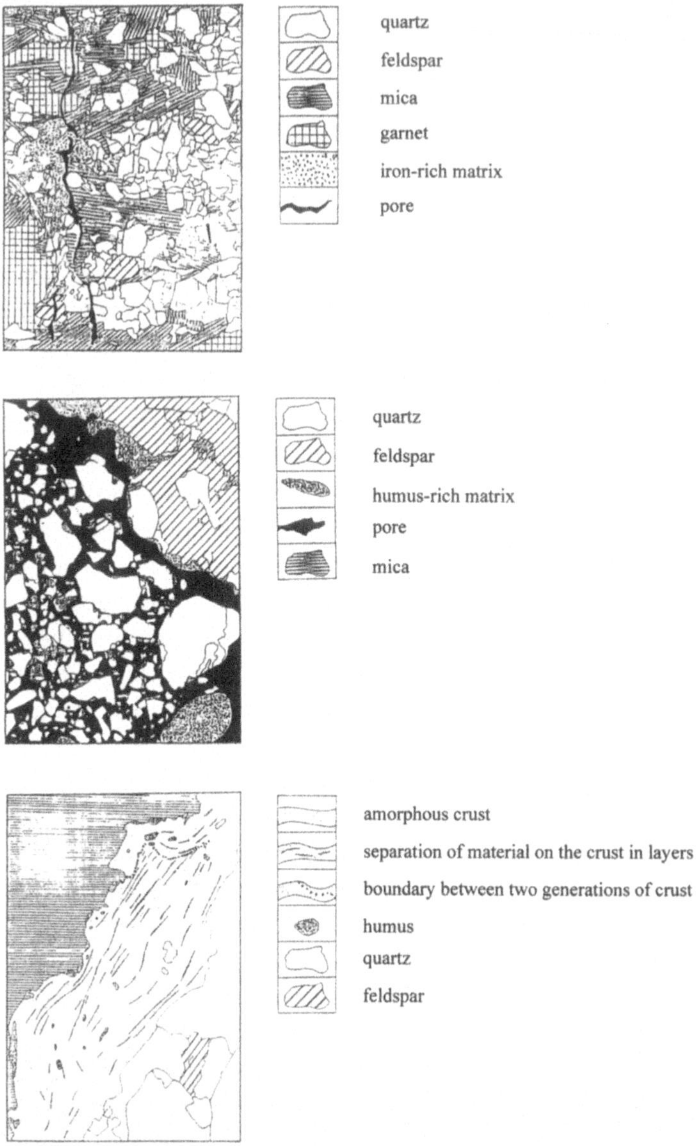

Fig. 2.7a–c. Illustrations of thin sections. **a** sample of rock from the Stations inselberg in the Taï National Park. Initial stages of weathering can be observed because of the beginning of exfoliation (cracks parallel to the rock surface) and the exposure of iron compounds out of the mica (longitudinal side around 5 mm). **b** Soil out of an oriçanga on inselberg in the Comoé National Park: the picture shows the grain fabric in the quartz-rich sand as well as a flake with fresh feldspars from an inselberg. The formation of matrix coatings around the grains is hinted at (longitudinal side around 5 mm). **c** Crust from an oriçanga on inselberg in the Comoé National Park: the formation of a crust out of amorphous (silicate) material which can enclose quartzes and humus-rich material was observed on the edge of oriçangas in particular. In the example shown the crust is multi-phase (longitudinal side around 2 mm)

2.5.2 Microforms

The study of the genesis of microforms on outcrops of unstratified rock has yielded a comprehensive amount of literature. Wilhelmy (1981) and Twidale (1982) have produced overviews of the topic. They show clearly that processes and conditions of the formation of microforms are, in part, still controversial today. The influence of climate and rock structure is judged differently, just as is the question of which of these forms were situated under a particular soil cover and how they develop subaerially. For a discussion of the connection between the formation of meso and microforms and the development of inselbergs see Bremer (1965, 1967, 1981).

In order to find out the botanical conditions, it is particularly important to find out whether the formation of microforms differentiates the abiotic conditions on the inselbergs. Stress factors that are typical for inselbergs – edaphic aridity, lack of fine-textured soil material which can be penetrated by roots – can be diminished on a small scale. In addition, however, a lack of unweathered rock as a source of cations, which is a disadvantage for the soil in the surrounding areas, can also be balanced out on the inselbergs by a thin layer of substratum and replenishment of fresh grus particles. Thus the change in microforms leads to the small-scale differentiation of habitats.

Two groups of microforms can be differentiated:

1. Forms which can at least partly compensate for the lack of soil and water which is typical for inselbergs. This applies in cases where the removal of fine-textured soil material and water is hindered or slowed down or where fine-textured soil material and water are supplied. These forms include oriçangas, joints/fissures, and in the case of migmatites, forms of selective weathering.
   Water can be present either because of the direct influence of rainfall or because it runs down off rock lying above, as well as due to overflowing oriçangas.
   Oriçangas. Closed depressions were found ranging in size from a few decimeters to several meters in width and 2–50 cm in depth. They may already have been formed under a weathering cover and have spread subaerially through granular disintegration and chemical denudation. Small oriçangas can be used as grinding holes and may therefore have been superimposed to become regular oval depressions (often without being filled with fine-textured soil material). In remote regions there are no such superimposed forms. As far as the non-anthropogenically superimposed oriçangas are concerned, the deeper ones are more often (but not necessarily always!) filled with fine-textured soil material than

the flatter ones. In the case of the latter, only a thin layer of humus or grus is to be found.

Tectonic Joints and Fissures. In contrast to the oriçangas, the tectonic joints have less space for loose material and penetration by roots. They can be enlarged through weathering and thus some oriçangas are oriented towards joints along their longitudinal axis. Joints and fissures can show a more regular soaking in the lower part of inselbergs in particular.

Joints and Fissures Which Arise Through Exfoliation. It is usually assumed that the cause of exfoliation parallel to the surface is the expansion of the rock due to a release of pressure. As both horizontal and vertical joints arise due to exfoliation, lack of space for roots, which is the normal disadvantage of joints, can be partly compensated for due to the network created.

Forms of Selective Weathering. In the case of homogeneous rocks, as for example the granites from which samples were taken, the denudation or flaking of individual grains leads to pitting, so that it is easier for a root system to form and be preserved. As far as the migmatites are concerned, however, a series of rises and depressions of a few centimeters arise following the veining in the rock structure. The depressions have an effect similar to the oriçangas in that they collect fine-textured soil material.

2. The other microforms on inselbergs, in contrast, show signs of unfavorable conditions for higher plants since fine-textured soil material, water, and seeds are carried away. These forms include, amongst others, steep surfaces or surfaces which are exposed to the wind as well as quartz veins that have been cleared out. It is not absolutely clear how runnels are to be classified. On the one hand, there is a large supply of water available for a short time after rainfall, on the other, fine-textured soil material and seeds find it very difficult to remain under such circumstances. Thus, most runnels had no higher plants. Runnels on Mt. Niénokoué, where a root system with fine-textured soil material remained, were an exception here.

2.5.3 Soil Formation

Until now, hardly anything was known about soil formation on inselbergs, apart from Bakker's research (1957).

2.5.3.1 Soil type and Substratum Characteristics: Findings

Three different soil types were found on slight inclines on inselbergs in the Côte d'Ivoire. Their distribution is at least in part regulated by the microrelief (Table 2.2).

The lithosols are mostly sandy-gritty weathered material which has been washed off the inselbergs as well as rock fragments (up to 1 × 3 × 4 cm in diameter). In the case of thicker (5–20 cm) lithosols, often a 1–2 cm layer of grus was found on top of the sand. An accumulation of humus can be made out here: a moderately humus-rich $A_{(h)}$-horizon changes abruptly into the bedrock or into a layer of exfoliated sheets. In the perhumid region the accumulation of humus in the oriçangas can be very significant, so that the $A_{(h)}$-horizon acquires a black-gray color. Bakker (1957) records humus contents of up to 18.2 %. Our own research confirms this finding.

In addition, in the deeper oriçangas in the Comoé National Park, cambic soils 15–50 cm thick occurred. As a rule they show a 5–10 cm thick $A_{(h)}$-horizon which differs little in color from the rest of the soil but is penetrated extensively by roots. Underneath that there is a 10-45 cm-thick cambic horizon which changes into the unweathered parent rock or rather sheet without a transition phase. The assemblage of soil types on slopes is somewhat more variable. They are differentiated according to the slope and intensity of the weathering of the sheets. Also, on the slopes of some inselbergs (Mt. Niénokoué, Sénéma) dense layers of peaty soil with little mineral content occurred. On Mt. Niénokoué this was above all observed in trough-shaped runnels in the slopes. Thick plant growth and a thick layer of the humus here protects them against erosion.

Table 2.2. Soil type on slight slopes on the surface of inselbergs

Soil type	Horizontal sequence	Most important soil formation processes	Situation in microrelief
Lithosol (without greatly differing humus content)	$A_{(h)}$–C	Accumulation of humus	Joints, oriçangas, forms of selective weathering, fractures formed by exfoliation
Cambisols	$A_{(h)}$–$B_{(v)}$–C	Accumulation of humus Grain coating	In oriçangas and fractures formed by exfoliation if the substratum is thick enough
Peaty soil	O/$A_{(h)}$–C	Extensive accumulation of humus	Hardly adapted to microforms

Two different forms of soil genesis are discernible on inselbergs:

1. On the one hand, the substratum may be unaltered fine-textured soil material, flakes or grus from the inselberg which has collected in depressions.
2. In the case of shield inselbergs it is also possible that parts of the fine-textured soil material are a soil relict which used to cover the now exposed basal knob. Pisoliths found in the oriçangas in the Comoé National Park may be an indication of this. A recent formation of pisoliths seems unlikely in the seasonal climate, since neither their thin soil profiles, nor their morphological position allows such an iron accumulation. There is no doubt, though, that these soils also experienced a fresh input of grus afterwards. It is difficult to prove, although very probable, that fresh material was brought in by the Harmattan dust. Herrmann (1996), who investigated the dust deposition in soils in West Africa on the basis of a transect from southwest Niger to south Bénin, found that 156 g of dust $m^{-2} a^{-1}$ was imported in the most northerly location; in the most southerly location the figure was 14 g. The main components were found to be quartz, feldspar, kaolinte, mica, and calcite. Nutrients which were imported were, above all, K and Ca, in smaller quantities Mg and N as well as, through the importing of plant ash from bush fires, P (mostly < 1 kg $ha^{-1} a^{-1}$).

2.5.3.2 Substratum Characteristics: Laboratory Findings

Methods:

Statements in Bremer (1995) regarding the methods used in the geomorphological laboratory apply. The only change was that a new diffractometer (Siemens D 5000) was used in the X-ray diffraction analysis. The water analyses were carried out by the geo-ecological laboratory at the Geographical Institute of the University of Cologne.

Results:

1. Grain Size
   In the oriçangas, loosely layered, well-sorted sands up to loamy sands (sand content 60–90 %, degree of sorting according to TRASK 2–5) with a highly variable proportion of soil skeleton were observed. A single-grain fabric or rather only slight aggregation was found in connection with this.

2. Micromorphology
The following components were observed in the oriçangas soils:

a) *Grains from the parent rock.* Quartz grains, ranging from angular to slightly rounded at the edges are greatly enriched in comparison with the parent rock, dominate the thin section. In all the cases investigated the quartz content of the soil is at least twice that of the parent rock. In individual cases the grains show solution cavities; in pisoliths fracturing of quartz grains can be observed.
The number of feldspar grains in the oriçangas is clearly lower than in the parent rock. Since hardly any neoformations can be made out (hardly clayish matrix in the thin section; weak reflexes in the X-ray diffraction samples), the feldspar grains have practically totally dissolved. Micas, which are also found in the parent rock, were found in the soil of the oriçangas even less frequently. On the edge of the oriçangas the proportion of unweathered minerals is increased due to the importing of grus.
b) *Coating.* The color of the soil substantially depends on whether the grains are present with or without a coating. These coatings can comprise iron compounds (red-brown), clay (pale brown) or humus (black-gray). The color of the soil changes according to this. If there is no coating, then shades of gray occur. In the case of the cambic soils of the oriçangas, an iron-rich clay plasma coating was usually observed.
c) *Pisoliths.* In some Oriçanga fillings in the shield inselbergs, pisoliths of mm to cm size were found, as well as in the surrounding soil. They occur in a rounded form, but also as angular debris (see above for their probable genesis). Bakker (1957) assumes that in perhumid regions it is possible that iron ore can form in oriçangas where peat has formed. This was not found to be the case in the inselbergs in the Taï National Park.
d) *Matrix.* The matrix covers only a small percentage of the thin section. It usually comprises a mixture of humus, quartz chips and some clay. In addition, humus was also found outside the matrix in the form of remains of roots.
In so far as it was possible to maintain and analyze aggregates in the very loose material, a single-grain fabric with typical primary voids (cf. Sander 1995) was visible. The volume of pores with a diameter of more than 10 μm is 25–30 % in these cases. Therefore, one must proceed on the assumption that permeability is very good. However, due to the build-up of water in the oriçangas, the erosion is greatly restricted. Solutions may even be supplied due to the overflowing of oriçangas that lie upslope. Current soil-forming processes can hardly be made out in the thin section. Bioturbation due to root growth is probable. Biogenetic pores are,

nevertheless, seldom preserved because there is little cohesion in the substratum. In addition, there is usually too little fine-textured soil material in the oriçangas to permit termite nests, so that these cannot be the reason for bioturbation. It is only possible to prove that the grains have a thin coating of iron oxides, clay, and humus. This result leads us to call them cambic soils. Furthermore, humus has accumulated in the top soil.

3. X-ray Diffraction Analysis
   The X-ray diffraction analyses were carried out on ground, unaltered rock on the one hand and on silt and clay fraction of the soils on the other. They were carried out both on soils in oriçangas and in the area surrounding the inselbergs (plains and pediments).

   a) *Examples from the Comoé National Park.* By far the most dominant mineral is quartz (Table 2.3). Feldspars occurred with considerably less frequency, although it could be demonstrated that they can occur in small quantities in the oriçanga fillings. Only in two cases was it possible to prove that mica, which occurs in the parent rock, exists in small amounts in the oriçangas. The low peaks for the primary minerals reflect the impoverishment of the oriçanga fillings. The oriçanga soils are, in comparison to the parent rocks, clearly impoverished.
   It is usual to find that the weathering of primary minerals in soils is contrasted with an accumulation of secondary minerals. This can hardly be detected in the samples tested. The peaks in the clay minerals are – if they exist at all – very small.
   In comparison to the soils in the surrounding areas, the oriçanga soils showed slightly less weathering. As far as the former are concerned – in most cases these are cambic soils – the feldspar is completely decayed and slight kaolinite formation can be observed.
   b) *Examples from the Taï National Park.* An obvious dominance of quartz in the fraction <63 μm is also evident in the samples taken in the Taï National Park (Table 2.4). It is nearly always possible to prove the existence of residual minerals. It is surprising that more residual minerals can be proved than in the Comoé National Park, which is considerably drier today. This may be an indication of the fact that these

Table 2.3. Mineral content in soils and substrata in the Comoé National Park

	Quartz	Feldspar	Mica	Kaolinite	Smectite
Unaltered rock	++++	++	+	–	–
Soil in the area	++++	–	–	(+)	(+)
Oriçangas	++++	(+)	–	–	(+)

Table 2.4. Mineral content in soils and substrata in the Taï National Park

	Quartz	Feldspar	Mica	Garnet	Gibbsite	Kaolinite	Smectite
Unaltered rock	++++	++	+++	+	–	–	–
Pediment	++++	+	+	-	(+)	+	–
Soil in the area	++++	+	(+)	-	(+)	–	–
Oriçangas	++++	+	(+)	-	–	+	+

materials in the oriçanga fillings in the Comoé National Park are in part remainders of old, intensively weathered soils. Kaolinite, which is normally present, as well as smectite were found in the oriçangas. If one compares the oriçanga soils with the soils in the surrounding area - for example with the soil of the inselberg's surroundings or with that of the pediment - then two differences are noticeable:

- in the oriçangas in the Taï National Park, smectite could regularly be detected. This is absent in the surrounding area, and
- in the area surrounding the inselberg it was possible in isolated cases to detect gibbsite, which is a sign of great leaching of silica.

The differences are due to the young age, the slowing down of erosion, (see above) and the fact that the oriçanga soils have a fresh supply of material.

c) *Significance of the Mineral Content as Recorded by the X-rays.* Clay minerals, along with humus, are temporary repositories for nutrients which are present in the soil solution in the form of cations. A lack of clay minerals can lead to alkali and alkaline earth ions, which are released by the weathering, not being available for an exchange, but which are rather washed out with the drainage water. The kaolinite has a lower ion exchange capacity than 2:1 clays.

By recording the residual minerals one gains an impression of which quantities of the substances mentioned above can still be released by weathering processes. Quartz, being a pure SiO_2 mineral, can offer no nutrients; feldspar and mica, on the other hand, release alkali and alkaline earth ions during hydrolytic weathering.

If one assumes the two criteria

- the presence of clay minerals, and
- the residual mineral content

as a standard for the nutrient balance, then the soils in the oriçangas are, in fact, impoverished compared with the parent rock and also occasionally show that the unfavorable kaolinite is present. However, they are richer in terms of residual minerals and the 2:1 clays than the soils surrounding the inselberg.

Table 2.5. Median of the pH value measured in the various climate zones

	Comoé National Park		Taï National Park	
	All samples	Oriçangas	All samples	Oriçangas
Rainfall (mm a^{-1})	1150	1150	1900	1900
Median	4.6	5.0	4.3	4.6
n=	45	7	15	3

4. pH Value
   Values of between 3.8 and 6.8 occurred in a total of 69 pH tests (KCI) carried out on soil samples from the Côte d'Ivoire. The tests included both soils from the area surrounding the inselbergs as well as soils from oriçangas. Table 2.5 presents the values from 70 tests from the Comoé National Park and the Taï National Park. It proved that the pH value is dependent on the annual rainfall, which has already been described in other studies (Jenny and Leonhard 1934; Fitzpatrick 1980; Sander 1996 amongst others).
   For the samples taken from oriçangas the following applies: in both the Comoé National Park and the Taï National Park the values for the substances taken from the oriçangas are a little higher than is usual for the entire sample collection. The median of the seven pH values measured in the soils taken from oriçangas in the Comoé National Park was 5. One value even reached 6.8. The three tests taken from oriçanga soils in the Taï National Park showed values of between 4.6 and 4.7.
   The better supply with bases in the oriçanga soils may be due to:
   - a higher proportion of unaltered rock in the oriçangas' substrata, which is due to pitted rock particles or Harmattan-dust which has blown in, and
   - incomplete removal of solutions, which evaporate on the spot and which leave behind the substances dissolved in them.
5. Water
   Water samples from several oriçangas were analysed after rainfall. The results are shown in Table 2.6.
   Neither Mn nor Si are recorded in the table, since Mn only shows up values under 0.5 ppm (usually 0–0.2 ppm) with little variation. Furthermore, the values of Si usually lie on or rather under the level at which they can be detected (which is, in this case, around 10 ppm).
   In the oriçangas it is evident that as a rule (the exception being inselberg no. 4) the number of cations which form bases (alkali and earth alkali ions) is significantly higher than the amount of iron and aluminum.

Table 2.6. Solubles in the water samples (in ppm)

	Na	K	Ca	Mg	Fe	Al
Oriçangas in the Comoé National Park						
Inselberg no. 13, 3 samples	2– 5	1–4	2–3	1	1	1
Inselberg no. 7, 2 samples	1–11	2–7	1–3	1	1	1–2
Inselberg no. 4, 1 sample	1	3	3	1	7	6
Oriçangas in the Taï National Park						
Stations inselberg, 2 samples	4– 6	4–5	2–3	1	1	1–2
Oriçangas on Mt. Niangbo						
Mt. Niangbo, 1 sample	1	2	2	1	0	0

Since the oriçanga soils have higher pH values than those of the surrounding area, it can be assumed that alkali and earth alkali ions are released during hydrolytic weathering of the parent rock and Harmattan dust (abrasion pH, cf. Loughnan 1969). It is, however, not necessary to deduce that a large amount of chemical denudation takes place, since this occurs only if water flows off when the oriçangas overflow and does not evaporate on the spot. The occasionally reported formation of a smooth crust up to 3 mm thick on the upper edges of oriçangas may be due to the latter process.

6. Crustification
The above-mentioned crust appears in the thin-section analyses in thin layers. It can totally enclose grains or humus particles deposited on what was formerly the surface of the rock. Following the microscopic analysis of the thin section, calcite and probably also iron compounds can be ruled out as crust formers. This makes it probable that silica compounds (as also assumed by Bakker 1957) are present. Microscopic unconformities within the crust lead us to assume that – in micromorphological terms! – granular disintegration has recently taken place.

2.5.3.3 The Influence of Geomorphological Factors for Habitat Boundaries

The formation of microforms leads to a wealth of greatly differing habitats with respect to the thickness of the loose material and the water balance. These habitats sometimes have clear boundaries, sometimes they gradually merge into one another. The gradual transition of forms and substrata can, amongst others, be observed along the edges of oriçangas and shield inselbergs. Here, the change in the soil thickness and the characteristics of the substrata also correspond to a change in the makeup of the vegetative covering. The formation of microforms of the relief may thus be seen as a factor which influences the differentiation of the vegetative covering. This is clearly evident along the edges of the shield inselbergs, where the thickness of the substratum, height of growth and composition of the plant cover correspond (Table 2.7): Illustration no. 12 in Bakker (1957) gives an example of the conditions in a rainforest climate, which shows the transition between an inselberg plateau and the rainforest in Suriname.

2.6 Discussion

Inselbergs are practically always very old, highly stable land forms. This applies even to block inselbergs or forms with large amounts of exfoliation sheets. Landslides are hardly to be expected, even where there is soil cover, as this is usually not very thick. Since it is normal for the soil to be very porous, the water can run off very quickly, so that the soil does not become oversaturated. Even on steep slopes plants are thus hardly threatened by natural processes of erosion.

On inselbergs the formation of microforms leads to significant small-scale differentiation of habitats (with highly variable amounts of water and soil which can be penetrated by roots) than is the case in the surrounding area. Therefore, there are habitats for very specialized plants which may not be present in the surrounding area.

Extremely weathered soils may have developed in the area surrounding inselbergs in dry (Ayers Rock) and subhumid (Comoé National Park) regions. Recent climatic conditions cannot explain these. They are either very impoverished in terms of primary minerals (Comoé National Park) or have neoformations which occur only in extremely leached soils (Central Australia) or show signs of multiphase pore fillings (Central Australia). This leads one to conclude that the inselbergs which occur in these zones have times of humid climate to thank for their formation and that they outlived at least one change in climate. There is only small subsequent superimposition: the maintenance of different microforms of weathering on Ayers Rock and of hollows (e.g., Mt. Mafa) and only slightly tafonized blocks (e.g., Mt. Niangbo) on the surface of inselbergs can hardly be explained in any other way. This only slight superimposition allows one to record early stages of chemical weathering in tropical regions by analyzing products of weathering on inselbergs. This only partly applies to shield inselbergs, where remainders of the former weathering cover can be present in oriçangas.

Table 2.7. Connection between soil thickness and vegetative covering along the edge of a shield inselberg in the Comoé National Park (classification of the plants and level of cover by R. Seine)

Thickness of the substratum (cm)	0–1	1–4	5–15	15–20	20–50
substrata and soils	Grus and angular weathering debris	Grus and angular weathering debris with little accumulation of humus	Lithosol in the weathering debris with $A_{(h)}$ horizon and grus scattered on the surface	Transition from lithosol to cambic soil	Cambic soil ($A_{(h)}$–$B_{(v)}$–C profile; with particles of grus on the surface and in the subsoil
Maximum plant height (cm)	0	<10	<20	<60	<200
total cover (%)	0	25	35	10	65
Plants:					
Cyanotis lanata		2	2		
Microchloa indica		1	1		
Sporobolus pectinellus		1	1		
Isoetes nigritiana			1	1	
Loudetia togoensis				+	3
Riccia sp.				+	
Loudetiopsis kagerensis					3
Chamaecrista mimosoides					+
Fimbristylis dichotoma					+
Oldenlandia lancifolia					r
Heliotropium strigosum					r
Pandiaka heudelotii					r

The beginnings of weathering are of interest for the supply of nutrients to plants. Water analyses taken from inselbergs in the Côte d'Ivoire show that more nutrients are present than are to be expected in view of the velocity of weathering of the bedrock. At least in the case of the oriçangas, it cannot be excluded that material is brought by the wind. In Australia this can be verified using Ayers Rock as an example, although apparently the humidity is enough to set the weathering processes in motion. The transformation of dust into gel and the fixing of the dust are probably above all due to iron compounds. Moreover, easily soluble substances in cracks in the rock show that even in climatic as well as edaphic arid positions, conversion of substances takes place. In West Africa it is very probable that dust is carried in by the Harmattan. Together with the arrival of fresh minerals through exfoliation on the inselberg, this is one of the reasons for the increased pH values in oriçanga soils and the high proportion of solubles in oriçanga waters.

References

Bakker JP (1957) Zur Entstehung von Pingen, Oriçangas und Dellen in den feuchten Tropen, unter besonderer Berücksichtigung des Voltzberggebietes (Surinam). Abh Geogr Inst FU Berlin 5 (Festschrift Maull): 7–20

Birot P (1978) Évolution des conceptions sur la genèse des inselbergs. Z Geomorphol NF Suppl 31:42–63

Bornhardt W (1900) Zur Oberflächengestaltung und Geologie Deutsch-Ostafrikas. Reimer, Berlin

Bremer H (1965) Ayers Rock, ein Beispiel für klimagenetische Morphologie. Z Geomorphol NF 9:249–284

Bremer H (1967) Zur Morphologie von Zentralaustralien. Heidelb Geogr Arb 17

Bremer H (1971) Flüsse, Flächen- und Stufenbildung in den feuchten Tropen. Würzb Geogr Arb 35

Bremer H (1981) Inselberge – Beispiele für eine ökologische Geomorphologie. Geogr Z 69:199–216

Bremer H (1982a) Abtragungsgeschwindigkeiten in den feuchten Tropen. Z Geomorphol NF Suppl 43:19–27

Bremer H (1982b) Verwitterungsformen als Stabilitätszeugen. Geogr Z 70:69–80

Bremer H (1989) Allgemeine Geomorphologie. Borntraeger, Berlin

Bremer H (1993) Etchplanation, review and comments on Büdel's model. Z Geomorphol NF Suppl 92:189–200

Bremer H (1994) Soils in tropical geomorphology. Z Geomorphol NF 38:257–265

Bremer H (1995) Boden und Relief in den Tropen: Grundvorstellungen und Datenbank. Relief Boden Paläoklima 11

Brook GA (1978) A new approach to the study of inselberg landscapes. Z Geomorphol NF Suppl 31:138–160

Büdel J (1957) Die 'doppelten Einebnungsflächen' in den feuchten Tropen. Z Geomorphol NF 1:201–228

Büdel J (1977) Klimageomorphologie. Borntraeger, Berlin

Büdel J (1978) Das Inselberg-Rumpfflächenrelief der heutigen Tropen und das Schicksal seiner fossilen Altformen in anderen Klimazonen. Z Geomorphol NF Suppl 31:79–110

Büdel J (1982) Climatic geomorphology. Princeton University Press, Princeton

Eldin M (1971) Le climat. Mém ORSTOM 50:73–108

Fitzpatrick EA (1980) Soils. Their formation, classification and distribution. Longman, London

Herrmann L (1996) Staubdeposition auf Böden Westafrikas. Hohenh Bodenk Hefte 36

Hövermann J (1978) Untersuchungen und Darlegungen zum Inselbergproblem in der deutschen Literatur der 1. Hälfte des 20. Jahrhunderts. Z Geomorphol NF Suppl 31:64–78

Howard JA, Mitchell CW (1985) Phytogeomorphology. Wiley, New York

Jenny H, Leonhard CD (1934) Functional relationships between soil properties and rainfall. Soil Sci 38:363–381

Jessen O (1936) Reisen und Forschungen in Angola. Reimer, Berlin

Loughnan FC (1969) Chemical weathering of the silicate minerals. Elsevier, New York

Sander H (1995) Untersuchungen zur Porosität tropischer Böden mittels dünnschliff-mikroskopischer Erfassung von Grobporen. Regensb Geogr Schr 25:143–155

Sander H (1996) Relief- und Boden in Nordost- und Zentralaustralien. Relief Boden Paläoklima 12

Späth H (1981) Bodenbildung und Reliefentwicklung in Sri Lanka. In: Bremer H, Schnütgen A, Späth H (eds) Zur Morphogenese in den feuchten Tropen. Verwitterung und Reliefbildung am Beispiel von Sri Lanka. Relief Boden Paläoklima 1:185–238

Thomas MF (1978) The study of inselbergs. Z Geomorphol NF Suppl 31:1–41

Thomas MF (1994) Geomorphology of the tropics. Wiley, Chichester

Twidale CR (1978) On the origin of Ayers Rock, Central Australia. Z Geomorphol NF Suppl 31:177–206

Twidale CR (1982) Granite landforms. Elsevier, Amsterdam

Twidale CR, Bourne JA (1975) Episodic exposure of inselbergs. Bull Geol Soc Am 86:1473–1481

Wilhelmy H (1981) Klimamorphologie der Massengesteine. 2nd edn. Westermann, Braunschweig

3 Xeric Islands: Environmental Conditions on Inselbergs

J. SZARZYNSKI

3.1 Introduction

Environmental conditions on inselbergs such as climate, moisture, or nutrient availability drive important ecosystem processes and modify the structure and composition of flora and fauna (Cantlon 1953; Phillips 1982). Especially microclimatic conditions often differ considerably from regional climate and steep gradients may occur within very short distances (Porembski et al. 1996). Beyond these fundamental observations, however, very little is known about diurnal and seasonal patterns of microclimate. As a contribution to the interdisciplinary framework of this book, the present chapter explores some of the important climatic interactions between inselbergs, the vegetation, and the atmosphere. Main aspects are the interception, retention, modification, and conductance of energy and water. Parallel to botanical and geological observations, the required micrometeorological data have been quantified by a long-term study in the Côte d'Ivoire, where special regard was paid to inselbergs in semi-deciduous rainforest. In order to emphasize the contrasting environmental conditions of these directly neighboring ecosystems, measurements were made on top of an inselberg, as well as within the surrounding rainforest during both wet and dry seasons under a variety of weather conditions. Following supplemental short-term investigations were accomplished in Cameroon, Nigeria, Bénin, and South America (Venezuela).

In general, environmental conditions on inselbergs are extremely harsh for plant life. Soil cover is restricted to gentle slopes including shallow soil-filled depressions or plain surfaces bearing small forest patches. Therefore, rainfall is largely lost to runoff at the bare rock and the retention and storage of water is limited. These features characterize a structural edaphic drought that primarily restricts the amount of water available to plants. Furthermore, the water budget of inselbergs is governed by the strong absorption of solar radiation at the rock surface causing high temperatures and appreciable rates of evapotranspiration. As a consequence of extreme

Ecological Studies, Vol. 146
S. Porembski and W. Barthlott (eds.) Inselbergs

environmental circumstances, inselbergs provide habitats for a highly adapted flora differing almost completely from the surrounding vegetation.

3.2 The Study of Microclimate on Inselbergs in the Côte d'Ivoire

The study was conducted in the western part of the Taï National Park (5°52'N, 7°27'W), at an altitude of 180–220 m in southwestern Côte d'Ivoire. The vegetation is classified as Forêt dense humide semi-decidue (Guillaumet 1967). Throughout the year the regional climate is variably influenced either by dry northeast trade winds (Harmattan) or the southwest monsoon, a humid airstream blowing from the Gulf of Guinea (Nieuwolt 1982). According to data from the meteorological station of Taï (period: 1945–1975) the mean annual rainfall in the study area is about 1.830 mm with a mean annual temperature of 26.2 °C (Casenave et al. 1981). There is usually a main wet season from March–July and a short wet season from September–November interrupted by a short dry period in August. The main dry season predominates from December–February. Detailed information about regional climate are published by Cardon (1978), Monteny (1983), and UNESCO/MAB (1984).

Field work was done during the 1992 dry season (January–February) and the 1992 short wet season (September–October). The measurement equipment contained two electronical data logger (GRANT) with different external sensors. On the inselberg 10-min averages of air temperature, relative humidity, global radiation, and wind velocity were continuously monitored in 2 m height, as well as temperatures of a plant cushion and the rock surface. In order to minimize errors due to irradiation while measuring the surface temperature, a fine-wire thermocouple thermometer and heat-conductive paste were used. Several walking transects were undertaken with the help of portable tools recording photosynthetic photon flux density (PPFD) in the waveband 400–700 nm, and air temperature and humidity. In combination with comprehensive data concerning regional climate and water run-off provided by the Department of Meteorology and Hydrology (Ministry of Environmental Affairs, Côte d'Ivoire), it was possible to integrate the local investigation into a spatially and temporarily broader range.

Two sites were selected for detailed measurements: the exposed bare rock of a representative inselberg and heavily shaded understory within the adjacent rainforest (Fig. 3.1). The total size of the inselberg was determined as about 10 ha and the slope's angle of inclination vary between 10° and 15°. Apart from forest patches on the summit, irregularly shaped rock pools, ephemeral flush vegetation, and small cushions of *Cyanotis arachnoidea* occurred on the inselberg, representing characteristic habitats for this region. The rock mainly consists of Precambrian migmatites and has a dark appearance due to a high portion of pyroxenes and amphiboles (Papon 1973). Furthermore the apparent bare rock is almost completely covered with lichens (e.g., *Peltula* spp.) also causing a darkish color (Porembski and Barthlott 1992; Barthlott et al. 1993). The rainforest station was installed approximately 120 m west of the inselberg, measuring air temperature, relative humidity, global radiation, and wind velocity in 2 m height and soil temperature in 20 cm depth. Within mature forest three different layers with diverse microclimatic conditions were perceptible (cf. Allen et al. 1972; Kira and Yoda 1989). The understory was sparsely developed because of the density of canopy which reaches up to 35 m in height, discontinuously interrupted by emergent trees over 40 m in height.

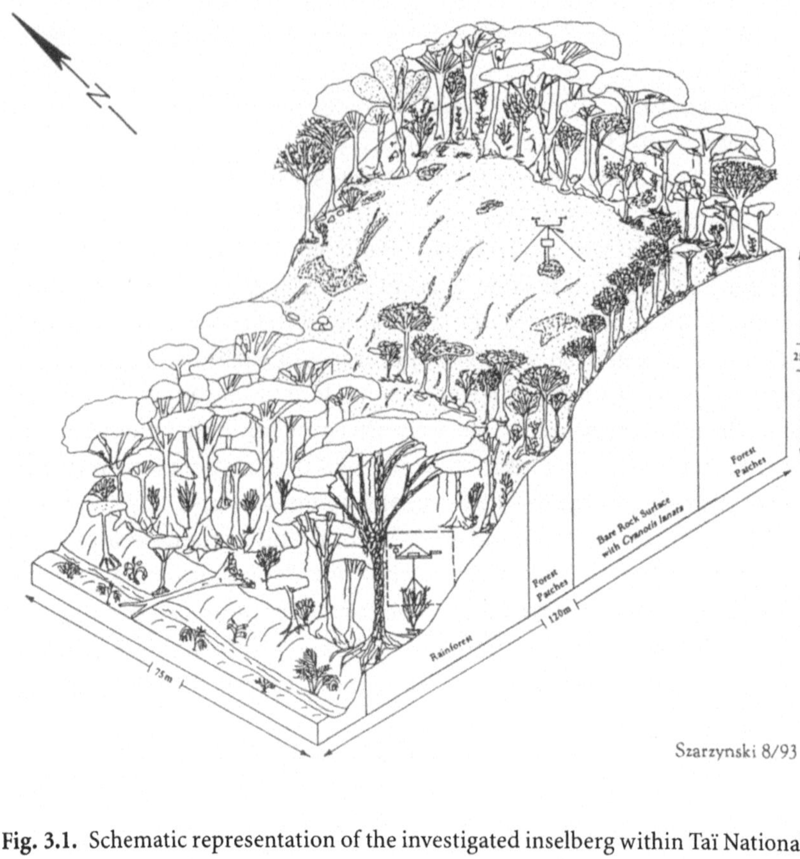

Fig. 3.1. Schematic representation of the investigated inselberg within Taï National Park

3.2.1 Microclimatic Observations During the Dry Season (January–February 1992)

During the dry season the atmosphere in the study area usually contains a high amount of Saharian dust transported by northeastern trade winds. When considering regional climate, the dimming effect of aerial particles must be taken into account. Therefore results from radiation measurements are a useful introduction to the general knowledge of climatic conditions found at the two investigation sites. The maximum of global radiation on the inselberg was 830 Wm^{-2}, while the intensity within the forest hardly exceeded 40 Wm^{-2}. In other words, maximal 5% of the incident radiation have reached the forest floor corresponding to the results of other authors (Longman and Jenik 1974; Endler 1993). Photosynthetic photon flux density was temporarily measured with a portable sensor during the same period and ranged from less than 10 $\mu mol\ m^{-2} s^{-1}$ in the understory to over 1600 $\mu mol\ m^{-2} s^{-1}$ on the inselberg (cf. Brown et al. 1994).

Hygrothermal conditions of both sites are presented by the graphs of an exemplary 5-day recording period including measured and calculated parameters (Fig. 3.2a–d). On the inselberg diurnal amplitudes of air temperature (18–37 °C) and relative humidity (100–28 %) were extremely high (Fig. 3.2a). In the rainforest, air temperature differed only between 18–26.5 °C and the minimum relative humidity was about 70 %. Soil temperature can be regarded as nearly isothermal (Fig. 3.2 c). A striking similarity between the temperature graphs of the plant cushion and the rock surface (Fig. 3.2a) can be noticed. Even during times of maximal temperatures at the rock surface (55 °C) only a slight difference of 5 K was measured inside the cushion (50 °C).

The described hygrothermal conditions on the inselberg are caused by the specific radiation and heat balance. In turn, these budgets are influenced by physical attributes of the rock surface. The mentioned dark coloration of the rock primarily increases the amount of absorbed irradiation relatively to the decrease in reflected shortwave radiation (albedo). Further important aspects are the heat capacity and thermal conductivity of the rock, summarized by the expression thermal admittance, i.e., the physical property of a surface that governs the ability to take up or release heat (Oke 1987). According to these attributes, the high amount of absorbed energy leads to a strong heating of the rock surface and an adequate emission of longwave thermal radiation warming the air near the ground (sensible heat flux).

As a matter of course, atmospheric humidity is also affected and the transfer of water vapor has important energetic (latent heat flux), hydro-

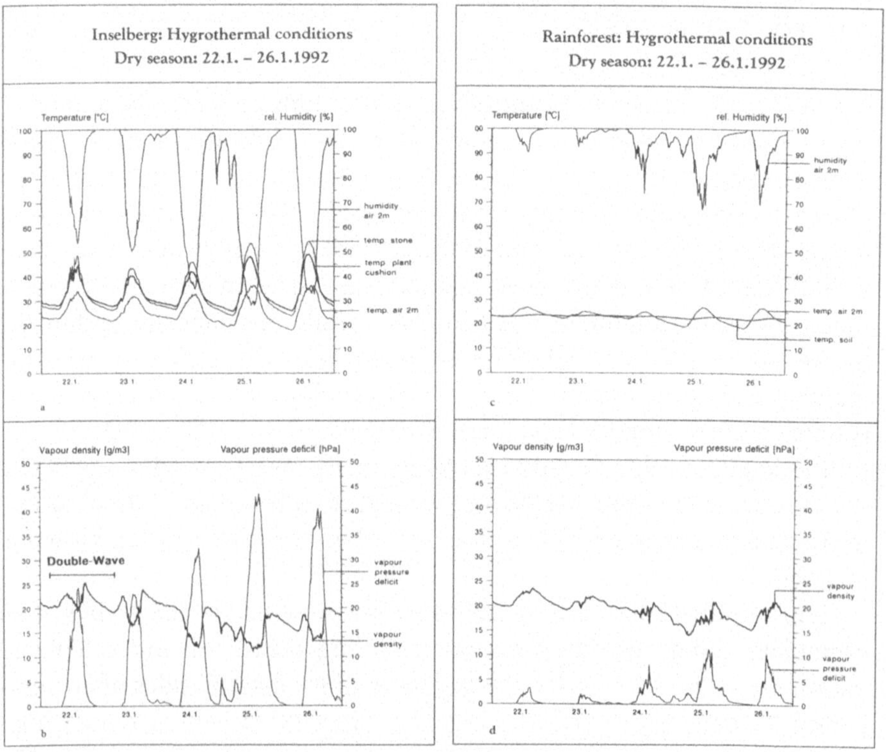

Fig. 3.2a–d. Measured (**a,c**) and calculated (**b,d**) parameters for the two investigation sites during the dry season

logical, and biological implications. Two basic parameters are illustrated to represent diurnal variations of air humidity: vapor pressure saturation deficit and vapor density, also referred to as absolute humidity (Fig. 3.2b,d). The vapor pressure deficit is considered as an important eco-physiological factor because it largely determines plant transpiration and indirectly inhibits photosynthesis by reducing stomatal aperture (Larcher 1994). In consequence of high air temperature and low humidity, the vapor pressure deficit increases corresponding with the evaporative demand at the leaf surface. During the illustrated recording period no precipitation occurred and the small reservoir of available water was considerably exhausted by evapotranspiration (Fig. 3.2b). Numerically, the recorded daily maximum was four times higher on the inselberg (44 hPa) compared with the value obtained within the rainforest (11 hPa) (Fig. 3.2b,d).

Focusing on the progress of vapor density or absolute humidity on the inselberg, the graph shows a typical diurnal double wave (Fig. 3.2b). After sunrise, the curve ascends due to increasing temperature and evaporation.

During midday, however, the absolute humidity decreases with a minimum in the early afternoon. Responsible for the change in water vapor content is the occurrence of convection, the major process of transporting the energy surplus of the surface into the atmosphere (Roedel 1994). The transport itself is mostly connected with thermals. These are rising masses of warm air heated at the rock surface and carrying temperature as well as water vapor or other properties of the air. Especially the dry and sun-exposed slopes of the inselberg act as sources for the temporal development of a thermal air rise. It starts with a bubble of hot air at the rock surface until the buoyancy becomes sufficient due to instability. The air bubble begins to rise initially with expanding tendencies and increasing velocity (Fig. 3.3; cf. Oke 1987). In greater heights, the mixing with cooler air and condensation into clouds prevent a further rise. In order to compensate for the air deficit near the ground, descending air from higher altitudes and air volumina from the surrounding rainforest close the thermally induced circulation. When insolation is reduced at late afternoon the buoyancy effects of convection are weakened and once again the content of absolute humidity increases until nocturnal condensation and dewfall (Fig. 3.2b). The two peaks of decreasing humidity during the night from 24.1.–25.1. can be explained by advective dry air (Harmattan) also influencing microclimatic conditions (Fig. 3.2b).

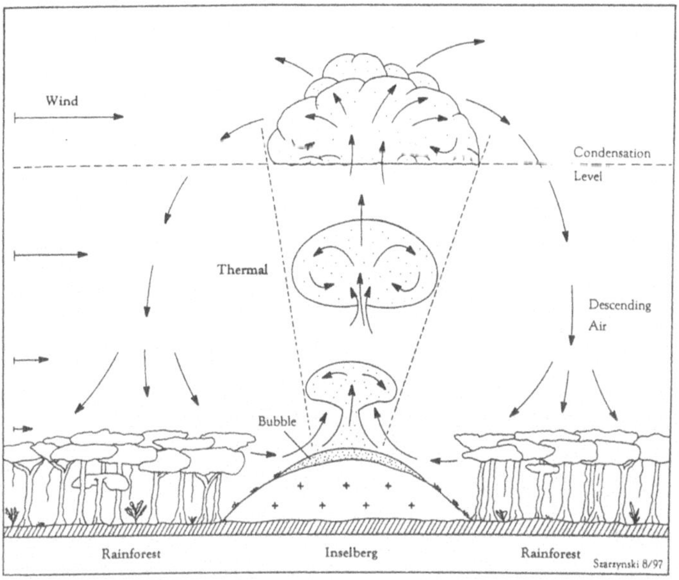

Fig. 3.3. Schematic depiction of different stages in the temporal development of a thermal over an inselberg in tropical rainforest. (After Oke 1987)

The velocity of convective air movements normally ranges between 2–5 m s^{-1}; however, within wind spouts or extensive cumulonimbus clouds it may increase to 10–30 m s^{-1} associated with high atmospheric instability (Malberg 1994). Over inselbergs convective upcurrents are often recognizable by the thermal soaring of birds of prey. While these birds use the vertical lift actively, tiny seeds of anemochorus plants are carried passively. In the case of the studied inselberg this transport mechanism may be an important aspect for plant dispersal because the tree crowns of the surrounding rainforest rise above the rock surface where, in turn, the horizontal wind velocity is strongly reduced (95 % of the measured values ranged below 1 m s^{-1}). For this reason, plants like *Lindernia exilis, Hymenodictyon floribundum*, or *Stereospermum acuminatissimum* (S. Porembski pers. comm.) are separated from the airstream above the rainforest canopy. However, convective thermals, even if small-scaled, enable tiny seeds to pass the obstructive distance up to the horizontal airstream by which they are finally distributed to a broader circle.

Certainly convective fluxes also occur over the rainforest, where the canopy represents the principal interface with the atmosphere (Lee 1993). Within the canopy the biomass and the interstitial air are heated by incident radiation. Yet the retention and conductance of heat are different compared with the rock surface of the inselberg which definitely appears like a thermal spot within the rainforest. Furthermore, the decay of radiative energy from the canopy to the forest floor produces a stable density stratification because warmer and lighter air is located above cooler and denser air below which partially isolates the understory or lower canopy from climatic fluctuations of the crown region and the atmosphere above (Fitzjarrald and Moore 1995). Thus the diurnal double wave of vapor density as described for the inselberg could not be measured within the forest, whereas the already mentioned influence of advective air (Harmattan) was perceptible mainly due to the proximity of the uncovered rock outcrop causing a gap where dry air could enter into the forest (Fig. 3.2d).

3.2.2 Microclimatic Observations During the Wet Season (September–October 1992)

The short wet season in the study area is caused by the equatorward movement of the Intertropical Convergence Zone (ITCZ) (Ojo 1977). In September and October 1992 the meteorological station in Taï recorded 537 mm rainfall, i.e., 30 % of the total annual precipitation (1.757 mm). Especially the amount of rainfall in September (348 mm) describes extraordinary humid conditions. Due to the absence of dust particles within the atmosphere, the intensity of irradiation was higher compared to the situation in January–February. The maximum 10-min average reading on the inselberg was 1100 Wm^{-2} and photosynthetic photon flux density was about 2150 $\mu mol\ m^{-2}\ s^{-1}$. In the understory of the rainforest the radiation that reached the ground was reduced to 3 %, caused by an increased leaf area during the wet season.

With regard to the question of how seasonal patterns modify the microclimate at both study sites, again data are illustrated by a selected 5-day sampling period (Fig. 3.4a–d). The greater supply of water and moisture during the wet season is particularly reflected by the graphs of the rainforest station (Fig. 3.4 c,d). Air within rainforest was almost constantly saturated (90–100 %) while air temperature ranged between 22–27 °C (Fig. 3.4 c). As a result, the vapor pressure deficit was low with a maximum of 5 hPa and vapor density varied between 19–24 $g\ m^{-3}$ (Fig. 3.4d). In comparison, the microclimate on the inselberg was obviously less affected by

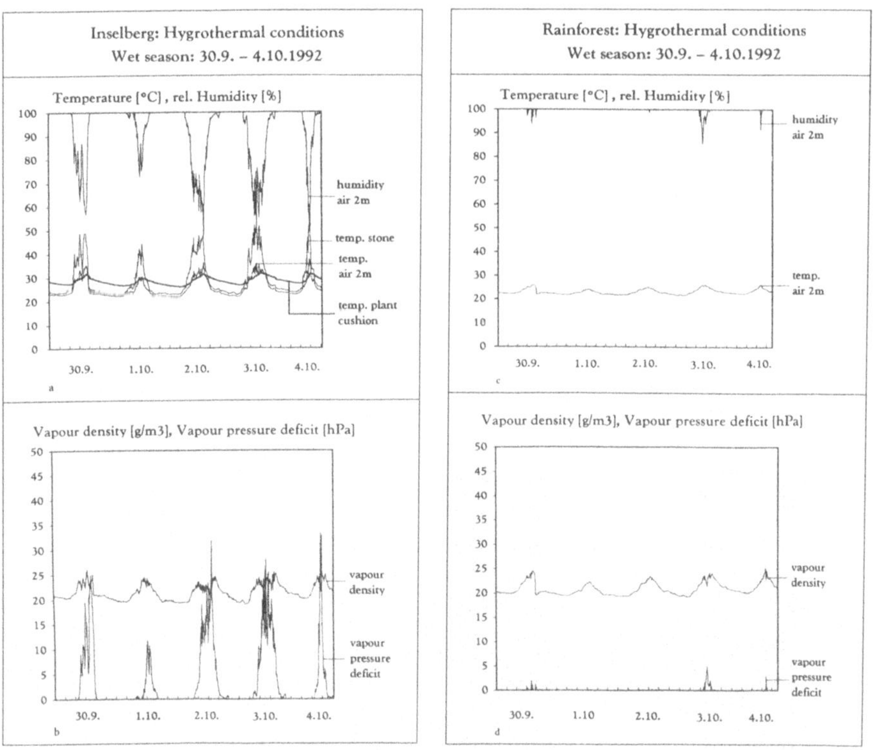

Fig. 3.4a–d. Measured **(a,c)** and calculated **(b,d)** parameters for the two investigation sites during the wet season

regional climatic conditions (Fig. 3.4a,b). Although relative humidity hardly subsided below 50% due to the higher moisture potential, the thermal conditions of the air and the rock surface were quite similar to those observed during the dry season (cf. Fig. 3.2a and 3.4a). Therefore the vapor pressure deficit occasionally reached maximum values of 34 hPa (Fig. 3.4b).

In order to draw a more precise picture of the interrelation between regional climate, microclimate, and resulting effects on the vegetation, it is useful to operate with a parameter such as evapotranspiration (ET), which integrates meteorological and physiological aspects. As far as climatological and biological implications are concerned, actual evapotranspiration (AET) is considered a more relevant factor than potential evapotranspiration (PET), yet its accurate determination requires a high technical measurement support (Lockwood 1976). In contrast, PET, defined as the amount of water loss that would occur with unlimited water supply, can be computed from meteorological samplings and the use of empirical formulas.

Based on regional climatic data from the meteorological station in Taï and recorded measures from the inselberg, PET was calculated according to the equation of Papadakis (1965). In Fig. 3.5 the results are summarized for the dry season (17.1.–9.2.92) and the wet season (28.9.–21.10.92). The total sum of PET determined for the inselberg (298 mm) was considerably higher than PET referring to regional climate (182 mm). Moreover, the individual terms of the inselberg were characterized by similar amounts during dry and wet season (147–151 mm). The corresponding values of the Taï station, however, differed between 105 mm in the dry season and 77 mm in the wet season (Fig. 3.5). As a conclusion, microclimate on the

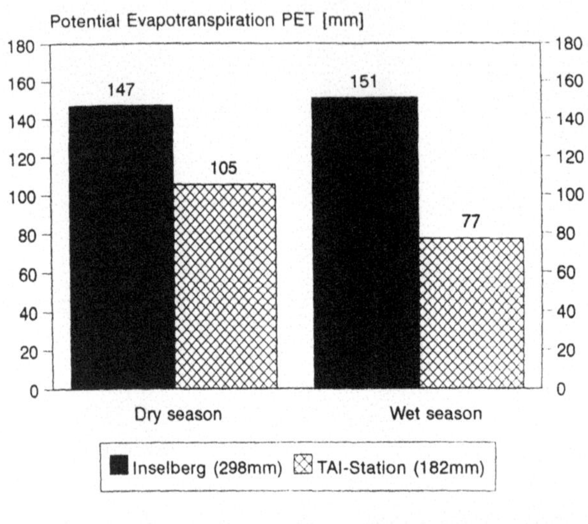

Fig. 3.5. Comparison of potential evapotranspiration (PET) computed for the inselberg (representing microclimate) and for Taï-Station (representing regional climate) according to the formula of Papadakis (1965)

inselberg seemed to be rather unaffected by variations in regional climate, and steady conditions predominate. Virtually, the calculations signify that the actual loss of water on the inselberg (AET) is directly related to its availability. Therefore in October the amplitudes of vapor density were smaller due to sufficient moisture supply and the values differed only between 18 and 27 g m^{-3} (2.10.–3.10.92, Fig. 3.4b). Nevertheless, in drier periods, especially when the reservoir of soil water is exhausted, high evapotranspiration leads to climatic drought that decisively influences the water balance of plants.

Finally, attention is focused on the illustrated temperature trend within the cushion of *Cyanotis arachnoidea* (Fig. 3.4a). In the dry season the curve nearly coincided with the temperature of the rock surface, characterized by high daily amplitudes (25 K) and maxima of about 50 °C (cf. Fig. 3.2a). On the other hand, during the wet season the graph was well-balanced and measures ranged from 27–32 °C (Fig. 3.4a). These completely different thermal conditions were based simply on the structure of the cushion. During the dry season the parched leaves were thin and barren. Insolation was able to reach the ground, greatly increasing the temperature within the cushion. In the wet season the leaves were more voluminous due to stored water and because of the greater density of the cushion, consequently an internal microclimate was produced, mitigating the environmental temperature fluctuations. A further aspect worth mentioning is the seasonal change of color. Under hot and dry climatic conditions the leaves of *Cyanotis arachnoidea* were red due to anthocyanin, which seemingly protects against a high degree of insolation. The humid conditions in October furthered the activity of growth and photosynthesis expressed by the prevalent green color of chlorophyll. These observations may contribute to the well-known fact that extreme environmental conditions favor the selection of morphological and physiological adaptations of plants such as tolerance strategies or avoidance mechanisms (cf. Lange 1959; Kluge and Brulfert, Chap. 9, this Vol.). Vice versa, they emphasize the ecological significance of edaphic and microclimatic features on inselbergs.

3.3 Conclusions

Summarizing the results obtained in the Côte d'Ivoire, inselbergs can be regarded as xeric islands or microenvironmental deserts in the sense of Phillips (1982). Due to edaphic features and the specific principles of heat exchange, the environmental conditions are generally dry and characterized by high degrees of temperature and evapotranspiration. In addition, xeromorphic and even succulent adaptations of plants reflect the extreme ecophysiological situation. Further investigations on inselbergs throughout the tropics and extratropics have proven that microclimate as well as the physiognomy of vegetation are quite similar wherever studied, and both aspects differ more or less considerably from the surrounding environment. Thus, inselbergs represent terrestrial ecosystems with insular attributes. Depending on the general character of the landscape they provide habitats for an azonal type of vegetation associated with azonal climatic conditions. For example, in the northern savanna zone of the Côte d'Ivoire where mean annual precipitation is about 1000 mm, inselbergs are often surrounded by an extrazonal forest belt caused by the local water runoff. The number of species found on those northern rock outcrops was higher than in the southern rainforest zone, where annual rainfall exceeds 2000 mm. In other words, biodiversity conducts contrary to the climatic gradient of precipitation (Barthlott et al. 1996). However, the number of species in the savanna zone is largely increased by ephemeral flush vegetation and plant communities that occur only during wet seasons, which finally emphasizes the significance of meso- and macro-scaled seasonal disturbances.

Acknowledgements. The presented investigations were financed by the Deutsche Forschungsgemeinschaft (DFG) within the program Mechanisms of maintainance of tropical biodiversity. The possibility of field research was offered by W. Barthlott (Bonn) and S. Porembski (Rostock) within the frame of the inselberg project. Their help is gratefully acknowledged. We are indebted to several institutions of the Côte d'Ivoire (ASECNA, ANAM, IET) for providing comprehensive climatic and hydrologic data. For discussions and support we have to express our thanks to D. Anhuf (Mannheim), W. Schmiedecken (Bonn), M. Winiger (Bonn), and a large number of colleagues in various countries. Their collaboration contributed decisively to the success of this project. Special thanks for field assistance are directed to J.-P. Mund (Mainz) and T. Austel (Bonn), U. Becker (Cologne), G. Radl (Taï), H. Sander (Cologne) and R. Seine (Bonn).

References

Allen LH, Lemon E Jr, Müller L (1972) Environment of a Costa Rican forest. Ecology 53:102–111

Barthlott W, Gröger A, Porembski S (1993) Some remarks on the vegetation of tropical inselbergs: diversity and ecological differentiation. Biogéographica 69:105–124

Barthlott W, Porembski S, Szarzynski J, Mund J-P (1996) Phytogeography and vegetation of tropical inselbergs. In: Guillaumet J-L, Belin M, Puig H (eds) Actes du colloque international de phytogéographie tropicale. ORSTOM, Paris, pp 15–24

Brown MJ, Parker GG, Posner NE (1994) A survey of ultraviolet-B radiation in forests. J Ecol 82:843–854

Cantlon JE (1953) Vegetation and microclimate on north and south slopes of Cushetunk Mountain, New Jersey. Ecol Monogr 23:241–270

Cardon D (1978) État des connaissanses climatiques dans le sud-ouest de la Côte d'Ivoire. ORSTOM, Adiopodoumé

Casenave A, Flory J, Ranc N, Simon JM (1981) Étude hydrologique des bassins de Taï. Campagne 1980. ORSTOM, Adiopodoumé

Endler JA (1993) The colour of light in forests and its implications. Ecol Monogr 63:1–28

Fitzjarrald DR, Moore KE (1995) Physical mechanisms of heat and mass exchange between forests and the atmosphere. In: Lowman M, Nadkarni NM (eds) Forest canopies. Academic Press, San Diego, pp 45–72

Guillaumet J-L (1967) Recherche sur la végétation et la flore de la région du Bas-Cavally (Côte d'Ivoire). Mém ORSTOM 20, Paris

Kira T, Yoda K (1989) Vertical stratification in microclimate. In: Lieth H, Werger N (eds) Tropical rain forest ecosystems. Elsevier, Amsterdam, pp 55–71

Lange OL (1959) Untersuchungen über Wärmehaushalt und Hitzeresistenz mauretanischer Wüsten- und Savannenpflanzen. Flora 147:595–651

Larcher W (1994) Ökologie der Pflanzen. Ulmer, Stuttgart

Lee R (1993) Forest microclimatology. Columbia University Press, New York

Lockwood JG (1976) The physical geography of the tropics. Oxford University Press, London

Longman KA, Jenik J (1974) Tropical forest and its environment. Longman, London

Malberg H (1994) Meteorologie und Klimatologie. Springer, Berlin Heidelberg New York

Monteny BA (1983) Observations climatiques à la station écologique de Taï dans le sud-ouest Ivoirien. 1978–1982. ORSTOM, Adiopodoumé

Nieuwolt S (1982) Tropical climatology. Wiley, London

Ojo O (1977) The climates of West Africa. Heinemann, London

Oke TR (1987) Boundary layer climates. Methuen, London

Papadakis J (1965) Potential evapotranspiration. Buenos Aires

Papon A (1973) Géologie et minéralisations du sud-ouest de la Côte d'Ivoire. Synthèse des travaux de l'operation Sacsa 1962–1968. SODEMI, Abidjan

Phillips DL (1982) Life-forms of granite outcrop plants. Am Midl Nat 107:206–208

Porembski S, Barthlott W (1992) Struktur und Diversität der Vegetation westafrikanischer Inselberge. Geobot Kolloq 8:69–80

Porembski S, Mund J-P, Szarzynski J, Barthlott W (1996) Ecological conditions and floristic diversity of an inselberg in the savanna zone of the Ivory Coast - Mt. Niangbo. In: Guillaumet J-L, Belin M, Puig H (eds) Actes du colloque international de Phytogéographie tropicale, ORSTOM, Paris, pp 251–261

Roedel W (1994) Physik unserer Umwelt - die Atmosphäre. Springer, Berlin Heidelberg New York

UNESCO/MAB (1984) Recherche et aménagement en milieu forestier tropical humide: le projet Taï de Côte d'Ivoire. UNESCO, PNUE, ORSTOM, IET, Note Tech. MAB 15, Paris

4 Islands on Islands: Habitats on Inselbergs

S. Porembski, U. Becker, and R. Seine

4.1 Introduction

When seen from a distance, inselbergs seem to form homogeneous landscape features which mainly consist of large expanses of bare, dark-colored rock (Fig. 4.1). However, a closer look reveals that they have to be considered as an ecosystem which comprises an unexpected amount of clearly distinguished vegetational habitat types representing "islands on islands" (Ornduff 1987). In the 19th century, the unusual appearance of inselbergs attracted the attention of famous naturalists, who provided the first descriptive accounts of their vegetation. Carl Friedrich Philipp von Martius (1842) and Alexander von Humboldt (1819) were amongst them. The latter discussed in detail the dark color of rocks in the Orinoco riverbed and of inselbergs in the same area. Humboldt assumed that the black color of the rocks was due to a sheath of manganese oxide and carbon. Today, however, we know that the occurrence of epi- and endolithic cyanobacteria is responsible for the dark coloration of these rocks. Among the first floristic studies concentrating on tropical inselbergs, one has to mention the work of Willis (1906) concerning vascular plants on Ceylonese rock outcrops. For a remarkably long period, the vegetation of inselbergs was simply denoted as lithophytic without any further differentiation, or it was described as "savanes-roche" due to a certain amount of physiognomic similarity with savanna vegetation (de Granville 1978). For granite outcrops in the USA, Burbanck and Platt (1964) provided a classification of inselberg plant communities according to maximum soil depth.

An essential result of our own studies has been the observation that inselbergs are not randomly assembled ecosystems. Despite fundamental differences in their floristic composition between distant geographic and climatic regions, they bear a typical set of physiognomically defined habitats that occur almost identically on inselbergs all over the world. However, there is a lack of a generally applicable terminology that facilitates com-

Ecological Studies, Vol. 146
S. Porembski and W. Barthlott (eds.) Inselbergs

Fig. 4.1. Large expanses of seemingly bare rock are typical features of inselbergs. The dark to grayish coloration of the rock is due to a cryptogamic crust with cyanobacteria and lichens as main components. (Brazil, Minas Gerais, photograph N. Biedinger)

parisons. A first preliminary enumeration of the major habitat types on inselbergs was given by Barthlott et al. (1993).

In the following, a concise description of the individual habitats on inselbergs is given. We have focused on non forest communities that occur under open, fully exposed conditions because these constitute the most typical vegetation units. Forest-type vegetation is therefore largely omitted from this enumeration. The classification of habitats is based on both geomorphological characteristics of the sites and physiognomic criteria of the vegetation. The distinction between individual habitats is surprisingly clear, and only in a few cases can transgressions be observed. For detailed information on floristics we refer to Chapter 10 (this Vol.).

4.2 Habitat Types on Inselbergs

4.2.1 Vegetation of Rock Surfaces

Typical is the lack of a soil cover and the occurrence of microclimatic extremes (e.g., temperatures over sun-exposed rock surfaces regularly exceed 60 °C, see Szarzynski Chap. 3, this Vol.) A biofilm consisting of cryptogams is the most important characteristic of the habitats included here which can be subdivided as follows:

4.2.1.1 Cryptogamic Vegetation of Rock Surfaces

On average, inselbergs are almost completely covered by a mosaic of either cyanobacterial lichens (typically *Peltula* spp., Fig. 4.2) or nitrogen-fixing cyanobacteria (frequently *Stigonema* spp. and *Scytonema* spp.; see Büdel et

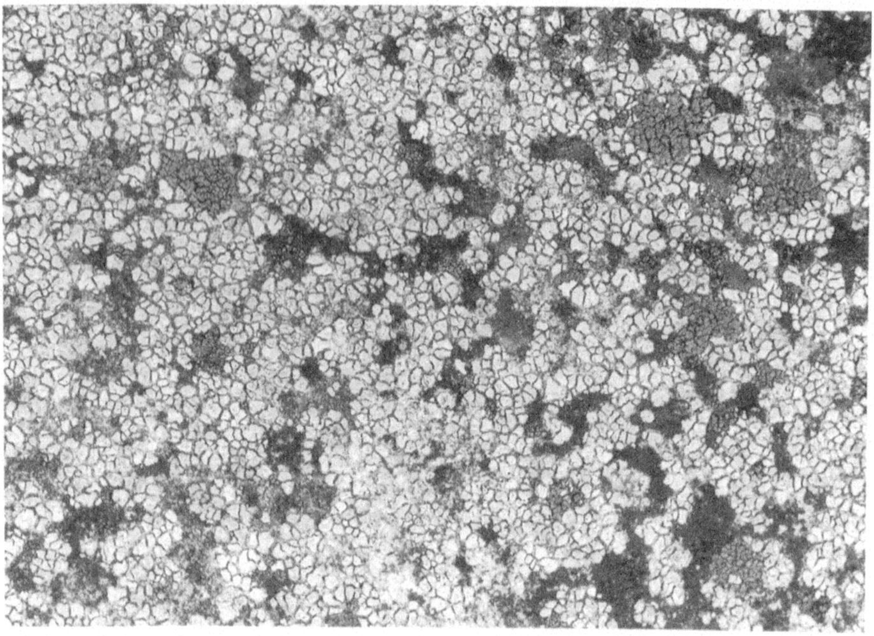

Fig. 4.2. Cyanobacterial lichens belonging to the genus *Peltula* form the most important component of the cryptogamous crust on exposed rocks of inselbergs situated in savanna regions. Usually, several *Peltula* species occur together, rendering a mosaic-like appearance to the cryptogamous crust. (Côte d'Ivoire, Sénéma, photograph S. Porembski)

al. Chap. 5, this Vol.), which are responsible for the mostly brownish or grayish color of inselbergs. Lichens and cyanobacteria, however, not only form epilithic surface layers but also grow endolithically. Cyanobacteria and certain cyanobacterial lichens living on light-exposed rocks are protected against high levels of ultraviolet radiation by the presence of ultraviolet-absorbing substances (Büdel et al. 1997). In some respects, the biofilm of inselbergs resembles microphytic crusts which are a common sight in desert regions (Lange et al. 1992; Danin 1996). They consist of cyanobacteria, green algae, lichens, and bryophytes. Among the cyanobacterial component of the microphytic crusts, several genera (*Schizothrix*, *Scytonema*) also contribute to the cryptogamic crust on inselbergs. In seasonally dry savanna regions, cyanobacterial lichens dominate on granite outcrops (lichen inselbergs), whereas under rainforest climates the rocky slopes are commonly covered by cyanobacteria (cyanobacteria inselbergs). In many respects, this habitat shows close affinities to the Tintenstrich communities which are widespread on natural and anthropogenic rock surfaces in temperate regions (Jaag 1945; Lüttge 1997). In particular, on temperate zone inselbergs extensive moss swards have been reported to cover low-angled rocky slopes (e.g., in Australia, Smith 1962). Due to the considerable water-holding capacity of the mosses, small herbs and geophytes colonize the swards.

It should be mentioned that the investigation of this habitat is not without risk if conducted on steep slopes during rainfall. Immediately after the onset of rain, the rocks become slippery due to the slimy sheaths of the cyanobacteria, which leads to considerable problems when descending.

4.2.1.2 Cryptogamic Vegetation of Boulders

Frequently, more or less rounded boulders of different sizes (up to several meters in height) occur which are often concentrated on the highest parts of inselbergs. On lichen inselbergs they are mainly covered by chlorophytic lichens and thus differ from the level rock faces with cyanobacterial lichens dominating (Fig. 4.3; see Büdel et al. Chap. 5, this Vol.). Preliminary observations indicate that the lichen species composition of individual boulders may change drastically over short distances (i.e., a few meters). On cyanobacteria inselbergs basal and overhanging parts of boulders form preferential growing sites for lichens.

Fig. 4.3. Isolated boulders occurring on sun-exposed rocky slopes bear a cryptogamous crust which mainly consists of chlorophytic lichens and thus differs completely from the surrounding level rock faces, which are covered by cyanobacterial lichens and cyanobacteria. (Brazil, Minas Gerais, photograph N. Biedinger)

4.2.1.3 Cryptogamic Vegetation of Drainage Channels

The scarce vegetation cover on inselbergs retains only a small amount of the precipitation. Most of it runs downwards through more or less steeply inclined drainage channels which were cut into the rock due to the erosive power of the water. Drainage channels may attain a length of several hundred meters, with their usual width ranging between 20 and 30 cm. Occasionally, due to the erosive power of the water, small but relatively deep rock pool-like Opferkessel have developed which may retain water over long periods. Cyanobacteria, and to a lesser extent cyanobacterial lichens, form a dense cover on the bottom of the drainage channels. Mosses and vascular plants are found only very rarely in drainage channels. Frequently, drainage channels are bordered by a zone (2–10 cm in width) of bare rock (Fig. 4.4).

Fig. 4.4. During rainfall runoff water flows downhill via drainage channels which are densely covered by crusts of cyanobacteria and certain lichens (in particular *Peltula lingulata*). (Côte d'Ivoire, Boundiali, photograph S. Porembski)

4.2.1.4 Wet Flush Vegetation

Occurring on inclined, bare rocky slopes where water flows continuously during the rainy season (Fig. 4.5). Typical is a cyanobacterial crust which in the wet state may attain a thickness of nearly 1 cm. This crust provides a substrate for colonizing vascular plants, with Xyridaceae, Eriocaulaceae, and the genus *Utricularia* being most important. These are small-sized annuals which are firmly attached by their roots to the cyanobacterial substrate. Occasionally, small patches of mosses can be found which provide establishment sites for other vascular plants. This community develops best under humid tropical climates and seems to be absent from temperate regions.

4.2.1.5 Vascular Epilithic Vegetation

Only relatively few species of vascular plants are able to colonize freely exposed bare rock. Typical constituents of this habitat, which is best developed on level rock faces of tropical inselbergs, are succulents or

Fig. 4.5. Wet flush vegetation develops on inclined rocky slopes with continuous seepage water during the rainy season. Higher plants (frequently *Xyris* spp.) root in dense cyanobacteria mats. (Côte d'Ivoire, Mt. Niangbo, photograph N. Biedinger)

xerophytic species which frequently occur as isolated individuals. Obviously, because of similar ecological constraints, vascular lithophytes (in particular in the neotropics) show pronounced systematic affinities to epiphytic relatives. Several taxa (mostly within Bromeliaceae and Orchidaceae) occur both epilithically and epiphytically.

4.2.2 Vegetation of Rock Crevices

On steep and bare rocky slopes, crevices provide opportunities for the establishment of vascular plant species. According to physical properties (e.g., width and depth), the following habitats are distinguished.

4.2.2.1 Vegetation of Horizontal and Vertical Crevices

Crevices may attain a considerable length, but are mostly less than 2 cm in width and allow for only small amounts of soil to accumulate. Therefore, plant cover is usually very sparse, with short-lived plants dominating on both tropical and temperate inselbergs. Frequently, crevices provide shaded conditions which result in the sometimes abundant occurrence of rhizomatous ferns and mosses.

4.2.2.2 Vegetation of Clefts

Due to the broader width and depth of clefts, they accumulate more substrate which renders them better growing sites than the preceding habitat type (Fig. 4.6). Perennial shrubs and trees (mostly deciduous) dominate. With increasing soil accumulation, species possessing underground storage organs or pachycaulous trees may become established. Under arid climatic conditions (e.g., in parts of the Sahara and in the Namib desert) clefts provide advantageous relatively moist sites for shrubs and trees which preferentially prosper there.

Fig. 4.6. Rock crevices form a preferred habitat for shrubs and trees due to better soil conditions. On West African inselbergs in particular legumes (here *Tephrosia mossiensis*) are richly represented in crevices. (Côte d'Ivoire, Bouaké, photograph N. Biedinger)

4.2.2.3 Vegetation around boulder bases

Large boulders provide shelter against high light intensities and desiccating winds. Occasionally, a humose substrate has accumulated. Vegetation cover is mostly sparse and frequently consists of mosses, shade-tolerant ferns, and small trees.

4.2.3 Vegetation of Depressions

Mostly on flat parts of rock outcrops, numerous depressions occur covering a wide range of sizes, forms, and substrate depths. All habitats associated with depressions are included here. Differentiation between individual habitat types is made by using characters such as duration of waterlogging, substrate depth, and physiognomy of the plant cover.

4.2.3.1 Vegetation of Seasonal Rock Pools

A wealth of names exists for seasonally water-filled rock pools on inselbergs, e.g., ephemeral pools, grinding holes, Opferkessel, oriçangas, weathering pits (cf. Freise 1938; Wilhelmy 1958; Bakker 1970). Despite the lack of exact data, it can be assumed that rock pools are relatively old habitats possessing an estimated age of at least several hundred years. In the following, the term rock pool is used. Depending on outline and development, two rock pool types can be distinguished: (1) irregularly shaped depressions of variable depth, covering up to several square meters (Fig. 4.7) possibly solutional in origin; (2) round to oval hollows mainly 10–15 cm deep, covering less than 0.5 m^2 (Fig. 4.8), of anthropogenic origin due to their usage as grinding holes in the past. The latter type occurs mainly throughout tropical Africa and is today no longer used for grinding purposes.

Best developed on level sites or only moderately inclined slopes, seasonal rock pools bear sometimes highly specialized vascular plants (even poikilohydric species occur) that depend on rainfall for their development. Rock pools form a very unreliable habitat for vascular plants because they may dry out even in the rainy season during longer rainless intervals. Generally, vegetation cover is not dense, and frequently vascular plants do not occur at all. In most cases, epilithic and endolithic cyanobacteria and certain lichens that form a dense cover on the rock-pool walls are of widespread occurrence. The vascular plants encountered in this habitat are mostly short-lived herbs that root in sandy substrate, but accidentally, free-floating water plants (*Lemna* spp.) can be seen. Apart from rock-pool specialists, a regionally variable amount of species typical for ponds (e.g., *Ludwigia* spp.) and marshy ground (e.g., *Cyperus* spp.) is present. In certain regions (e.g., West Africa) therophytes dominate which rely on seed banks for the survival of the dry season.

4.2.3.2 Vegetation of Permanently Water-Filled Rock Ponds

Only very rarely, permanently water-filled rock ponds occur. They are of natural origin and are mostly characterized by being relatively deep (occasionally several meters) whereas their surface area is comparatively small (e.g., gnammas on Australian inselbergs). Mostly, establishment conditions for vascular plants are difficult due to the absence of shallow zones. Therefore vascular plants can be found only occasionally with plants that prefer to grow in wet places dominating.

Fig. 4.7. Seasonally water-filled rock pools are particularly frequent on inselbergs possessing gently inclined or level rocky slopes. The Cyperaceae *Cyperus submicrolepis* is specialized to this habitat. (Côte d'Ivoire, Comoé National Park, photograph S. Porembski)

4.2.3.3 Vegetation of Rock Debris

Typical is a thin soil cover that usually does not exceed 2 cm in depth. The substrate is mainly sandy with only very low amounts of humose material. Cryptogams are almost always present with mosses, liverworts, and cyanobacteria, which under humid conditions may form dense swards. Therophytes dominate among the higher plants but poikilohydric vascular plants (e.g., *Selaginella* spp., *Lindernia* spp.) may also play an important role.

4.2.3.4 Herbaceous Vegetation of Soil-Filled Depressions

Shallow soil-filled depressions occur in large numbers where the rock is relatively flat. They occur in a wide range of sizes but their average size is 3 to 5 m^2 (Fig. 4.9). After heavy downpours, shallow depressions can be inundated for a couple of days. The soil is coarse-textured and dries out rapidly. Soil depth ranges between 2 and 20 cm and increases from the

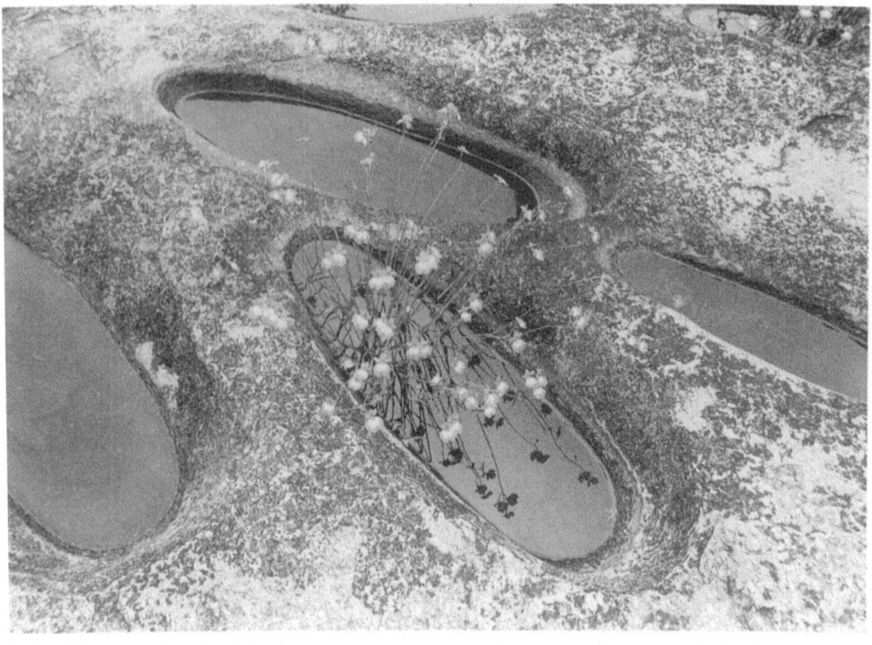

Fig. 4.8. Man-made grinding holes are subject to considerable variations in water level. Typical rock-pool specialists, like *Dopatrium longidens*, are rarely observed as colonizers of this habitat. (Bénin, Savé, photograph N. Biedinger)

Fig. 4.9. The vegetation of shallow soil-filled depressions mainly consists of short-lived species, e.g., *Cyanotis lanata, Aeollanthus pubescens.* (Côte d'Ivoire, Mt. Niangbo, photograph N. Biedinger)

periphery towards the center. This is accompanied by an increasing importance of longer-lived species. Cryptogams, therophytes, and to a lesser extent geophytes are abundant colonizers of this habitat. Occasionally, shrubs and small trees can be encountered.

4.2.3.5 Woody Vegetation of Soil-Filled Depressions

Depressions containing deeper soils offer favorable conditions for the establishment of trees (usually deciduous) and shrubs, whereas cryptogams and herbs are only rarely present. The latter groups sometimes form fringe-like vegetation belts at the periphery of this habitat.

4.2.4 Ephemeral Flush Vegetation

This habitat occurs on both tropical and temperate zone inselbergs, where it is best developed at the base of steeply inclined slopes. During the rainy season, the ephemeral flush vegetation has a meadow-like physiognomic appearance whereas the dry season aspect is almost desolate with the bare exposed soil covered with sparse desiccated plant remnants. In locations like this, the mostly tiny (Diels 1906 used the appropriate term Zwergflora) and ephemeral constituents of this habitat benefit during the rainy season from the continuous supply of seepage water. This community is considerably less profusely developed on dome-shaped inselbergs (bornhardts) where slopes are too steep and water runoff is too fast for the establishment of typical elements. Under these conditions, the ephemeral flush vegetation forms at best small rims fringing monocotyledonous mats. In this case, the ephemeral flush vegetation forms a highly dynamic and species-rich ecotone.

As a consequence of low nutrient availability, carnivorous plants (Lentibulariaceae, *Drosera*, Fig. 4.10) and other taxa (e.g., Burmanniaceae, Eriocaulaceae, Xyridaceae) which are indicative of poor soils are abundantly represented in this habitat, which is frequently rich in species. Moreover, grasses and sedges thrive in abundance. Remarkable are the ecological and physiognomical affinities of the ephemeral flush vegetation

Fig. 4.10. Tiny annuals dominate the ephemeral flush vegetation. Carnivorous plants (e.g., *Drosera indica*) belong to the most prominent species of this community. (Bénin, Parakou, photograph N. Biedinger)

with therophyte-rich pioneer communities in the temperate zone, like the class Isoeto-Nanojuncetea (Pietsch 1973).

4.2.5 Mat Vegetation

Almost the whole rock surface of inselbergs is covered by a cryptogamic crust, whereas vascular plants occur only exceptionally under these extreme environmental conditions. Remarkably, however, in certain independent lineages vascular plants have developed adaptations (most of all poikilohydry) in order to colonize this inhospitable environment. In contrast to other rock outcrop types, the relatively smooth rocky slopes of inselbergs offer advantageous conditions for the establishment of mat-forming plants. Vascular mat-forming plants on inselbergs firmly cling with their roots directly to the rock and do not have to rely on soil cover for establishment. Most mats attain a considerable age, i. e., probably hundreds of years (Bonardi 1966). During this long period the mats form their own substrate, that mainly consists of decaying plant material.

4.2.5.1 Monocotyledonous Mats

On tropical inselbergs, even fully sun-exposed steep rocky slopes are covered by carpet-like mats formed by xerophytic Bromeliaceae or poikilohydric Cyperaceae and Velloziaceae. Typically, the representatives of the latter two families possess pseudostems with the species-specific heights ranging from 10 cm up to more than 3 m. Surrounded by rock, these mats appear like islands of vegetation when seen from a distance. Frequently, individual mats which are circular in outline attain a size of between 5 and 10 m^2. On level rock surfaces, mats may cover much larger areas. The ability of effectively colonizing open rock faces is certainly related to the fact that the majority of the mat-forming species possess the ability to spread horizontally by means of vegetative propagation (via stolons or by basal branching), leading to the formation of clonal populations.

Occasionally, mats form a community very poor in species (even monospecific, Porembski et al. 1996). In tropical Africa and Madagascar, Cyperaceae are dominating (*Afrotrilepis* in West Africa, *Coleochloa* in East Africa and Madagascar), and Velloziaceae (*Xerophyta*) sometimes attain the status of a codominant. On neotropical inselbergs, Bromeliaceae (e.g., *Deuterocohnia*, *Orthophytum* spp., *Pitcairnia*, *Vriesea* spp., Fig. 4.11) and Velloziaceae dominate, whereas Cyperaceae (*Trilepis* spp.) are of minor

Fig. 4.11. Monocotyledonous mats occur on inselbergs throughout the tropics. In the neotropics the Bromeliaceae (here *Encholirium* sp.) belong to the most important mat-formers. (Brazil, Minas Gerais, photograph N. Biedinger)

importance. Mat-forming Cyperaceae and Velloziaceae possess convergently developed morphological (treelet-like habit), anatomical (velamen radicum; Porembski and Barthlott 1995), and physiological (poikilohydry) adaptations. Monocotyledonous mats are most diverse on neotropical, in particular Brazilian, inselbergs. Monocotyledonous mats provide establishment sites for a number of other plant life-forms. For example, succulents (e.g., Aloaceae, Cactaceae, Crassulaceae) may be present in considerable abundance and a moderate number of therophytes occurs at the edge of the mats. Often the exposed periphery of the mats is covered by mosses and lichens and transgresses into ephemeral flush vegetation.

Throughout the tropics the most striking component among the accompanying flora of the monocotyledonous mats is formed by several orchid genera which are strictly limited to this community. These orchids can be found as "epiphytes" growing on the pseudostems of both Cyperaceae and Velloziaceae or as colonizers of the humose mat substrate (i.e., in a sense forming humus epiphytes). The reasons for the host specificity of these orchids are unknown.

4.2.5.2 *Selaginella* Mats

Poikilohydric species of the fern genus *Selaginella* form small and relatively flat cushion-like mats on inselbergs in the palaeotropics, neotropics and in temperate regions (Fig. 4.12). *Selaginella* mats frequently form a monospecific community. Field observations suggest that they form a pioneer stage which subsequently will be replaced by mat-forming monocots or woody dicots which outcompete *Selaginella*. In a later stage of this succession, *Selaginella* is restricted to the margin of the mat.

4.2.5.3 Other Types of Mats

Only rarely do dicots occur as mat formers on inselbergs. For example cacti (*Echinopsis hammerschmidii* on Bolivian inselbergs, Ibisch et al. 1995) have been recorded to have built up mats as well as the Apiaceae *Eryngium histrix*, which vegetatively resembles a bromeliad (southern Brazil, own unpubl. data).

Fig. 4.12. Being of moss-like appearance, the poikilohydric *Selaginella sellowii* forms extensive mats on Brazilian inselbergs. Frequently, these mats provide establishment sites for other species. (Brazil, Minas Gerais, photograph N. Biedinger)

4.2.6 Other Types of Vegetation

4.2.6.1 Forest Islands

On deeper soil (>30 cm in depth) forest-like stands develop under moist climatic conditions. Most species occurring here are not specific for inselbergs but are widespread in the surroundings. Inselbergs situated in humid tropical regions bear a higher percentage of drought-resistant trees than the surrounding forest. On larger inselbergs summital forests may develop on less inclined slopes. Due to the fact that the highest parts of inselbergs are frequently covered by clouds or mist, these forests are relatively rich in epiphytes. Inselbergs situated in savanna are frequently surrounded by a forest belt that is maintained by runoff water. Today, in many regions where human influence has led to the disappearance of forest, the woody vegetation on inselbergs has become an important refuge for forest-bound species.

On inselbergs the ecotone between forest and open rock underlies relatively dry conditions (e.g., high radiation, shallow soil). Deciduous trees and shrubs are therefore frequently encountered in this transitional zone.

4.2.6.2 Savanna Islands

On deeper soil (>30 cm in depth) a grass-dominated, herbaceous vegetation may develop. Just as in the case of forest islands, most species occurring here are not specific for inselbergs but are common in the surroundings. Occasionally, this vegetation type occurs on inselbergs surrounded by rainforest, where it may represent a relict of drier climatic conditions in the past.

4.3 Floristic Affinities to Other Rock Outcrop Ecosystems

In comparison to their temperate zone counterparts (cf. Quarterman et al. 1993 for a survey about limestone, sandstone, and granite outcrop communities in southeastern USA), tropical rock outcrops belong to the least-known ecosystems. Apart from inselbergs consisting of Precambrian rocks, there are a number of other rock outcrop assemblages in the tropics which are characterized by similar environmental conditions. In this connection one has to mention, for example, sandstone outcrops and flat ferricretes. Although our knowledge is still very preliminary, it can be stated that in certain regions inselbergs share many species with other rock outcrop localities (Porembski et al. 1994). This is true in particular for species which are adapted to the specific ecological constraints on rock outcrops. Therefore, species that are able to cope with nutrient deficiency (carnivorous plants) or water shortage (e.g., succulents and poikilohydric vascular plants) belong to the most prominent components of tropical rock outcrop ecosystems. Concerning the distribution of habitats, there is also a large degree of superficial similarity between different outcrop types. For example, this can be seen from the almost ubiquitous distribution of cryptogamic crusts on exposed rock surfaces. However, one has to note the conspicuous fact that the occurrence of monocotyledonous mats is largely restricted to inselbergs. Only very occasionally can mat-forming species (e.g., *Afrotrilepis pilosa* in West Africa and *Trilepis ihotzkiana* in Brazil) be seen on other rock outcrop types. However, in these cases, carpet-like mats are usually not formed.

Apart from rock outcrop communities, the vegetation of inselbergs shows floristic relationships to a number of other ecosystems. In the neotropics there is a close floristic correspondence between inselbergs, other rock outcrop vegetation types, epiphyte communities, and white sand savannas (Porembski et al. 1998). In contrast to this, the link between tree canopies and inselbergs is only weakly developed on African inselbergs, where a large percentage of species is shared with savanna vegetation.

Acknowledgements. The Deutsche Forschungsgemeinschaft (grant no. Ba 605/4-3 to W. Barthlott) is thanked for supporting our research. Predoctoral fellowships to R. Seine (Studienstiftung des deutschen Volkes) and R. Becker (Graduiertenförderung des Landes Nordrhein-Westfalen) are gratefully acknowledged.

References

Bakker JP (1970) Zur Ökologie einer Blaualgenkruste in den feuchten Tropen auf dem Biotitgranit einer Inselberglandschaft (Voltzberggebiet-Surinam). Colloq Geogr 12:152–176

Barthlott W, Gröger A, Porembski S (1993) Some remarks on the vegetation of tropical inselbergs: diversity and ecological differentiation. Biogéographica 69:105–124

Bonardi D (1966) Contribution à l'étude botanique des inselbergs de Côte d'Ivoire forestière. Diplome d'études supérieures de sciences biologiques, Université d'Abidjan

Büdel B, Karsten U, Garcia-Pichel F (1997) Ultraviolet-absorbing scytonemin and mycosporine-like amino acid derivatives in exposed, rock-inhabiting cyanobacterial lichens. Oecologia 112:165–172

Burbanck MP, Platt RB (1964) Granite outcrop communities of the Piedmont Plateau in Georgia. Ecology 45:292–306

Danin A (1996) Plants of desert dunes. In: Cloudsley-Thompson JL (ed) Adaptations of desert organisms. Springer, Berlin Heidelberg New York

Diels L (1906) Die Pflanzenwelt von West-Australien südlich des Wendekreises. In: Engler A, Drude O (eds) Die Vegetation der Erde. VII. Engelmann, Leipzig

Freise FW (1938) Inselberge und Inselberg-Landschaften im Granit- und Gneisgebiete Brasiliens. Z Geomorphol 10:137–168

Granville JJ de (1978) Recherches sur la flore et la végétation guyanaises. Thesis, Université Montpellier

Humboldt A v (1819) Relation historique du voyage aux regions équinoxiales du Nouveau Continent, ait en 1799–1804 par A. de Humboldt et A. Bonpland, vol II. Maze, Paris

Ibisch PL, Rauer G, Rudolph D, Barthlott W (1995) Floristic, biogeographical, and vegetational aspects of Pre-Cambrian rock outcrops (inselbergs) in eastern Bolivia. Flora 190:299–314

Jaag O (1945) Untersuchungen über die Vegetation und Biologie der Algen des nackten Gesteins in den Alpen, im Jura und im schweizerischen Mittelland. Beitr Kryptogamenflora Schweiz 9:1–560

Lange OL, Kidron GJ, Büdel B, Meyer A, Kilian E, Abeliovich A (1992) Taxonomic composition and photosynthetic characteristics of the "biological soil crust" covering sand dunes in the western Negev Desert. Funct Ecol 6:519–527

Lüttge U (1997) Cyanobacterial tintenstrich communities and their ecology. Naturwissenschaften 84:526–534

Martius CFP v (1842) Tabulae physiognomicae. Flora Brasiliensis I–IX. Fleischer, München

Ornduff R (1987) Islands on islands: plant life on the granite outcrops of Western Australia. University of Hawaii, H. L. Lyon Arboretum Lecture No 15. University Press of Hawaii, Honolulu

Pietsch W (1973) Beitrag zur Gliederung der europäischen Zwergbinsengesellschaften (Isoeto-Nanojuncetea Br.-Bl. & Tx. 1943). Vegetatio 28:401–438

Porembski S, Barthlott W (1995) On the occurrence of a velamen radicum in Cyperaceae and Velloziaceae. Nord J Bot 15:625–629

Porembski S, Barthlott W, Dörrstock S, Biedinger N (1994) Vegetation of rock outcrops in Guinea: granite inselbergs, sandstone table mountains and ferricretes - remarks on species numbers and endemism. Flora 189:315–326

Porembski S, Brown G, Barthlott W (1996) A species-poor tropical sedge community: *Afrotrilepis pilosa* mats on inselbergs in West Africa. Nord J Bot 16:239–245

Porembski S, Martinelli G, Ohlemüller R, Barthlott W (1998) Diversity and ecology of saxicolous vegetation mats on inselbergs in the Brazilian Atlantic rainforest. Divers Distrib 4:107–119

Quarterman E, Burbanck MP, Shure DJ (1993) Rock outcrop communities: limestone, sandstone, and granite. In: Martin WH, Boyce SG, Echternacht AC (eds) Biodiversity of the southeastern United States. Upland terrestrial communities. Wiley, New York, pp 35–86

Smith GG (1962) The flora of granite rocks of the Porongorup Range, South Western Australia. J R Soc West Aust 45:18–23

Wilhelmy H (1958) Klimamorphologie der Massengesteine. Westermann, Braunschweig

Willis JC (1906) The flora of Ritigala, an isolated mountain in the North-Central Province of Ceylon; a study in endemism. Ann R Bot Gard Peradeniya 8:271–302

5 Algae, Fungi, and Lichens on Inselbergs

B. BÜDEL, U. BECKER, G. FOLLMANN, and K. STERFLINGER

5.1 Introduction

Bare calcareous and quartzitic rock surfaces in arid and semiarid regions, as well as dry spots like inselbergs in humid savannas, are generally inhabited by a complex community of actinomycetes, cyanobacteria, algae, dematiaceous fungi [= Hyphomycetes (imperfect fungi) having dark-colored hyphae and/or spores], and lichens (Danin and Garty 1983; Eppard et al. 1996; Friedmann 1980; Krumbein and Jens 1981; Staley et al. 1992). Especially cyanobacteria are known to occur also on man-made substratum like the walls of buildings (Garty 1990), older concrete walls, or the marble of historic temples (Anagnostidis et al. 1983), where they can initiate deterioration of the material, often together with fungi. Diels (1914) and later Jaag (1945) published extensive works on the composition and ecology of what is called Tintenstriche, a community basically created by different species of cyanobacteria, cyanobacterial lichens, and a few green algae, that stain vertical rock faces in the European Alps ink-blue. However, this phenomenon is not restricted to the European Alps, it also occurs in many high mountain ridges of the world (e.g., Wessels and Büdel 1995). Many species creating these Tintenstriche can be found in cyanobacteria-dominated biofilms all over the world. Particularly inselbergs are the habitat where prokaryotic algae (blue-green algae, cyanobacteria), fungi, and lichens show luxuriant growth (Hambler 1964; Scott 1967; Büdel and Wessels 1991; Büdel et al. 1994, 1997a; Sarthou et al. 1995; Seine et al. 1997). However, certain distribution patterns on the surface of inselbergs and differences among inselbergs of geographically different savannas exist. A first attempt to classify different inselberg types can be made on the basis of their overall color. In the dry and thornbush savannas, inselbergs are typically yellowish brown-colored (Fig. 5.1A), while inselbergs of humid savannas are black-colored (Fig. 5.1B; Humboldt 1849; Hambler 1964). Inselbergs at higher altitudes are more colorful, showing a spectrum from green, yellow to red (Fig. 5.1E,F; Scott 1967; Seine et al. 1997). The

Ecological Studies, Vol. 146
S. Porembski and W. Barthlott (eds.) Inselbergs

C
D
E
F

G
H
I
J
K

Fig. 5.1.A Brown granite inselberg in dry savanna, North Transvaal, South Africa. Cyanolichen growth indicated by *arrows.* **B** Black granite inselberg in the humid savanna around Pto. Ayacucho, Venezuela. Dense forest surrounds the inselberg. **C** *Scytonema crassum*, a major species on the rock surface of black inselbergs in the humid tropics; yellow sheath pigment is scytonemin, a UV sunscreen; *scale* 10 µm. **D** *Lichenothelia* sp., microcolonies of the black fungus, quartz from granite of an inselberg, Côte d'Ivoire; *scale* 0.5 mm. **E** Surface of the green-red vegetation pattern, showing beards of *Usnea* sp., green-gray *Dimelaena* sp., red *Caloplaca* sp., and yellow *Dermatiscum thunbergii*; *scale* 5 cm. **F** Surface of the red-yellow vegetation pattern with red *Caloplaca* sp., yellow *Acarospora* sp., and *Dermatiscum thunbergii* and few thalli of *Dimelaena* sp.; *scale* 5 cm. **G** Surface of a sandstone inselberg in South Africa, at the border of Botswana and Zimbabwe. Vertical faces of the boulder covered by *Dermatiscum thunbergii*, the surrounding rock surface is covered by gray *Peltula umbilicata* (*arrow*) and brown *P. placodizans* (*arrowheads*); *red scale* on boulder 2 cm. **H** *Peltula placodizans*, Venezuela; *scale* 10 mm. **I** *Peltula tortuosa*, Brazil, *scale* 10 mm. **J** *Peltula euploca*, Australia, *scale* 1 mm. **K** Cross fracture of granite rock from a South African inselberg, showing the endolithic *Chroococcidiopsis* sp. and *Peltula umbilicata* thalli on the rock face; *scale* 1 mm

differences in coloration are caused by characteristic combinations of microorganisms and will be dealt with in the following sections.

5.2 Algae

The vast majority of algae on the naked rock surface of inselbergs in all geographical regions are cyanobacteria. The few green algae that occasionally occur belong to the genera *Zygogonium* cf. *ericetorum* (Zygnematales, Charophyta; Büdel et al. 1994), a common soil alga, *Klebsormidium* cf. *montanum* (Charophyceae, Charophyta, Weber 1997), *Mougeotia* (Zygnematales, Charophyta), and *Oedogonium* (Oedogoniales, Chlorophyta, Sarthou 1992). Little is known about the algal communities of temporarily water-filled rock pools. In this context, an exciting result was the discovery of a new species and genus in a seasonal rock pool of a Zimbabwean granite inselberg: *Starria zimbabweënsis* (Oscillatoriales, Cyanophyceae/Cyanobacteria), with its peculiar triradiate morphology in cross sections (Lang 1977). A species of *Haematococcus* sp. (Volvocales, Chlorophyta) was also found in this rock pool (Lang 1977).

Although cyanobacteria constitute a prominent part of the biofilm of exposed rocks in the tropics and subtropics, their diversity and vegetation patterns are poorly investigated. Thus, the knowledge concerning their worldwide distribution is very limited and basically restricted to inselbergs occurring in Africa and South America. A preliminary summary of cyanobacterial genera and species occurring on inselbergs in Africa and America is given in Table 5.1. Members of the genus *Chroococcidiopsis* grow mostly inside the rock (= endolithic in 1–4 mm depth; Fig. 5.1K), while the genera *Gloeocapsa, Schizothrix, Scytonema* (Fig. 5.1C), and *Stigonema* grow epilithic and are very common in these habitats (Table 5.1). From temple wall surfaces (sandstone?) in India, the cyanobacteria *Calothrix parietina, Gloeocapsa* sp., *Nostoc* sp., *Scytonema ocellatum,* and *Tolypothrix byssoidea* have been reported, forming dark-colored crusts (Adhikary and Satapathy 1996). These species might occur also on inselbergs in the surrounding region.

Almost all species of cyanobacteria listed in Table 5.1 are characteristic inhabitants of rock faces and are not restricted to the tropical regions.

Table 5.1. Preliminary list of cyanobacteria found on inselbergs

Species	Africa	America
Amphithrix janthiana[a]	+	
Calothrix gypsophila[b]		+
Calothrix parietina	+	
Chroococcidiopsis sp. (spp ?)	+	+
Chroococcus turgidus[b]		+
Dichothrix baueriana[a]	+	
Dichothrix sp.	+	
Doliocatella formosa[b]		+
Entophysalis sp.	+	
Fortiea caucasica[c]		+
Geitleribactron sp.		+
Gloeocapsa biformis[d]		+
Gloeocapsa cf. *dermochroa*	+	
Gloeocapsa compacta[d]		+
Gloeocapsa granosa[b]		+
Gloeocapsa itzigsohnii[b]		+
Gloeocapsa kützingiana[c]		+
Gloeocapsa montana[b]		+
Gloeocapsa quaternaria		+
Gloeocapsa sabulosa[c]		+
Gloeocapsa sanguinea	+	+
Gloeocapsa sp.	+	+
Gloeothece confluens[b]		+
Gloeothece palea[b]		+
Hapalosiphon fontinalis[b]		+
Hapalosiphon luteolus[b]		+
Nostoc muscorum[a]	+	
Nostoc sp.	+	
Plectonema tomasinianum	+	

Table 5.1 (*continued*)

Species	Africa	America
Pleurocapsa minor[c]		+
Porphyrosiphon kaernbachii	+	
Porphyrosiphon notarisii	+	
Rivularia sp.	+	
Schizothrix cf. *calcicola*		+
Schizothrix cf. *epiphytica*[c]		+
Schizothrix cf. *rubra*	+	
Schizothrix heufleri[d]		+
Schizothrix lutea		+
Schizothrix luteola[b]		+
Schizothrix purpurascens	+	
Schizothrix telephoroides		+
Scytonema amplum		+
Scytonema crassum	+	+
Scytonema densum[b]		+
Scytonema guyanense		+
Scytonema hofmanni	+[a]	+
Scytonema minus[d]		+
Scytonema multiramosum	+	+[b]
Scytonema myochrous	+	+
Scytonema ocellatum	+	+
Scytonema stuposum	+	
Scytonema subcoatile[b]		+
Starria zimbabweënsis[e]	+	
Stigonema flexuosum[b]		+
Stigonema hormoides	+	+
Stigonema mammilosum	+	+
Stigonema minutissimum		+
Stigonema minutum	+	+
Stigonema ocellatum	+	+
Stigonema panniforme	+	+
Stigonema tomentosum	+	
Symphyonema sinense[b]		+
Symploca muscorum	+	
Xenococcus sp.		+

[a] Hambler (1964). [b] Sarthou et al. (1995). [c] Weber (1997). [e] Lang (1977). [d] Golubic (1967).

5.3 Fungi

While several studies deal with the growth of microcolonial fungi on exposed rock surfaces in arid regions (Krumbein and Jens 1981; Henssen 1987; Staley et al. 1992), it seems that the peculiar fungal growth on inselbergs has not been looked for specifically. Thus, the results presented here come from a hitherto unpublished study on granite rocks from inselbergs at the Côte d'Ivoire and a few observations from other continents. Rock surfaces of inselbergs in the Côte d'Ivoire are inhabited by fungi forming black microcolonies via meristematic growth. The fungal colonies are found between lichen thalli and cyanobacterial colonies, respectively, but also on surfaces free of other organisms (Fig. 5.1D). On the latter, the fungal colonies are more dense, growing mainly in contact zones between quartz crystals like pearls on a chain. Where the granitic surface is covered by a calcitic crust, this crust appears to be extremely carsted with pits that are filled with fungal colonies.

Several genera of dematiaceous fungi have been isolated from granite samples of Ivorian inselbergs: e.g., *Cladosporium, Scolecobasidium, Chaetomium, Coniosporium,* and some strains of meristematically growing black fungi, the identity of which is unclear due to the lack of morphological differentiation. However, isolation of fungi by transferring single colonies on agar plates has shown that most microcolonies on the granite samples are formed by members of the genus *Lichenothelia* (Fig. 5.1D), species of which had been reported as inhabitants of bare rocks on a worldwide scale (18 species, Henssen 1987). *Lichenothelia intermedia*, for example, was isolated and described from a granite inselberg in the thornbush savanna of South Africa and occurs on almost every inselberg in Africa, while, for instance, the species *L. gigantea* and *L. radiata* are restricted to Australia and *L. globulifera* was found on granite outcrops of the island La Digue, Seychelles (Henssen 1987).

Although the fungal microcolonies sometimes resemble initial stages of lichen thalli, they can be clearly distinguished by light and electron microscopy. The fungal thalli have a size from 10–100 µm in diameter and a single cell measures 3–5 µm. On granite samples, vegetative cells of *Lichenothelia* are densely aggregated to form flat colonies with erythrocyt-like shape. Microcolonial fungi produce small, single hyphae, able to spread over the rock surface and penetrate it where possible.

Rock-inhabiting fungi are poikilohydric organisms like many other, but not all, fungi. These fungi overcome periods of high temperatures in the desiccated state. Thus, desiccation tolerance is a prerequisite for temperature tolerance (Sterflinger and Krumbein 1995; K. Sterflinger unpubl. data).

Moreover MCF (= microcolonial fungi) can grow with very low nutrient concentrations, absorbing volatiles as carbohydrates and aromates from the atmosphere. This special feature allows them to grow – albeit slowly (several months for a thallus) – on rock surfaces devoid of nutrients, and makes them independent of primary production by autotrophic organisms (Krumbein and Gorbushina 1995). The up to 0.8 μm-thick cell walls are encrusted by melanins which provide shelter from high UV radiation on surface areas exposed to the sun (Diakumaku et al. 1995). Rock-inhabiting fungi are known to contribute quite extensively to the biodeterioration of stone monuments and buildings (Krumbein 1969; Diakumaku et al. 1995).

5.4 Lichens

In contrast to the situation faced with the cyanobacteria and fungi, our knowledge of the diversity of lichens, especially those with cyanobacterial photobionts (called cyanolichens throughout the text) and their distribution patterns is much better (e.g., Hambler 1964; Scott 1967; Büdel 1987; Wessels and Büdel 1989; Büdel et al. 1997a; Seine et al. 1997). The genera and species of cyanolichens known to occur on inselbergs are listed in Table 5.2.

Swinscow and Krog (1988) mention that cyanobacterial species of the genera *Sticta*, *Pseudocyphellaria*, and *Leptogium* typically occur on trees of fog-influenced slopes on inselbergs in Kenya. Hambler (1964) reported lichens with green phycobionts (called green algal lichens throughout the text) like *Anaptychia leucomelaena* (= *Heterodermia leucomelos*), *Placodium zahlbruckneri* (= *Lecanora* subgen. *Placodium*), *Graphis hylinella*, *Usnea ledienii*, etc. from the bark of *Ficus abutilifolia* on an inselberg in western Nigeria. However, since this occurrence is not restricted to inselbergs and is also common on many other mountain ridges, this type of flora and vegetation will not be dealt with here.

The most common lichen genus on inselbergs throughout the tropics and subtropics seems to be the genus *Peltula*, belonging to the lichen family Peltulaceae (some 40 species). So far, 20 *Peltula* species have been found on African inselbergs, 12–13 on Australian, 7 on American, and 4–5 on Asian inselbergs. However, this comparison is not very reliable, since the degree of investigation is quite different between the continents, with Africa being best documented, followed by Australia and South America. The knowledge of the lichen flora of Asian inselbergs is still very poor.

Peltula tortuosa is the only species occurring on all continents in the tropical and subtropical region (Fig. 5.1I), and at the same time represents

Table 5.2. Preliminary list of saxicolous cyanobacterial lichens on inselbergs

Species/Continent	Africa	America	Asia	Australia
Anema decipiens[a]			+	
Collema ryssoleum[a]			+	
Gloeoheppia turgida	+			+
Heppia egentissima[b]	+			
Heppia lutosa	+		[+]?	[+]?
Heppia nigra[b]	+			
Heppia trichophora[a]			+	
Metamelanea sp.		+		
Paulia myriocarpa	+	+		
Paulia tesselata	+			
Paulia sp.	+			
Peltula sp. II		+		
Peltula sp. nov. ined.		+		
Peltula bolanderi	+	+		+
Peltula boletiformis	+			
Peltula clavata	+	+		+
Peltula congregata	+			
Peltula coriacea	+			
Peltula cylindrica	+	+		+
Peltula euploca	+		+	+
Peltula farinacea	+	[+] ?		
Peltula impressa	+			+
Peltula leptophylla ined.	+	+		
Peltula lingulata	+			
Peltula marginata	+			
Peltula obscurans	+		+	+
Peltula omphaliza			[+]?	[+]?
Peltula patellata	+			+
Peltula placodizans	+			+
Peltula rodriguesii	+			+
Peltula santessonii	+			
Peltula tortuosa	+	+	+[d]	+
Peltula umbilicata	+			+
Peltula zahlbruckneri	+		+[d]	+
Phylliscum abuense[a]			+	
Phylliscum monophyllum[c]		+		
Phylliscum testudineum	+	+		
Psorotichia sp.[e]	+			
Pterygiopsis sp.	+			
Pyrenopsis sp. I	+			
Synalissa sp.		+		
Synalissa symphorea	+			
Thyrea indica[a]			+	
Zahlbrucknerella indica[a]			+	

[a] Awasthi and Singh (1979). [b] Hambler (1964). [c] Krempelhuber (1876). [d] Upreti and Büdel (1990). [e] Scott (1967).

a typical inhabitant of inselbergs. *Peltula bolanderi, P. clavata, P. euploca* (Fig. 5.1J), *P. obscurans*, and *P. zahlbruckneri* have been found on inselbergs of three continents (Table 5.2). All *Peltula* species found in Australia also occur in Africa, and only South America has two species not shared with the African savannas and rainforests so far (Table 5.2). South and East Africa seem to be the center of radiation of the family (Büdel 1987). Widespread on African inselbergs is *Peltula umbilicata* (Fig. 5.1G,K), a species found only twice in tropical/subtropical Australia and missing in all other regions. *Peltula tortuosa* is very common on inselbergs in the neotropics. In addition, a new species of *Peltula* not yet described was found on several inselbergs distributed over a large area in Venezuela. The new species seems to be restricted to that geographical region.

Apart from the genus *Peltula*, a number of lichens occur regularly on the rock surface of inselbergs in tropical/subtropical southern Africa. The green algal lichens *Acarospora* (yellow, 3 species; Fig. 5.1F), *Caloplaca* (usually only one species; Fig. 5.1E,F), *Dermatiscum thunbergii* (Fig. 5.1G), *Toninia* (often *T. bumamma*), *Dimelaena* (usually *D. oreina*), *Usnea* (several species; Fig. 5.1E), and *Xanthoparmelia* (several species) are the most common in the tropical southern African region. In shallow depressions of the rock, where debris accumulates and soil formation is initiated, the green algal lichens *Catapyrenium lacinulatum, Eremastrella crystallifera*, and the cyanolichens *Heppia lutosa* and *Peltula patellata* often can be found. In Zimbabwe 53 inselbergs, distributed all over the country, were investigated and some 180 lichen species belonging to 24 families and 61 genera have been found, the 10 most important of which are *Peltula* (16 spp.), *Usnea* (16 spp., e.g., *U. pulvinata, U. welwitschiana, U. maculata*), *Buellia* (15 spp.; e.g., *B. spuria*), *Xanthoparmelia* (15 spp.; e.g., *X. plitii, X. lavicola*), *Rinodina* (12 spp.; e.g., *R. oxydata* s.l., *R. detecta, R. substellulata*), *Parmotrema* (6 spp.; e.g., *P. andinum, P. grayanum, P. soyauxii*), *Heterodermia* (8 spp.; e.g., *H. leucomelos, H. diademata, H. albicans*), *Caloplaca* (6 spp.), *Lecanora* (6 spp.; *L. bicincta, L. pseudistera, L. farinacea*), and *Acarospora* (4 spp.), and *Pyxine* (4 spp.; e.g., *P. petricola*; Seine et al. 1997 and unpubl. results). In western Nigeria, *Parmelia conturbata* (= *Neofuscelia conturbata*) and *Heppia nigra* were observed growing on shaded vertical rock faces, *Buellia subalbula* and *Lepraria* sp. on overhanging damp ledges and cave roofs, and *Buellia* sp., *Anaptychia leucomeleana* (= *Heterodermia leucomelos*), and the orange *Placodium zahlbruckneri* on boulders (Hambler 1964). From an inselberg in French Guiana, Sarthou (1992) reported the green algal lichens *Parmotrema cristiferum, Cladonia corallifera, Cladonia didyma*, and *Trypethelium aeneum*.

While most lichen species found on inselbergs are fertile and produce ascospores, some species are only known in the sterile state (e.g., *Peltula clavata*, *P. lingulata*). The production of vegetative diaspores like isidia or soredia is frequently observed in the inselberg lichen flora [e.g., *Peltula euploca*, *P. placodizans* (Fig. 5.1H), *Usnea* spp., *Parmotrema* spp., *Pertusaria* spp.].

5.5 Habitat Separation

As mentioned before, the three different types of inselbergs are defined by a characteristic combination of microorganisms, roughly following the rule that cyanolichens dominate on the brown, free living cyanobacteria dominate on black-colored inselbergs, and green algal lichens on the green, yellow, and red ones (Fig. 5.1A,B,E,F). At the lighter parts of the rock surface, the ocher-brown type (Fig. 5.1A) is caused by oxidation processes with iron substances inside the rock, while the darker parts are brought about by a dense cover of cyanolichens. The black-colored inselbergs (Fig. 5.1B) are mainly caused by a very dense cover of free-living cyanobacteria (Fig. 5.1C). The third type are inselbergs of higher altitudes, which are characterized by green algal lichens (Scott 1967; Seine et al. 1997).

5.5.1 The Large Scale: Differences Between Dry and Humid Savannas and Low and High Altitudes

In general, the variation found between inselberg types can roughly be correlated with the degree of latitude and with altitude. Consequently, the number of arid months, precipitation rates, and the occurrence of fog play a major role in the establishment of the different inselberg types.

5.5.1.1 Dry Savanna Versus Humid Savanna

Brown inselbergs are characterized by the dominating presence of cyanolichens, which in total can reach 90 % coverage of the rocks along rainwater tracks, and up to 65 % on rock surfaces with minor water supply (Scott 1967; Wessels and Büdel 1989; Seine et al. 1997). The brown type is found in the dry savanna and the thornbush savanna and in semideserts of the arid and semiarid zones of Africa (Fig. 5.1A) and, to a lesser extent, Australia. However, intermingled with the lichens, brown inselbergs also accommodate quite a large number of cyanobacterial species, but they never form dense crusts or come to dominance there.

In contrast to the brown inselbergs, those of the humid savanna are mostly black-colored and this coloration is caused by dominating, free-living cyanobacteria, accompanied by scattered thalli of cyanolichens (Büdel et al. 1994, 1997a). The cyanobacterial mats often reach values close to 100 % coverage. Usually, values of 50–90 % coverage of the exposed rock surface are reached. Cyanolichens occur as single thalli or have a very scattered distribution only. Black inselbergs occur in the humid savannas and rainforest zones of Africa (e.g., Cameroon, Côte d'Ivoire, Guinea, Nigeria), South America (e.g., Brazil, French Guiana, Suriname, Venezuela; Fig. 5.1B), the Seychelles, and probably also in India.

The limit of precipitation for the formation of black, cyanobacteria-dominated inselbergs seems to be about 900–1200 mm a^{-1}. In addition, the duration of the dry season seems to be of importance. Annual precipitation rates clearly below these values obviously result in the formation of brown, cyanolichen-dominated inselbergs. However, rather than the mean annual amount of rain, the number of rain events are responsible for the establishment of the brown or black inselberg type. Differences in species composition of different geographical regions can be taken from Tables 5.1 and 5.2.

5.5.1.2 High Altitudes

Since data from other continents are not available, we focus here on the situation in Africa. While both cyanolichen and cyanobacterial inselbergs occur in the lower range of altitude (up to 1300 m), green algal lichen inselbergs in Africa are related to higher altitudes (above 1300–1400 m). Where inselbergs are situated at altitudes above ca. 1400 m asl, high air humidity and frequent fog events occur and green algal lichens dominate (Fig. 5.1E,F). However, the occurrence of this inselberg type is not always uniform regarding their color, and different patterns of coloration can sometimes be observed, due to the different composition of green algal lichen communities.

According to observations of S. Porembski (pers. comm.), green algal lichens seem to dominate on inselbergs of high latitudes in Australia.

5.5.2 The Small Scale: Vegetation Patterns

Among the cyanolichens of inselbergs, a clear correlation between water regime and growth form could be observed (Fig. 5.2; Scott 1967; Büdel et. al.

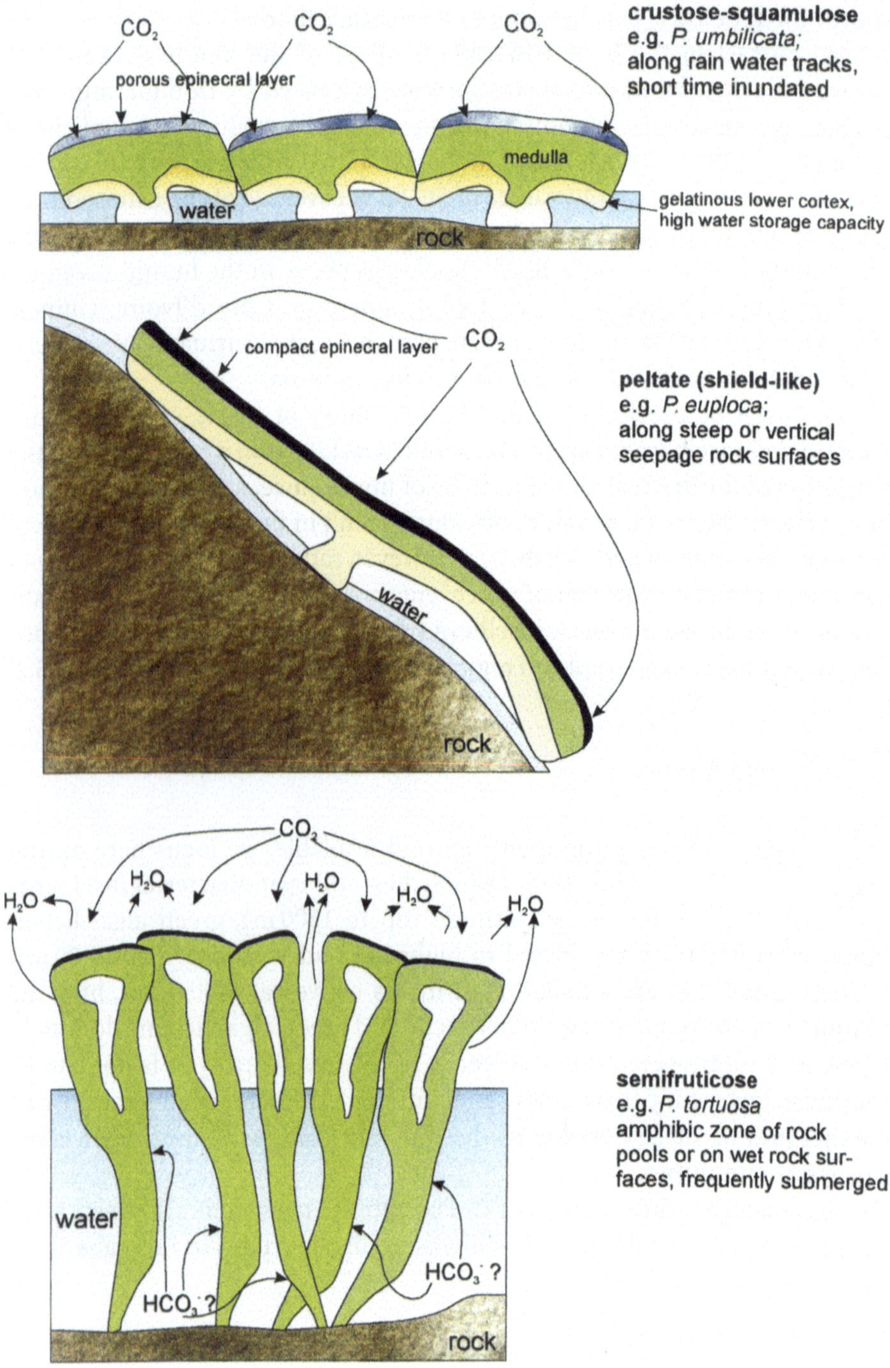

Fig. 5.2. Three main growth types of *Peltula*, properties of thallus structure, and habitat preferences

1997a; Wessels and Büdel 1989). While crustose-squamulose species like *Peltula marginata, P. placodizans, P. impressa*, and *P. obscurans* are restricted to fully exposed areas of rock which only become wet during rainfall, peltate species like *P. euploca* or *P. coriacea* preferably grow on steep or vertical rock surfaces with an even shorter inundation time for the thalli (Fig. 5.2). The semifruticose and fruticose type, e. g., *P. clavata, P. cylindrica, P. lingulata*, and *P. tortuosa*, however, is always associated with temporarily water-filled rock pools (Fig. 5.2) and drainage channels. Here, *P. cylindrica* and *P. tortuosa* grow along the amphibic zone, and *P. clavata*, often together with *P. lingulata*, extends more towards the center of the pool, thus being completely submerged. Another habitat are drainage channels, fed with water for a certain period after rain, rock pools or phanerogamic vegetation islands (Fig. 5.1G). In these cases, the crustose-squamulose *Peltula umbilicata* (Fig. 5.2) or, more rarely, *P. santessonii* occur.

Ecophysiological reasons – related to the growth form – are most probably responsible for the dominance of certain species in a certain habitat (see Sect. 5.6.1). The patchwork of different cyanolichen species is only to a lesser extent influenced by stochastic processes. The following communities of cyanolichens, correlated with a certain habitat, can be distinguished in the case of African inselbergs (South Africa, Namibia, Zimbabwe, Côte d'Ivoire, and partly Kenya): (dominant species underlined)

1. Broad rain water runoffs (drainage channels; Fig. 5.1G), wet for a longer period after rains: *P. congregata, P. cylindrica, P. obscurans, P. santessonii, P. umbilicata*.
2. Fully exposed, slanted rock surfaces, quickly dry after rain: *P. obscurans, P. placodizans, P. umbilicata, P. zahlbruckneri, Pyrenopsis* spp.
3. Fully exposed rock faces not or only slightly inclined, driest part of the rock: *P. impressa, P. marginata, P. placodizans, P. umbilicata, P. zahlbruckneri*.
4. Seasonal rock pools, wet: *P. boletiformis, P. clavata, P. cylindrica, P. lingulata, P. santessoni, P. tortuosa* (rarely *P. euploca*).
5. debris, detritus, beginning soil formation: *P. patellata, Heppia lutosa*.
6. Steeply inclined rock surfaces, large boulders: *P. euploca*.
7. Furrows, crevices: *P. congregata, P. obscurans*.

On more elevated parts in the microrelief of the rock, green algal lichen species form a typical element of the communities no 1 and 2. Many of them grow preferably along vertical or steeply inclined faces of rocks and boulders (Fig. 5.1G), where they are not in contact with flowing water, thus being activated by damp air after rainfall. Typical lichens of elevated microrelief areas and boulders are *Acarospora* spp., *Caloplaca* spp.,

Dermatiscum thunbergii (Fig. 5.1G), and *Xanthoparmelia* spp. A very unusual, grayish-white-colored stripe free of any cyanobacterial, fungal, or lichen growth is always associated with drainage channels or rock pools, indicating the region between the amphibic and the dry zone along them (Scott 1967; Bakker 1970; Wessels and Büdel 1989). Scott (1967) discussed a local concentration effect of substances by the cycle of ebb and flow of capillary water throughout each rain season, while Bakker (1970) hypothesized that the abrasion activity of flowing water along the edges of water channels might be responsible. However, a convincing explanation for this phenomenon is still missing.

Three types of vegetation patterns on green algal lichen inselbergs could be distinguished in the case of a study in Zimbabwe. So far, the ecological reasons why these three types are differentiated are not clear: (1) the bright green type, (2) the red-yellow type, and (3) the green-red type. The bright green type was dominated by 95 % *Dimelaena* spp. (often *D. oreina*) with additional *Usnea* spp. (often *U. welwitschiana*) and rarely *Caloplaca* spp. (Fig. 5.1E). Tufts of *Usnea pulvinata* and *U. welwitschiana* are sometimes formed. The red-yellow type (Fig. 5.1F) was dominated by a cover of more than 80 % *Caloplaca* spp. with *Dermatiscum thunbergii* and yellow *Acarospora* spp., occasionally together with *Dimelaena* spp. The green-red type finally had 30 % *Dimelaena* spp. together with 40 % red *Caloplaca* spp. and scattered *Dermatiscum thunbergii* thalli (Seine et al. 1997).

5.6 Ecophysiology

What are the reasons for the limited growth of cyanolichens in the humid savanna? Why do cyanobacteria dominate there? Seemingly, principal photosynthetic properties of cyanobacteria, cyanolichens, and green algal lichens can explain at least in part the different distribution patterns.

5.6.1 Photosynthetic Parameters of Different Growth Forms

Water, light, and temperature are the basic factors influencing the assimilatory carbon gain of algae and lichens. Two basic questions are to be answered here: (1) Does the degree of water availability constitute a mechanism influencing the assortment of growth forms found along transects on inselbergs? (2) In which way does the growth form influence the photosynthetic behavior and subsequently the carbon gain under different water regimes in lichens?

To answer these questions, we have to consider the following facts: green algal lichens and many eukaryotic algae, in contrast to cyanobacteria and cyanolichens, can make use of water vapor alone to gain positive net photosynthesis (Büdel and Lange 1991; Lange et al. 1993a). Supersaturation with water results in many lichens, especially in green algal lichens like *Dermatiscum thunbergii*, in a limitation of photosynthetic carbon gain (Fig. 5.3) due to extremely high diffusion resistance for CO_2 (Cowan et al. 1992; Lange et al. 1993b). These facts help to explain the preference of green algal lichens for habitats with influence of fog or damp air and reduced influence of rainwater runoff, for instance, large boulders lying on the surface (Fig. 5.1G). In these places, a zone of high air humidity develops in the direct vicinity of the hot rock surface after rainfall. Most species of the genus *Peltula* can overcome supersaturated phases by the activity of a carbon dioxide-concentrating mechanism, which is also found in other cyanolichens (Badger et al. 1993), thus creating a steeper gradient between internal and external CO_2-concentrations and or being able to make use of HCO_3^- as an additional carbon source. Although different growth forms like the crustose-squamulose *Peltula umbilicata*, the peltate *P. euploca*, or

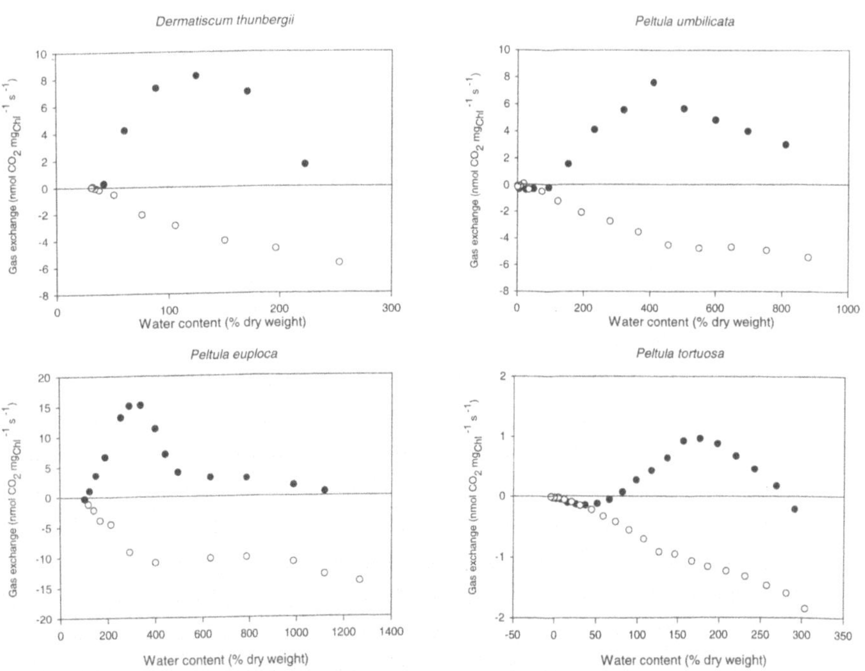

Fig. 5.3. Net photosynthesis (*NP*) related to thallus water content of the three main *Peltula* growth types, and the green algal lichen *Dermatiscum thunbergii*. Variability of optimal NP is not reached by cyanobiont diversity but by diversity in growth form

the semifruticose *P. tortuosa* share the same type of photobiont – an unicellular cyanobacterium (*Chroococcidiopsis* sp.) – their optimum for CO_2 uptake is reached at quite different water contents in the field (Fig. 5.3). This different behavior is most probably due to the growth form itself and might help to explain the preferences for habitats with differentiated water regimes. Nonlichenized, endolithic *Chroococcidiopsis* sp. is regularly associated with *Peltula* species (Fig. 5.1K). It is a typical low light-adapted cyanobacterium, in which light saturation for net photosynthesis is reached at about 200 $\mu mol\ m^{-2}\ s^{-1}$ photon flux density (Weber et al. 1996). However, the light climate inside the rock and *Peltula* species is almost comparable (Büdel 1990; Büdel and Lange 1994).

5.6.2 Nutrient Input into the Ecosystem

Since the biofilm of inselbergs experiences a certain degree of abrasion by desquamation, water, and biodeterioration of lichens and cyanobacteria, a considerable amount of cyanobacteria and cyanolichens are brought to the soil at the base of these rock outcrops. With their decay, important nutrients like nitrogen compounds are brought down into the soil.

Measurements carried out by E. Medina (unpubl.) of the nitrogen contents of the upper soil layers along a gradient at the base of an inselberg at Galipero in the Venezuelan savanna revealed an increased nitrogen content of the upper soil layers (Fig. 5.4). Compared with the surrounding area, values of N were roughly three times higher in the vicinity of the inselberg than in the surrounding soil. At least for the savanna region in the Amazon State of Venezuela, where tree growth is not limited by precipitation, the observed N input might be responsible for the luxuriant growth of gallery-like forests around the inselbergs (Fig. 5.1B).

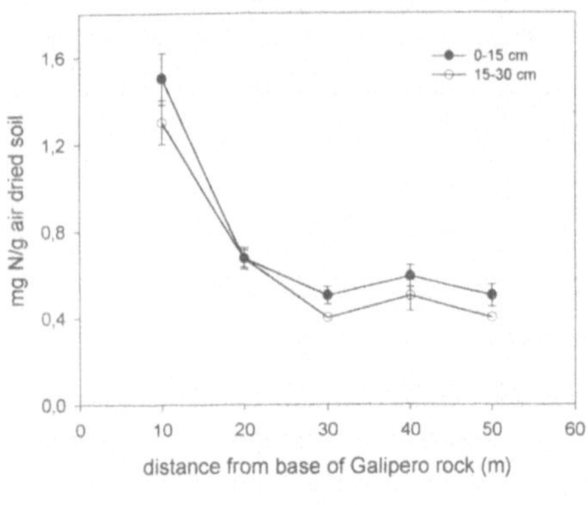

Fig. 5.4. Nitrogen content of the soil from different depth, taken along a transect beginning from the base of an inselberg at Galipero, Venezuela (Original data of E. Medina, with permission of the author)

5.6.3 Soil Formation/Weathering

Desquamation is the typical process of weathering that occurs on the granite of the inselbergs worldwide. Weathering activity of cyanobacteria and lichens on granitic rocks in the humid tropics of Suriname was studied extensively by Bakker (1970), who attributed the main effect of weathering activity to short period alternations of pH and temperature. Bakker concluded his studies with the following remark (translated): "...blue-green algal and lichen-weathering rates on granite in the humid tropics are different, but nevertheless relatively fast". The strong influence of cyanobacteria on microdesquamation (1–1.5 cm deep) was shown not only for the granite of inselbergs in the humid tropics (Bakker 1970), but also on sandstone inselbergs in semiarid regions (Wessels and Büdel 1995; Weber et al. 1996). On the basis of Indian engravings, Bakker (1970) gives an age of at least 200–300 years for the cyanobacterial biofilm.

Microcolonial fungi are known to have a strong impact on the surface relief of rocks (Krumbein 1969; Sterflinger and Krumbein 1997). SEM photographs of granite samples from the Côte d'Ivoire show that the fungal thalli are immersed not only in lesions of carbonate crusts but also in lesions of the quartz grains of the granite. Both pitting structures can be produced by the physiological and mechanical growth activities of black fungi penetrating the rock with thin hyphae, decohesing the grains of granite and mechanically removing quartz crystals as well as mica. In this manner the fungi contribute to the formation of a carsted surface relief of the granite. Together with coryneform bacteria and actinomycetes, black fungi are now assumed to be real specialists of rock decay.

The cyanobacterial genera *Nostoc, Porphyrosiphon*, and *Schizothrix* form large amounts of biomass in ephemeral flush communities and dominate completely in wet flush communities, thus, together with the decaying material, serving as substrate for the establishment of higher plants.

5.6.4 Pioneers

Which are the primary colonizers of bare rock surfaces after their exposure? The either chasmolithic or cryptoendolithic unicellular cyanobacterium *Chroococcidiopsis* belongs to the first colonizers of a freshly exposed rock surface. However, it obviously takes years before the rock is colonized. In the dry savanna of northern Transvaal we exposed rock surfaces by chiseling away scales of rock colonized with the cyanolichen *Peltula placodizans* in 1983. After reexamination in 1994, only first stages of recolonization by the lichen species *P. placodizans* could be observed (B. Büdel and D.C.J. Wessels, unpubl.). *Peltula placodizans*, together with *P. impressa*, were observed first colonizing the center of small exposed rock faces of about 20 × 20 cm on inselbergs in Zimbabwe, created by desquamation, while *P. umbilicata* and *P. zahlbruckneri* started to colonize the exposed rock faces beginning from the margins (U. Becker, unpubl.).

As slow-growing pioneers in a highly UV-exposed environment, cyanobacteria and cyanolichens depend on a very effective protection against UV radiation. In cyanobacteria, the indol alkaloid scytonemin, located in the outer sheath envelope, was found to provide shelter against UV-A radiation (Garcia-Pichel and Castenholz 1991). In addition, the nonglycosylated mycosporine-like amino acid compounds (MAAs), occurring intracellularly (Garcia-Pichel and Castenholz 1993), and an extracellularly located oligosaccharide MAA were found (Scherer et al. 1988), protecting against UV-B. It has been conclusively shown that the production of scytonemin is strongly induced by near UV-A, and that of MAA exclusively by UV-B (Ehling-Schulz 1997). In cyanolichens of the families Collemataceae, Lichinaceae, and Peltulaceae, the UV sunscreens scytonemin and nonglycosylated MAAs were detected in high amounts (up to 8.1 % of dry weight, Büdel et al. 1997b), where scytonemin is obviously supplied by the cyanobiont, and MAAs might also be produced by the mycobiont. In lichen thalli, scytonemin is concentrated in the sheath along thallus margins (Büdel et al. 1997b).

5.7 Conclusion

It is not clear whether the biofilms of the bare rock of inselbergs are old communities, but there are many reasons to believe so. First of all, cyanobacteria are the oldest organisms on earth, possessing oxygenic photosynthesis, which, in many species, is combined with the ability to fix atmospheric nitrogen. The second is the ability of cyanobacteria, as well as of cyanolichens, to withstand long periods of drought in a dehydrated state, showing no measurable signs of activity or energy consumption. Since the discovery of Taylor et al. (1995), we know that cyanolichens are at least 400 Ma old. It might therefore have well been that cyanolichens, together with cyanobacteria, dominated rock surfaces for a long period of time before they became out-competed by more efficient eukaryotic photosynthesizers. As a result, we find them today in habitats which are not convenient for modern algae, mosses, and higher plants. Ancient cyanobacteria and cyanolichens, with their ability to weather rock, might have contributed to soil formation, a necessary step in accommodating larger terrestrial plants.

Many rock faces which are thought to be completely bare of life are thought to be extreme in conditions for life; but they are not naked; instead, they are inhabited by a community of historically "old" primary producers, of which we do not know how much they contribute to the Earth's CO_2 sink. The discovery of new species shows us that even the species number of the biofilms and the vegetation patterns are only poorly understood. Further investigations will focus on this topics. In the present work, we have tried to give a first overview on species composition and ecology of biofilms of rocks on inselbergs, and with this we hope to bring the vividly living community to the awareness of the readers.

Acknowledgements. We would like to express our sincere thanks to Dr. E. Medina (IVIC, Caracas, Venezuela) for providing the data of the soil nitrogen contents. This work was supported by the Deutsche Forschungsgemeinschaft by grants to B. Büdel (Grant no. Bu 666/5-1, 5-2, 5-3) and U. Becker (Grant no. Ba 605/4-1. 4-2, 4-3 to W. Barthlott) as well as grants of the Graduiertenförderung of Nordrhein-Westfalen and the Richard-Winter-Stiftung to U. Becker.

References

Adhikary SP, Satapathy DP (1996) *Tolypothrix byssoidea* (Cyanophyceae/Cyanobacteria) from temple rock surfaces of coastal Orissa, India. Nova Hedwigia 62: 111–116

Anagnostidis K, Economou-Amili A, Roussomoustakaki M (1983) Epilithic and chasmolithic microflora (Cyanophyta, Bacillariophyta) from marbles of the Parthenon (Acropolis-Athens, Greece). Nova Hedwigia 38:227–287

Awasthi DD, Singh SR (1979) New or otherwise interesting lichens from Mt. Abu, Rajasthan, India. Norw J Bot 26:91–97

Badger MR, Pfanz H, Büdel B, Heber U, Lange OL (1993) Evidence for the functioning of photosynthetic CO_2-concentrating mechanisms in lichens containing green algal and cyanobacterial photobionts. Planta 191:57–70

Bakker JP (1970) Zur Ökologie einer Blaualgenkruste in den feuchten Tropen auf dem Biotitgranit einer Inselberglandschaft (Voltzberggebiet-Surinam). Colloq Geograph 69:152–176

Büdel B (1987) Zur Biologie und Systematik der Flechtengattungen *Heppia* und *Peltula* im südlichen Afrika. Bibl Lichenol 23:1–105

Büdel B (1990) Anatomical adaptations to the semiarid/arid environment in the lichen genus *Peltula*. Bibl Lichenol 38:47–61

Büdel B, Lange OL (1991) Water status of green and blue-green phycobionts in lichen thalli after hydration by water vapor uptake: do they become turgid? Bot Acta 104:361–366

Büdel B, Lange OL (1994) The role of cortical and epinecral layers in the lichen genus *Peltula*. Cryptogam Bot 4:262–269

Büdel B, Wessels DCJ (1991) Rock inhabiting blue-green algae/cyanobacteria from hot arid regions. Algol Stud 64:385–398

Büdel B, Lüttge U, Stelzer R, Huber O, Medina E (1994) Cyanobacteria of rocks and soils of the Orinoco lowlands and the Guayana uplands, Venezuela. Bot Acta 107:422–431

Büdel B, Becker U, Porembski S, Barthlott W (1997a) Cyanobacteria and cyanobacterial lichens from inselbergs of the Ivory Coast, Africa. Bot Acta 110:458–465

Büdel B, Karsten U, Garcia-Pichel F (1997b) UV-absorbing scytonemin and mycosporine-like amino acid derivatives in exposed, rock inhabiting cyanobacterial lichens. Oecologia 112:165–172

Cowan I, Lange OL, Green TGA (1992) Carbon-dioxide exchange in lichens: determination of transport and carboxylation characteristics. Planta 187:282–294

Danin A, Garty J (1983) Distribution of cyanobacteria and lichens on hillsides of the Negev Highlands and their impact on biogenic weathering. Z Geomorph NF 27: 423–444

Diakumaku E, Gorbushina AA, Krumbein WE, Panina L, Soukharjevski S (1995) Black fungi in marbles and limestones – an esthetical, chemical, and physical problem for the conservation of monuments. Sci Total Environ 167:295–304

Diels L (1914) Die Algen-Vegetation der Südtiroler Dolomitriffe. Ein Beitrag zur Ökologie der Lithophyten. Ber Dtsch Bot Ges 32:502–526

Ehling-Schulz M, Bilger W, Scherer S (1997) UV-B-induced synthesis of photoprotective pigments and extracellular polysaccharides in the terrestrial cyanobacterium *Nostoc commune*. J Bacteriol 179:1940–1945

Eppard M, Krumbein WE, Koch C, Rhiel E, Staley, JT, Stackebrand E (1996) Morphological, physiological and molecular characterisation of actinomycete isolates from dry soil, rocks and monument surfaces. Arch Microbiol 166:12–22

Friedmann EI (1980) Endolithic microbial life in hot and cold deserts. Origins Life 10:223–235

Garcia-Pichel F, Castenholz RW (1991) Characterization and biological implications of scytonemin, a cyanobacterial sheath pigment. J Phycol 27:395–409

Garcia-Pichel F, Castenholz RW (1993) Occurrence of UV-absorbing, mycosporine-like compounds among cyanobacterial isolates and an estimate of their screening capacity. Appl Environ Microbiol 59:163–169

Garty J (1990) Influence of epilithic microorganisms on the surface temperature of building walls. Can J Bot 68:1349–1353

Golubic S (1967) Die Algenvegetation an Sandsteinfelsen Ost-Venezuelas (Cumaná). Int Rev Gesamten Hydrobiol 52:693–699

Jaag O (1945) Untersuchungen über die Vegetation und Biologie der Algen des nackten Gesteins in den Alpen, im Jura und im schweizerischen Mittelland. Beitr Kryptogamenflora Schweiz 9:1–560

Hambler DJ (1964) The vegetation of granitic outcrops in western Nigeria. J Ecol 52: 573–594

Henssen A (1987) *Lichenothelia*, a genus of microfungi on rocks. In: Progress and problems in Lichenology in the eighties. Bibl Lichenol 25:257–293

Humboldt A v (1849) Ansichten der Natur. Quoted after the edition of Greno Verlagsgesellschaft, Nördlingen (1986)

Krempelhuber A (1876) Lichenes brasilienses, colecti a D.A. Glaziou in provincia brasiliensi Rio de Janeiro. Flora 59:56–63

Krumbein WE (1969) Über den Einfluß der Mikroflora auf die exogene Dynamik (Verwitterung und Krustenbildung). Geol Rundsch 58:333–363

Krumbein WE, Gorbushina AA (1995) On the interaction of water repellent treatments of building surfaces with organic pollution, microorganisms, and microbial communities. In: Wittman F, Siemes T, Verhoef L (eds) Surface treatment of building material with water-repellent agents. Delft University of Technology, Delft, pp 29/1–29/10

Krumbein WE, Jens K (1981) Biogenic rock varnishes of the Negev Desert (Israel), an ecological study of iron and manganese transformation by cyanobacteria and fungi. Oecologia 50:25–38

Lang NJ (1977) *Starria zimbabweënsis* (Cyanophyceae) gen. nov. et sp. nov.: a filament triradiate in transverse section. J Phycol 13:288–296

Lange OL, Büdel B, Meyer A, Kilian E (1993a) Further evidence that activation of net photosynthesis by dry cyanobacterial lichens requires liquid water. Lichenologist 25:175–189

Lange OL, Büdel B, Heber U, Meyer A, Zellner H, Green TGA (1993b) Temperate rainforest lichens in New Zealand: high thallus water content can severely limit photosynthetic CO_2 exchange. Oecologia 95:303-313

Sarthou C (1992) Dynamique de la végétation pionière sur un inselberg en Guyane Française. Thèse de Doctorat d'Etat, Université Paris

Sarthou C, Thérézien Y, Couté A (1995) Cyanophycées de l'inselberg des Nourages (Guyane Française). Nova Hedwigia 61:85–109

Scherer S, Chen TW, Böger P (1988) A new UV-A/B-protecting pigment in the terrestrial cyanobacterium *Nostoc commune*. Plant Physiol 88:1055–1057

Scott GD (1967) Studies of the lichen symbiosis: 3. The water relations of lichens on granite kopjes in Central Africa. Lichenologist 3:368–385

Seine R, Becker U, Porembski S, Follmann G, Barthlott W (1997) Vegetation of inselbergs in Zimbabwe. Edinb J Bot 55:267–293

Staley JT, Adams JB, Palmer TE (1992) Microcolonial fungi: common inhabitants on desert rocks? Science 215:1093–1095

Sterflinger K, Krumbein WE (1995) Multiple stress factors affecting growth of rock inhabiting fungi. Bot Acta 108:490–496

Sterflinger K, Krumbein WE (1997) Dematiaceous fungi as a major agent of biopitting for Mediterranean marbles and limestones. Geomicrobiol J 14:219–231

Swinscow TDV, Krog H (1988) Macrolichens of East Africa. British Museum (Natural History), London

Taylor TN, Hass H, Remy W, Kerp H (1995) The oldest fossil lichen. Nature 378:244

Upreti DK, Büdel B (1990) The lichen genera *Heppia* and *Peltula* in India. J Hattori Bot Lab 68:279–284

Weber B, Wessels DCJ, Büdel B (1996) Biology and ecology of cryptoendolithic cyanobacteria of a sandstone outcrop in the Northern Province, South Africa. Algol Stud 83:565–579

Weber H-M (1997) Ein Biofilm auf freiem Fels: Cyanobakterien der Inselberge Brasiliens. MSc Thesis, University of Bonn

Wessels DCJ, Büdel B (1989) A rockpool lichen community in northern Transvaal, South Africa: composition and distribution patterns. Lichenologist 21:259–277

Wessels DCJ, Büdel B (1995) Epilithic and cryptoendolithic cyanobacteria of Clarens Sandstone cliffs in the Golden Gate Highlands National Park, South Africa. Bot Acta 108:220–226

6 Bryophytes

J.-P. FRAHM

6.1 Introduction

Hornworts (Anthocerotae), hepatics (Marchantiatae), and mosses (Bryatae) are usually classified as bryophytes. Although apparently polyphyletic and without common origin, they share similar morphological and anatomical structures and a similar life cycle and make use of similar life strategies.

Bryophytes show strong contrasts to flowering plants in many respects:

- Bryophytes belong to the oldest green land plants. They existed already in the Paleozoic in forms which were hardly different from the extant species. In spite of the strong environmental changes on earth during the past geological periods and in contrast to lycopods and horsetails, they have always successfully adapted to new habitats such as Paleozoic swamps, Mesozoic deserts, or Tertiary forests.
- Many bryophytes possess several cormophytic structures such as a cuticle, stomata, or conducting tissues; however, they make no use of them. Thus, it can be postulated that they derived from cormophytic anchestors by reducing cormophytic structures and adapting to a poikilohydric life, thus choosing an alternative pathway in evolution.
- Most bryophytes have ranges which are much wider than those of phanerogams. For example, 40 % of the moss species of Central and South America have a wide neotropical distribution from Mexico to SE Brazil; more than 10 % of the moss species of the Neotropics occur also in tropical Africa. Therefore, some bryophyte species of inselbergs in the neotropics and the paleotropics are identical, and several bryophyte species from inselbergs in western Africa and Zimbabwe are the same, which is not the case to the same extent in flowering plants.

Because of these important differences, it can be expected that bryophytes show different reactions as compared with vascular plants concern-

Ecological Studies, Vol. 146
S. Porembski and W. Barthlott (eds.) Inselbergs

ing morphological adaptations, physiology, distribution, and dispersal, especially on such a harsh habitat as inselbergs.

The information about inselberg bryophytes is, however, small. Collections on inselbergs were occasionally made; however, with no special focus on this group of plants. The first intensive collections of bryophytes with special reference to this habitat were made during the inselberg project of the Botanical Institute of the University of Bonn. Within this project, inselberg bryophytes were collected by S. Porembski in the Côte d'Ivoire, Senegal, and Guinea (Frahm and Porembski 1994). The collections revealed a total of 67 species, 31 from inselbergs. Forty-three of them were new records to these countries, which shows that not only the countries but especially this kind of habitat are undercollected.

Additional bryophyte collections from the Côte d'Ivoire made by S. Porembski and also from Zimbabwe made by R. Seine were published by Frahm et al. (1996). Twenty-five bryophyte species were recorded from inselbergs in Zimbabwe. A preliminary evaluation of the floristic records from western Africa and Zimbabwe was published by Frahm (1996). It gave a preliminary idea of the diversity, life strategies, fertility, structural adaptations, dispersal, and ranges of inselberg bryophytes.

Additional collections of bryophytes on inselbergs were made by Porembski and Biedinger in the Seychelles. The floristic results of these collections were incorporated in a moss flora of the Seychelles (O'Shea et al. 1996). Most recently, a bryophyte collection from inselbergs in Bénin revealed seven species of hepatics and ten species of mosses, most of them new records for this country (Frahm and Porembski 1997).

The results on inselberg bryophytes must be interpreted with some care, because the inselberg collections were made by non-bryologists. Although the numbers of collections, especially in western Africa, seem to be sufficient for an evaluation, the fact that different species were found in different years on the inselbergs indicates that collecting was generally not complete. Also the fact that less numbers of bryophyte species were found at higher altitudes on inselbergs in Zimbabwe as compared with the inselbergs in western Africa may not be generalized and is probably an effect of incomplete collecting, since bryophyte diversity usually rises worldwide with elevation.

6.2 Floristic Diversity

A list of all bryophyte species collected on inselbergs within the frame of the inselberg project performed by the Botanical Institute, University of

Bonn, is given in Table 6.1. The inselbergs investigated are all from tropical Africa and can thus be easily compared. In contrast to a previous evaluation (Frahm 1996), the data from the Seychelles as well from Bénin have been incorporated. Furthermore, epiphytic species from inselberg forests were omitted since it can be supposed that these are identical with those from other forest types.

In general, the species composition in the vegetation surrounding the inselbergs was not determined. Therefore we do not know which bryophyte species are confined to this kind of habitat.

Species diversity of inselberg bryophytes seems to be very low as compared with other habitats, although no exact data are available for comparison. The highest species diversity of inselberg bryophytes is found in the Côte d'Ivoire with 39 species, the lowest with 16 and 17 species in the Seychelles and Bénin, respectively. The inselbergs in the Côte d'Ivoire are, however, situated in the rainforest as well as in the savanna belt, which both have a different floristic inventory. In contrast, the inselbergs in Bénin are all situated in the savanna. Their bryoflora shows a strong relation to the savanna inselbergs in the Côte d'Ivoire.

The majority of inselberg bryophytes consist of mosses, almost all of which are acrocarpous species. The only pleurocarpous species, *Sematophyllum fulvifolium* and *Erythrodontium squarrosum*, which were found in the Côte d'Ivoire, are mainly growing on bark and seem to occur on sand and rock just by chance and can therefore be excluded from further consideration. Thus mosses represent 66% of the species in the Côte d'Ivoire and 68% in Zimbabwe. Liverworts are less represented. They consist only exceptionally of foliose species and mainly of thalloid species, predominantly *Riccia* species with xeromorphous thalli, which seems to be the reason that liverworts are lacking on rainforest inselbergs (e.g., on the Seychelles). This composition indicates that bryophytes are apparently adapted to strong desiccation and that dry-resistant species dominate.

There is only one species common to all inselbergs studied in Africa, *Brachymenium exile. Hyophila involuta, Brachymenium acuminatum* and *Bryum arachnoideum*, were found in three of four regions studied. These are, interestingly, the species with the widest ranges. *Hyophila* is of pantropical distribution, *Bryum arachnoideum* occurs disjunct in South America, and *Brachymenium acuminatum* occurs disjunct in SE Asia.

The Seychelles and the Côte d'Ivoire have five species in common. The closest affinities between inselberg bryofloras exist between Bénin and the Côte d'Ivoire, which have 11 species (or 64% of the known inselberg species in Bénin) in common. In contrast, Zimbabwe and the Côte d'Ivoire have 5 species in common, Zimbabwe and Bénin only 3 species.

Table 6.1. Floristic diversity of bryophytes from inselbergs in tropical Africa

	Seychelles	Côte d'Ivoire	Bénin	Zimbabwe
Mosses				
Anoectangium aestivum		x		
Archidium ohioense		x	x	
Brachymenium acuminatum		x	x	x
B. dicranoides				x
B. exile	x	x	x	x
B. philonotula				x
Braunia arbuscula				x
Bryum arachnoideum		x	x	x
B. argenteum			x	
B. depressum		x	x	
Calymperes erosum	x	x		
C. palisotii	x	x		
C. tenerum	x			
Campylopus brevirameus	x			
C. decaryi				x
C. flaccidus				x
C. nanophyllus	x	x		x
C. pilifer				x
C. robillardei	x			
Campylopus savannarum		x		
Clastobryophilum rufo-viride	x			
Ectropothecium seychellarum	x			
Erythrodontium squarrosum		x		
Fissidens basicarpus		x		
F. crispulus		x		
F. reflexus		x		
F. sciophilus		x		
F. zollingeri		x		
Funaria calvescens		x		
Garckea moenkemeyeri		x	x	
Haplodontium cf. *ovale*		x		
Hedwigia albicans				x
Himantocladium seychellarum	x			
Hyophila involuta	x	x	x	
Leucobryum madegassum				x
Octoblepharum albidum	x	x		x
Philonotis hastata	x	x		
P. mniobryoides		x	x	
P. sp.	x			
P. strictula		x		
Plagiobryum zierii		x		x
Pohlia cf. *elongata*				x
Polytrichum piliferum				x
Rhacocarpus purpurascens				x
Schizomitrium seychellense	x			
Sematophyllum fulvifolium		x		

Table 6.1 (*continued*)

	Seychelles	Côte d'Ivoire	Bénin	Zimbabwe
Mosses				
Syrrhopodon mahensis	x			
S. planifolius		x		
Trichostomum brachydontium		x		x
Vesicularia sp.		x		
Weissia cf. *edentula*		x	x	
Total mosses	16	30	10	18
Hepatics				
Acrolejeunea emergens			x	
Caudalejeunea hanningtonii		x		
Exormotheca holstii				x
Fossombronia sp.			x	
Lejeunea caespitosa		x		
Riccia atropurpurea			x	x
R. cf. *fluitans*		x		
R. cf. *radicosa*		x		
R. congoana			x	x
R. discolor		x	x	
R. lanceolata		x	x	
R. microciliata				x
R. moenkemeyeri		x	x	
R. okahandjana				x
R. rosea				x
Total hepatics	0	7	7	6
Hornworts				
Anthoceros sp.		x	x	
Notothylas aff. *javanica*		x		
Total hornworts	0	2	1	0
Σ	16	39	18	24

The low number of species corresponds with a low number of genera and families (Table 6.2). The 30 species of mosses from the Côte d'Ivoire belong to 13 families, the 19 mosses from Zimbabwe to only 6 families.

The hepatic and hornwort flora consists mainly of thalloid species, in Zimbabwe exclusively, whereas in the Côte d'Ivoire only two foliose hepatics were collected, but seven thalloid liverworts.

The closest relationship between the inselberg bryofloras exists between the Côte d'Ivoire and Bénin in the sense that all bryophytes which were found in Bénin on inselbergs are also present on inselbergs of the Côte d'Ivoire. The percentage similarity between the inselberg bryofloras of the

Table 6.2. Species numbers of families of inselberg bryophytes in tropical Africa

	Seychelles	Côte d'Ivoire	Bénin	Zimbabwe
Hepatics and hornworts				
Ricciaceae		6	5	5
Exormothecaceae				1
Condoniaceae		1	1	
Lejeuneaceae		1	1	
Anthocerotaceae		2	1	
Acrocarpous mosses				
Fissidentaceae	1	5		
Archidiaceae		1	1	
Ditrichaceae		1	1	
Dicranaceae	3	1		4
Leucobryaceae	1			2
Calymperaceae	4	2		
Pottiaceae	1	4	2	1
Funariaceae		1		
Bryaceae	1	6	5	7
Bartramiaceae	2	4	1	
Hedwigiaceae				2
Rhacocarpaceae				1
Polytrichaceae				1
Pleurocarpous mosses	3	2		

Côte d'Ivoire and Bénin is lower, because the inselbergs in Bénin were all situated in the savanna belt whereas those in the Côte d'Ivoire were situated in savannas as well as in the rainforest belt.

A comparison between the bryophyte flora of inselbergs in Zimbabwe and the Côte d'Ivoire shows similarities but also differences in the composition of species and families, which probably reflect the different altitude (20–400 m in the Côte d'Ivoire but 400–2500 m in Zimbabwe) and the different latitude (the length of the rainy season and the precipitation with 900–1700 mm is comparable).

Nevertheless, there is some correspondence in the fact that Dicranaceae (with Leucobryaceae), Pottiaceae, and Bryaceae are found in both regions and that Dicranaceae and Bryaceae are the families with the highest species numbers in both regions. Twenty-three percent of the families found in the Côte d'Ivoire but 50 % of the families found in Zimbabwe are the same.

The floristic affinities between the Côte d'Ivoire and Zimbabwe on the genus level includes seven genera. On the species level, 25 % of the species from inselbergs in the Côte d'Ivoire are the same as in Zimbabwe and 42 % of the inselberg bryophytes from Zimbabwe also occur in the Côte d'Ivoire. This relatively high floristic correspondence results from the large ranges of bryophytes and is essentially different in flowering plants.

The floristic diversity of inselbergs in the Côte d'Ivoire was studied in correlation with their size, their location in different vegetation belts, and their elevation (Frahm 1996). The species numbers were not correlated with the size of the inselbergs or their elevation. Some of the inselberg bryophytes were confined to the rainforest region, others to the savanna region. The highest species numbers were therefore found in the transition zone. Rainforest inselbergs had less species than savanna inselbergs, which seems paradox, because annual precipitation of 1500–2000 mm should favor bryophyte growth more than the 1100–1500 mm in savanna regions. The reduced species number in rainforest inselbergs could be a result of higher competition of flowering plants or the stronger isolation of inselbergs.

A basic question is how the floristic composition of inselberg bryophytes is determined. Generally, there are two possibilities. First the species numbers and species composition can be related to the number of available ecological niches. A second possibility is that species numbers are determined by the chance of spore dispersal. Since high similarities between comparable study areas indicate deterministic processes and low-similarity stochastic processes, the calculation of the similarity index could solve this question. In a comparison of the bryoflora of eight inselbergs in the Côte d'Ivoire, the Søerensen index was generally <0.5 and thus an indication of stochastic processes in the species composition. Generally, the β-diversity was very low. This seems, however, not surprising with regard to the harsh ecological conditions, which reduce the species numbers.

6.3 Habitats

Inselbergs provide numerous habitats for bryophytes such as forests, open rocks, wet flush areas, rock pools, and ephemeral flush communities. With the exception of Zimbabwe and also Australia, where the moss *Grimmia laevigata* typically colonizes open rocks (Keever 1957), bryophytes are not very characteristic for open rocks in equatorial latitudes, where lichens and cyanobacteria seem in general to be more competitive. This suggests that bryophytes cannot tolerate the strong desiccation in tropical lowlands in exposed habitats with perhaps too high temperatures or too high respiration. In contrast, the inselbergs studied in Zimbabwe are situated in elevations up to 1700 m, where bryophyte growth in open habitats seems physiologically to be unproblematic.

The main minimum factor for bryophytes seems to be water. Therefore species richness in different habitats is determined by the availability of water. Typical bryophyte habitats are mats of phanerogams, episodic water flushes, and shallow depressions. A short analysis of the habitats of all bryophytes from inselbergs in Bénin (Frahm and Porembski 1997) revealed that there is apparently not much habitat specifity. Almost the same species but in different frequencies showed up in wet flush areas, *Afrotrilepis* mats, and rock pools. These are all habitats which are supplied with water. Especially the margins of phanerogam mats in western Africa show a typical border of bryophytes such as *Bryum arachnoideum* and *Brachymenium exile.* Wet flush areas with the low amount of water available for bryophytes have the lowest species numbers; phanerogam mats which retain humidity for a longer time are richer in species and richest are rock pools. Rock pools are filled periodically with water. As a consequence, rock pools harbor annual species which germinate in the rainy season, finish their life cycle in the rainy season, and survive dry seasons as spore in the soil. Examples for annual hepatics are all species of *Riccia.* The spores develop within the thallus and are released when the thallus decomposes. *Riccia* species are a characteristic element of rock pools in western Africa, where they can cover large areas with dense mats. These mats are a result of the spore release into the soil, where a large spore bank is built up, which results in a sudden germination and establishment of dense mats after the first rain falls in the rainy season. The same life cycle is found in the moss *Archidium ohioense*, which is often associated with species of *Riccia. Riccia* species and *Archidium* are therefore found only during rainy seasons and their existence on inselbergs can easily be overlooked in dry seasons. Thus, the species diversity found on inselbergs in semihumid regions depends very much on the collecting season.

Inselbergs situated in savannas are not affected by savanna fires and have presumably an effect on the conservation of bryophyte populations in such areas. It can be hypothesized that inselbergs may function as refugia for savanna species, and that burnt savannas can be recolonized from diaspores produced by plants on inselbergs.

6.4 Structural Adaptations

Bryophytes are known for a rich plasticity of anatomical and morphological structures as adaptations to their environment, e.g., hyalocysts and water sacks as water-storing structures, alar cells, cilia, papillae on the leaf surfaces, concave leaves or tomentose stems as water-conducting structures, and the presence of a central strand for internal water conducting. The number of structural adaptations varies between inselbergs in rainforest areas (e.g., southern Côte d'Ivoire, the Seychelles) and inselbergs in savanna regions. Surprisingly, there are hardly any structural adaptations to be observed in the case of savanna inselberg bryophytes, in spite of their extreme ecological conditions. Most of the mosses look quite "normal", e.g., most species of *Bryum* and all species of *Brachymenium, Garckea*, and *Fissidens.* Some of the *Riccia* species have xeromorphic thalli. The Pottiaceae (*Trichostomum, Weissia*) have involute leaves in the dry state, the species of *Philonotis* a scarce tomentum and papillae on the leaf surface; however, any specialized adaptations are present in only a minority of the species. Thus, papillae on leaf surfaces are found in only 2 of 30 species of mosses from inselbergs in the Côte d'Ivoire (*Trichostomum brachydontium* and *Weissia* cf. *edentula*), and in Zimbabwe in 4 of 19 species (*Trichostomum brachydontium, Hedwigia ciliata, Braunia arbuscula, Rhacocarpus purpurascens*, the latter with a reticulate leaf surface). Thus, structurally adapted species are surprisingly more common in the more humid climate at higher elevation in Zimbabwe as compared with the Côte d'Ivoire. Vermiculate and silvery foliation are found in the common *Bryum arachnoideum* and in the rare *B. argenteum* and *Plagiobryum zierii*, but these species are usually associated with other species without any special adaptations, which can apparently easily compete with the better-adapted species. In contrast, the mosses from rainforest regions belonging to the families Leucobryaceae and Calymperaceae have highly developed structures for storing water in hyalocysts. Both families are, in general, highly characteristic for the humid tropics, especially in low elevations; however, it remains unclear why especially species from humid environments have water-storing structures, where water supply is not a problem, and species from dry habitats have not.

The lack of structural adaptations and special life-forms indicates that the bryoflora on inselbergs consists of generalists and not of specialists.

6.5 Dispersal

A speciality of the inselberg bryophytes is their apparent sterility. From all species collected on inselbergs in Zimbabwe, only two species were found with sporophytes. All mosses collected on inselbergs in the Côte d'Ivoire and Bénin were sterile. Regular spore production is found only in all hepatic species of *Riccia* and in the moss *Archidium ohioense*, both annual species. The spores of *Riccia* are, however, 60–90 µm in diameter, and the spores of *Archidium*, with a diameter between 120 and 250 µm, are the largest bryophyte spores in the world. The size of these spores and the fact that the spores are only released in the soil when the spore capsules are decayed suggests that these spores are not wind-dispersed. The only possibility for dispersal seems to be by birds, which is supported by the habitat of these species in shallow depressions.

The mosses lack not only sporophytes but also any methods of vegetative propagation. Rhizoidal gemmae were found only in one specimen of *Brachymenium exile* from Zimbabwe; however, this is not a sufficient method for long-distance dispersal. The lack of sporophytes could be explained by the fact that most of the species (e.g., all species of Dicranaceae and Bryaceae, which are the dominant families on inselbergs) are dioiceous. However, the lack of means for vegetative propagation can hardly be explained. Although species of *Bryum*, *Brachymenium*, or *Philonotis* can principally produce small buds, these were never observed in inselberg specimens. Thus the inselberg bryophytes are mostly "achor", without any means of propagation. Dispersal within an inselberg is possible by plant fragments. The lack of any effective means of propagation is in strong contrast to the wide ranges of the species. It can only be postulated that the inselbergs were colonized by these species at a past climatic period, when the species produced sporophytes and that these species are now "trapped" in sterile condition on inselbergs, or that these species are not confined to inselbergs but produce sporophytes elsewhere, colonize inselbergs, but remain sterile on this kind of habitat.

The lack of mechanisms of effective dispersal seems also to be the reason that the inselberg bryofloras are relatively inhomogenous, and chance seems to be the determining factor in the colonization of inselbergs.

Table 6.3. Phytogeographic elements of inselberg bryophytes

	Seychelles	Bénin	Côte d'Ivoire	Zimbabwe
Cosmopolitan		13	13	12
Pantropical	42	25	35	28
Tropical African	21	38	32	28
Zambian				20
Sudanian		13	6	
Sudano-Zambian		13	9	4
Holarctic-African			3	8
Endemic	36			

6.6 Distribution

An analysis of the phytogeographic composition of African inselberg bryophytes (Table 6.3) shows that cosmopolitan, pantropical or tropical African species represent the majority of bryophyte species. More than one third of the species found on inselbergs in the Seychelles is endemic, in contrast to continental Africa, where no endemics are found on inselbergs. The inselbergs in Bénin and the Côte d'Ivoire are linked with those in Zimbabwe by a Sudano-Zambian element and both harbor representatives of the Sudanian viz. Zambian element. In contrast to flowering plants, there are no species endemic to inselbergs. This result is in sharp contrast to the fact that most inselberg bryophytes are sterile or have no means of long-distance dispersal, if they produce spores. This counts only for the present climatic conditions, and it could be that fertility will increase under changing climatic conditions.

6.7 Comparison with Flowering Plants

Bryophytes and flowering plants have different life strategies. The main difference concerns the poikilohydric life of bryophytes, which suits them especially for such harsh ecological conditions as on inselbergs. Interestingly, this – apparently very effective – adaptation is copied by some flowering plants, e.g., *Afrotrilepis pilosa* (Cyperaceae), *Coleochloa setifera*, *Tripogon minimus* (Poaceae), *Myrothamnus flabellifolia* (Myrothamnaceae), and *Xerophyta* spp. (Velloziaceae). Poikilohydric life avoids water stress and allows direct reaction to sudden water supply.

A comparison with the data for flowering plants reveals the following differences between bryophytes and flowering plants:

1. The phytogeographic distribution of flowering plants is exactly opposite to bryophytes. The distribution types with the largest extension have the highest percentages in bryophytes, whereas the local element dominates in flowering plants. More than 30 % of the phanerogamic inselberg flora in the Côte d'Ivoire is Guinean, however, there is absolutely no correspondence in bryophytes; 17 % of the phanerogams on inselbergs in the Côte d'Ivoire are Sudano-Zambian in distribution, however only 9 % of the bryophytes.
2. There are about 24 species of flowering plants endemic to inselbergs of the Guinean region, but none in bryophytes.
3. Forty % of the flowering plants from inselbergs in the Côte d'Ivoire are endozoochorous, 25 % anemochorous, 10 % are epizoochorous. However, only 17.5 % of the bryophytes in the same area are (presumably) epizoochorous, all others are "achorous", without any methods of dispersal, as outlined above.
4. The phanerogamic flora of inselbergs is highly influenced by deterministic processes as indicated by the dominance of highly adapted species. Bryophytes react just opposite. They are structurally unspecialized generalists; their composition is mainly influenced by stochastic processes.
5. An average of 30–40 % of phanerogams is confined to inselbergs, whereas the inselberg bryophytes are also found in other types of habitats.

Similar effects are found in bryophytes and flowering plants only in the respect that most of the species belong to few families.

References

Frahm J-P (1996) Diversity, life strategies, origins and distribution of tropical inselberg bryophytes. An Inst Bot Univ Nac Autónn México Ser Bot 67:73–86

Frahm J-P, Porembski S (1994) Moose von Inselbergen aus Westafrika. Trop Bryol 9:59–68

Frahm J-P, Porembski S (1997) Moose von Inselbergen in Benin. Trop Bryol 14:3–9

Frahm J-P, Porembski S, Seine R, Barthlott W (1996) Moose von Inselbergen aus Elfenbeinküste und Zimbabwe. Nova Hedwigia 62:177–189

Keever C (1957) Establishment of *Grimmia laevigata* on bare granite. Ecology 38: 422–429

O'Shea BJ, Frahm J-P, Porembski S (1996) Die Laubmoosflora der Seychellen. Trop Bryology 12:169–191

7 Vascular Plants on Inselbergs: Systematic Overview

W. Barthlott and S. Porembski

7.1 Introduction

To date, only incomplete information is available about details of the systematic composition of most tropical ecosystems or particular functional types. There are still relatively few published works dealing with the taxonomic composition of major components of tropical ecosystems (lianas Gentry 1991; epiphytes Madison 1977 and Kress 1989), which is apparently due to their unparalled diversity and complexity.

This chapter is intended as an introduction to the systematic distribution of vascular plants occurring on inselbergs. In forming an ecosystem that is considerably less complex than most forest types, an analysis of its sytematic composition on a global scale presents no insuperable obstacle. Species numbers will be given for the taxa occurring on inselbergs to illustrate their respective importance. Facing similar growth conditions, epiphytes can be considered to represent ecological counterparts of rock outcrop plants. Many valuable data are available for vascular epiphytes as far as species numbers and their systematic distribution are concerned (Kress 1989). How many vascular plants occur on inselbergs and what about their systematic occurrence? This question can only be answered with caution. In contrast to other ecologically circumscribed groups of vascular plants, there exists no comprehensive compilation of plants that grow on inselbergs: not even rough estimations have been provided hitherto about the numbers of families, genera, and species.

Over the past years a number of studies have been published which provided insights into the floristic composition of plant communities on inselbergs throughout different geographic regions. When species lists were compiled, usually no distinction was made between inselberg specialists (i.e., mainly found on inselbergs) and accidental colonizers. In the following account only inselberg specialists are included, whereas accidentals are omitted. The classification of species occurring on inselbergs as either specialist or accidental is mainly based on long-term experience

Ecological Studies, Vol. 146
S. Porembski and W. Barthlott (eds.) Inselbergs

gained from our own field surveys and observations over a broad geographic spectrum (South America: Bolivia, Brazil, French Guiana, Venezuela; Africa: Bénin, Cameroon, Côte d'Ivoire, Equatorial Guinea, Guinea, Madagascar, Malawi, Seychelles, Zimbabwe; Australia). In the following synthesis all species which preferentially occur on inselbergs are regarded as inselberg specialists.

7.2 Data Availability

The data presented in this chapter are based on extensive fieldwork, the analysis of floras (e.g., for tropical Africa: *Flora of West Tropical Africa*, Keay and Hepper 1954–1972; *Flora of Tropical East Africa*, Polhill 1973; *Flora Zambesiaca*, Pope 1960) and of published floristic surveys and herbarium studies (e.g., in Berlin-Dahlem, Kew, Paris). The extent to which these sources of information were available varied between different regions. Within the tropics the vegetation of inselbergs is relatively well known in West and East Africa, in southern Venezuela, and in other parts of the Guayana shield. However, for many regions, only sparse information exists concerning the floristic composition of inselberg plant communities. For this reason the systematic overview given here is preliminary and only after the completion of hitherto existing gaps of knowledge can a more detailed account be provided. In particular there is almost no published information about the vegetation of inselbergs from the Indian subcontinent. Likewise, there is a need for more data to be obtained from certain parts of Brazil, Madagascar, and central to southern Africa (e.g., Angola).

7.3 Systematic Distribution and Species Numbers

7.3.1 Systematic Distribution of Vascular Plants on Inselbergs

Species which are characteristic for inselbergs are widespread throughout the major divisions of vascular plants (Table 7.1, Fig. 7.1). However, they are clearly concentrated on the most derived groups within the angiosperms. The following list of taxa is based on the classification of Kramer and Green (1990) for ferns and gymnosperms. For angiosperms we follow a modified system of Ehrendorfer (1998) which considers modern molecular data (Chase et al. 1993). Only groups containing inselberg specialists

Table 7.1. Systematic distribution of vascular plants on inselbergs

Major groups	Class	Order	Family	Genera
Pteridophyta	Lycopodiopsida	Selaginellales	Selaginellaceae	*Selaginella*
		Isoetales	Isoetaceae	*Isoetes*
	Pteridopsida	Ophioglossales	Ophioglossaceae	*Ophioglossum*
		Schizaeales	Schizaeaceae	*Anemia*
				Mohria
				Schizaea
		Pteridales	Pteridaceae	*Actiniopteris*
				Afropteris
				Cheilanthes
				Doryopteris
				Notholaena
				Pellaea
		Aspidiales	Aspleniaceae	*Asplenium*
				Ceterach
Gymnosperms	Cycadopsida	Cycadales	Zamiaceae	*Encephalartos*
				Zamia
Angiosperms	Piperopsida	Piperales	Piperaceae	*Peperomia*
	Liliopsida	Potamogetonales	Aponogetonaceae	*Aponogeton*
		Arales	Araceae	*Anthurium*
				Caladium
		Areacales	Arecaceae	*Syagrus*
		Pandanales	Pandanaceae	*Pandanus*
		Velloziales	Velloziaceae (70)	*Barbacenia*
				Pleurostima
				Vellozia
				Xerophyta
		Commelinales	Commelinaceae (30)	*Cyanotis*
				Murdannia
		Bromeliales	Bromeliaceae (110)	*Aechmea*
				Dyckia
				Encholirium
				Navia
				Pitcairnia
				Tillandsia
				Vriesea
		Juncales	Eriocaulaceae (80)	*Eriocaulon*
			Xyridaceae (30)	*Xyris*
			Cyperaceae (340)	*Afrotrilepis*
				Coleochloa
				Bulbostylis
				Cyperus

Table 7.1 (*continued*)

Major groups	Class	Order	Family	Genera
				Microdracoides
				Rhynchospora
				Scleria
				Trilepis
		Poales	Centrolepidaceae	*Aphelia*
				Centrolepis
			Poaceae (390)	*Axonopus*
				Eragrostis
				Microchloa
				Panicum
				Sporobolus
				Thrasya
		Burmanniales	Burmanniaceae	*Burmannia*
		Asparagales	Amaryllidaceae	*Worsleya*
			Iridaceae	*Cipura*
			Aloaceae	*Aloe*
			Anthericaceae	*Anthericum*
				Chlorophytum
		Orchidales	Boryaceae	*Borya*
			Orchidaceae (180)	*Angraecum*
				Caladenia
				Cyrtopodium
				Epidendrum
				Polystachya
	Ranunculopsida	Proteales	Proteaceae	*Hakea*
	Rosopsida	Vitales	Vitaceae	*Cissus*
		Saxifragales	Crassulaceae (30)	*Crassula*
				Kalanchoe
		Myrtales	Myrtaceae (60)	*Eucalyptus*
				Eugenia
			Melastomataceae (110)	*Acanthella*
				Antherotoma
				Comolia
				Dissotis
				Ernestia
				Graffenrieda
			Lythraceae	*Rotala*
		Rutales	Burseraceae	*Commiphora*
		Malvales	Malvaceae	*Hibiscus*
			Bombacaceae	*Bombax*
				Pseudobombax
		Capparanae	Capparaceae	*Cleome*
		Fabales	Polygalaceae (30)	*Polygala*
			Mimosaceae	*Acacia*
			Caesalpiniaceae	*Chamaecrista*
			Fabaceae (330)	*Indigofera*
				Tephrosia

Table 7.1 (*continued*)

Major groups	Class	Order	Family	Genera
		Cucurbitanae	Begoniaceae	*Begonia*
		Casuarinales	Myrothamnaceae	*Myrothamnus*
		Urticales	Moraceae (60)	*Dorstenia*
				Ficus
		Violales	Erythroxylaceae	*Erythroxylum*
		Euphorbiales	Euphorbiaceae (140)	*Croton*
				Euphorbia
				Manihot
		Nepenthales	Droseraceae	*Drosera*
		Caryophyllales	Portulacaceae (25)	*Anacampseros*
				Calandrinia
				Portulaca
			Cactaceae (40)	*Coleocephalocereus*
				Melocactus
				Rhipsalis
		Hypericales	Clusiaceae	*Clusia*
			Balsaminaceae	*Impatiens*
		Rubiales	Rubiaceae (190)	*Anthospermum*
				Hymenodictyon
				Tocoyena
				Virectaria
			Apocynaceae (60)	*Mandevilla*
				Pachypodium
				Plumeria
			Asclepiadaceae (30)	*Ceropegia*
			Gentianaceae (40)	*Irlbachia*
				Sebaea
		Solanales	Boraginaceae	*Cordia*
		Lamiales/Scrophulariales	Lamiaceae (30)	*Aeollanthus*
				Plectranthus
				Solenostemon
			Bignoniaceae	*Tabebuia*
			Scrophulariaceae (100)	*Craterostigma*
				Dopatrium
				Lindernia
			Gesneriaceae (40)	*Sinningia*
				Streptocarpus
			Lentibulariaceae (40)	*Genlisea*
				Utricularia
		Pittosporales	Apiaceae	*Steganotaenia*
				Trachymene
		Asterales	Stylidiaceae	*Stylidium*
			Asteraceae (200)	*Brachycome*
				Kleinia
				Mikania
				Oyedaea

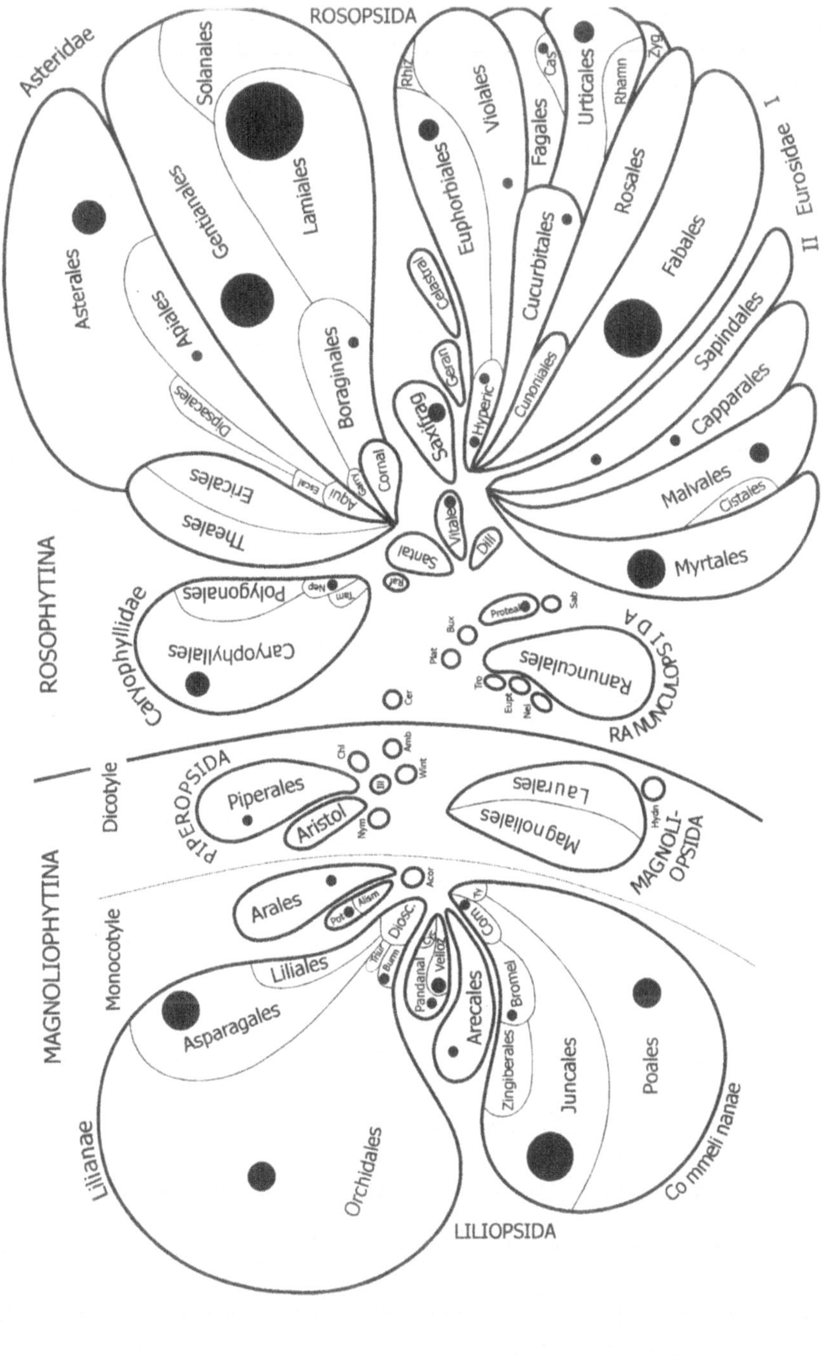

Fig. 7.1. Distribution of inselberg specialists among the major divisions of angiosperms (graphic design of the angiosperm system according to Barthlott et al. 2000). Approximate species numbers are given for orders comprising many or very characteristic inselberg specialists

are considered. On the family level the most important genera (i.e., containing more than five inselberg specialists or comprising a few but highly typical or widespread species) are mentioned. For a number of families (e.g., Xyridaceae) no exact data were available, therefore we had to estimate the approximate number of species present on inselbergs.

Despite the widespread systematic distribution of inselberg specialists, some general principles can be derived. Largely absent from inselbergs are groups that mainly comprise large forest trees (e.g., Fagaceae, Lauraceae). Richly represented are xerotolerant taxa which can be assigned to two different life-strategy types: (1) k-strategists are represented by slow-growing, centuries-old, poikilohydric species and (2) r-strategists, which consist of ephemeral plants. Concerning pollination and dispersal biology, no particular types are dominant among inselberg plants. However, in regard to dispersal, tiny dust-like anemochorous or epizoochorous diaspores are preponderant.

7.3.1.1 Ferns and Gymnosperms

The Pteridophyta and gymnosperms account for only a small percentage of inselberg specialists. Most prominent in regard to the number of species are the fern genera *Anemia*, *Asplenium*, *Cheilanthes*, and *Pellaea*, each including more than 20 species. The tiny fern spores are wind-dispersed and are therefore highly suited for the colonization of isolated habitat patches. Rock outcrop specialists among the ferns are adapted to xeric conditions by being either poikilohydric or geophytic (i.e., *Isoetes* and *Ophioglossum*).

Within the gymnosperms only the cycads (i.e., the Zamiaceae) include true inselberg specialists. For example, the genus *Encephalartos* includes rock outcrop species which are centered in central and southern Africa. Most species of *Encephalartos* are ornithochorous and form extremely slow-growing caulescent rosette trees.

7.3.1.2 Angiosperms

The angiosperms account for the vast majority of inselberg plants. The Poaceae (comprising 390 species as inselberg specialists), Cyperaceae (340 spp.), and Fabaceae (330 spp.) form the most speciose families. Of those angiosperm families that include inselberg specialists, the systematically isolated Myrothamnaceae are the only family which is more or less restricted to rock outcrops, with granitic and gneissic inselbergs being the most important growing sites. Families showing particularly high percentages of inselberg specialists are the Velloziaceae (24.3% of all species), Lentibulariaceae (16.3%), Xyridaceae (11.5%), Cyperaceae (7.7%), and Eriocaulaceae (7.3%). Altogether, 57 families of angiosperms contain a significant number of species that can be classified as inselberg specialists. In total, ca. 3500 species of angiosperms can be considered to be inselberg specialists, with monocotyledons and dicotyledons being roughly equally represented in regard to species number. In being generally less speciose than the dicotyledons, the monocotyledons are thus relatively overrepresented on inselbergs. The relatively high percentage of monocotyledons is a consequence of a wide range of vegetative forms which are especially successful in open, xeric environments. An important role is played by poikilohydric mat-formers and therophytes with both life-forms being well represented within Poaceae and Cyperaceae. Specialized annuals which are concentrated in seasonally wet localities are also found elsewhere, e.g., within Centrolepidaceae, Eriocaulaceae, and Xyridaceae. Certain families (e.g., Bromeliaceae, Orchidaceae) are notable for the possession of adaptational traits characteristic for epiphytes which are also of importance on inselbergs. The bromeliads on inselbergs are either closely related to terrestrial species (e.g., *Pitcairnia*) or derive from epiphytic relatives (e.g., *Aechmea*). Extremely slow-growing, poikilohydric caulescent trees occur within Velloziaceae, Cyperaceae, and Boryaceae, where a velamen radicum evolved independently in each of these families.

Within the dicotyledons the good representation of both legumes and Asteraceae seems to be the result of the sheer size of these families. Apart from a few exceptions (e.g., succulence), no specific adaptations for life on inselbergs has been developed. In contrast to this the Euphorbiaceae comprise a large number of taxa which are highly adapted to cope with arid conditions and their rich representation on inselbergs is therefore no surprise. Among the dicotyledons, several mostly herbaceous families are particularly speciose. Apparently the possession of specific adaptational traits, such as carnivory (Lentibulariaceae) or poikilohydry (Scrophulariaceae) promotes their rich presentation on inselbergs.

The largest number of inselberg specialists is present in the following genera: *Vellozia* (ca. 30 spp.), *Rhynchospora* (ca. 30 spp.), *Utricularia* (ca. 30 spp.), *Cyperus* (ca. 25 spp.), *Chamaecrista* (ca. 20 spp.), *Indigofera* (ca. 20 spp.), *Polygala* (ca. 20 spp.), and *Xyris* (ca. 20 spp.). Larger genera with high shares of inselberg specialists are best represented within the monocotyledons. Approximately two thirds of all species of *Xerophyta* (paleotropic Velloziaceae) and nearly one quarter of all species of the genus *Vellozia* (neotropical Velloziaceae) is more or less bound to inselbergs. Within the Bromeliaceae, the genera *Orthophytum* (25%), *Aechmea* (23.5%), and

Dyckia (18.7 %) are particularly well represented on inselbergs. Among the dicotyledons the genera *Lindernia* (24 %, Scrophulariaceae), *Dissotis* (20 %, Melastomataceae), and *Utricularia* (14 %, Lentibulariaceae) contain many inselberg specialists.

Within the angiosperms, the systematic distribution of inselberg specialists is not limited to particular groups. However, the basic groups of the angiosperms are depauperate in rock outcrop plants. Inselberg specialists are widely scattered throughout the more derived angiosperms with a considerable number of families containing at least a few species. Despite this fact it is remarkable that a high percentage of the total (more than 30 %) is concentrated to just three families (Poaceae, Cyperaceae, Fabaceae). Another third is contributed by the seven families Asteraceae, Rubiaceae, Orchidaceae, Euphorbiaceae, Melastomataceae, Bromeliaceae, and Scrophulariaceae. In summary, only ten angiosperm families account for two thirds of all inselberg specialists.

7.3.2 Phytogeographical Aspects

As can been seen from Table 7.2, there are considerable region-specific differences in the systematic composition of the inselberg vegetation (Porembski et al. 1997). Only a relatively small number of families, e.g., Poaceae and Cyperaceae, are important on inselbergs throughout temperate and tropical regions.

Apart from families which are important concerning the number of species, each region is characterized by a set of typical families which dominate particular habitats or are prominent in regard to life-form or adaptational trait. For example, on South American inselbergs, Cactaceae (e.g., *Melocactus*, *Coleocephalocereus*) and Gesneriaceae (*Sinningia*) comprise succulents and cryptophytes respectively, which are frequently encountered. In tropical Africa and Madagascar, Crassulaceae, Droseraceae, Lamiaceae, and Myrothamnaceae contain typical inselberg elements. The monotypic Medusagynaceae and a single representative of the Nepenthaceae are specific for inselbergs on the Seychelles. Australian inselbergs bear short-lived Crassulaceae (also on rock outcrops in the USA) and Portulacaceae (present also on South American rock outcrops).

Concerning the most species-rich genera occurring on inselbergs, there are large differences between South America and Africa. A preliminary survey of the most important genera from South American and African inselbergs is given in Table 7.3.

The monocotyledons account for the great majority among the most speciose genera on South American inselbergs. Of the plant families that

Table 7.2. Regional differences in the systematic distribution of vascular plants occurring on inselbergs. For each region the ten most species-rich families are listed. The data are confined to species which are more or less restricted to inselbergs. For South America and Africa estimations of species numbers are given in parentheses

South America	Africa	Madagascar	Seychelles	Australia	USA
Cyperaceae (120)	Fabaceae (200)	Asteraceae	Apocynaceae	Asteraceae	Poaceae
Poaceae (120)	Poaceae (180)	Poaceae	Arecaceae	Poaceae	Cyperaceae
Bromeliaceae (110)	Cyperaceae (150)	Orchidaceae	Cyperaceae	Stylidiaceae	Asteraceae
Rubiaceae (110)	Scrophulariaceae (70)	Asclepiadaceae	Euphorbiaceae	Centrolepidaceae	Liliaceae
Melastomataceae (90)	Asteraceae (60)	Cyperaceae	Rubiaceae	Boryaceae	Commelinaceae
Orchidaceae (90)	Rubiaceae (60)	Euphorbiaceae	Orchidaceae	Apiaceae	Caryophyllaceae
Fabaceae (80)	Euphorbiaceae (50)	Aloaceae	Pandanaceae	Droseraceae	Rosaceae
Apocynaceae (50)	Orchidaceae (40)	Scrophulariaceae	Poaceae	Orchidaceae	Hypericaceae
Euphorbiaceae (50)	Commelinaceae (30)	Gentianaceae	Fabaceae	Cyperaceae	Portulacaceae
Myrtaceae (50)	Lentibulariaceae (20)	Fabaceae	Sapotaceae	Fabaceae	Rubiaceae

Table 7.3. The ten most speciose genera of vascular plants occurring on South American and African inselbergs. Approximate species numbers are given in parentheses

South America	Family	Africa	Family
Rhynchospora (30)	Cyperaceae	*Cyperus* (25)	Cyperaceae
Vellozia (30)	Velloziaceae	*Dissotis* (20)	Melastomataceae
Vriesea (30)	Bromeliaceae	*Indigofera* (20)	Fabaceae
Pitcairnia (20)	Bromeliaceae	*Utricularia* (20)	Lentibulariaceae
Polygala (20)	Polygalaceae	*Xerophyta* (20)	Velloziaceae
Utricularia (20)	Lentibulariaceae	*Eragrostis* (15)	Poaceae
Xyris (20)	Xyridaceae	*Eriocaulon* (15)	Eriocaulaceae
Aechmea (20)	Bromeliaceae	*Euphorbia* (15)	Euphorbiaceae
Dyckia (20)	Bromeliaceae	*Ficus* (15)	Moraceae
Bulbostylis (20)	Cyperaceae	*Tephrosia* (15)	Fabaceae

include genera which occur on South American inselbergs, the Bromeliaceae are best represented with four genera. Amongst them the genus *Dyckia* includes the highest percentage of inselberg specialists (more than 20 % out of ca. 100 species) and is only exceeded by *Vellozia* (25 % inselberg specialists). In contrast to this, no such dominance of monocotyledons exists among the important genera on African inselbergs. Moreover, there is no particular family dominating among the most speciose genera. Despite the lack of detailed data, it can be concluded that the genus *Xerophyta* (Velloziaceae) contains the largest percentage of inselberg specialists (probably more than 80 %).

7.3.3 Systematic Distribution Within Inselberg Habitats

Species-rich families like Poaceae and Cyperaceae occur over a wide range of inselberg habitats. Certain families, however, show close ecological affinities to particular inselberg habitats. Rock-pool specialists are particularly well represented within the Scrophulariaceae and in the genus *Isoetes*. In the southeastern USA, throughout many parts of tropical Africa, and in Australia, rock pools are colonized by highly specialized and mostly tiny species of the Scrophulariaceae genera *Amphianthus*, *Glossostigma*, and *Lindernia*, which are frequently accompanied by *Isoetes* species. Monocotyledons make up the bulk of the mat-forming species. Apart from a few exceptions, nearly all Bromeliaceae and Velloziaceae occur as mat-formers, as do most poikilohydric Cyperaceae (e.g., *Afrotrilepis*, *Coleochloa*) whereas dicotyledons as mat-formers are largely absent. Ephemeral flush communities are characterized by the presence of a large number of families which are relatively poor in species. Families like Burmanniaceae, Eriocaulaceae, Gentianaceae, Polygalaceae, and Xyridaceae are largely restricted to this habitat, which is characterized by low nutrient availability.

7.4 Comparison with Vascular Epiphytes

The abiotic conditions on inselbergs are characterized by severe microclimatic and edaphic conditions. In many respects, epiphytic vascular plants have to overcome very similar stressors. Thus, certain adaptive traits (e.g., the common occurrence of CAM) are shared between rock outcrop plants and epiphytes. In the tropics it is therefore not surprising to observe floristic affinities between the vegetation of inselbergs and those of tree canopies. These relationships are especially well developed in South America. Here, certain Bromeliaceae, Orchidaceae, and ferns can be found growing either epilithically or epiphytically.

Judging from the data reported by Kress (1989) and from the numbers given above, it is evident that vascular epiphytes (23 466 species = more than 10% of all vascular plants) are by far more numerous than rock outcrop species (ca. 3500 species = ca. 1.5% of all vascular plants). Apart from this huge difference in quantity, there are also large discrepancies in the systematic composition between both groups. Within the angiospermous epiphytes the monocotyledons clearly dominate (67%), with the orchids alone accounting for 60% of all epiphytic species. In contrast to this, monocotyledons and dicotyledons are equally well represented on inselbergs. One can only speculate about the factors which have led to the higher species richness of epiphytes in contrast to those of inselberg species. However, it seems to be clear that the explosive radiation of epiphytic orchids has been decisive (for discussion of reasons see Benzing 1981; Dressler 1981; Benzing and Atwood 1984; Gentry and Dodson 1986).

When comparing the most species-rich families on inselbergs with those containing the largest numbers of epiphytes, significant differences are revealed. Only Orchidaceae, Melastomataceae, and Bromeliaceae constitute important components on both rock outcrops and in tree canopies. In regard to the systematic distribution, the dominance of Poaceae, Cyperaceae, and Fabaceae on rock outcrops is the main difference between inselberg plants and epiphytes. This difference is largely due to the structural attributes of inselbergs, which allow for the development of large-sized, individuum-rich plant communities favoring the establishment of frequently wind-pollinated, short-lived herbaceous to shrubby species. Since short-lived, anemophilous species are almost completely lacking among epiphytes, these structural differences between inselbergs and canopies may explain the considerable amount of dissimilarity in systematic distribution.

Inselberg plants and epiphytes are not only distinguished concerning the most species-rich families. Similarly, there are sharp contrasts in regard to several other families occurring mutually exclusive with Velloziaceae, Eriocaulaceae, and Xyridaceae comprising many inselberg specialists on the one hand and Araceae, Ericaceae, and Gesneriaceae including large numbers of epiphytes on the other.

According to Benzing (1989), the large number of epiphytic vascular plants, and especially of orchids, is due to the fact that they possess a large set of specializations. Poaceae, Cyperaceae, and Fabaceae, as being the numerically most important families on inselbergs, do not show unique adaptations to this ecosystem. Most inselberg specialists belonging to these families are not substantially different from close relatives outside inselbergs. The origin of inselberg specialists thus can be located among closely related taxa in grasslands which were already predisposed to cope with drought and high solar irradiation. Only within a small number of clades did inselberg specialists evolve which developed a set of morphological, anatomical, and physiological traits that are rarely encountered outside inselbergs. Most specialized in this regard are tree-like poikilohydric Velloziaceae and Cyperaceae, which are characterized by the possession of velamentous roots.

Acknowledgements. The Deutsche Forschungsgemeinschaft (grant no. Ba 605/4-3 to W. Barthlott) is thanked for supporting research on the biodiversity of inselbergs.

References

Barthlott W, Borsch T, Neinhuis C (2000) Vermutete Verwandtschaftsbeziehungen zwischen den Angiospermengruppen. In: Weberling F, Schwantes HO (Hrsg), Pflanzensystematik, 7. Aufl. Ulmer, Stuttgart

Benzing DH (1981) Why is Orchidaceae so large, its seeds so small and its seedlings mycotrophic? Selbyana 5:241–242

Benzing DH, Atwood JT Jr (1984) Orchidaceae: ancestral habitats and current status in forest canopies. Syst Bot 9:155–165

Chase MW, Soltis DE, Olmstead RG, Morgan D, Les DH, Mishler BD, Duvall MR, Price RA, Hills HG, Qiu Y-L, Kron KA, Rettig JH, Conti E, Palmer JD, Manhart JR, Systma KJ, Michaels HJ, Kress WJ, Karol KG, Clark WD, Hedren M, Gaut BS, Jansen RK, Kim K-J, Wimpee CF, Smith JF, Furnier GR, Strauss SH, Xiang O, Plunkett GM, Soltis PS, Swensen SM, Williams SE, Gadek PA, Quinn CJ, Eguiarte LE, Golenberg E, Learn GHJ,

Graham SW, Barrett SC, Dayanandan S, Albert VA (1993) Phylogenetics of seed plants: an analysis of nucleotide sequences from the plastid gene rbcL. Ann Miss Bot Gard 80:528–580

Dressler RL (1981) The orchids: natural history and classification. Harvard University Press, Cambridge

Ehrendorfer F (1998) Angiospermae, Bedecktsamer. In: Strasburger's Lehrbuch der Botanik. 34 edn. Gustav Fischer, Stuttgart

Gentry AH (1991) The distribution and evolution of climbing plants. In: Putz FE, Mooney HA (eds) The biology of vines. Cambridge University Press, Cambridge, pp 3–49

Gentry AH, Dodson CH (1986) Diversity and biogeography of neotropical vascular epiphytes. Ann Mo Bot Gard 74:205–233

Keay RJW, Hepper FN (1954-72) Flora of West Tropical Africa. Revised edn. Crown Agents, London

Kramer KU, Green PS (eds) (1990) Pteridophytes and gymnosperms. In: Kubitzki K (ed) The families and genera of vascular plants, vol 1. Springer, Berlin Heidelberg New York

Kress WJ (1989) The systematic distribution of vascular epiphytes. In: Lüttge U (ed) Vascular plants as epiphytes. Ecological Studies 76. Springer, Berlin Heidelberg New York, pp 234–261

Madison M (1977) Vascular epiphytes: their systematic occurrence and salient features. Selbyana 2:1–13

Polhill RM (1973-) Flora of tropical East Africa. A A Balkema, Rotterdam

Pope G (1960) Flora Zambesiaca. Flora Zambesiaca Managing Committee, London

Porembski S, Seine R, Barthlott W (1997) Inselberg vegetation and the biodiversity of granite outcrops. In: Withers PC, Hopper SC (eds) Granite outcrops symposium. J R Soc West Aust 80:193–199

8 Vascular Plants on Inselbergs: Vegetative and Reproductive Strategies

N. Biedinger, S. Porembski, and W. Barthlott

8.1 Introduction

As being shown in some detail by Szarzynski in Chapter 3, this Volume, the environmental conditions on inselbergs are stressful. Whereas detailed information is available about the ecophysiological characteristics of rock outcrop cryptogams (for survey see Büdel et al. Chap. 5, this Vol.) there are only sparse data concerning the adaptive traits of vascular plants. It could, however, be demonstrated that most rock outcrop specialists are adapted to high light environments, with their growth and photosynthetic rates being maximal under these conditions. Closer insights into their ecophysiological traits are given by Kluge et al. (Chap. 9, this Vol., see also Lüttge 1997 for a concise survey).

Various morphological, anatomical, and reproductive adaptations are presumed to be advantageous for plant survival on inselbergs. They serve in one way or the other to overcome ecological constraints of the environment where the major environmental stresses are drought, high temperatures and light intensities, and low nutrient availability. We focus on inselberg-specific plant types or functional types, which are reviewed and discussed here concerning their relative importance within the inselberg vegetation. The following deliberate selection of examples cannot replace a detailed account of the ecological properties of plants growing on inselbergs. This chapter rather tries to provide a brief introduction to this field.

8.2 Adaptational Aspects of Plant Life- and Growth-Forms on Inselbergs

Life-forms of plants are vivid testimony of the growth conditions within communities. A wide spectrum of different habits and life-forms can be found among species that grow on inselbergs. None of them is restricted to

Ecological Studies, Vol. 146
S. Porembski and W. Barthlott (eds.) Inselbergs

granite outcrops or other rocky outcrops but they often are more frequent – in terms of percentages of the whole inventories – than in the surrounding habitats.

As is mentioned in the regional accounts on inselberg vegetation (see Chap. 10, this Vol.), drastic differences exist in life-form spectra between different geographic regions and between individual habitats. For instance, therophytes are the predominant life-form on inselbergs in tropical Africa, whereas in the neotropics hemicryptophytes, chamaephytes, and phanerophytes are by far more important. In the following, the life-forms and habits are distinguished following the definition of Raunkiaer (1934) and the most frequent and typical will briefly be described.

Additionally, an overview of the most important morphological and anatomical traits of typical inselberg plants (i.e., species preferentially occurring on inselbergs) is given. It has to be emphasized that most morphological and anatomical characters mentioned here are not exclusively restricted to inselbergs. The vegetation of inselbergs shares the stressful environmental conditions with plant communities of other ecosystems. Consequently, adaptive traits related to desiccation or high temperature are widespread outside inselbergs and will not be discussed here in detail. We rather focus on those features which possess a specific adaptive value for survival in the inselberg environment.

8.2.1 Caulescent Rosette Trees

Island woodiness, i.e., the tendency of predominantly herbaceous dicot plant families to form trees and shrubs on oceanic islands, is a well-known phenomenon (Carlquist 1974). Woody representatives of predominantly herbaceous families mainly occur on islands, whereas they can be encountered in only a few continental ecosystems. Prominent examples are the giant rosette plants of the genera *Lobelia*, *Senecio*, and *Espeletia* which form character plants of the paramos of high tropical mountains (Rauh 1988).

Besides the paramos, inselbergs form another important continental ecosystem where woody representatives of otherwise herbaceous families occur. However, this fact has received little attention hitherto (e.g., Chevalier 1933; Mora-Osejo 1989). In contrast to oceanic islands, the phenomenon of island woodiness is restricted to poikilohydric monocotyledonous plants on inselbergs. These shrubby to arborescent species may attain a height of up to 4 m and possess woody-fibrous pseudostems (Fig. 8.1) with leaves crowded at the top. The pseudostems consist of persistent leaf bases and adventitious roots. It could be demonstrated that the adventitious

Fig. 8.1. The poikilohydric *Xerophyta splendens* (Velloziaceae) attains a tree-like habit with pseudostems up to 4 m in height. (Mt. Mulanje, Malawi, photograph S. Porembski)

roots of tree-like Cyperaceae and Velloziaceae possess a one- or multi-layered velamen radicum (Porembski and Barthlott 1995) which shows the same structural features as the velamen of epiphytic orchids (Fig. 8.2). Functionally, a velamen could be of considerable adaptive value for arborescent monocots on inselbergs where immediate water uptake is of crucial importance. As could be observed in different regions, arborescent monocots from inselbergs are highly resistant to fire, which is due to the isolating effect of a thick sheath consisting of leaf bases and roots that protects the central portions.

Most Cyperaceae and Velloziaceae which form caulescent rosette trees on rock outcrops have convergently developed xerophytic leaves that are characterized by similar morphological and anatomical traits. For example, the stomata are placed in abaxial furrows which are closed by inrolling of the leaf as a response to water loss during desiccation (see also Ayensu 1973), which reduces transpirational water loss. Anatomically, the leaves are characterized by the presence of a considerable amount of sclerenchyma around the vascular bundles (Fig. 8.3).

Vegetative propagation by means of stolons or by basal branching is not uncommon in Cyperaceae (for instance in arctic species of the genus *Carex*, Jonsson et al. 1996) and Velloziaceae. According to Morawetz (1983),

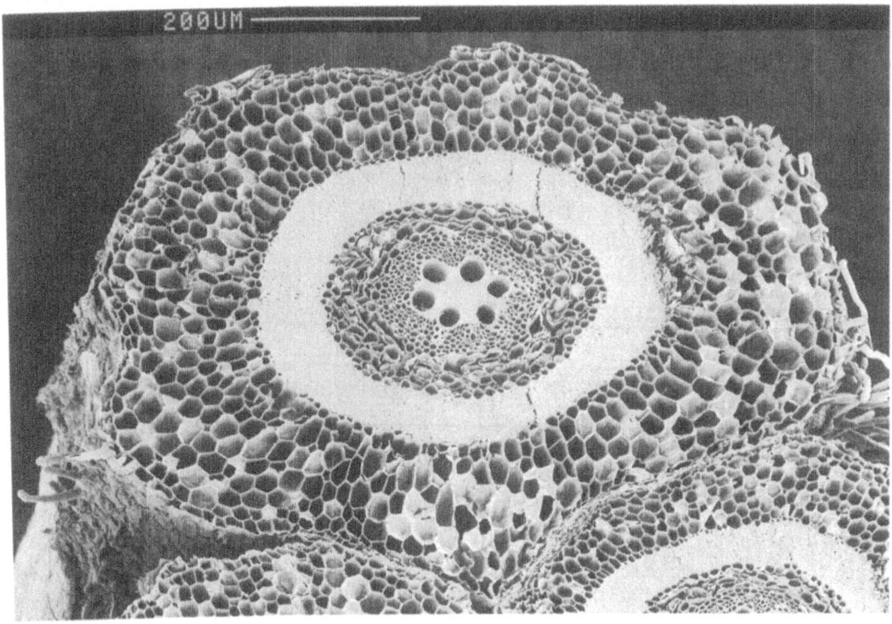

Fig. 8.2. SEM micrograph of transversal section through adventitious root of *Xerophyta pinifolia* showing the multilayered velamen radicum

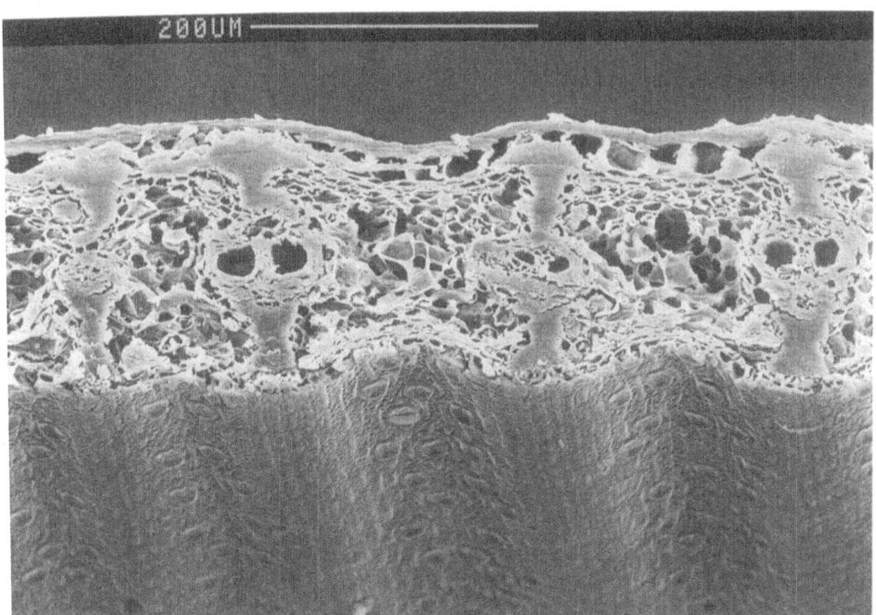

Fig. 8.3. SEM micrograph of a cross section through a leaf of *Lophoschoenus hornei* (Cyperaceae) showing large amounts of sclerenchyme

Vellozia cf. *glochidea*, which occurs in coastal sands in Bahia (Brazil), forms large colonies by vegetative propagation. Many rock outcrop-dwelling Velloziaceae and Cyperaceae likewise form clonal colonies which have an estimated age of several hundred years. Clonal growth of mat-forming species may provide substantial advantages:

- in isolated localities large populations can be built up after the establishment of a single individual;
- because of the potentially considerable age (i.e., hundreds or even thousands of years) of clonal populations long-term persistence is promoted;
- the stressful conditions on inselbergs may enhance seedling mortality; vegetative reproduction may reduce the risk of local extinction;
- the ability to spread horizontally possibly leads to the rapid occupation of suitable sites.

Remarkable is the fact that both mat-forming Velloziaceae and Cyperaceae bear a highly specific set of epiphytic orchids. The highest diversity of orchids restricted to caulescent rosette trees is found in tropical Africa. Here, several species of the genus *Polystachya* occur on *Afrotrilepis pilosa*, *Coleochloa setifera*, and different *Xerophyta* species. No data are available yet which could explain the dependency of these orchids on their specific phorophytes.

The dominant mat-former on West African inselbergs, *Afrotrilepis pilosa*, is characterized by great phenotypic plasticity. Whereas *Afrotrilepis pilosa*-mats are phenotypically uniform on an individual inselberg, there is frequently a high degree of morphological variation when compared with populations from geographically separated sites. Large amounts of phenotypical plasticity are already seen over relatively short distances (i.e., several kilometers) with environmental conditions (e.g., climate) being the same. We thus conclude that the observed patterns of morphological variation do not reflect specific adaptive traits.

Table 8.1 gives an overview of the systematic and geographic distribution of arborescent or shrub-like species on inselbergs. It becomes clear that most species are restricted to the tropics. In Australia the genus *Borya* (Boryaceae) occurs on both tropical and temperate zone inselbergs. Moreover one has to note the absence of caulescent rosette trees from inselbergs in North America.

Apart from inselbergs, caulescent Cyperaceae and Velloziaceae are also known from other tropical rock outcrop types and, moreover, form typ-ical elements of the white sand savanna vegetation (e.g., in the Brazilian cerrado). In South America, azonal sites are particularly rich in caulescent

Table 8.1. Caulescent rosette trees on inselbergs. The geographical distribution is roughly outlined

Species/genus	Family	Distribution
Afrotrilepis	Cyperaceae	West Africa
Barbacenia	Velloziaceae	South America
Borya	Boryaceae	Australia
Bulbostylis	Cyperaceae	South America
Coleochloa	Cyperaceae	East Africa, Madagascar
Lophoschoenus hornei	Cyperaceae	Granitic Seychelles
Microdracoides squamosus	Cyperaceae	West Africa
Pleurostima	Velloziaceae	South America
Trilepis	Cyperaceae	South America
Vellozia	Velloziaceae	South America
Xerophyta	Velloziaceae	East Africa, Madagascar

monocotyledons. Obviously, these localities have played a major role in their evolution, as is documented by the occurrence of tree-like species within Cyperaceae (apart from the above-mentioned taxa also in *Cephalocarpus, Everardia*) and Velloziaceae but moreover also within Bromeliaceae (*Ayensua*), Eriocaulaceae (*Paepalanthus*), and Xyridaceae (*Orectanthe*). In Australia, inselbergs in both tropical and temperate zones are colonized by the genus *Borya*. This genus is endemic to Australia with most species being resurrection plants (Gaff 1978) which possess a cushion-like habit. Several species form micro stilt plants (Pate 1989). In contrast to arborescent monocotyledons such as Pandanaceae and Arecaceae, which possess massive aerial prop roots, the plants which are aptly called microstilt plants are microchamaephytes and usually less than 10 cm in height. Microstilt plants are centered in southwestern Australia and comprise both monocotyledons and dicotyledons (in particular the genus *Stylidium*). The ecological significance of the microstilt habit lies in lessening the effect of high temperatures by elevating the shoots several centimeters above the surface.

An interesting morphological parallel to the caulescent monocotyledons on inselbergs is provided by woody Iridaceae (i.e., *Klattia, Nivenia*), which are restricted to the Cape Floral region of southern Africa. These plants are characterized by their shrubby growth form (sometimes cushion-forming) and their affinity to rock outcrops and nutrient-poor soils derived from Table Mountain sandstone (Goldblatt 1993).

An explanation for the existence of unusual shrubby and arborescent monocotyledons on inselbergs is difficult to provide. In our eyes the evolution of a tree-like habit out of a short-lived herbaceous ancestry was enabled by the fact that the tree niche on exposed slopes of inselbergs was not occupied. Moreover, there is an ecological advantage in becoming tree-like and long-lived, as the competitive ability for the long-term occupation of space increases considerably.

8.2.2 Poikilohydric Vascular Plants

The most spectacular adaptation among vascular plants on inselbergs which is related to the stressful water conditions is desiccation tolerance (for details on their physiology see Chap. 9, this Vol.; Lüttge 1997; Hartung et al. 1998; Tuba et al. 1998). These species are poikilohydric and can withstand considerable losses of their water content (more than 90 %) and may survive in the desiccated state for months or even years (Ayo-Owoseye and Sanford 1972; Gaff 1977, 1987). According to the behavior during desiccation, it is possible to distinguish between homoiochlorophyllous (i.e., keeping their photosynthetic pigments) and poikilochlorophyllous (i.e., losing their pigments) species. In particular, arborescent monocotyledonous species that occur as mat-formers on inselbergs belong to the so-called resurrection plants but, remarkably, certain dicotyledonous species which occur in seasonally water-filled rock pools are also desiccation-tolerant. The latter are best represented within the Scrophulariaceae with species which preferentially occur in seasonally water-filled rock pools, like the famous *Chamaegigas intrepidus* (*Lindernia*), endemic to Namibia. The vegetative organs of this species can survive dehydration for at least 8 months. During desiccation, large amounts of abscisic acid are accumulated in the vegetative organs (Schiller et al. 1997). After rewatering, they regain their metabolic activity in less than 2 h (see Hickel 1967 for details). The African genus *Craterostigma* comprises the most thoroughly studied poikilo-hydric species thus far. In particular the molecular biology (i.e., the gene expression) of *C. plantagineum* has been investigated in detail (Ingram and Bartels 1996).

Monocotyledons and ferns clearly dominate among the poikilohydric vascular plants (cf. survey in Fahn and Cutler 1992), whereas dicotyledons are rarely desiccation-tolerant. Examples among the dicotyledons are provided by Myrothamnaceae (comprising two species of *Myrothamnus* in Africa and Madagascar), Scrophulariaceae, and Gesneriaceae (e.g., *Streptocarpus* in East Africa and Madagascar). Among poikilohydric plants, *Myrothamnus* is unique in possessing woody stems which dehydrate with xylem vessels completely embolized in the air-dry state. Sherwin et al. (1998) assume that xylem refilling is due to root pressure which provides water at the base of the stem, and subsequent capillary rise in the vessels results in the reestablishment of hydraulic continuity in the xylem.

In particular the cyperaceous resurrection plants on inselbergs are remarkable for their change in leaf color from green to grayish between the wet and the dry state. This phenomenon was already noted long ago, but not before Hambler (1961) was a detailed description of a poikilohydric and poikilochlorophyllous species (i.e., *Afrotrilepis pilosa*) given. Today it clearly can be stated that inselbergs form a center of diversity for poikilohydric vascular plants (see Table 8.2 for an overview of the most important taxa) which recently attracted enhanced interest by molecular biologists (e.g., Bartels and Nelson 1994).

Poikilohydric vascular plants are of widespread occurrence on inselbergs throughout the tropics and to a lesser extent in temperate regions (Porembski and Barthlott, in press). Although our knowledge about their exact geographic distribution is not complete it appears that resurrection plants are most prominent in seasonally wet tropical regions whereas their importance decreases towards constantly humid or arid zones. Whereas inselbergs situated in perhumid rainforest usually bear at least a few poikilohydric vascular species, these are mostly absent from inselbergs situated in deserts.

8.2.3 Pachycaulous and Caudiciformous Plants

Pachycaulous trees (i.e., possessing massive water-storing, mostly unbranched, relatively soft trunks) and caudiciform succulents are widespread in seasonally dry tropical regions (see Table 8.3). Pachycaulous or caudiciformous species occur over a broad range of inselberg habitats. However, a certain preference for clefts where the development of deep roots is possible can be observed. In contrast to many savanna trees, which are fire-resistant due to the possession of an insulating corky bark, certain pachycaulous trees and caudiciformous species are characterized by a papery thin bark and grow preferentially in rocky sites sheltered from fire. In this case the weak development of the bark is related to the photosynthetic activity of the stems of these mostly leaf deciduous species.

Remarkably broad is the size range covered by pachycaulous and caudiciformous species on inselbergs. The latter are frequently more or less spherical and may attain a diameter of more than 1 m. Apart from caudiciformous species with aboveground storage organs there also exist geophytes possessing an underground caudex. Striking examples are provided by several species of the genus *Raphionacme* (Periplocaceae) on West African inselbergs, which often occur in shallow soil with their caudex laterally flattened.

Table 8.2. Important poikilohydric vascular plant genera on inselbergs. Data on geographic distribution and habitat preference are mainly based on personal observation. Indications of geographical region and habitat type refer to poikilohydric species

Genus	Family	Distribution	Habitat on inselbergs
Actiniopteris	Pteridaceae	E Afr	Crevices, mats
Afrotrilepis	Cyperaceae	W Afr	Mats
Anemia	Schizaeaceae	S Am	Crevices, mats
Asplenium	Aspleniaceae	Trop Afr	Crevices, mats
Barbacenia	Velloziaceae	S Am	Mats
Boea	Gesneriaceae	NE Austr	Crevices
Borya	Boryaceae	Austr	Shallow depressions, mats
Bulbostylis	Cyperaceae	S Am	Shallow depressions, crevices
Chamaegigas	Scrophulariaceae	Namibia	Rock pools
Cheilanthes	Sinopteridaceae	E Afr, S Am, Austr	Crevices
Coleochloa	Cyperaceae	E Afr, Madagascar	Mats
Craterostigma	Scrophulariaceae	E Afr	Shallow depressions
Doryopteris	Pteridaceae	SE Brazil	Crevices, mats
Eragrostiella	Poaceae	Austr	Shallow depressions
Eragrostis	Poaceae	E, S Afr	Shallow depressions
Fimbristylis	Cyperaceae	Trop Afr, Austr	Crevices, shallow depressions
Hemionitis	Pteridaceae	S Am	Crevices, mats
Limosella	Scrophulariaceae	S Afr	Rock pools
Lindernia	Scrophulariaceae	Trop Afr	Shallow depressions, rock pools
Micraira	Poaceae	Austr	Shallow depressions
Microchloa	Poaceae	Trop Afr, S Am	Shallow depressions
Microdracoides	Cyperaceae	W Afr	Mats
Myrothamnus	Myrothamnaceae	S, E Afr, Madagascar	Crevices, mats
Nanuza	Velloziaceae	Brazil	Mats
Notholaena	Sinopteridaceae	N, S Am	Crevices
Oropetium	Poaceae	S Afr	Shallow depressions
Pellaea	Sinopteridaceae	N, S Am, Afr	Crevices, mats
Pleurostima	Velloziaceae	S Am	Mats
Schizaea	Schizaeaceae	E Afr, Sey	Crevices, mats
Selaginella	Selaginellaceae	Pantrop	Mats
Sporobolus	Poaceae	Trop Afr	Shallow depressions
Streptocarpus	Gesneriaceae	E Afr, Madagascar	Crevices
Talbotia	Velloziaceae	S Afr	Mats
Trilepis	Cyperaceae	S Am	Mats, crevices
Tripogon	Poaceae	Trop Afr, S Am, Austr	Shallow depressions
Vellozia	Velloziaceae	S Am (mainly Brazil)	Mats, shallow depressions
Xerophyta	Velloziaceae	Trop Afr, Madagascar	Mats, shallow depressions

Table 8.3. Selected genera occurring with pachycaulous and caudiciformous species on inselbergs. Distribution data are restricted to species found on inselbergs

Genus	Family	Distribution
Acanthella	Melastomataceae	Northern S Am
Adenia	Passifloraceae	Trop Afr
Adenium	Apocynaceae	Trop Afr, S Arabia
Aloe	Aloaceae	Trop Afr, Madagascar
Brachystelma	Asclepiadaceae	Trop Afr
Ceropegia	Asclepiadaceae	Trop Afr
Cochlospermum	Cochlospermaceae	Trop Afr, S Am
Dorstenia	Moraceae	Trop Afr
Encephalartos	Zamiaceae	Trop and S Africa
Euphorbia	Euphorbiaceae	Paleotrop
Ficus	Moraceae	Pantrop
Graffenrieda	Melastomataceae	S Am
Hildegardia	Sterculiaceae	W Afr
Himatanthus	Apocynaceae	S Am
Mandevilla	Apocynaceae	S Am
Othonna	Asteraceae	E, S Afr
Pachira	Bombacaceae	S Am
Pachycormus	Anacardiaceae	Baja California
Pachypodium	Apocynaceae	E, S Afr, Madagascar
Plumeria	Apocynaceae	S Am
Pseudobombax	Bombacaceae	S Am
Pyrenacantha	Icacinaceae	E Afr
Raphionacme	Periplocaceae	Trop Afr
Tabebuia	Bignoniaceae	S Am

8.2.4 Succulents

Succulence (for a definition see von Willert et al. 1992) is a widespread desiccation-avoidance strategy in desert habitats where water storage in either leaves, stems, or roots permits survival. Mainly in the tropics, both leaf and stem succulents can be observed on inselbergs throughout but even in perhumid regions. Succulent species that occur on inselbergs have not developed specific structural modifications that distinguish them from their relatives in other ecosystems. On inselbergs succulents prefer exposed sites and frequently occur as lithophytes. Their ecophysiological traits are summarized by Kluge and Brulfert (Chap. 9, this Vol.).

Frequently the common occurrence of succulents representing different life-strategy types can be observed. Little information is available yet about the functional differentiation between them, so up to now there are no unequivocal data indicating a correlation between frequency of certain life strategies and environmental constraints on inselbergs. Particularly diverse both systematically and in respect to life strategy-types of succulents are inselbergs in East Africa and Madagascar (Barthlott and Porembski 1996).

In regard to life strategy, succulents with above-ground perennating organs dominate. Prominent examples are found within Aloaceae, Cactaceae, and Euphorbiaceae. Another life strategy is expressed by succulents with deciduous leaves. Included here are plants with perennating succulent stems, e.g., the West African Lamiaceae *Solenostemon graniticola*. Remarkable is the presence of annual leaf succulents on inselbergs in both tropical and temperate regions. Examples are *Cyanotis lanata* (tropical Africa), *Sedum smallii* (southeastern USA), and several species of the genera *Crassula* und *Calandrinia* which are common on rock outcrops in southwestern Australia. Preferably, these species occur in shallow depressions, where they appear in large quantities during the rainy season. Obviously, succulent annuals benefit from their enhanced water-storage capacity which allows the survival of droughts which may even occur during the rainy season. Most succulents are not exclusively bound to inselbergs but generally prefer dry, open sites. Today, fire plays a fundamental role in many seasonally dry ecosystems, leaving inselbergs and other types of rock outcrops as refugia for fire-sensitive plants, such as succulents.

Succulents on inselbergs have different ecological origins, with most of them being widespread terrestrial colonizers of dry localities. However, in the Neotropics a considerable number of succulents on inselbergs are either derived from close epiphytic relatives or occur both epilithically and epiphytically. Examples for succulents on inselbergs with a strong epiphytic background are mainly found in Orchidaceae, Bromeliaceae, and Araceae, which mostly show the crassulacean acid metabolism (CAM) mode of photosynthesis. In this regard it is worth mentioning the neotropical shrubby to tree-like genus *Clusia* (Clusiaceae) which comprises, apart from forest trees, many epiphytic and epilithic species possessing succulent leaves. Remarkably high is the physiological plasticity of this genus, with rapid changes between CAM and C_3 photosynthesis, which is an important advantage in localities underlying climatic stressors (Lüttge 1991).

An alternative strategy of water storage has been developed by tank-forming Bromeliaceae. These mostly epiphytic bromeliads (e.g., *Alcantarea* spp., *Vriesea* spp.) are particularly well represented, with epilithic species on inselbergs located in the Brazilian Atlantic rainforest, and possess tanks containing up to 20 l of water (incl. organic debris) which is absorbed via characteristic trichomes.

Data on temperature resistance of succulents growing on inselbergs are virtually non-existent. It is conceivable, however, that adaptations against overheating exist. A striking example is provided by certain Cactaceae on Brazilian inselbergs. The cereoid genus *Coleocephalocereus* comprises species with procumbent to prostrate stems creeping over exposed rock where surface temperatures regularly exceed 50 °C. Remarkably prostrate species of *Coleocephalocereus* possess two morphologically very different types of spines. Areoles in the upper parts of the stem have normal pungent spines whereas areoles of stem portions directly overlaying rock bear a dense cover of long (more than 10 cm, Fig. 8.4), hair-like spines which may have an insulating effect (Porembski et al. 1998). A strategy that recalls the Australian microstilt plants is exhibited by certain epilithic Orchidaceae (e.g., *Angraecum* spp. on Madagascar) which possess water-storing pseudobulbs or leaves. These species possess robust aerial stilt roots that fix the plant body in an upright position, thereby avoiding the high temperatures near the rock surface.

Fig. 8.4. Prostrate stems of *Coleocephalocereus buxbaumianus* (Cactaceae) bear long, hair-like spines which may have an insulating effect. (Brazil, Minas Gerais, photograph N. Biedinger)

8.2.5 Sclerophylly

Species possessing evergreen sclerophyllous leaves are typical for mediterranean-climate regions, with South Africa being an evolutionary center of sclerophyllous plants. The functional importance of sclerophylly is mainly attributed to water scarcity and nutrient-deficient soils. Both factors are of major significance for plants growing on inselbergs, where sclerophyllous trees and shrubs can be observed throughout the tropics. Inselbergs in South America and Australia are particularly rich in sclerophyllous species (see Table 8.4). In particular in southwestern Australia it is probably the high regional richness in sclerophyllous species that acts as major determinant of diversity on an inselberg scale. However, it is still not clear why this adaptational trait is so richly represented on inselbergs in South America too.

Preferred habitats of sclerophyllous species on inselbergs are clefts and the narrow ecotone between freely exposed rock habitats and forest. Within this floristically rich transition zone sclerophylls frequently occur side by side with deciduous trees.

Table 8.4. Selected genera occurring with sclerophyllous species more or less restricted to inselbergs. Distribution data are confined to species found on inselbergs

Genus	Family	Distribution
Byrsonima	Malpighiaceae	S Am
Clusia	Clusiaceae	S Am
Elaeophorbia	Euphorbiaceae	Trop Afr
Erythroxylon	Erythroxylaceae	S Am, Seychelles
Eucalyptus	Myrtaceae	Australia
Eugenia	Myrtaceae	S Am
Euphorbia	Euphorbiaceae	Madagascar, Seychelles
Ficus	Moraceae	S Am, trop Afr
Gastonia	Araliaceae	Seychelles
Graffenrieda	Melastomataceae	S Am
Hakea	Proteaceae	Australia
Licania	Chrysobalanaceae	S Am
Mandevilla	Apocynaceae	S Am
Manilkara	Sapotaceae	W Afr
Memecylon	Melastomataceae	Seychelles
Mimosa	Mimosaceae	S Am
Mimusops	Sapotaceae	Trop Afr, Seychelles
Ouratea	Ochnaceae	S Am

8.2.6 Other Life- and Growth-forms

The fact that inselbergs can be considered as microenvironmental deserts is reflected by the mostly considerable importance of therophytes among their vegetation. Annuals in particular dominate over shallow soils, but as soon as soil depth increases, they are accompanied by geophytes, which are likewise well adapted to this seasonal habitat. Remarkably, on inselbergs in Western Australia several annual species of the genus *Calandrinia* behave like part-time geophytes due to the possession of below-ground water-storage organs (Pate and Dixon 1982), which could be an advantage in surviving drought during the growing season.

Morphological adaptations of annuals on inselbergs are mainly related to water availability and are not specific to this ecosystem. Like their counterparts in deserts, they develop a shallow and wide-ranging root system. Many annuals from inselbergs are capable of producing fairly extensive root systems characterized by a large specific root length. For instance, the root system of the Commelinaceae *Cyanotis lanata* is characterized by a large amount of superficial (i.e., occurring in the upper 2 cm of the soil) roots which achieve considerable lengths, which allows for the depletion of water over large areas.

Another adaptive trait of annuals that is frequently observed on inselbergs and which is shared with plants growing in deserts is the broad range of individual body sizes. Under extremely dry conditions the individuals stay minute with aerial parts sometimes less than 1 cm high, in contrast to plants growing under moist conditions attaining a height of more than 15 cm, as is the case in, e.g., *Lindernia exilis* (Scrophulariaceae, see Porembski et al., Chap. 12, Fig. 12.13, this Vol.).

With regard to germination ecology, data are available mainly for therophytes from rock outcrops in southeastern USA (e.g., Baskin and Baskin 1982). Characteristically, seeds of winter annuals germinate in early autumn, plants overwinter with rosettes and flower in spring. Summer dryness is survived in the seed stage. Detailed information on germination ecology of plants growing on tropical inselbergs is practically nonexistent. However, it can be assumed that similar adaptations concerning the germination ecology (e.g., afterripening of the seeds) are present and that the same set of abiotic factors (e.g., temperature, rainfall) determines their germination behavior.

Remarkable among the geophytes on inselbergs are those species which possess below-ground organs that are absent among their closest relatives. Very interesting in this regard are the fern genera *Ophioglossum* and *Isoetes*. On West African inselbergs the geophytes *Ophioglossum costatum* and *Isoetes nigritiana* are widespread as specialized colonizers of seasonally wet depressions. They survive the dry season by means of a bulbous rhizome (*O. costatum*) or with a corm-like organ (*I. nigritiana*, Fig. 8.5). Relatively frequent on neotropical inselbergs are lianas with below-ground storage organs. Mostly sites with deep soil are preferred with Apocynaceae (e.g., *Mandevilla*), Convolvulaceae, Dioscoreaceae, and Smilacaceae as important families.

Despite the lack of detailed measurements, it can be assumed that the presence of nitrogen-fixing cyanobacteria is of considerable importance for the nitrogen supply within the inselberg ecosystem (see Kluge and Brulfert, Chap. 9, this Vol.). In general, however, the nutrient content of the substrate in inselberg habitats is very low (in particular N and P). Carnivorous plants as typical colonizers of extremely nutrient-poor sites rely on trapping and subsequently digesting insects as their primary source of nitrogen and other elements (Juniper et al. 1989). On inselbergs, carnivorous plants (Droseraceae and Lentibulariaceae) are richly represented in particular habitats, i.e., ephemeral and wet flush communities (Seine et al. 1995) where they grow in considerable densities in seasonally wet soils. Species of the genus *Genlisea* (Lentibulariaceae) show a preference for nutrient-poor white sand areas and rock outcrops, like inselbergs in South America, Africa, and Madagascar. The plants possess small rosettes appressed to the ground and achlorophyllous subterranean leaves. The long basal part of the leaves opens into a hollow utricle which contracts into a tubular channel. The latter divides into two helically contorted arms which bear slit-like openings (width 400 µm, height 180 µm). It could be proved that these specialized leaves are traps for catching protozoa which attract their prey chemotactically (Barthlott et al. 1998). Another remarkable carnivorous species from inselbergs is *Utricularia nelumbifolia*, which is known from several localities in the Brazilian states of Rio de Janeiro and Minas Gerais. This relatively large species is restricted to the water-filled tanks of large Bromeliaceae (*Vriesea* spp.) which grow on exposed steep rocky slopes (Ule 1898). Like the closely related *U. humboldtii* (distributed throughout the Guayana Highland region), *U. nelumbifolia* possesses long aerial stolons which descend into the tanks of nearby bromeliads, thus forming effective means of vegetative propagation.

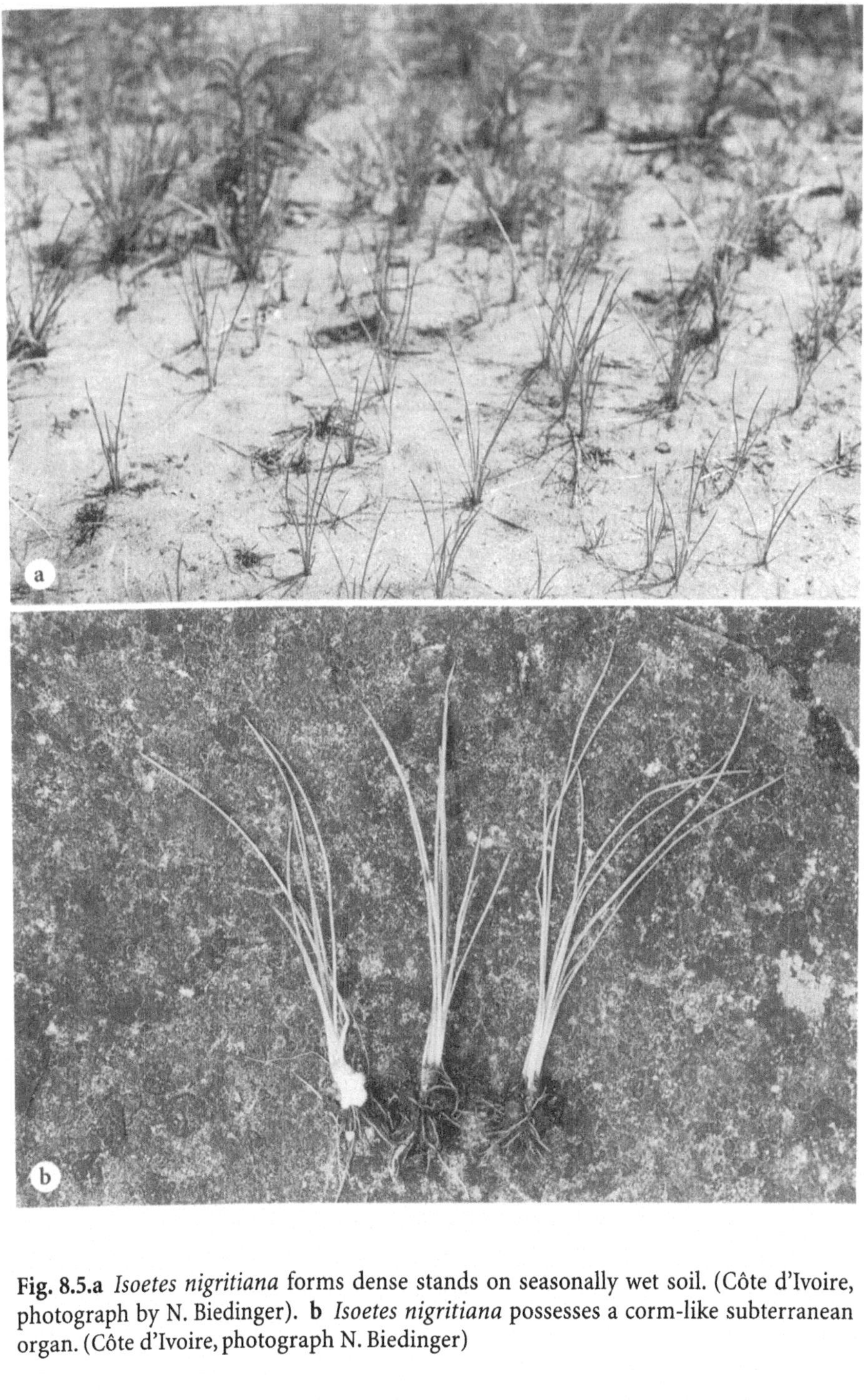

Fig. 8.5.a *Isoetes nigritiana* forms dense stands on seasonally wet soil. (Côte d'Ivoire, photograph by N. Biedinger). **b** *Isoetes nigritiana* possesses a corm-like subterranean organ. (Côte d'Ivoire, photograph N. Biedinger)

8.3 Reproductive Strategies

8.3.1 Pollination

In regard to pollination ecology, detailed observations on inselbergs have mainly been made in temperate regions. A detailed survey of the reproductive ecology of inselberg plants in the southeastern USA was accomplished by Wyatt (1997). According to this author, many granite outcrop species in the southeastern USA are outcrossers which are self-incompatible or more frequently have evolved self-fertilization. In contrast to this, inbreeding is reported for the Lobeliaceae *Isotoma petraea* which occurs with isolated populations on rocky outcrops in arid Australia (Bussell and James 1997). Populations of this species differ in regard to their stigma protrusion frequencies and thus vary in the extent of inbreeding. Highly inbreeding populations of *I. petraea* form complex hybrids. Detailed investigations of the genetic properties and competitive abilities of *I. petraea* rock outcrop populations were conducted by James (1965, 1970). Hopper (1981) provided insights into the relationships between honeyeaters (Meliphagidae) and their winter food plants (mostly Myrtaceae and Proteaceae), which almost exclusively rely on bird pollination on rock outcrops in Western Australia. It can be supposed that the great mobility of the honeyeaters is of importance for maintaining adequate levels of outcrossing in species occurring in small isolated populations (Hopper 1981). Small flies (especially syrphids) and bees were reported by Wyatt (1986) as primary pollinators of the winter annuals *Arenaria uniflora* and *A. glabra* on rock outcrops in southeastern United States. According to Wyatt (1981), both species of *Arenaria* compete for pollinators, with *A. glabra* possessing much showier flowers. Wherever both species are found sympatrically *A. uniflora* occurs with self-pollinated morphs, which assures reproduction despite the presence of a superior competitor for pollinators. The rare case of ant-pollination was demonstrated by Wyatt (1981) for the granite outcrop endemic *Diamorpha smallii* occurring in southeastern USA. In accordance with Hickman (1974), he proposed characters of an ant-pollination syndrome which is typically found in very small plants with flowers borne in short pedicels occurring in hot, dry habitats. To date no data on ant-pollination on inselbergs are available from other regions. However, our own observations indicate the possible presence of this phenomenon on Western Australian inselbergs also. Flower-visiting ants were observed in shallow depressions which were characterized by the abundant occurrence of *Calandrinia granulifera* (Portulaceae). This species almost perfectly fits the above-mentioned criteria of ant-pollination and is morphologically very similar to *Diamorpha smallii*. More field observations, however, are needed to prove the assumption of ant-pollination for *Calandrinia granulifera*.

The spectrum of observed pollination syndromes on tropical inselbergs is broad, ranging from entomophily and anemophily to ornithophily and chiropterophily. The latter two syndromes are reported mostly from the neotropics. Apart from other features (e.g., smell, coloration) chiropterophily is frequently expressed by the very exposed position of the flowers (Faegri and van der Pijl 1979). Chiropterophily mainly occurs in large plants (i.e., trees, lianas) which possess adaptations (e.g., flagelliflory) that provide easy access to visiting bats, whereas herbaceous forest plants are almost never chiropterophilous due to difficult flight conditions for bats in the forest understory. The presence of bat-pollinated herbaceous species on South American inselbergs is therefore testimony to the open character of this ecosystem with Bromeliaceae (e.g., *Alcantarea*), Cactaceae (e.g., *Coleocephalocereus*), and Gesneriaceae (e.g., *Sinningia*) being most prominent. Similarly, the bat-pollinated Gentianaceae *Irlbachia alata*, which is a typical element of the inselberg flora of the Guayana region, is another example that underlines the importance of inselbergs for the evolution of chiropterophilous herbs. Since bats are efficient pollinators over great distances, they should promote genetic exchange between plant populations situated on isolated inselbergs. Detailed studies on the pollination of extremely long-spurred Malagasy orchids revealed that the epilithic *Angraecum sororium* is pollinated by the hawkmoth *Coelonia solani* (Wasserthal 1997). Otherwise hawkmoth pollination has rarely been demonstrated to occur among plants growing on inselbergs.

Concerning the number of species, in most cases entomophily is dominant on inselbergs, followed by anemophily (Fig. 8.6). As can be expected, the latter mode of pollination is preponderant when the number of individuals is considered, whereas the population densities of, e.g., chiropterophilous taxa are much lower. Wind-pollinated trees are rare on tropical inselbergs. In this regard, one has to mention several species of the rubiaceous genus *Anthospermum* (East Africa, Madagascar) and both species of the genus *Myrothamnus* (*M. flabellifolia*, South and East Africa; *M. moschata*, Madagascar) that are wind-pollinated.

In particular, in the rainy season the vegetation of tropical inselbergs offers relatively more flowers for potential pollinators than the surrounding vegetation. In the Côte d'Ivoire this is mainly due to the presence of large numbers of annuals which are present in high abundance in certain habitats (e.g., shallow depressions). On Ivorian inselbergs the peak in number of flowering species is reached relatively soon after the onset of the rainy period (Fig. 8.7). Numerous species are already in flower 3 to 4

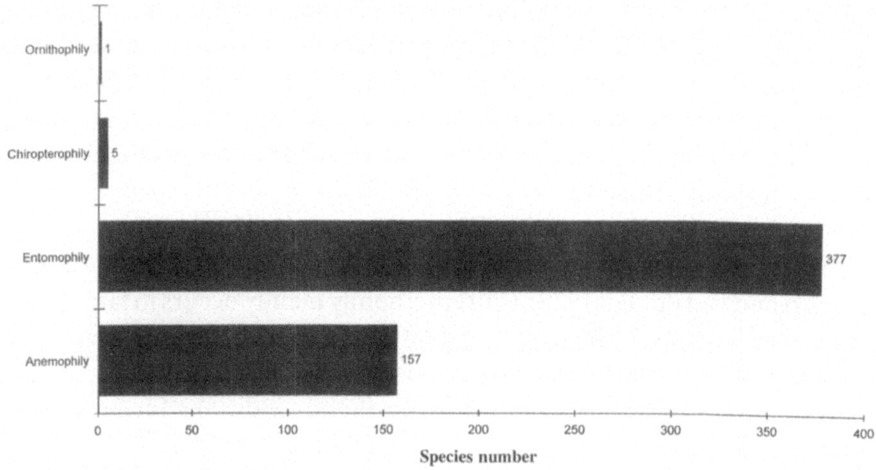

Fig. 8.6. Pollination syndrome for the vegetation of inselbergs in the Côte d'Ivoire. Entomophily and anemophily are clearly dominating (own unpubl. data)

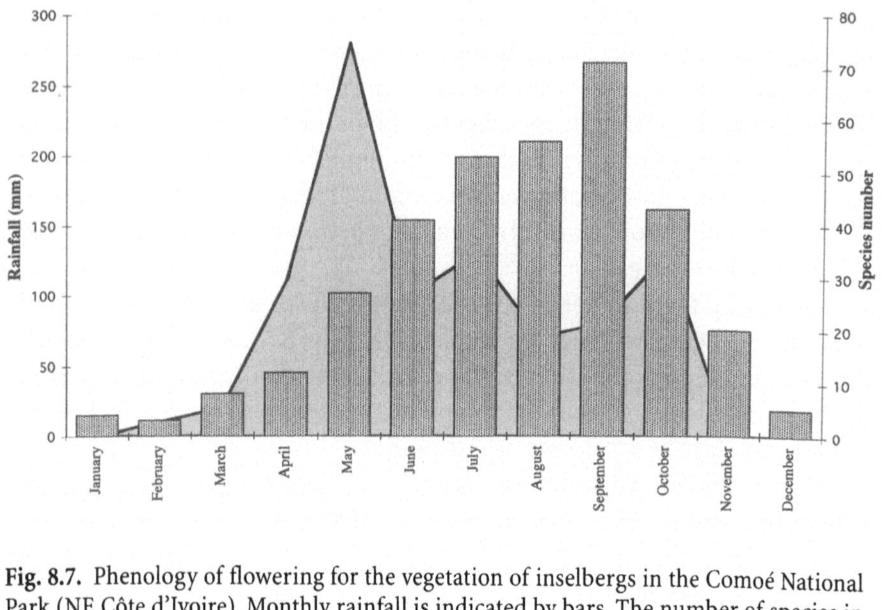

Fig. 8.7. Phenology of flowering for the vegetation of inselbergs in the Comoé National Park (NE Côte d'Ivoire). Monthly rainfall is indicated by bars. The number of species in flower reaches a peak at the begin of the rainy season in May. (own unpubl. data)

weeks after germination. There exist, however, habitat specific differences regarding the timing of the flowering peak, with the ephemeral flush vegetation attaining a maximum of flowering species at the height of the wet season. The number of flowering species is low during the dry season, with certain cryptophytes and phanerophytes being in flower. In particular, Anthericaceae (i.e., *Chlorophytum*) grow vegetatively in the rainy season and flower in the dry season. According to Cowling et al. (1997), many geophytes in the fynbos biome of southern Africa also have uncoupled flowering and growth periods. In general, flowering of inselberg plants is highly seasonal throughout most regions (cf. Gröger 1995 with data for Venezuela).

On inselbergs situated in the savanna zone of the Côte d'Ivoire certain species are characterized by gregarious flowering (i.e., nearly all individuals of a population are in flower at the same time). In particular, several species of *Utricularia* demonstrated this phenomenon with most individuals (occasionally thousands) being in flower over a relatively long (i.e., several weeks') time span. Remarkably, however, almost no visits of potential pollinators could be observed within these extended stands of *Utricularia*.

Studies on the pollination biology of the orchid *Polystachya rosea* revealed that this species, which occurs on inselbergs on the central plateau of Madagascar, exhibits a deceptive pollination system (Petterson and Nilsson 1993). This nectarless terrestrial or epilithic species is characterized by a large degree of flower color polymorphism. In contrast to most other species of the genus *Polystachya*, which offer pseudopollen (Dressler 1981), no such reward is provided by *P. rosea*. This species obviously exploits male solitary bees that perform mate-seeking patrolling behavior in which potential food plants are included.

8.3.2 Dispersal

In contrast to oceanic islands, where many plant species have lost their dispersibility, the majority of plants on inselbergs show dispersal abilities that allow for transport over longer distances. However, Wyatt (1997) assumes that plants from rock outcrops in the southeastern USA possess adaptations against long-distance dispersal in order to avoid inhospitable sites in the surroundings. It should be mentioned that exact information concerning the effectiveness of dispersal of inselbergs plants is largely lacking.

Data obtained in tropical Africa indicate that on inselbergs the percentage of endozoochorous species (incl. ornithochorous taxa) is highest, followed by anemochorous and epizoochorous plants. Endozoochory is frequent among Poaceae, Cyperaceae, and certain phanerophytes whose diaspores are swallowed by granivorous birds (pers. observ.). Anemochory is mainly by dust diaspores. In particular, the ephemeral flush vegetation is rich in families with spore-like diaspores: Burmanniaceae, Droseraceae, Gentianaceae, Lentibulariaceae, and Xyridaceae. Winged diaspores are characteristic for inselberg specialists such as *Hymenodictyon floribundum* and the genus *Myrothamnus*. Plumed diaspores occur in Asclepiadaceae, e.g., *Ceropegia*. Since thermal turbulence is a frequently observed phenomenon on inselbergs (see Szarzynski, Chap. 3, this Vol.), it can be assumed that dispersal by wind is an effective mode of transport on freely exposed rock outcrops. Epizoochory on inselbergs is probably mainly due to birds which carry diaspores either in their plumage or sticking to their feet. Adhesion to birds is affected by a myxospermous testa, as in *Aeollanthus pubescens* and some other Lamiaceae. In other cases, as in *Cyanotis lanata*, unadapted diaspores are lying on the ground and may stick to the feet of visiting birds. Hydrochory is assumed to be the mode of dispersal for a few Lamiaceae and Eriocaulaceae.

According to Gröger (1995), in the vegetation of inselbergs in southern Venezuela zoochorous species are more frequent than anemochorous taxa. Among the zoochorous species bird dispersal is of major importance. Data obtained on Brazilian inselbergs indicate a broad dispersal spectrum with anemochory (e.g., in Velloziaceae and most Bromeliaceae) dominating (Porembski et al. 1998). Endozoochory (e.g., in Cactaceae like *Coleocephalocereus* spp. and *Melocactus* spp. with reddish berry-like fruits) seemed to be far less important; epizoochory seems to be lacking altogether.

Species possessing diaspores without obvious means of dispersal (barochory) were recorded on both tropical African and South American inselbergs. Usually, their seeds simply fall to the soil. Possibly most plants belonging to this category are unintentionally dispersed by birds to whose feet the seeds stick, intermingled with mud. On paleotropical inselbergs the genus *Euphorbia* is richly represented. The lack of effective modes of dispersal within this genus could be the reason for the considerable amount of divergence between isolated populations of certain species. This is particularly well expressed by Malagasy species of *Euphorbia* (e.g., within the *E. milii* group). Their relatively large barochorous seeds do not allow for interinselberg transport, thus resulting in reproductive isolation of individual populations.

Due to the open character of their vegetation cover and because of abundant climatic irregularities, inselbergs are susceptible to new colonists. Today, however, apart from native species, weeds, which are usually good dispersers invade from adjacent disturbance prone sites (e.g., roads, pastures). The tradeoff between good dispersability and low competitiveness which is often postulated does not seem to exist for certain invasive weeds on inselbergs. Instead, one frequently encounters a combination of effective dispersal with strong competitiveness, as, for example, in *Furcraea foetida* and *Ananas comosus*. The latter two species have become a serious threat to the indigenous vegetation of inselbergs in certain regions of tropical Africa and Madagascar.

8.4 Plant Functional Groups and Redundancy in the Vegetation of Inselbergs

Ecological classifications of plants which are based on physiognomy and life-strategy characters date back a long time ago (e.g., Theophrastus about 300 B.C.). One of the most widely utilized systems is those of Raunkiaer (1934), who used the position of perennating buds during unfavorable periods as major criterion for his classification. In recent years, an increasing amount of functional classifications have been developed in order to describe those biotic components of ecosystems which perform the same function within the ecosystem. According to Gitay and Noble (1997), one attempt in this direction is to describe plant functional types (PFT) which are based on their response to certain perturbations. The circumscription of PFTs is somewhat arbitrary and should possibly be restricted to the feedback of plants to particular environmental constraints. Plant species belonging to a certain PFT share a set of covarying parameter ranges for attributes related to resource acquisition, growth, reproduction, dispersal, and response to environmental stress (Scholes et al. 1997). The vegetation of inselbergs is subject to a high degree of disturbance due to climatic perturbations. Here, it is not the aim to present a detailed PFT classification for inselbergs, instead, we focus on key environmental factors, such as water or nutrient availability, which exert major impacts on functional attributes of inselberg plants, in order to examine to what extent species redundancy contributes to the observed patterns of diversity for inselbergs in geographically distinct regions. Redundant species are considered to perform the same role in ecosystem processes as in primary production, and are believed to be present in most ecosystems (Lawton and Brown 1993). With respect to water-use efficiency, the following PFTs are of particular interest on inselbergs: mat-forming monocotyledons, succulents, and annuals. Concerning their species richness, these PFTs are extraordinarily rich in certain regions (e.g., mat-forming monocotyledons and succulents in Brazil, annuals in West Africa) whereas some areas are extremely impoverished (e.g., mat-forming monocotyledons and succulents in West Africa, annuals in South America). Undoubtedly, inselbergs in certain regions bear more species with parallel function than in others. Functional redundancy is strongly developed among mat-forming species on Brazilian inselbergs (Porembski et al. 1998) and among annuals in West Africa. Many attempts have been made to describe how increasing diversity enhances ecosystem stability (for survey see Peterson et al. 1998). By using the example of poikilohydric respectively annual species on inselbergs, it is possible to demonstrate that stability in the sense of resistance to invasions increases with species richness (i.e., redundancy) within these PFTs. Our hitherto available data indicate that invasive plants are of no importance in communities bearing locally species-rich PFTs (i.e., mats on Brazilian inselbergs, annual-dominated communities in West Africa). However, introduced plants have successfully invaded inselberg communities with low degrees of redundancy. This is shown by annual weeds which are becoming widespread on inselbergs in Brazil and on the granitic Seychelles or by the establishment of xerophytic species in mats in West Africa where particular PFTs are impoverished in species. In contrast to this, species-poor mats in West Africa are easier to invade than their more diverse counterparts in Brazil whereas species-rich annual-dominated communities in West Africa are more resistant to annual invaders than less speciose therophyte communities in Brazil. This inselberg example may provide support for the view that there is a general relationship between diversity and the invasibility of ecosystems.

In regions where the vegetation of inselbergs is characterized by low redundancy, those species which represent the sole member of certain PFTs may determine important ecosystem functions (e.g., water transfer). Species like these can be considered keystone species (Paine 1966). On inselbergs mat-forming monocotyledons play particularly important roles in several ecosystem processes, such as water availability or providing establishment sites for other species. On West African inselbergs the presence of whole communities (most of all ephemeral flush vegetation) strictly depends on the presence of mats formed by the Cyperaceae *Afrotrilepis pilosa*. Frequently situated on steep rocky slopes, these mats store a considerable amount of rainwater which otherwise would be completely lost by runoff. During the rainy season a continuous supply of seepage water out of the mats results in moist conditions at the feet of *Afrotrilepis*-covered slopes and allows for the development of species-rich ephemeral flush communities. Thus, the extent of *Afrotrilepis* mats on inselbergs is critical for the control of the amount of seepage water available for other plant communities. Any reduction in the size of the mats unavoidably causes an increase in direct water runoff during rainfall, which subsequently leads to a decrease in the amount of seepage water during dry periods. With *Afrotrilepis pilosa* being the only representative of poikilohydric mat-forming species throughout West Africa, the consequences of increasing human disturbance (e.g., fire) which cause a reduction in mat-covered area will result in decreasing species numbers within the whole ecosystem inselberg.

Acknowledgements. For discussions and various help we thank in particular G. Brown (Rostock) and R. Seine (Bonn). The research was supported by the Deutsche Forschungsgemeinschaft (grant no. Ba 605/4-3 to W. Barthlott).

References

Ayensu ES (1973) Biological and morphological aspects of the Velloziaceae. Biotropica 5:135–149

Ayo-Owoseye J, Sanford WW (1972) An ecological study of *Vellozia schnitzleinia*, a drought-enduring plant of northern Nigeria. J Ecol 60:807–817

Bartels D, Nelson D (1994) Approaches to improve stress tolerance using molecular genetics. Plant Cell Environ 17:659–667

Barthlott W, Porembski S (1996) Biodiversity of arid islands in tropical Africa: the vegetation of inselbergs. In: van der Maesen LJG, van der Burgt XM, van Medenbach de Rooy JM (eds) The biodiversity of African plants. Proc XIVth AETFAT Congress. Kluwer, Dordrecht, pp 49–57

Barthlott W, Porembski S, Fischer E, Gemmel B (1998) *Genlisea* – first protozoa-trapping plant found. Nature 392:447

Baskin JM, Baskin CC (1982) Germination ecophysiology of *Arenaria glabra*, a winter annual of sandstone and granite outcrops of southeastern United States. Am J Bot 69:973–978

Baskin JM, Baskin CC (1988) Endemism in rock outcrop plant communities of unglaciated eastern United States: an evaluation of the roles of the edaphic, genetic and light factors. J Biogeogr 15:829–840

Bussell JD, James SH (1997) Rocks as museums of evolutionary processes. In: Withers PC, Hopper SD (eds) Granite Outcrops symposium. J R Soc West Aust 80:221–229

Carlquist S (1974) Island biology. Columbia University Press, New York

Chevalier A (1933) Deux cypéracées arbustiformes remarquables de l'ouest africain. Terre Vie 3:131–141

Cowling RM, Richardson DM, Mustart PJ (1997) Fynbos. In: Cowling RM, Richardson, DM, Pierce, SM (eds) Vegetation of Southern Africa. Cambridge University Press, Cambridge, pp 99–130

Dressler RL (1981) The orchids. Natural history and classification. Harvard University Press, Cambridge

Faegri K, van der Pijl L (1979) The principles of pollination ecology. Pergamon Press, Oxford

Fahn A, Cutler DF (1992) Xerophytes. In: Braun HJ, Carlquist S, Ozenda P, Roth I (eds) Handbuch der Pflanzenanatomie. Spezieller Teil, vol 13, part 3. Borntraeger, Berlin
Gaff DF (1977) Desiccation-tolerant flowering plants of southern Africa. Oecologia 31:95–109
Gaff DF (1978) The occurrence of resurrection plants in the Australian flora. Aust J Bot 26:485–492
Gaff DF (1987) Desiccation tolerant plants in South America. Oecologia 74:133–136
Gitay H, Noble IR (1997) What are functional types and how should we seek them? In: Smith TM, Shugart HH, Woodward FI (eds) Plant functional types. Cambridge University Press, Cambridge, pp 3–19
Goldblatt P (1993) The woody Iridaceae: *Nivenia*, *Klattia*, and *Witsenia*: systematics, biology, and, evolution. Timber Press, Portland
Gröger A (1995) Die Vegetation der Granit-Inselberge Südvenezuelas. PhD Thesis, University of Bonn, Germany
Hambler DJ (1961) A poikilohydrous, poikilochlorophyllous angiosperm from Africa. Nature 191:1415–1416
Hartung W, Schiller P, Dietz KJ (1998) Physiology of poikilohydric plants. Prog Bot 59: 299–327
Hickel B (1967) Zur Kenntnis einer xerophilen Wasserpflanze: *Chamaegigas intrepidus* Dtr. aus Südwestafrika. Int Rev Gesamten Hydrobiol 52:361–400
Hickman JC (1974) Pollination by ants: a low-energy system. Science 184:1290–1292
Hopper SD (1981) Honeyeaters and their winter food plants on granite rocks in the central wheatbelt of Western Australia. Aust Wildl Res 8:187–197
Ingram J, Bartels D (1996) The molecular basis of dehydration tolerance in plants. Annu Rev Plant Physiol Plant Mol Biol 47:377–403
James SH (1965) Complex hybridity in *Isotoma petraea*. I. The occurrence of interchange heterozygosity, autogamy and a balanced lethal system. Heredity 20:341–353
James SH (1970) Complex hybridity in *Isotoma petraea*. II. Components and operation of a possible evolutionary mechanism. Heredity 25:53–78
Jonsson BO, Jónsdottir IS, Cronberg N (1996) Clonal diversity and allozyme variation in populations of the arctic sedge *Carex bigelowii* (Cyperaceae). J Ecol 84:449–459
Juniper BE, Robins RJ, Joel DM (1989) The carnivorous plants. Academic Press, London
Lawton JH, Brown VK (1993) Redundancy in ecosystems. In: Schulze E-D, Mooney HA (eds) Biodiversity and ecosystem function. Springer, Berlin Heidelberg New York, pp 255–270
Lüttge U (1991) *Clusia*: Morphogenetische, physiologische und biochemische Strategien von Baumwürgern im tropischen Wald. Naturwissenschaften 78:49–58
Lüttge U (1997) Physiological ecology of tropical plants. Springer, Berlin Heidelberg New York
Mora-Osejo LE (1989) La bioforma de *Bulbostylis leucostachya* Kunth (Cyperaceae) y de otras monocotiledoneas arboriformes tropicales. Rev Acad Colomb Cienc 17: 215–230
Morawetz W (1983) Dispersal and succession in an extreme tropical habitat: coastal sands and xeric woodland in Bahia (Brazil). Sonderb naturwiss Verein Hamburg 7:359–380
Paine RT (1966) Food web complexity and species diversity. Am Nat 100:65–75
Pate JS (1989) Australian micro stilt plants. TREE 4:45–49
Pate JS, Dixon KW (1982) Tuberous, cormous and bulbous plants. University of Western Australia Press, Perth
Peterson G, Allen CR, Holling CS (1998) Ecological resilience, biodiversity, and scale. Ecosystems 1:6–18

Petterson B, Nilsson LA (1993) Floral variation and deceit pollination in *Polystachya rosea* (Orchidaceae) on an inselberg in Madagascar. Opera Bot 121:237–245

Porembski S, Barthlott W (1995) On the occurrence of a velamen radicum in Cyperaceae and Velloziaceae. Nord J Bot 15:625–629

Porembski S, Barthlott W (in press) Granitic and gneissic outcrops (inselbergs) as centers of diversity for desiccation-tolerant vascular plants. Plant Ecol

Porembski S, Martinelli G, Ohlemüller R, Barthlott W (1998): Diversity and ecology of saxicolous vegetation mats on inselbergs in the Brazilian Atlantic rainforest. Divers Distrib 4:107–119

Rauh W (1988) Tropische Hochgebirgspflanzen. Wuchs- und Lebensformen. Springer, Berlin Heidelberg New York

Raunkiaer C (1934) The life forms of plants and statistical plant geography. Oxford University Press, Oxford

Schiller P, Heilmeier H, Hartung W (1997) Abscisic acid (ABA) relations in the aquatic resurrection plant *Chamaegigas intrepidus* under naturally fluctuating environmental conditions. New Phytol 136:603–611

Scholes RJ, Pickett G, Ellery WN, Blackmore AC (1997) Plant functional types in African savannas and grasslands. In: Smith TM, Shugart HH, Woodward FI (eds) Plant functional types. Cambridge University Press, Cambridge, pp 255–268

Seine R, Porembski S, Barthlott W (1995) A neglected habitat of carnivorous plants: inselbergs. Feddes Repert 106: 555-562

Sherwin HW, Pammenter NW, February E, Vander Willigen C, Farrant JM (1998) Xylem hydraulic characteristics, water relations and wood anatomy of the resurrection plant *Myrothamnus flabellifolius* Welw. Ann Bot 81:567–575

Tuba Z, Proctor MCF, Csintalan Z (1998) Ecophysiological responses of homoiochlorophyllous and poikilochlorophyllous desiccation tolerant plants: a comparison and an ecological perspective. Plant Growth Regul 24:211–217

Ule E (1898) Ueber Standortsanpassungen einiger Utricularien in Brasilien. Ber Dtsch Bot Ges 16:308–314

Wasserthal LT (1997) The pollinators of the Malagasy star orchids *Angraecum sesquipedale*, *A. sororium* and *A. compactum* and the evolution of extremely long spurs by pollinator shift. Bot Acta 110:343–359

Willert DJ von, Eller BM, Werger MJA, Brinckmann E, Ihlenfeldt H-D (1992) Life strategies of succulents in deserts with special reference to the Namib desert. Cambridge Universiy Press, Cambridge

Wyatt R (1981) Ant-pollination of the granite outcrop endemic *Diamorpha smallii* (Crassulaceae). Am J Bot 68:1212–1217

Wyatt R (1986) Ecology and evolution of self-pollination in *Arenaria uniflora* (Caryophyllaceae). J Ecol 74:403–418

Wyatt R (1997) Reproductive ecology of granite outcrop plants from the southeastern United States. In: Withers PC, Hopper SD (eds) Granite outcrops symposium. J R Soc West Aust 80:123–129

9 Ecophysiology of Vascular Plants on Inselbergs

M. KLUGE and J. BRULFERT

9.1 Introduction

As described in Chapter 4, this Volume (Porembski et al.) in more detail, inselbergs represent by no means uniform ecosystems but rather show clear fragmentation in subhabitats such as exposed rock surfaces, drainage channels, crevices between rocks, humus-filled depressions on top or the slopes of the rocks, or even wet flushes and seasonal rock pools (Barthlott et al. 1996). These habitat types differ largely in the constellation of edaphic and microclimatic factors (see Szarzynski, Chap. 3, this Vol.) and thus in the ecophysiological demands with which the plants colonizing such ecological units have to cope. As a consequence, inselbergs show a high floristic β-diversity, i.e., they are covered by mosaics of plant communities consisting of species highly adapted to the environmental conditions of the given habitat type. For this reason, inselbergs provide promising models for comparative studies on mechanisms and effectivity of ecological adaptation in plants, but the ecophysiological investigation of inselberg vegetation is still at its beginning (for review see Lüttge 1997). The aim of this chapter is first to analyze some mechanisms of ecophysiological adaptation in vascular plants inhabiting inselbergs. In the second part we will apply these considerations to the vegetation of Mt. Angavokely, an inselberg of the central high plateau of Madagascar. Finally, possible consequences of adaptation for the diversity of life-forms in vascular inselberg plants will be discussed. The ecophysiological adaptations in the cryptogams of the biofilm covering the rocks of inselbergs are treated by Büdel et al. (Chap. 5, this Vol.).

Ecological Studies, Vol. 146
S. Porembski and W. Barthlott (eds.) Inselbergs

9.2 Environmental Factors as Stressors

The recent biological stress concept (Levitt 1980; Larcher 1987) applies the term stress to any state of life performance outside of optimum. Any biotic and abiotic factor can push an organism into the state of stress, i.e., can become a stressor, if its dosage is sub- or hyperoptimal. As far as the inselbergs are concerned, the plants have to cope mainly with the following potential stressors:

- short supply of water or drought,
- high irradiance,
- high temperature,
- limitation in the availability of mineral nutrients.

In the analysis of plant response to stressors two aspects are important. One of them concerns the time patterns, i.e., the dynamics of stress situations (Larcher 1994; Lüttge 1997). This aspect is beyond the scope of this chapter. Rather, we will focus on the other aspect, which is ecophysiologically and increasingly molecular-biologically orientated and aims to uncover the causal chain between perception of the signal set by a given stressor and the reaction of the plant. In other words, we will analyze some examples of ecophysiological adaptation of inselberg plants towards the stressors typical for the specific environment.

In order to understand mechanisms of ecological adaptation, it is important to note that the primary reaction of the organism to the stressor takes place in the protoplasm of the cells. This means that only that component of the stressor which really reaches the protoplasm is capable of provoking stress (Larcher 1994). Depending on the plant, the protoplasm can to a certain extent resist the action of the stressor before a significant reaction occurs (tolerance towards the stressor), or mechanisms have been developed to avoid the stressor reaching the level of the protoplasm with the full intensity that the outer surfaces of the plant receive (avoidance of effects due to the potential stressor). We will later see that inselberg plants provide examples both for such tolerance and avoidance strategies.

The extent of ability to tolerate or to avoid stress is, within a certain variability, genotypically determined but, signaled by the stressor, can also undergo phenotypic alterations (acclimation, deacclimation). It has been pointed out by Solbrig (1993) and Lüttge (1996) that finally both genotypical variability and phenotypic plasticity allow the plants ecophysiological functioning along environmental gradients, and thus are important factors which determine the diversity of species and growth forms in a habitat.

Causal analyses of the response of plants to a given potential stressor are impeded by the fact that in situ environmental stressors do not act independently of each other, but rather synergistically in a kind of network. For instance, high irradiance acts not only per se on the photosynthetic machinery but rather increases the temperature in the plant tissues, thus establishing overheating as a further potential stressor; or high temperature increases evapotranspiration and thus may induce water-deficiency stress. Moreover, it has to be taken into account that the intensity of environmental stressors and with it their ecophysiological demand is not constant, but rather varies along the diurnal and seasonal cycles.

9.3 Adaptation of Inselberg Plants to Scarcity of Water

Driven by the water vapor saturation deficit of the atmosphere with respect to the plant body, the latter unavoidably loses water by evapotranspiration. On inselbergs, where full exposure to the sun creates very hot microclimates, during the day this deficit may reach extreme values, thus increasing the danger of excessive water loss from the plant. Moreover, on the bare rocks and shallow soils of the inselbergs, water supply is inadequate or, during cessation of precipitation, even totally missing. Therefore, apart from seasonally water-filled rock pools and water runoffs, the vegetation of inselbergs consists entirely of xerophytes, which are in different ways adapted to the scarcity of water. These adaptations comprise both tolerance and avoidance mechanisms and follow the same principles as those holding true also for the plants in deserts (Nobel 1988; von Willert et al. 1992) or otherwise dry habitats, for instance epiphytic sites (Lüttge 1989; Benzing 1990).

9.3.1 Desiccation Tolerance

A classic example of the tolerance strategy in plants is desiccation tolerance. This refers to the ability to survive cycles of dehydration and subsequent rehydration and represents one of the most spectacular ecophysiological adaptations shown by inselberg plants (Lüttge 1997; Porembski and Barthlott, in press). During the state of dehydration, the life processes are reduced to such extent that they are practically no longer measurable, whilst upon remoistening, these processes stepwise recover completely (see below). Thus, desiccation-tolerant plants (resurrection plants) are poikilohydric, in contrast to homoiohydric plants, which keep their protoplasmic hydrature, and with it the performance of life processes, constant.

Physiological, biochemical, and molecular-biological aspects of desiccation tolerance have been reviewed repeatedly (Bewley and Krochko 1982; Ingram and Bartels 1996; Lüttge 1997; Hartung et al. 1998) so that it is sufficient to focus here on some more general aspects of the phenomenon. Desiccation tolerance is mainly a protoplasmic property (Gaff 1980). The affected plants are able to survive for months (duration of the dry season) or, in extreme cases, even for years, equilibration of the cell water with low ambient air humidity (often in the range near zero %). Equilibration with dry air leads to dehydration of the protoplasm to relative leaf water contents lower than 2%. For comparison, nontolerant plants suffer already severe or even deadly stress if hydration of the protoplasma slightly deviates from saturation.

During the cycles of de- and rehydration the plants undergo drastic structural, physiological, and biochemical alterations (for details consult the quoted reviews). It is important to note that not only in the desiccation-sensitive but also in the tolerant plants, dehydration of the protoplasma leads to certain degradation on the level of biomembranes and protein structures (see e.g., Vieweg and Ziegler 1969; Ziegler and Vieweg 1970; Bewley and Krochko 1982; Hartung et al. 1998). This can be concluded from the observation that during the process of rehydration of previously desiccated tissues solutes leak unspecifically from the cells (Levitt 1980) or, even more strikingly, that during dehydration most (but not all) resurrection plants lose their chlorophyll and other photosynthetic pigments (Tuba et al. 1993a,b, 1994, 1996; for reviews see Lüttge 1997; Hartung et al. 1998), suggesting damage of the thylacoid membranes. However, due to synthesis of protective compounds (compatible solutes, protective proteins, special membrane lipids), in the resurrection plants the extent of protoplasmatic damage is limited to an extent which can be tolerated by the cell (Bewley and Krochko 1982; Ingram and Bartels 1996; Mullet and Whitsitt 1996; Müller et al. 1997; Hartung et al. 1998). Additionally, desiccation-tolerant plants possess effective mechanisms to repair, during the process of rehydration, the protoplasmatic destruction brought about the previous desiccation. There is evidence that both the protection and reparation mechanisms include up- and down-regulation of genes, with the phytohormone abscisic acid being involved as messenger in the signal transduction chain (Ingram and Bartels 1996; Hartung et al. 1998).

Hartung et al. (1998) proposed that one of the key factors bringing about desiccation tolerance is the ability of the plant cells to perform cytorrhysis. In contrast to plasmolysis, where upon dehydration the protoplast, not, however, the cell wall, contracts, cytorrysis is characterized by concomitant shrinkage of both the protoplast and the adjacent cell wall. That is, cytorrhysis avoids extensive plasmolysis during desiccation which otherwise would irreversibly disrupt the plasmodesmatal connections between the cells and with it the essentially important symplastic continuum in the tissue. In the drought-sensitive plants, where the ability of cytorrhysis is missing, dehydration leads to damaging plasmolysis.

A more macroscopic consequence of cytorrhysis is that, due to the contraction of the leaf tissue in the resurrection plants, the laminae of the leaves curl or fold during desiccation. This reduces the surfaces exposed to the often extremely high irradiation of the habitats. It is worth mentioning that among the poikilohydric angiosperms the monocotyledons clearly dominate. One of the reasons for this dominance might be the parallel leaf nervature typical of these plants, which provides a better preadaptation with respect to advantageous leaf folding compared with the net-shaped nervature of the dicotyledons.

During the cycle of de- and rehydration, desiccation-tolerant plants undergo a characteristic sequence of physiological events (Fig. 9.1) which

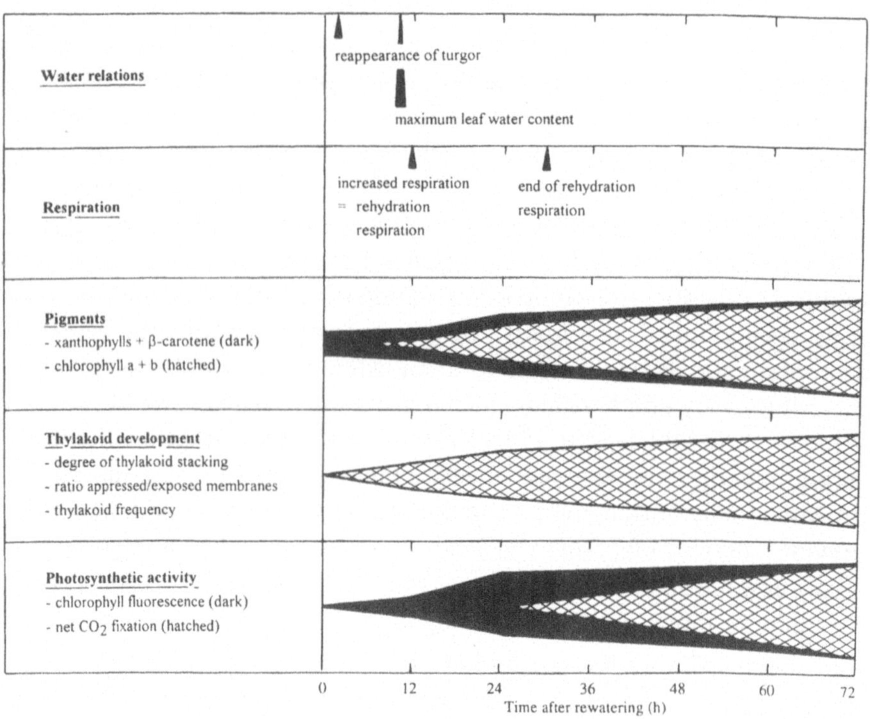

Fig. 9.1. Recovery of a poikilochlorophyllous desiccation-tolerant plant (*Xerophyta scabrida*) upon rehydration. The graph compiles data of Tuba (1993a,b, 1994) and indicates in the *lower three panels on the right* the relative increase of the structures and functions described in the *left column*. (Lüttge 1997)

have recently been studied by Tuba et al. (1993a, b, 1994; see also the review by Lüttge 1997) in the African inselberg plant *Xerophyta scabrida* (Velloziaceae). This investigation showed that during desiccation active respiration is maintained even at cell water potentials clearly more negative than –3.2 MPa. Upon rehydration, the physiological features of the cells recover in the following sequence in time (quoted from Lüttge 1997; see also Fig. 9.1): turgor → maximum leaf water content → respiration → resynthesis of carotenoids and → chlorophylls (accompanied by restoration of thylacoid structure) → chlorophyll fluorescence yield (indicating full recovery of the photosynthetic light reaction) → net CO_2 fixation. Also in the African resurrection plant *Myrothamnus flabellifolia* (Myrothamnaceae) upon rehydration the onset of respiration precedes that of photosynthesis (Vieweg and Ziegler 1969; Ziegler and Vieweg 1970). The delay in the reappearance of photosynthesis can most easily be explained in terms of repair of damage suffered by the photosynthetic apparatus when the protoplasma is dehydrated.

During long-lasting drought, most desiccation-tolerant plants lose their functional roots so that after rewetting water has to be trapped initially by capillar forces at the stem's bark and the invaginations of the curled leaves. The capillarly trapped water is then resorbed via the leaf surface. Later, new adventitious roots develop allowing normal water uptake. It has been found (Porembski and Barthlott 1995) that, as is also true for epiphytic orchids, in many desiccation-resistant Cyperaceae and Velloziaceae, the adventitious roots possess a *velamen radicum* which facilitates rapid water uptake.

Desiccation tolerance is a feature not only of cryptogamic plants but occurs also in vascular plants (Bewley and Krochko 1982; Ingram and Bartels 1996; Lüttge 1997; Hartung et al. 1998). There they comprise mainly ferns and monocotyledons, but also some dicotyledons (mainly members of the Scrophulariaceae). No poikilohydric gymnosperm is known. According to Gaff (1977), in Africa resurrection plants represent one of the most important components of the inselberg flora. Particularly important are species of the monocotyledonous families Poaceae (e.g., with the genus *Tripogon*), Cyperaceae (with the poikilohydric genera *Afrotrilepis* and *Coleochloa;* Porembski et al. 1996), and Velloziaceae (in America with the genus *Vellozia*, in Africa with the genera *Xerophyta* and *Talbotia*). In fact, according to Lüttge (1997), all species of Velloziaceae are desiccation-tolerant.

Particularly fascinating exponents of desiccation-tolerant inselberg plants are the dicotyledonous species *Myrothamnus flabellifolia* (mainly southeast Africa), *Myrothamnus moschata* (Madagascar), and *Chamaegigas intrepidus* (Namibia). The *Myrothamnus* species represent small bushes where, upon desiccation, the leaves dry out and fold completely so that only the lower brownish leaf surface remains visible. In nature, already 30–60 min after rainfall the leaves are fully rehydrated and spread their green laminae. The structural and physiological alterations occurring in *Myrothamnus flabellifolia* during de- and rehydration have been investigated by Ziegler and Vieweg (1970). *Chamaegigas intrepidus* is a desiccation-resistant water plant growing on inselbergs in shallow ephemeral rock pools which dry out completely during the dry season. The behavior of this species has been described by Heil (1925), Ziegler and Vieweg (1970), and Hartung et al. (1998). *Chamaegigas intrepidus* roots in a thin layer of sand usually gathered at the bottom of the pool, and possesses both ephemeral leaves swimming on the water surface and small perennial bulb-like underwater leaves. When the pools dry out, the swimming leaves die, whereas the underwater leaves and roots desiccate and survive in this state for up to 11 months of severe drought and extreme heat. Upon rewatering, within a few hours the previously strongly shrunken underwater leaves retain their previous volume and physiological activity. Later also new swimming leaves are developed.

Although on a first view desiccation tolerance represents a simple and effective mode of ecophysiological adaptation to drought, it has the serious selective disadvantage that the concerned plants can perform photosynthesis and thus increase their biomass only during the time when the plant body is sufficiently hydrated, i.e., as far as vascular plants under natural conditions are concerned, in extreme cases only during a few weeks of the year. For this reason poikilohydric plants are thrust aside to habitats with extreme ecological conditions which exclude other plants as competitors. As will be shown later, this trend can be observed even within the microhabitats of inselbergs. On the other hand, desiccation-tolerant plants of inselbergs can grow also in other dry habitats which provide similar growth conditions.

9.3.2 Desiccation Avoidance

Considering the mentioned selective disadvantage of the poikilohydric way of life, it is not surprising that the majority of the cormophytes has evolved structural and physiological adaptations which avoid desiccation and guarantee homeostasis of plant water status even under very arid environmental conditions. This holds true also for inselberg plants. Desiccation avoidance in cormophytes is based on the possession of an epidermis covered with a cuticula which is practically impermeable to water vapor and CO_2. However, the cuticula is perforated by the adjustable stomata mediating controlled diffusional gas exchange between the plant and the ambient atmosphere. More than 95%, i.e., practically the whole transpiratory water loss, proceeds through the stomata and therefore can be controlled by regulation of the stomatal aperture in such a way that homoeostasis of cell water status is maintained. However, the fact that also the CO_2 exchange proceeds via the stomata throws the plants into the dilemma of suffering either desiccation or starvation. Namely, in order to effectively throttle transpiration, the resistance to water vapor diffusion out of the leaves has to be increased by stomatal closure. This, however, would hinder or even prevent uptake of external CO_2, thus starving the photosynthetic production of biomatter. Because of the naturally small downhill CO_2 concentration gradient between ambient atmosphere and the sites of photosynthesis inside the leaves, effective diffusion of CO_2 into the leaves requires that the diffusion resistance is kept as low as possible, i.e., the stomata are kept wide open. Under the arid conditions of the inselbergs, this gas exchange dilemma becomes the most severe ecophysiological challenge for homoiohydric cormophytes and requires specific adaptation.

9.3.2.1 Structural Avoidance Mechanisms: Xeromorphism

Although there are practically no in situ ecophysiological investigations of inselberg plants (see, however, Sect. 9.5), the following generally accepted interpretation of the adaptive value of xeromorphism should apply also for the xerophytes of this ecosystem. Xeromorphic adaptation is mainly governed by two principles, namely (1) protection of the photosynthetic tissue from excessive evaporation and (2) temporary storage of water in succulent tissues and organs. Apart from succulence, which will be discussed later, the following xeromorphic avoidance mechanisms are particularly important: thickened outer walls of epidermal cells furnished with thick cuticles lower the cuticular transpiration which cannot be regulated by the plant. Dense covering of the leaf surfaces by dead hairs, or possession of sunken stomata help to maintain high boundary layer resistance to evaporation by stabilizing an unstirred air layer at the leaf surface. Stomatal transpiration is kept low by low stomatal density, often drastic reduction of leaf surface, and small stature of the plants.

In succulents (for definition of succulence see Kluge and Ting 1978; von Willert et al. 1992), specialized tissues of stems, leaves and, less frequently, of roots serve as reservoirs for utilizable water. These reservoirs are filled during the wet season via water uptake by the roots (in some specialists, e.g., many bromeliads, also by special hairs on the leaf surface) and emptied during the dry season by internal transfer of water to those sites of the plant body where it is important to keep the cell water potential constant even if external water is no longer available. Once a succulent organ has been completely emptied of water, it will definitively die. The morphological, anatomical, and physiological fundaments of the repeated filling and emptying of succulents organs with utilizable water has been discussed in detail by von Willert et al. (1992) and will thus not be considered in this chapter. However, it is worth mentioning here that there are two types of succulent vascular plants: in the all-cell succulent type (denotation after von Willert et al. 1992) water storage takes place in all cells of the water storing organs, whereas in the partially succulent type water storage takes place either in peripherous (mainly epidermal) tissues or in central parenchyma. As will be outlined below, many succulents possess, additionally to their ability for water storage, also other xeromorphic characteristics such as mentioned above, and the crassulacean acid metabolism (CAM) type of photosynthesis (see Sect. 9.3.2.2) as a metabolic means to avoid rapid loss of the stored water.

Concerning succulents of the deserts, von Willert et al. (1992) showed that they exhibit a large diversity of life strategies as part of the adaptation to the specific demands of the given habitat. Some leaf succulents are annuals, but most succulents are above-ground persistent perennials and include species without green leaves (stem succulents), species with evergreen leaves, and such with seasonally deciduous leaves.

The different types of succulents are important constituents of inselberg vegetation. Worldwide there are examples of inselberg annual leaf succulents, mainly members of the families Crassulaceae, Portulacaceae, and Commelinaceae (S. Porembski, pers. comm.). At least for *Crassula sieberiana*, an annual leaf succulent growing on granitic outcrops in New South Wales (Australia), crassulacean acid metabolism (CAM, see Sect. 9.3.2.2) has been reported (Brulfert et al. 1991), but it is very likely that CAM occurs also in other annual inselberg succulents.

In Madagascar some of the most interesting communities of succulents can be found on the inselbergs of the Central High Plateau (Rauh 1973) and comprise both leaf succulents (mainly species of the genera *Aloe*, *Kalanchoe*, and *Senecio*) and stem succulents (mainly species of the genera *Pachypodium*, *Euphorbia* and *Cynanchum*). On inselbergs of South Madagascar *Rhipsalis baccifera* ssp. *horrida*, a subspecies of the only paleotropic cactus *R. baccifera*, can be found (Barthlott et al. 1996). Also in the tropical South Americas the inselbergs carry a rich vegetation of succulent species, mainly of Agavaceae, Araceae, Bromeliaceae, and Cactaceae (Lüttge 1997). In both Africa and South America inselbergs are inhabited by orchids (Rauh 1973; Lüttge 1997) which are capable of storing water either in the thick leaves and/or pseudobulbs made up of a single or several thickened internodes.

According to Lüttge (1997), on the granitic rock vegetation in Brazil hemicryptophytes (plants having rosettes and rhizomes close to the ground, with only small parts of the plant body protruding) represent with a relative abundance of 32% by far the most prominent life-form. It is conceivable, but remains to be investigated in detail, whether or not the hemicryptophytic life-form helps to maintain homeostasis in the plant water budget under the semiarid conditions of the inselbergs.

9.3.2.2 Metabolic Avoidance Mechanism: C_4 Photosynthesis and Crassulacean Acid Metabolism

In vascular plants there are two modifications of the photosynthetic carbon assimilation pathway representing metabolic means to reduce or avoid desiccation stress: C_4 photosynthesis and crassulacean acid metabolism (CAM). Although biochemically quite similar, the two pathways follow different ecophysiological strategies to escape the mentioned gas exchange dilemma that terrestrial plants face in nature.

C_4 photosynthesis and CAM have in common that they consist of two carboxylation steps switched in series. The primary CO_2 fixation is mediated by phosphoenolpyruvate carboxylase (PEPC) and leads to synthesis of a C_4 dicarboxylic acid (mainly malate) which is afterwards again decarboxylated. The resulting CO_2 is refixed by the secondary carboxylation step mediated by ribulose-1,5-bisphosphate carboxylase/ oxygenase (Rubisco), the enzyme starting the photosynthetic C_3 pathway (Calvin cycle). In C_4 photosynthesis (for review see, e.g., Hatch and Osmond 1976), primary CO_2 fixation by PEPC and refixation of the CO_2 by Rubisco occur simultaneously in time, however, separated in space, i.e., in leaf mesophyll and bundle sheath, respectively. C_4 photosynthesis represents basically a mechanism which concentrates CO_2 in the bundle sheath where Rubisco is located and the Calvin cycle operates. At first the primary CO_2 fixation by PEPC with its high CO_2-affinity creates a low CO_2 concentration in the substomatal intercellulars of the mesophyll and with it a steeper gradient for inward diffusion of CO_2 from the ambient atmosphere. Then the transport of the intermediary C_4 product into the bundle sheath and its decarboxylation there leads to a massive CO_2 enrichment at the site where the Calvin cycle operates. This suppresses photorespiration (which otherwise would lead to loss of carbon), thus increasing the effectivity of the carbon assimilation. As a consequence, the C_4 pathway allows photosynthesis to proceed with partially closed stomata, which, on the other hand, decreases transpiratory water loss.

Baskin and Baskin (1988) report C_4 photosynthesis to occur in certain species (*Portulaca smallii* and *Cyperus granitophilus*) inhabiting rock outcrops of the unglaciated eastern United States. However, altogether the present knowledge on the abundance and ecophysiological meaning of C_4 photosynthesis in inselberg plants is still quite fragmentary.

Compared with C_4 photosynthesis, in inselberg plants the crassulacean acid metabolism (CAM) appears to be ecophysiologically more significant. The denotation of that mode of photosynthesis refers to its discovery in a species of the Crassulaceae (*Bryophyllum calycinum* = *Kalanchoë pinnata*). Up to now, CAM plants are known from 33 families distributed over the whole phylogenetic tree of the cormophytes (Smith and Winter 1996). This broad taxonomic distribution suggests that CAM is of polyphyletic origin. In the vegetation of inselbergs, species of the families Asclepiadaceae, Cactaceae, Crassulaceae, Euphorbiaceae, Agavaceae, Bromeliaceae, and Orchidaceae are quite frequent. These families are rich in CAM-performing species, which suggests that in inselberg plants CAM must be an important phothosynthetic option. Direct evidence supporting this assumption will be discussed later in Sect. 9.5. Recently, it has been reported that the poikilohydric plants *Haberlea rhodopensis* and *Ramonda serbica* (Markovska et al. 1997), which are abundant on rocky habitats in the eastern mediterranean, are capable of CAM performance in the well hydrated state. Up to now this is the only reported case of such a combination of desiccation tolerance and CAM.

In contrast to C_4 photosynthesis, in CAM (for reviews see, e.g., Kluge and Ting 1978; Osmond 1978; Winter 1985; Lüttge 1987, 1997) the primary and secondary CO_2 fixation proceed in the same cell but separated in time. That is, CO_2 is taken up and fixed by PEPC during the night when opening of stomata does not endanger the water balance of the plant, because the leaf/air water-vapor saturation difference, i.e., the driving force for transpiration, is much lower than during the day. The CO_2 fixed via PEPC is stored overnight in form of malic acid in the vacuoles. In order to avoid too high concentrations of malic acid, the vacuole stores also large amounts of water. For this reason, CAM plants show a succulent habit.

During the day, the malic acid synthetized and stored during the previous night is released from the vacuoles, decarboxylated, and the resulting CO_2 finally assimilated via the Calvin cycle. By providing the Calvin cycle with the malate-derived endogenous CO_2 as substrate, the photosynthetic machinery becomes independent of direct supply with CO_2 from the atmosphere. This allows photosynthesis to operate behind closed stomata, without the otherwise unavoidable transpiratory water loss.

A simplified scheme of the carbon flow pathway in CAM is given in Fig. 9.2. For more details, the mentioned reviews on CAM should be consulted.

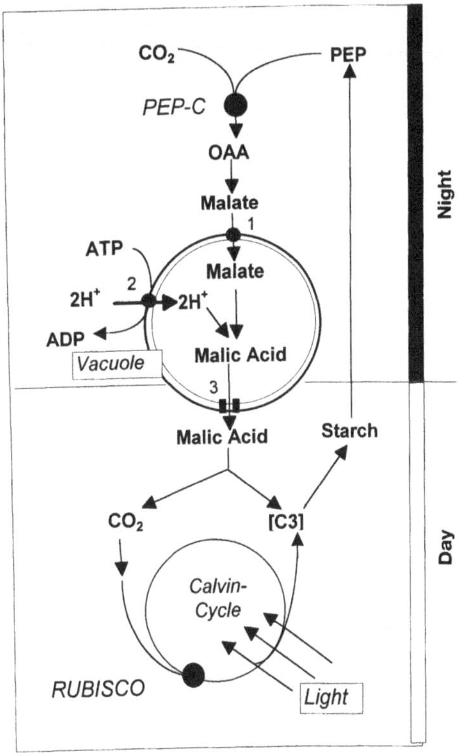

Fig. 9.2. Largely simplified scheme of the CAM pathway. The *black bar on the right* indicates the reactions proceeding during the night, the *open bar* those during the day. *PEP* Phosphoenolpyruvate; *OAA* oxaloacetic acid; *PEP-C* phosphoenolpyruvate carboxylase, the enzyme which catalyzes the primary (nocturnal) CO_2 fixation in CAM; *Rubisco* ribulose-1,5-bisphosphate carboxylase/oxygenase, the enzyme which catalyzes the refixation of the malate-derived endogenous CO_2; *1* the malate carrier of the tonoplast; *2* the ATP-driven vacuolar proton pump (V-H^+-ATPase); *3* the malate export channel of the tonoplast

It is worth mentioning that certain CAM plants, in particular species of the genus *Clusia*, store overnight, together with malic acid, large amounts of citric acid (e.g., Wolf 1960; Franco et al. 1992; Haag-Kerwer et al. 1992; von Willert et al. 1992). The biological meaning of nocturnal citric acid accumulation in CAM is still under discussion. It has been proposed that the breakdown of storage carbohydrates during the night by glycolysis and following generation of acetyl coenzyme A for the citrate synthase reaction produces ATP which could help to cover the high demand on energy required to pump organic acids into the vacuoles during the night (Lüttge 1988; Feng et al. 1994). Moreover, breakdown of citric instead of malic acid during the day may possibly result in the production of more internal CO_2 which, on the other hand, could prevent photoinhibition of photosynthesis when light intensity is high (Lüttge 1997).

In its classical performance, the net CO_2 exchange resulting from CAM shows four phases (denotation according to Osmond 1978): besides the nocturnal CO_2 fixation (phase I) there is to some extent uptake of external CO_2 also during the day, mainly during a short transient phase at the onset of the light period (phase II) and towards the end of the day (phase IV) after the vacuoles have been depleted of the previously stored malate and the photosynthetic apparatus is no longer supplied with internal CO_2. The phases II and IV are separated by phase III, which consists of a depression in net CO_2 uptake, with the stomata being closed.

Depending on the plant species (i.e., genotypically determined), but even more on the environmental conditions and on the physiological state of the plant (i.e., phenotypically determined), the relative contribution of these phases to the total CO_2 uptake and with it the water-saving effect of CAM can vary between CO_2 uptake taking place exclusively during the night and taking place mainly during the day. In the latter case, the gas exchange pattern is close to that of a C_3 plant. The water saving effects linked to a given gas exchange pattern can be quantified by the transpiration quotient [H_2O (g) lost by transpiration/carbon (g) gained by photosynthesis]. Figure 9.3 correlates transpiration quotients with gas exchange

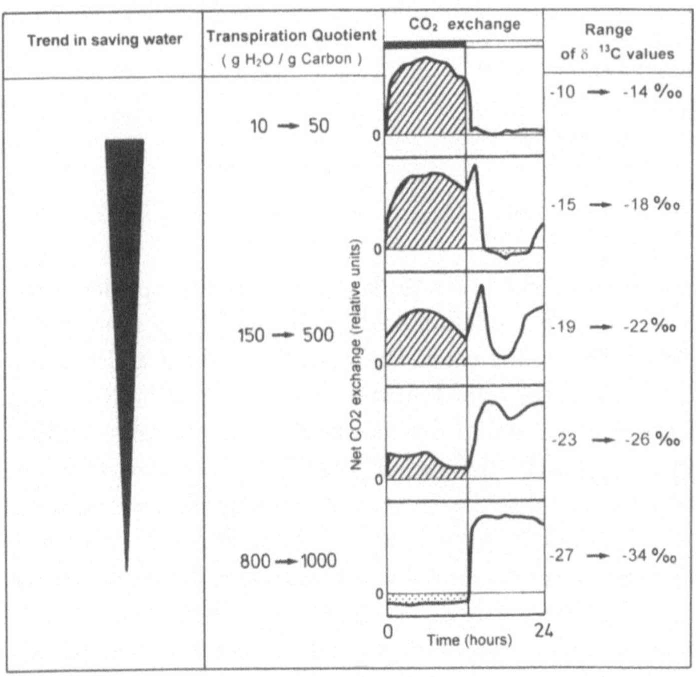

Fig. 9.3. Scheme illustrating the trends in the interrelationship between the various types of CAM gas exchange patterns, the corresponding approximative range of $\delta^{13}C$ values and the water-saving effect (trend in the transpiration quotient) related to the gas exchange patterns and $\delta^{13}C$ values. It has to be mentioned that in reality the transients between the ranges are smooth. *Hatched areas above zero line* Net CO_2 uptake in the night mediated by PEP-carboxylase; *open areas above zero line* Net CO_2 uptake in the light mediated mainly by Rubisco; *dotted areas below zero line* Net CO_2 output. (After Kluge and Vinson 1995)

patterns shown by CAM plants. The lowest transpiration quotients are achieved when CO_2 is harvested entirely during the night, and the opposite holds true with CO_2 fixation taking place only during the day.

In fact, water availability is the most important external factor governing the CAM gas exchange. As a rule, under drought conditions, CAM plants reduce and finally cancel first the CO_2 uptake during the day, while the water-saving nocturnal CO_2 uptake persists much longer. However, under long-lasting drought, finally also the nocturnal CO_2 uptake may be reduced and even drop to zero. Then CAM operates even for months by recycling respiratory CO_2 (CAM idling; see the quoted reviews on CAM).

On the other hand, the so-called C_3/CAM intermediate plants may perform C_3 photosynthesis when well watered and shift to the water saving CAM when water becomes in short supply.

Some of the most striking examples of C_3/CAM intermediate plants are certain species of the Clusiaceae where, depending on the local conditions, simultaneously some parts of the same tree can acquire external carbon by CAM and others by C_3 photosynthesis, respectively (for review see Lüttge 1997). Such CAM/C_3 indermediate *Clusia* species are abundant also on South American inselbergs (Lüttge 1995, 1997).

There is evidence that CAM is not only a water saving mechanism but can also faciliate osmotic uptake via the leaf surface of water from mist, dew or rain (Lüttge 1987; von Willert et al. 1992). In many plants, for instance species of the Bromeliaceae, water uptake by leaves is supported by specialized foliar trichomes. The driving force for the CAM-related water diffusion is a strong osmotic gradient created by the massive nocturnal malic acid accumulation in the vacuoles.

On first view, it is surprising that CAM occurs also in aquatic species of *Isoetes* (Lycopodioposidae; Isoetaceae) and *Crassula* (Crassulaceae) inhabiting oligotrophic ephemeral pools on inselbergs. However, in this case CAM is not an adaptation to scarcity of water but to temporary short supply of CO_2 (Keeley 1982, 1996). Due to the photosynthetic activity of the photoautotrophic organisms living in the pools, during the day the CO_2 concentration in the water decreases to such extent that it becomes the limiting factor for photosynthetic carbon acquisition. In contrast, during the night the concentration of CO_2 solved in the water is relatively high because of the respiratory activity of the organisms living in the pool. The CAM-performing aquatic *Isoetes* and *Crassula* species take advantage of this situation by performance of CAM with fixation of CO_2 during the night when it is available in higher concentrations.

Direct measurements in situ on performance of CAM in inselberg plants are still missing. However, valuable information on behavior, abundance, and ecophysiological relevance of CAM can be derived from the estimation of stable carbon isotope ratios ($\delta^{13}C$ values) in the plants of interest (for review see, e.g., Ehleringer and Rundel 1989; Ehleringer and Osmond 1989; Rundel et al. 1989; Griffiths 1992; Kluge et al. 1991a).

$\delta^{13}C$ values are measured by means of mass spectrometry and defined by the equation

$$\delta^{13}C\ (‰) = \left[\frac{^{13}C/^{12}C\,\text{sample}}{^{13}C/^{12}C\,\text{standard}} - 1 \right] \times 10^3.$$

The CO_2 of the air contains the stable carbon isotopes ^{13}C (1.11 %) and ^{12}C (98.89 %), together with traces of ^{14}C which are not relevant in the present context. As shown in Fig. 9.2, in CAM carbon is harvested during the night by phosphoenolpyruvate carboxylase and during the day by ribulose-1,5-bisphosphate carboxylase. When fixing CO_2, Rubisco strongly discriminates ^{13}C with respect to the lighter isotope ^{12}C, whilst PEPC shows no such discrimination. As a consequence of this different enzymatic behavior, biomatter produced by Rubisco-mediated primary CO_2 fixation (i.e., by CO_2 uptake during the day) is poorer in ^{13}C compared with biomatter produced via PEPC-mediated nocturnal CO_2 fixation. The more depleted a given sample in ^{13}C, the more negative is its $\delta^{13}C$ value. For this reason, C_3 plants, with CO_2 fixation exclusively by Rubisco, have the most negative $\delta^{13}C$ values with a range from about –26 to –34 ‰. In contrast, C_4-plants which fix external CO_2 entirely via PEPC, have less negative $\delta^{13}C$ values (ranging from about –10 to –14 ‰). As shown in the scheme of Fig. 9.3, CAM plants cover the whole mentioned range of $\delta^{13}C$ values, depending on the relative contribution of nocturnal (i.e., PEPC-mediated) and daytime (i.e., Rubisco-mediated) CO_2 fixation to total carbon gain. Since, as discussed previously, the mode of CO_2 fixation determines the capability of the plant to save water when harvesting carbon from the atmosphere, it is possible to conclude from the $\delta^{13}C$ values also as to the water economy of the CAM plants (Fig. 9.3). An application of $\delta^{13}C$ analysis to investigate options of photosynthesis in inselberg plants will be shown in Section 9.6.2.

9.4 Adaptation to High Irradiance and Temperatures

Together with the scarcity of water, high irradiance and temperature are the main stressors on inselbergs and require specific adaptation (Lüttge 1997).

According to our own measurements on Malagasy inselbergs, photosynthetic photon flux densities (PPFD) can exceed values of 2000 $\mu mol^{-2} s^{-1}$. Lüttge (1997) reports for an inselberg in Venezuela temperatures of the rock surface in the range of 60 °C (cf. Szarzynski, Chap. 3, this Vol.) and showed that in response to this rock temperature the leaves of *Ananas ananassoides* and *Pitcairnia pruinosa* overheat within a few hours by about 24 °C, finally reaching values around 47 °C. In the European Alps the succulent leaves of rock-inhabiting *Sempervivum* species at noon under full radiation temperatures may reach in the range of 60 °C (own observation) and easily survive such conditions. Such extreme values should apply also for inselberg plants.

The well-known structural avoidance mechanism (e.g., reflection of radiation by dead hairs on the leaf surface) can also be found in cormophytes on inselbergs. Basically, cooling the leaves by intense transpiration is an effective physiological means to avoid overheating of the plant body (Lange 1959; von Willert et al. 1992), but it requires the plant to have enough water available to replace the loss by water uptake via the roots. On the dry microhabitats of the inselbergs the latter is not the case, thus presumably cooling by transpiration is of minor importance, there.

In the dehydrated state poikilohydric vascular plants can tolerate temperatures of the plant body between 60 and 80 °C (Larcher 1994), but the fact that also homoiohydric species of hot habitats can tolerate strong overheating (confessedly to a lesser extent than dehydrated desiccation plants) implies the existence of protoplasmatic resistance to temperature stress. Moreover, there is evidence that, due to acclimation, the degree of temperature resistance in a given plant individual can change (Nobel 1988).

Temperature affects the fluidity and with it the proper functioning of biomembranes. On the other hand, microorganisms, animals, and plants have evolved mechanisms allowing them to maintain homeostasis of membrane fluidity while the environmental conditions change drastically (homeoviscous adaptation, HVA; Raison et al. 1980; Hazel 1988; Quinn 1988). HVA was found also in the tonoplast of the CAM-performing *Kalanchoe daigremontiana*, a leaf succulent of the dry bush of Madagascar (Kluge et al. 1991b; Kliemchen et al. 1993; Schomburg and Kluge 1994; Behzadipour et al. 1998), and there is evidence that in this plant HVA helps to stabilize the vacuolar membrane under temperature conditions which otherwise could disturb or even destroy it. It was also found by the mentioned authors that HVA of tonoplast fluidity is a multifaceted phenomenon which is based on alterations in the relative content and composition of membrane proteins, composition of the membrane lipids, and degree of saturation in the fatty acids. Since there is evidence that HVA concerns also biomembranes other than the tonoplast and plant species without CAM

(Raison et al. 1980; Berry and Raison 1981; Quinn 1988; Kluge et al. 1999), it should occur also in the inselberg plants and help there to guarantee optimal functioning of the cell membranes under the often extreme temperature conditions of the habitat.

There is increasing evidence that in plants short-term heating leads to synthesis of specific heat-shock proteins, often with the function remaining yet unclear. On the other hand, it is known that heat-shock proteins are involved in the prevention of thermotrope damage or in the repair of the tertiary protein structures (Vierling 1991). Nothing is known if specifically in inselberg plants heat-shock proteins contribute to bringing about resistance to heat stress.

Certain inselberg plants can adapt to the specific light conditions of that habitat. Kluge and Vinson (1995) studied the light response of the orchid *Angraecum sesquipedale*. This species grows in Madagascar both epilithically on inselbergs permanently exposed to the full sun and epiphytically in the shadow of rainforest trees. By means of chlorophyll fluorescence analysis (Schreiber and Bilger 1993; Lüttge 1997), we found that, at high photon flux density, individuals growing fully exposed on the rocks, compared with individuals grown in the shadow, showed higher effective quantum yield of photosystem II and higher rates of relative electron transport. This suggests adaptation on the level of the photosynthetic light reaction.

As pointed out by Lüttge (1997), in tropical habitats it can be often observed that individuals of a given plant species either growing in shade or fully exposed to the sun, not only show morphologically different phenotypes but are also strikingly different in pigmentation, with the shade plants exhibiting a deep green and the sun plants a bright yellow or red color. It was observed that such phenotypic adaptation to light occurs also in succulent rosette plants from Malagasy inselbergs, for instance in *Aloe capitata* and *Lomatophyllum prostratum* where, in contrast to the green shade plants, the individuals grown in the sun often show intense red coloration.

The mechanisms and the adaptational relevance of this phenotypic light adaptation are not yet fully understood. Fetene et al. (1990) showed that *Bromelia humilis* plants grown in weak light have higher chlorophyll contents and smaller chlorophyll a/b ratios as compared to plants grown in stronger light. Moreover, these authors found that the phenotypic adaptability depends on nitrogen nutrition. Diaz et al. (1990) observed that upon stress by high irradiance and drought the leaves of *Aloe vera* accumulate the xanthophylls rhodoxanthin and, to a lesser extent, zeaxanthin. The latter authors also found that reduced N availability enhances the xanthophyll accumulation, and interpret it as a mechanism of light-stress avoidance.

9.5 Adaptation to Deficiency of Mineral Nutrients: Carnivorous Plants

As shown by the detailed floristic analysis by Seine et al. (1995) and Dörrstock et al. (1996), carnivorous plants are a remarkable component of the inselberg vegetation. On the inselbergs themselves ephemeral and permanent wet flushes are the most important habitats of carnivorous plants. Occasionally, the vegetation of wet flushes consists entirely of such plants, for instance of *Drosera* species.

The many facets of carnivory in plants has been reviewed by Lüttge (1983). There is ample evidence supporting the view that carnivory in plants is an adaptation to scarcity of mineral nutrients in the root medium, in particular of N, P, and S. Carnivorous plants are extremely calcifuge and need acidic and wet soils. Exactly these conditions are provided by the microhabitats of inselbergs where carnivorous plants grow. The restriction of carnivors to wet microhabitats is certainly due to the fact that these plants are badly protected from water loss by transpiration, because digestion of the animal prey and resorption of the digestion products proceeds via special epidermal structures having tender cell walls and lacking cuticles. Such properties exclude survival at dry sites.

In fact, studies by Dörrstock et al. (1996) show that the soils and water of inselbergs and in particular of the runoffs are poor in mineral nutrients and thus provide ecological niches for carnivorous plants. As also true for peat bogs, where carnivorous plants are also quite abundant, on inselbergs scarcity of mineral nutrients is due to the circumstance that ground water is not available, thus the soil water available to the plants derives directly from the precipitation. Moreover, in the wet flushes of the inselbergs the permanent drainage creates a further pauperization in mineral nutrients.

Carnivorous plants can overcome the scarcity of N, P, and S in the soil by digestion and resorption of the various organic compounds of the prey caught, in the case of the inselberg plants mainly insects. Very interesting from the carnivory point of view are the species of the genus *Genlisea*. These plants grow on nutrient-poor white sand areas and moist sites on inselbergs and are abundant in South America and Africa, including Madagascar (Rauh 1973; Dörrstock et al. 1996; Barthlott et al. 1998). Although Darwin (1875) already postulated that root-like subterranean leaves typical for the *Genlisea* species function as traps for animal prey, there was no proof for this assumption, and the feeding habit of the *Genlisea* remained an unsolved puzzle. Now Barthlott et al. (1998) has shown that *Genlisea* traps and digests protozoa by its specialized subterranean leaves.

Altogether, to our knowledge up to now there are no systematic ecophysiological investigations on carnivorous plants of inselbergs. Investigations in this interesting field are highly desired.

9.6 Modes of Photosynthesis in Plants of Mt. Angavokely (Central High Plateau of Madagascar)

9.6.1 Description of the Site

In the following we will report on a study on options of photosynthesis in plants of an inselberg in Madagascar, Mt. Angavokely. The study was carried out in September 1993, i.e., during the dry season, and is based mainly on a survey by $\delta^{13}C$ analysis.

Mt. Angavokely is situated on the Central High Plateau, about 30 km east of the capital Antananarivo, and consists of a dome-shaped granitic outcrop surrounded by mainly secondary tropical forests and farmland. During the dry season (April to October) the top of the mountain, including the study site, is often covered by dense fog for several hours during the night and the early morning. The site where our observations were made is located about 1400 m asl and consists of smooth rock plates 20° steep and exposed to the southeast. The bare rock is nearly gaplessly covered by cyanobacteria and lichens.

Where the rock slope locally is less steep, humus amasses, forming islands ranging from about 0.5 to 25 m^2 in size. This humus islands are covered by a characteristic xerophytic vegetation (Fig. 9.4a), but there are also cormophytic pioneer plants which settle directly on the bare rocks (Fig. 9.5).

9.6.2 Adaptational Strategies in the Pioneer Plants of Mt. Angavokely

We have observed on Mt. Angavokely that at the site of study together with cryptogams mainly three species of higher plants appear as pioneers growing directly on the bare rocks: *Bulbophyllum* sp. (Orchidaceae), *Xerophyta dasylirioides* (Velloziaceae), and small rosettes of *Aloe capitata* (Aloaceae). As shown in Fig. 9.5, occasionally these plants grow closely together on an area of few dm^2, forming a kind of microcommunity of pioneer plants. The individuals shown in Fig. 9.5 had $\delta^{13}C$ values of –16.7 ‰ (*A. capitata*), –25.0 ‰ (*Bulbophyllum* sp.) and –27.5 ‰ (*X. dasylirioides*). Since, to our knowledge, none of the concerned genera

Fig. 9.4.a The border of a vegetation island on a southeast-exposed rock slope of Mt. Angavokely (Central High Plateau of Madagascar). *In the center* some of the impressive individuals of *Angraecum sororium* (Orchidaceae). **b** Schematic profile of the species distribution of the vegetation island shown in **a. 1** *Bulbophyllum* sp., epilithic on the bare rock (–24.9 ‰); **1a** *Bulbophyllum* sp., epiphytic in the center of the vegetation island (–29.3 ‰); **2** *Coleochloa setifera* (–25 .0 ‰); **3** *Xerophyta dasylirioides* (–28.0‰); **4** *Helichrysum* sp. (no data); **5** *Aloe capitata* (–16.7 ‰); **6** *Nematostylis anthophylla* (–25.8 ‰); **7** *Senecio melastomaefolia* (–27.9 ‰); **8** *Angraecum sororium* (–23.5 ‰); **9** *Cussonia bojeri* (–25.8 ‰); **10** *Pyrostria madagascariensis* (–25.7 ‰). On the surface of the ground protected by the vascular plants there is a rich vegetation of lichens, mosses, and ferns which were not analyzed with respect to the $\delta^{13}C$ values

comprises C_4 plants, it is reasonable to interpret the $\delta^{13}C$ values found in the vegetation of Mt. Angavokely in terms of CAM and C_3 photosynthesis. Thus, the data suggest that *A. capitata* performs pronounced CAM as water-saving mode of photosynthesis, with fixation of external CO_2 taking place mainly at night. In contrast, the other two species harvest external carbon via C_3 photosynthesis and follow other strategies of drought adaptation, namely water storage (*Bulbophyllum*) and desiccation tolerance (*Xerophyta*). The fact that these three species can grow spatially close together suggests that pronounced CAM, water storage, and desiccation tolerance are equally successful adaptations, allowing them to conquer the bare rock of the inselberg as habitat. In order to find out if throughout a longer time scale one of these three survival strategies is more successful than the others, it would be very interesting to trace in future work the succession in microcommunities such as shown in Fig. 9.5.

The adaptational functioning of these survival strategies is mirrored also by the distribution of species within the plant community developing on islands of humus amassing on flattenings of the otherwise steep slopes of the rocks. An example is shown in Fig. 9.4a and b. At the edges of such islands, still on the bare rock or on a very thin cover of soil, the vegetation is dominated by water-storing *Bulbophyllum* sp. (annotation: the species

Fig. 9.5. Cormophytic "pioneers" on the bare rocks of Mt. Angavokely. The community consists of *Xerophyta dasiliriodes* (*1*), *Bulbophyllum* sp. (*2*) and *Aloe capitata* (*3*)

could not be identified because the plants were not flowering when the investigation was carried out) and the desiccation-tolerants *Xerophyta dasylirioides* (Velloziaceae) and *Coleochloa setifera* (Cyperaceae). The $\delta^{13}C$ values in the samples of these species (see legend to Fig. 9.4b) are indicative of C_3 photosynthesis.

At sites where the humus layer reaches a thickness of some cm, the vegetation is dominated by large rosettes of *Aloe capitata*, by *Senecio melastomaefolius* (Asteraceae), and *Nematostylis anthophylla* (Rubiaceae). In the outer range of this zone also a clearly xerophytic *Helichrysum* sp. can be found. Although $\delta^{13}C$ values were not estimated in the latter plant, because of the nonsucculent habitus, occurrence of CAM most likely can be ruled out. Whether or not *Helichrysum* is desiccation-tolerant remains to be investigated.

As indicated by the $\delta^{13}C$ values (legend to Fig. 9.4b), also inside the larger vegetation complexes *Aloe capitata* performs strong CAM, whereas *Senecio melastomaefolius* and *Nematostylis anthophylla* harvest external CO_2 via C_3 photosynthesis. In the latter two species adaptation to drought is due to water storage by leaf succulence (*S. melastomaefolius*) and by root bulbs (*N. anthophylla*). In the context of the different photosynthetic behavior of *A. capitata, N. anthophylla* and *S. melastomaefolius*, it is interesting to note that at the vegetation island shown in Fig. 9.4a the three species are equally abundant and grow side by side. However, although the $\delta^{13}C$ values of *S. melastomaefolius* suggest that this species in situ acquires CO_2 mainly by C_3 photosynthesis, the possibility cannot be ruled out that potentially it is capable of CAM performance. In fact, the genus *Senecio* comprises also CAM plants (Kluge and Ting 1978; von Willert et al. 1992), and we will later become acquainted with *Kalanchoe campanulata* which *in situ* on Mt. Angavokely behaves as a C_3 plant, but under controlled conditions by application of drought can be shifted to CAM behavior (unpubl.). The fact that the two leaf succulents *A. capitata* (CAM) and *S. melastomaefolius* (C_3 photosynthesis) occupy the same inselberg habitat support the findings by Eller and Ferrari (1997) that CAM plants are not necessarily better water savers than C_3 plants if similar growth forms under the same environmental conditions are compared.

At sites where on the vegetation islands the humus layer is thicker than about 15 to 20 cm, the impressive individuals of the orchid *Angraecum sororium* can be found (Fig. 9.4a). For the specific site shown in Fig. 9.4a, the $\delta^{13}C$ values of *A. sororium* samples were in the range of −23.5‰. The mean $\delta^{13}C$ value of all *A. sororium* samples taken at Mt. Angavokely ($n = 13$) was 24.4 ± 2.0‰. Altogether, these data suggest that *A. sororium* in situ performs CAM with a high contribution of CO_2 uptake during the day to total carbon gain. This interpretation is fully consistent with the results of gas exchange measurements carried out in this plant under controlled conditions (Kluge et al. 1998a).

Sites of the vegetation islands where the soil reaches a thickness of about 0.5 m (i.e., mainly the central parts) are occupied by xeromorphic shrubs consisting of *Cussonia bojeri* (Araliaceae) and *Pyrostria madagascariensis* (Rubiaceae). At places where light is sufficient, the stems of these shrubs are covered by epiphytic individuals of *Bulbophyllum*. The $\delta^{13}C$ values of the two shrubs and the epiphytic *Bulbophyllum* (see legend to Fig. 9.4b) are clearly indicative of carbon acquisition by C_3 photosynthesis. Obviously, the deeper soil allows the shrubs to develop extended systems of roots enabling them to take up sufficient amounts of water so that CO_2 can be harvested from the atmosphere during the day without endangering their water balance. Moreover, in *Cussonia bojeri, Pyrostria madagascariensis,* and other C_3 photosynthesis-performing plants of the inselberg vegetation, excessive transpiration is prevented by their clearly xeromorphic habit.

The vegetation of Mt. Angavokely allows some informative comparisons elucidating the adaptative value of CAM. One comparison concerns the already-mentioned orchid *Angraecum sororium* and another species of the genus, *A. sesquipedale*. The latter is abundant in the lower regions of the east coast of Madagascar and grows there epiphytically in the coastal rain-forests and epilithically on rock outcrops (Fig. 9.6). As compared with the po-

Fig. 9.6. *Angraecum sesquipedale* growing epilithically on coastal rock outcrops near Taolanaro (southeast coast of Madagascar; photograph courtesy Prof. Dr. L.T. Wasserthal)

tential CAM plant *A. sororium*, which behaves in situ nearly as a C_3 plant ($\delta^{13}C$ = 24.4 ±2.0‰), in *A. sesquipedale* the $\delta^{13}C$ values are with −15.8 ± 3.7‰ (n = 11 independent samples) significantly less negative and indicate strong CAM with CO_2 uptake mainly at night (Kluge and Vinson 1995; Kluge et al. 1996; 1998a). This different photosynthetic behavior corresponds completely with the microhabitats naturally occupied by the two species. That is, *A. sesquipedale* roots practically direct on the bare bark of the phorophyte (in case of epiphytic growth) or rock (in the case of epilithic growth, Fig. 9.6). Thus, even relatively short cessations of precipitation endanger the acquisition of water so that photosynthetic adaptation provided by CAM is helpful to escape from the dilemma of suffering either starvation or desiccation. In contrast, for *A. sororium* the danger of drought stress is less relevant, because in these plants the roots are surrounded by humus and litter. Obviously, this substrate is capable to store sufficient amounts of water from rain and fog so that the plant has enough water available to allow opening of the stomata and uptake of external CO_2 also during the day.

Similar photosynthetic adaptation to the microhabitat might hold true also for another orchid of Mt. Angavokely, i.e., *Bulbophyllum* sp. The $\delta^{13}C$ values given in the legend to Fig. 9.7 suggest that the epiphytic individuals (or species ?) growing in the shadow and under protection of the dense shrubs acquire external CO_2 exclusively by C_3 photosynthesis, whereas the clearly less negative $\delta^{13}C$ values in the exposed plants (bare rock surface, edge of the vegetation island) indicate that in the latter case nocturnal CO_2 fixation of CAM contributes to some extent to the overall carbon gain. This fits very well with the finding that epiphytic *Bulbophyllum* species collected in humid montane forests of the High Plateau of Madagascar had a mean $\delta^{13}C$ value of −30.3 ± 2.6‰ (Kluge et al. 1995) indicative of C_3 photosynthesis, whereas the mean $\delta^{13}C$ value for samples collected at the study site at Mt. Angavokely was −24.8 ± 1.9‰ (n = 7), suggesting contribution of nocturnal CO_2 fixation by CAM. Since the identification of the species in the *Bulbophyllum* samples collected at Mt. Angavokely was not possible (not flowering when sampled), the question remains open if the differences in $\delta^{13}C$ values mirror inter- or intraspecific variability in the photosynthetic behavior. The occurrence of CAM in species of the genus *Bulbophyllum* has clearly been shown by Winter et al. (1983). On the other hand, it has to be mentioned that high temperatures and low internal CO_2 concentrations at the sites of photosynthesis in the cells can to some extent decrease the ^{13}C discrimination by Rubisco (Farquhar and Sharkey 1982; Ehleringer and Osmond 1989), thus shifting the $\delta^{13}C$ towards somewhat less negative values. However, the mentioned differences in the *Bulbophyllum* $\delta^{13}C$ values are too large to be explained solely in terms of this isotopic effect. Altogether, fur-

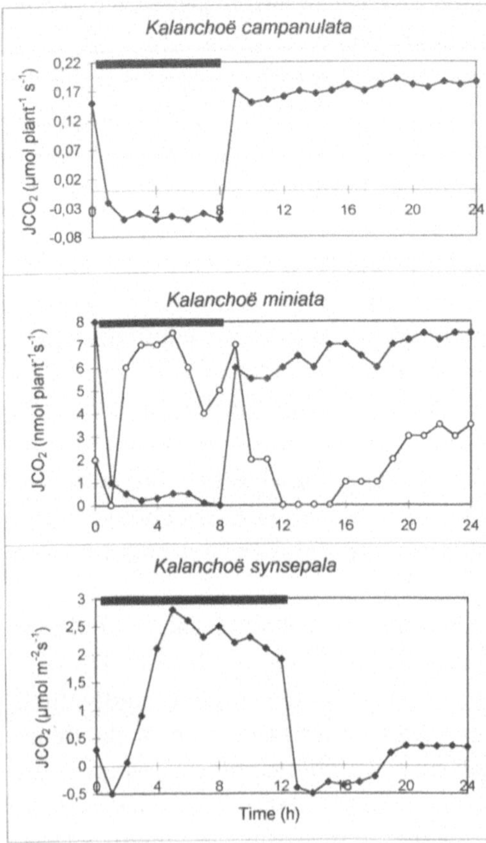

Fig. 9.7. Net CO_2 exchange (J_{CO_2}) patterns performed under controlled conditions by the *Kalanchoe* species of Mt. Angavokely discussed in the text. The temperature was 27 °C (day)/15 °C (night). Photosynthetic quantum flux density was in the range of 200 µmol m^{-2} s^{-1}. Gas exchange was measured in the well-watered state (*closed symbols*) except for *K. miniata* where also the gas exchange after 10 days of drought is shown (*open symbols*). The *black bar* indicates the duration of night. *Positive values of* J_{CO_2} indicate net CO_2 uptake, *negative values* net CO_2 output. It can be seen from the figure that, when watered, *K. campanulata* and *K. miniata* take up CO_2 entirely during the day, thus behaving as C_3 plants, whereas *K. synsepala* shows typical pronounced CAM with CO_2 uptake mainly during the night. Upon drought, *K. miniata* changes to performance of CAM. The same holds true also for *K. campanulata*; however, in this latter species already a few days of withholding water leads to stomatal closure throughout the whole day/night cycle (data not shown in the figure)

ther experimental work is required to answer definitively the question of weather the $\delta^{13}C$ values of the *Bulbophyllum* samples from Mt. Angavokely really mirror differences in the modes of photosynthetic carbon acquisition.

A third comparison of photosynthetic behavior concerns three species of the genus *Kalanchoe* (Crassulaceae) growing on Mt. Angavokely. It has been shown (Kluge et al. 1993; Kluge and Brulfert 1996; Gehrig et al. 1997) that all species of the genus are potentially CAM plants, and that the expression of CAM can be closely related on the one hand to the taxonomic position of the species within the genus and, on the other, to the ecophysiological demand of the habitat, thus to the biogeographic distribution of the species in Madagascar and Africa. Such coincidences between physiological behavior and habitat become strikingly evident also on the smaller scale of Mt. Angavokely. In the shadow of deep, humid gaps between the granitic outcrops, there grows the above-mentioned thin-leafed *K. campanulata*. As indicated by the $\delta^{13}C$ values of –27.0 ‰ shown by samples obtained at Mt. Angavokely, in situ these plants perform C_3 photosynthesis. This fits with the low ecophysiological demand the plant faces at its habitat as far as the water factor is concerned. In contrast, due to propagation by stolons, *K. synsepala* is capable of conquering bare rock surfaces as habitat. In coincidence with the extremely difficult water relations at this site, the plant in situ performs pronounced CAM, with CO_2 uptake proceeding entirely during the night ($\delta^{13}C$ = –13.4 ‰). Finally, *K. miniata* growing in the open bush under less demanding conditions than that of the rock surfaces showed $\delta^{13}C$ values of –18.0 ‰. This suggests CAM with some contribution of CO_2 uptake during the day. In fact, more detailed investigations in the latter plants by Brulfert et al. (1996) showed that, depending on the external conditions, in *K. miniata* CAM is extremely flexible, ranging from C_3 photosynthesis in the well-watered state to extreme CAM under drought.

Gas-exchange measurements under controlled conditions with the three mentioned *Kalanchoe* species (Fig. 9.7) are fully consistent with our interpretations of the $\delta^{13}C$ values of samples collected in situ at Mt. Angavokely. It can be seen from Fig. 9.7 that, when watered, *K. campanulata* and *K. miniata* take up CO_2 entirely during the day, and thus behave as C_3 plants. In contrast, even in the well-watered state, *K. synsepala* shows typically pronounced CAM with CO_2 uptake mainly during the night. Upon drought, *K. miniata* changes to CAM. The same holds true also for *K. campanulata*. However, in this latter species, already a few days of withholding water leads to stomatal closure throughout the whole day/night cycle (data not shown in the figure) so that there is no more net carbon gain.

9.7 Conclusions

The ecological success of a given taxon depends both on genotypically related diversity and intraspecific, i.e., functionally determined plasticity

(Lüttge 1996). This concept can also be applied to the vegetation conquering a given habitat, and thus holds true for the inselberg plants. High ecophysiological demands favor the selection of species equipped with highly specialized modes of adaptation, as exemplified by the various resistance and avoidance mechanisms discussed above. On the fundament and within the genotypic frame of a given mode of adaptation, the plants feature phenotypic plasticity. The best example is CAM, which represents by far the most flexible mode of photosynthesis. Phenotypic plasticity enables the plants to respond to stochastic environmental variations (i.e., variations which cannot be predicted in time and space) in such a way that the physiological performance, and with it the fitness, is always maintained as close as possible to the optimum. Lüttge (1997) has pointed out that the intensity in the ecophysiological demand brought about by stressors is of major importance. High intensity of the stressing demand selects mainly genotypically specialized modes of adaptation such as desiccation tolerance. Namely, low intensity is not sufficient to open ecological niches for such genotypes, because it still allows competition, thus success of several species at the same site. Thus, under weak or medium environmental demand phenotypic plasticity of adaptation is important.

The validity of this concept is clearly supported by the described spatial distribution in the vegetation island on Mt. Angavokely of plants representing different modes of adaptation to the water factor. Even more striking support comes from the coincidence between the different modes of CAM performance within the *Kalanchoe* species abundant on Mt. Angavokely and the microhabitat that the concerned plants occupy on the inselberg: under the low intensity of water deficiency stress in the humid gaps between the rocks, *K. campanulata,* although being potentially a CAM plant, performs C_3 photosynthesis and competes successfully with other herbaceous C_3 plants. In contrast, the hostile bare rocks are occupied by the highly specialized *K. synsepala,* which shows an extremely inflexible mode of CAM, with CO_2 fixation taking place only at night. The same was found in *K. beharensis*, a species native in the semiarid southwest of Madagascar. In this plant even artificial daily irrigation during the dry season could not shift the extreme CAM into a mode where more CO_2 is taken up during the day (Kluge et al. 1992). On the other hand, as mentioned above, *K. miniata* obviously growing at Mt. Angavokely under medium stress, shows very flexible CAM which can shift easily and reversibly in the direction of C_3 photosynthesis (Brulfert et al. 1996).

Altogether, irrespective of the still existing considerable deficits in the knowledge on ecophysiology of inselberg plants, the following general conclusion might be allowed: both diversity of adaptive strategies and intraspecific phenotypic plasticity in the performance of a given mode of adaptation are important categories determining the composition of the plant communities of inselbergs. In concerted action with the complex mosaic in the constellation of environmental factors provided by the microhabitats, these categories might by major reasons for the high β-diversity exhibited by the vegetation of a given inselberg.

Acknowledgements. We thank Dr. Eliane Deleens (Institut de Biotechnologie des Plantes, Université de Paris-Sud Orsay, France) for measuring the $\delta^{13}C$ values in the samples of Mt. Angavokely, Dr. Didier Ravelomanana (EES-Science, Université d'Antananarivo, Madagascar) for logistical support during our stays in Madagascar, and Monika Medina-Espana for valuable help in the preparation of the manuscript. Financial support of our work by the Deutsche Forschungsgemeinschaft, by the Heinrich Walther Foundation and the Deutscher Akademische Austauschdienst (Germany) for M.K., and of the CNRS (France) for J.B. is gratefully acknowledged.

References

Barthlott W, Porembski S, Szarzynski J, Mund JP (1996) Phytogeography and vegetation in tropical inselbergs. In: Guillaumet J-L, Belin M, Puig H (eds) Actes du colloque international de Phytogéographie tropicale, ORSTOM, Paris, pp 251–261

Barthlott W, Porembski S, Fischer E, Gemmel B (1998) First protozoa-trapping plant found. Nature 392:447

Baskin JM, Baskin CC (1988) Endemism in rock outcrop plant communities of unglaciated eastern United States: an evaluation of the roles of the edaphic, genetic and light factors. J Biogeogr 15:829–840

Behzadipour M, Ratajczak R, Faist K, Pawlitschek P, Trémolieres A, Kluge M (1998) Phenotypic adaptation of tonoplast fluidity to growth temperature in the CAM plant *Kalanchoe daigremontiana* Ham. et Per. is accompanied by changes in the phospholipid membrane and protein composition. J Membr Biol 166:61–70

Benzing DH (1990) Vascular epiphytes. Cambridge University Press, Cambridge

Berry JA, Raison JK (1981) Responses of macrophytes to temperature. In: Lange OL, Nobel PS, Osmond CB, Ziegler H (eds) Encyclopedia of plant physiology. New Series, vol I. Springer, Berlin Heidelberg New York, pp 277–338

Bewley JD, Krochko JE (1982) Desiccation tolerance. In: Lange OL, Nobel PS, Osmond CB, Ziegler H (eds) Physiological plant ecology. Encyclopedia of plant physiology. New Series, vol 2. Springer, Berlin Heidelberg New York, pp 325–378

Brulfert J, Güclü S, Kluge M (1991) Effects of abrupt or progressive drought on the photosynthetic mode of *Crassula sieberiana* cultivated under different day length. J Plant Physiol 138:685–690

Brulfert J, Ravelomanana D, Güclü S, Kluge M (1996) Ecophysiological studies in *Kalanchoe porphyrocalyx* (Baker) and *K. miniata* (Hils et Bojer), two species performing highly flexible CAM. Photosynth Res 49:29–36

Darwin C (1875) Insectivorous plants. John Murray, London

Diaz M, Ball E, Lüttge U (1990) Stress-induced accumulation of the xanthophyll rhodoxanthin in leaves of *Aloe vera*. Plant Physiol Biochem 28:679–682

Dörrstock S, Porembski S, Barthlott W (1996) Ephemeral flush vegetation on inselbergs in the Ivory Coast (West Africa). Candollea 51:407–419

Ehleringer JR, Osmond CB (1989) Stable isotopes. In: Pearcy RW, Ehleringer JR, Mooney HA, Rundel PW (eds) Plant physiological ecology: field methods and instrumentation. Chapman & Hall, London, pp 281–290

Ehleringer JR, Rundel PW (1989) Stable isotopes: History, units, instrumentation. In: Rundel PW, Ehleringer JR, Nagy KA (eds) Stable isotopes in ecological research. Ecological studies 68. Springer, Berlin Heidelberg New York, pp 1–15

Eller BM, Ferrari S (1997) Water use efficiency of two succulents with contrasting CO_2 fixation pathways. Plant Cell Environ 20:93–100

Farquhar GD, Sharkey TD (1982) Stomatal conductance and photosynthesis. Annu Rev Plant Physiol 33:317–345

Feng W, Ning L, Daley LS, Moreno Y, Azarenko A, Cridlle RS (1994): Theoretical fitting of energetics of CAM path to calorimetric data. Plant Physiol Biochem 32:591–598

Fetene M, Lee HSJ, Lüttge U (1990) Photosynthetic acclimation in a terrestrial CAM bromeliad, *Bromelia humilis* Jacq. New Phytol 114:399–406

Franco AC, Ball E, Lüttge U (1992) Differential effects of drought and light levels on accumulation of citric and malic acids during CAM in *Clusia*. Plant Cell Environ 15:821–829

Gaff DF (1977) Desiccation-tolerant vascular plants of Southern Africa. Oecologia 31:95–104

Gaff DF (1980) Protoplasmic tolerance of extreme water stress. In: Turner NC, Kramer PJ (eds) Adaptation of plants to water and high temperature stress. John Wiley, New York, pp 207–231

Gehrig HH, Rösike H, Kluge M (1997) Detection of DNA polymorphisms in the genus *Kalanchoe* by RAPD-PCR fingerprint and its relationships to infrageneric taxonomic position and ecophysiological photosynthetic behaviour of the species. Plant Sci 125:41–51

Griffiths H (1992) Carbon isotope discrimination and the integration of carbon assimilation pathways in terrestrial CAM plants. Plant Cell Environ 15:1051–1062

Haag-Kerwer A, Franco A, Lüttge U (1992) The effects of temperature and light on gas exchange and acid accumulation in the C_3-CAM plant *Clusia minor*. J Exp Bot 43:345–352

Hartung W, Schiller P, Dietz K-J (1998) Physiology of poikilohydric plants. Prog Bot 59:299–327

Hatch MD, Osmond CB (1976) Compartmentation and transport in C_4 photosynthesis. In: Stocking CR, Heber U (eds) Encyclopedia of plant physiology. New Series, vol 3. Springer, Berlin Heidelberg New York, pp 134–187

Hazel JR (1988) Homeoviscous adaptation in animal cell membranes. In: Aloia RC, Curtain CC, Gordon LM (eds) Advances in membrane fluidity, vol 3. AR Liss, New York, pp 149–189

Heil H (1925) *Chamaegigas intrepidus* Dr., eine neue Auferstehungspflanze. Beih Bot Zentralbl 41:41–50

Ingram J, Bartels D (1996) The molecular basis of dehydration tolerance in plants. Annu Rev Plant Physiol Plant Mol Biol 47:377–403

Keeley JE (1982) Distribution of diurnal acid metabolism in the genus *Isoetes*. Am J Bot 69:254–257

Keeley JE (1996): Aquatic CAM photosynthesis. In: Winter K, Smith, JAC (eds) Crassulacean acid metabolism: biochemistry, ecophysiology and evolution. Ecological Studies 114. Springer, Berlin Heidelberg New York, pp 281–295

Kliemchen A, Schomburg M, Galla H-J, Lüttge U, Kluge M (1993) Phenotypic changes in the fluidity of the tonoplast membrane of crassulacean acid metabolism plants in response to temperature and salinity stress. Planta 189:403–409

Kluge M, Brulfert J (1996) Crassulacean acid metabolism in the genus *Kalanchoe*: ecological, physiological and biochemical aspects. In: Winter K, Smith JAC (eds) Crassulacean acid metabolism: biochemistry, ecophysiology and evolution. Ecological Studies 114. Springer, Berlin Heidelberg New York, pp 325–335

Kluge M, Ting IP (1978) Crassulacean acid metabolism. Analysis of an ecophysiological adaptation. Ecological Studies 30. Springer, Berlin Heidelberg New York

Kluge M, Vinson B (1995) Der Crassulaceen-Säurestoffwechsel bei Orchideen Madagaskars. Analyse einer ökologischen Anpassung der Photosynthese. Rundgespräche der Kommission für Ökologie. Vol 10 Tropenforschung. H Pfeil, München, pp 163–175

Kluge M, Brulfert J, Ravelomanana D, Lipp J, Ziegler H (1991a) Crassulacean acid metabolism in *Kalanchoe* species collected in various climatic zones of Madagascar: a survey by $\delta^{13}C$ analysis. Oecologia 88:407–414

Kluge M, Kliemchen A, Galla H-J (1991b) Temperature effects on crassulacean acid metabolism: EPR spectroscopic studies on the thermotropic phase behaviour of the tonoplast membranes of *Kalanchoe daigremontiana*. Bot Acta 104:355–360

Kluge M, Razanoelisoa B, Ravelomanana D, Brulfert J (1992) In situ studies of crassulacean acid metabolism in *Kalanchoe beharensis* Drake del Castillo, a plant of the semiarid southern region of Madagascar. New Phytol 120:323–334

Kluge M, Brulfert J, Lipp J, Ravelomanana D, Ziegler H (1993) A comparative study by $\delta^{13}C$ analysis of crassulacean acid metabolism (CAM) in *Kalanchoe* (Crassulaceae) species of Africa and Madagascar. Bot Acta 106:320–324

Kluge M, Brulfert J, Rauh W, Ravelomanana D, Ziegler H (1995) Ecophysiological studies on the vegetation of Madagascar; $\delta^{13}C$ and δD survey for incidence of crassulacean acid metabolism (CAM) among orchids from montane forests and succulents from the xerophytic thorn-bush. Isot Environ Health Stud 31:191–210

Kluge M, Brulfert J, Vinson B (1996) Signification biogéographique des processus d'adaptation photosynthétique. II: l'exemple des orchidées malgaches. In: Lourenço WR (ed) Biogéographie de Madagascar. ORSTOM, Paris, pp 157–163

Kluge M, Vinson B, Ziegler H (1998a) Ecophysiological studies on orchids of Madagascar: incidence and plasticity of crassulacean acid metabolism in species of the genus *Angraecum* Bory. Plant Ecol 135:43–57

Kluge M, Nguyen B, Behzadipour M, Fischer-Schliebs E (1999) Phenotypic adaptation of membrane fluidity in the tonoplast and plasmalemma of the C_3 plant *Hordeum vulgare* var. Alexis. J Plant Physiol 154:185–191

Lange OL (1959) Untersuchungen über den Wärmehaushalt und Hitzeresistenz mauretanischer Wüsten-und Savannenpflanzen. Flora 147:595–651

Larcher W (1987): Streß bei Pflanzen. Naturwissenschaften 74:158–167

Larcher W (1994) Ökophysiologie der Pflanzen. Ulmer, Stuttgart

Levitt J (1980): Responses of plants to environmental stresses. Vol II, Water, radiation, salt and other stresses. Academic Press, New York

Lüttge U (1983) Ecophysiology of carnivorous plants. In: Lange OL, Nobel PS, Osmond CB, Ziegler H (eds) Encyclopedia of plant physiology. New Series, vol 3. Springer, Berlin Heidelberg New York, pp 489–517

Lüttge U (1987) Carbon dioxide and water demand: crassulacean acid metabolism (CAM), a versatile ecological adaptation exemplifying the need for integration in ecophysiological work. New Phytol 106:593–629

Lüttge U (1988) Day-night changes of citric acid levels in crassulacean acid metabolism: phenomenon and ecophysiological significance. Plant Cell Environ 13:977–982

Lüttge U (1989) Vascular epiphytes: Setting the scene. In: Lüttge U (ed) Vascular plants as epiphytes. Ecological Studies 76. Springer, Berlin Heidelberg New York, pp 1–12

Lüttge U (1995) Ecophysiological basis of the diversity of tropical plants: the example of the genus *Clusia*. In: Heinen HD, San José JJ, Caballero-Arias H (eds) Nature and human ecology in the neotropics. Sci Guaianae 5:23–26

Lüttge U (1996) *Clusia*: plasticity and diversity in a genus of C_3/CAM intermediate tropical trees. In: Winter K, Smith JAC (eds) Crassulacean acid metabolism. Ecological Studies 114. Springer, Berlin Heidelberg New York, pp 296–311

Lüttge U (1997) Physiological ecology of tropical plants. Springer, Berlin Heidelberg New York

Markovska Y, Kimenov T, Tsonev G (1997) Regulation of CAM and respiratory recycling by water supply in higher poikilohydric plants – *Haberlea rhodopensis* Friv. and *Ramonda serbica* Panc. at transition from biosis to anabiosis and *vice versa*. Bot Acta 110:18–24

Müller J, Sprenger N, Bortlik N, Boller T, Wiemken A (1997) Desiccation increases sucrose levels in *Ramonda* and *Haberlea*, two genera of resurrection plants in the Gesneriaceae. Physiol Plant 100:153–158

Mullet JE, Whitsitt MS (1996) Plant cellular responses to water stress. Plant Growth Regul 20:119–124

Nobel PS (1988) Environmental biology of agaves and cacti. Cambridge University Press, Cambridge

Osmond CB (1978) Crassulacean acid metabolism: a curiosity in context. Annu Rev Plant Physiol 29:379–414

Porembski S, Barthlott W (1995) On the occurrence of a velamen radicum in Cyperaceae and Velloziaceae. Nord J Bot 15:625–630

Porembski S, Barthlott W (in press) Granitic and gneissic outcrops (inselbergs) as centers of diversity for desiccation-tolerant vascular plants. Plant Ecol

Porembski S, Brown G, Barthlott W (1996) A species-poor tropical sedge community: *Afrotrilepis pilosa* mats on inselbergs in West Africa. Nord J Bot 16:239–245

Quinn PJ (1988) Regulation of membrane fluidity in plants. In: Aloia RC, Curtain CC, Gordon LM (eds) Physiological regulation of membrane fluidity. A R Liss, New York, pp 293–322

Raison JK, Berry JA, Armond PA, Pike CS (1980) Membrane properties in relation to adaptation of plants to temperature stress: In: Turner N, Kramer PJ (eds) Adaptation of plants to water and high temperature stress. Wiley, New York, pp 261–276

Rauh W (1973) Über die Zonierung und Differenzierung der Vegetation Madagaskars. Akademie der Wissenschaften und Literatur Mainz. Franz Steiner, Wiesbaden

Rundel PW, Ehleringer JR, Nagy KA (eds) (1989) Stable isotopes in ecological research. Ecological Studies 68. Springer, Berlin Heidelberg New York

Schomburg M, Kluge M (1994) Phenotypic adaptation to elevated temperatures of tonoplast fluidity in the CAM plant *Kalanchoe daigremontiana* is caused by membrane proteine. Bot Acta 107:328–332

Schreiber U, Bilger W (1993) Progress in chlorophyll fluorescence research: major developments during the past years in retrospect. Prog Bot 54:151–173

Seine R, Porembski S, Barthlott W (1995) A neglected habitat of carnivorous plants: inselbergs. Feddes Repert 106:555–562

Smith JAC, Winter K (1996) Taxonomic distribution of crassulacean acid metabolism. In: Winter K, Smith JAC (eds) Crassulacean acid metabolism: biochemistry, ecophysiology and evolution. Ecological Studies 114. Springer, Berlin Heidelberg New York, pp 427–434

Solbrig OT (1993) Plant traits and adaptive strategies: their role in ecosystem function. In: Schulze E-D, Mooney HA (eds) Biodiversity and ecosystem function. Ecological Studies 99. Springer, Berlin Heidelberg New York, pp 97–116

Tuba Z, Lichtenthaler HK, Csintalan Z, Pócs T (1993a) Regreening of desiccated leaves of the poikilochlorophyllous *Xerophyta scabrida* upon rehydration. J Plant Physiol 142:103–108

Tuba Z, Lichtenthaler HK, Maroti I, Csintalan Z (1993b) Resynthesis of thylakoids and function of chloroplasts in the desiccated leaves of the poikilochlorophyllous plant *Xerophyta scabrida* upon rehydration. J Plant Physiol 142:742–748

Tuba Z, Lichtenthaler HK, Csintalan Z, Nagy Z, Szente U (1994) Reconstitution of chlorophylls and photosynthetic CO_2 assimilation upon rehydration of the desiccated poikilochlorophyllous plant *Xerophyta scabrida* (Pax) Th. Dur. et Schinz. Planta 192: 414–420

Tuba Z, Lichtenthaler HK, Csintalan Z, Szente K (1996) Loss of chlorophyll, cessation of photosynthetic CO_2 assimilation and respiration in the poikilochlorophyllous plant *Xerophyta scabrida* during desiccation. Physiol Plant 96:383–388

Vierling E (1991) The roles of heat shock proteins in plants. Annu Rev Plant Physiol Plant Mol 42:579–620

Vieweg GH, Ziegler H (1969) Zur Physiologie von *Myrothamnus flabellifolia.* Ber Dtsch Bot Ges 82:29–36

von Willert DJ, Eller BM, Werger MJA, Brinkmann E, Ihlenfeldt HD (1992) Life strategies of succulents in deserts, with special reference to the Namib desert. Cambridge University Press, Cambridge

Winter K (1985) Crassulacean acid metabolism. In: Barber J, Barber NR (eds) Photosynthetic mechanisms and the environment. Elsevier, Amsterdam, pp 329–387

Winter K, Wallace BJ, Stocker GC, Roksandic Z (1983) Crassulacean acid metabolism in Australian vascular epiphytes and some related species. Oecologia 57:129–141

Wolf J (1960) Der diurnale Säurerhythmus. In: Ruhland W (ed) Encyclopedia of plant physiology, vol 12. Springer, Berlin Heidelberg New York, pp 809–889

Ziegler H, Vieweg GH (1970) Poikilohydre Pteridophyta (Farngewächse), Poikilohydre Spermatophyta (Samenpflanzen). In: Walter H, Kreeb K (eds) Die Hydratation und Hydratur des Protoplasmas der Pflanzen und ihre ökophysiologische Bedeutung. Protoplasmatologia, vol II. Springer, Wien New York, pp 88–108

10 Variations on One Theme: Regional Floristics of Inselberg Vegetation

10.1 West African Inselberg Vegetation

S. POREMBSKI

10.1.1 Geography and Geology

West Africa is defined here as the region which is covered by the *Flora of West Tropical Africa* (Keay and Hepper 1954–72). This work embraces Upper Guinea, a region extending from Lat. 18° in the north (southern Mauretania) to the coast, eastward to the boundary of Chad, and in the south to a line running through the northwestern parts of Cameroon. Cameroon, Equatorial Guinea and Gabon, which are either only partly or not at all included in the *Flora of West Tropical Africa*, will also be included in the following account. Moreover, the following description will also include data on the vegetation of inselbergs to the north and east of West Africa, i.e., those situated in the western parts of the Sahara and to the east of Cameroon into the Central African Republic.

Most of West Africa is underlain by Precambrian rocks. For detailed descriptions of structural patterns and rock types, refer to Falconer (1911), Leneuf (1959) and Wilson (1968). In particular, the latter work summarizes a large body of information about the Côte d'Ivoire, Nigeria and Cameroon. The landscape is dominated by extensive plains, interrupted occasionally by low mountain ranges and high plateaux (e.g., the Guinean Fouta Djalon). The altitude is almost consistently less than 1000 m asl, however, occasionally granitic rock outcrops that are part of larger mountain ranges reach 1200 m asl or more (e.g., Mt. Tonkoui near Man, Côte d'Ivoire; Loma Mts., Sierra Leone). Granitic and gneissic inselbergs are fairly widespread throughout the whole region and display an enormous range of sizes (from a few m^2 to several km^2). Apart from dome-shaped bornhardts with steep slopes, so-called shield inselbergs are widespread in particular in seasonally dry regions. Koppjes, which constitute heaps of large boulders, are relatively rare. With respect to their geomorphology, West African bornhardts situated in the savanna zone usually possess relatively smooth slopes, in comparison with their rainforest counterparts. In West Africa, inselbergs occur as hills that are isolated for hundreds of kilometers, or alternatively

Ecological Studies, Vol. 146
S. Porembski and W. Barthlott (eds.) Inselbergs

they form dense clusters of rock outcrops. Areas with concentrations of inselbergs occur in most West African countries, for example in Guinea (around Guéckédou and Macenta), Côte d'Ivoire (Déps. Séguéla and Mankono), Bénin (around Dassa-Zoumé and Savé), Nigeria (Jos Plateau), and Angola (e.g., Amboim Plateau).

Inselbergs form habitat complexes that show close ecological affinities to other azonal landform types, such as duricrusts, for instance, which form nearly impermeable pavements that can reach great age. In particular, ferricretes (i.e., iron sesquioxides which form a stony surface, in West Africa locally known as bowal) are widespread in rainforest and more frequently in savanna regions of West Africa, where they may be inundated for weeks or months during the rainy season.

10.1.2 Climate

According to Griffiths (1972), West Africa can be divided into the following climatic zones (Fig. 10.1.1): tropical wet, tropical wet with short dry season and tropical with long dry season. The wet climate type covers only a small part of the Upper Guinea Region and is surrounded to the north by a seasonally dry climate zone which, on the Gulf of Guinea, reaches the coast between Ghana and Bénin (Dahomey Gap).

Precipitation is controlled by the migration of the Intertropical Convergence Zone (ITCZ, i.e., converging air of low latitudes). During the Northern Hemisphere summer period the ITCZ migrates northward and produces rain in West Africa. In the Southern Hemisphere summer period the ITCZ moves back south of the equator and a dry northeasterly trade

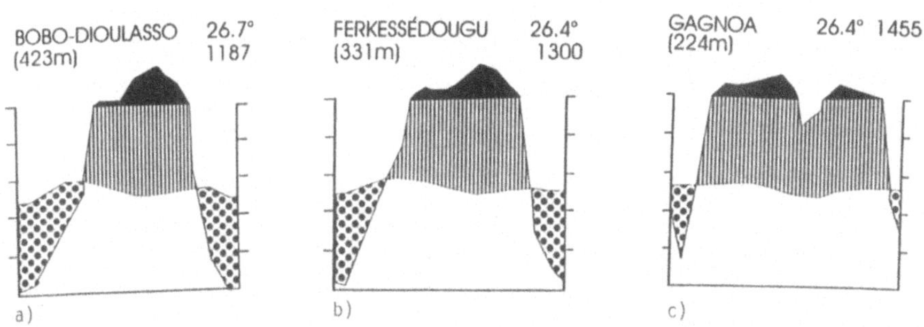

Fig. 10.1.1a–c. Typical Klimadiagramms for stations in West Africa (Walter and Lieth 1967). **a** Bobo-Dioulasso (Burkina Faso) 11°1'N/4°2'W. **b** Ferkessédougu (Côte d'Ivoire) 4°1'N/5°1'W. **c** Gagnoa (Côte d'Ivoire) 6°1'N/5°5'W

wind (Harmattan) prevails. The rainforest area in West Africa has a humid equatorial climate, and averages between 1500 and 2000 mm annual precipitation. Only very locally (e.g., coastal regions of Cameroon) does the annual rainfall exceed 3000 mm. Almost everywhere there is a pronounced dry season (between 1 and 3 months) with less than 100 mm of rain per month. Mean annual temperature is close to constant throughout the year.

West African woodland/grassland areas are characterized by a seasonal climate with a dry period of up to 6 months. Annual precipitation ranges between 700 and 1600 mm. Mean annual temperature (often >27 °C) is higher than in the rainforest zone. The severity of the dry season is enhanced by the desiccating effect of the Harmattan.

During the Quaternary, massive climatic changes caused a shift of all major West African vegetation belts. In particular the rainforests were disrupted (leading to their retreat to refugia) by the southward movement of the Sahara, as is indicated by pollen analyses from southern Ghana (Maley 1989). At the time of the last glacial maximum from 19 000 to 15 000 BP, arboreal pollen percentages reached minimum values, and herbaceous plants dominated. After ca. 8500 BP, arboreal pollen percentages steadily increased again. The aridity of West Africa during the late Quaternary may have been crucial for the impoverished nature of its forests (Richards 1973).

Since the following account on West African inselberg vegetation will focus mainly on the Côte d'Ivoire, a more detailed description of this country's climate is provided. Situated between 4°30'N and 10°30'N, the southern third of the Côte d'Ivoire has a humid equatorial climate (Af according to the Köppen classification), and receives an annual rainfall of between 1800 and 2300 mm. The north of the country has a seasonal climate (Aw) with a dry period of up to 6 months, mainly between October and April. Annual precipitation is lower than in the southern part of the country and usually ranges between 900 and 1600 mm. Moreover, the northern parts of the Côte d'Ivoire are characterized by a pronounced temporal and spatial variation in rainfall. A detailed description of the climate of the Côte d'Ivoire has been given by Eldin (1971) and Anhuf (1994). For data on the microclimate of West African inselbergs it is referred to Szarzynski Chapter 3, this Volume.

10.1.3 Vegetation

No detailed information exists concerning the number of species and endemics in West Africa. The *Flora of West Tropical Africa* (Hutchinson and Dalziel 1954–1972) comprises about 6000 species, a relatively small number compared to other tropical areas of equal geographical expansion. Only the Cameroon-Gabon center is considered to represent a global hot spot of phytodiversity (Barthlott et al. 1996).

According to White (1983, 1993), most of the West African vegetation is divided between two phytochoria. The lowland rainforest is part of the Guineo-Congolian regional centre of endemism (comprising 8000 spp., 80 % endemics), whereas woodlands and grasslands belong to the Sudanian regional center of endemism (ca. 3750 spp., ca. 33 % endemics). Rainforest – there exist a wide diversity of types – extends in West Africa from southern Senegal in the west to Ghana in the east. The Dahomey Gap in Ghana, Togo, and Bénin separates them from the Central African rainforests. Today, agriculture and other human activities have led to large-scale deforestation and fragmentation, leaving only small relict blocks of forest (e.g., Taï in southwestern Côte d'Ivoire). Separated from the rainforests by a transitional forest-savanna mosaic, the remainder of West Africa is covered by Sudanian woodlands and grasslands.

10.1.4 Inselberg Vegetation

Inselbergs occur in all vegetation and climate zones of West Africa. Frequently they can be found as large dome-shaped outcrops attaining an absolute height of more than 300 m, in drier regions shield inselbergs are particularly well represented (Fig. 10.1.2). Apart from a few exceptions (e.g., the Nigerian Jos Plateau), inselbergs of the koppje type do not occur. Inselbergs can be found in both a broad spectrum of sizes and in various degrees of isolation. In certain regions, groups of inselbergs exist which are characterized by huge numbers of rock outcrops occurring side by side, sometimes even forming continuous mountain ranges, such as those found in the northwestern part of the Côte d'Ivoire (Chaîne de Tiémé, Chaîne de Madinani).

The inselberg vegetation of certain West African countries is comparatively well known (Cameroon: Létouzey 1968, 1985; Mildbraed 1922; Villiers 1981; Central African Republic: Sillans 1958; Côte d'Ivoire: Adjanohoun 1964; Bonardi 1966; Miége 1955; Porembski and Barthlott 1992, 1993, Porembski and Brown 1995, Porembski et al. 1996b; Gabon: Reitsma et al. 1992; Guinea: Porembski et al. 1994; Nigeria: Hambler 1964; Kershaw 1968; Richards 1957; western Sahara: Quézel 1965). Most of these works present useful descriptions of the vegetation and flora, while studies concerned with ecological aspects of rock outcrop plant communities have only rarely been conducted (e.g., Ayo-Owoseye and Sanford 1972; Isichei

Fig. 10.1.2.a Dome-shaped inselberg with steeply rising flanks situated in lowland rainforest. (Equatorial Guinea, photograph N. Biedinger). **b** Inselbergs situated in the savanna zone are characterized by usually only slightly inclined slopes. (Côte d'Ivoire, photograph N. Biedinger)

and Longe 1984). However, up to now no comprehensive review of both floristical and ecological aspects has been available. For certain West African countries, such as Ghana, Equatorial Guinea and Sierra Leone, almost no accounts of the vegetation of inselbergs exist. Similarly, information on regions (for instance Central African Republic, Angola) linking West Africa with other phytochoria is sparse.

A survey of the characteristic elements of major habitat types on West African inselbergs is presented in the following (terminology according to Porembski et al., Chap. 4, this Vol.). The most prominent and well known are (1) cryptogamic crusts on open rock surfaces (consisting of lichens and cyanobacteria, but destitute of vascular plants); (2) drainage channels; (3) seasonally water-filled rock pools; (4) shallow depressions; (5) monocotyledonous mats, and (6) ephemeral flush vegetation. However, accounts of other less well-studied habitats are also provided. Not included in the following account are several types of forest vegetation. There are two main types of forest in association with inselbergs, which can be distinguished according to their location either at the foot of the outcrop or on the summit. The first type of forest benefits from runoff water (thus often containing rainforest trees even when situated in savanna) and an enhanced supply of mineral nutrients, whereas the summit forest is characterized by relatively shallow soils and a high percentage of drought-adapted trees. The latter type of forest, however, can be relatively rich in epiphytes due to occasional cloud cover, depending on the absolute height of the inselberg.

10.1.4.1 Vegetation of Rock Surfaces

Cryptogamic Vegetation of Rock Surfaces. The first descriptions of this community (cryptogamic crust) on West African inselbergs were provided by Zehnder (1953), Richards (1957), and Hambler (1964). A recent account of lichens and cyanobacteria found on Ivorian inselbergs is provided by Büdel et al. (1997). Among the lichens, species belonging to the family Peltulaceae (in particular *Peltula congregata*, *P. lingulata*, *P. tortuosa*, and *P. umbilicata*) are dominant. In addition, a few Lichinaceae occur. Of the cyanobacteria, an endolithic *Chroococcidiopsis* sp. and the epilithic *Gloeocapsa sanguinea* are frequent components of the cryptogamic crust. On inselbergs located in the seasonally dry Sudanian region, cyanobacterial lichens dominate, whereas in rainforest climates the rocky slopes are commonly covered by cyanobacteria. Büdel et al. (1997) report a gradient in species number for both cyanolichens and cyanobacteria, with a decrease from the savanna to the rainforest. So-called lichen mats (formed by *Usnea* spp.), which characteristically cover rock surfaces of inselbergs in

Malawi and Zimbabwe (Porembski 1996; U. Becker pers. comm.), occur in West Africa only on hills above an elevation of ca. 800 m. Nonlichenized microcolonial fungi (i.e., mostly dematiaceous Hyphomycetes), which are known from granitic outcrops in other tropical (e.g., Seychelles, Henssen 1987) and nontropical regions, have recently been reported to occur on Ivorian inselbergs as well (K. Sterflinger pers. comm.). Occasionally, small patches of mosses (most notably *Brachymenium exile* and *Bryum arachnoideum*, Frahm and Porembski 1994) occur on bare rock where seepage water is available. In arid regions, like in the western part of the Sahara, with annual rainfall below 100 mm, continuous cryptogamic crusts are missing.

Cryptogamic Vegetation of Boulders. Isolated boulders on exposed rocky slopes are covered by lichens. Floristically, boulders contrast sharply with the surrounding level rock faces. Cyanobacterial lichens are missing, whereas chlorophytic lichens such as *Caloplaca*, *Lecanora*, and *Toninia* dominate and are responsible for the orange, greenish, or whitish coloration of the boulders. Richards (1957) reports the presence of foliose lichens (e.g., *Parmelia tinctorum*) on boulders under moist, shaded conditions.

Cryptogamic Vegetation of Drainage Channels. Cyanobacteria (e.g., *Stigonema mamillosum*) and cyanobacterial lichens (most notably the black-colored *Peltula lingulata*) form a dense cover on the bottom of drainage channels (for details see Büdel et al. 1997). Frequently drainage channels are bordered by a zone of bare rock 2–10 cm in width.

Wet Flush Vegetation. Apart from a few hints (Hambler 1964; Porembski et al. 1994), information about this vegetation type is very scarce. This is probably due to the fact that this habitat is relatively small in extent and occurs in places where access is difficult, i.e., relatively steep slopes with water running continuously during the rainy season. Characteristic vascular plants are minute annuals, in particular *Xyris straminea*, *Utricularia subulata*, *U. andongensis*, and several Eriocaulaceae, which are attached to cyanobacterial crusts, forming slippery blackish brownish films. Frequently, dense mats of *Utricularia* species determine the physiognomy of this community. Occasionally, small patches of mosses can be found, which provide establishment sites for vascular plants. Though our knowledge of this habitat is at present only rudimentary, it can be assumed (based on personal field experience) that this community is more rich in species in the savanna zone, compared to the rainforest region.

Vascular Lithophytic Vegetation. On West African inselbergs only a few species occur as lithophytes on free exposed rock. Most prominent (with respect to frequency) throughout the wetter parts of West Africa is the Commelinaceae *Cyanotis arachnoidea* (Fig. 10.1.3). This perennial species possesses succulent leaves and forms cushions on steep slopes. Moreover, several other succulents occur regularly as colonizers of bare rock, for example acaulescent species of *Aloe* (e.g., *A. buettneri*, *A. schweinfurthii*). Less frequent are orchids (e.g., *Bulbophyllum* spp.), and the stem succulent *Cissus quadrangularis*. Under more shaded conditions, the fern *Platycerium angolense* can be found. One has to emphasize, however, that apart from *Cyanotis arachnoidea*, none of these species is restricted to this habitat. The occasional presence of invasive exotic weeds (e.g., *Ananas comosus*, *Bryophyllum pinnatum*) as lithophytes is remarkable. On Ivorian inselbergs situated in the rainforest zone, *Ananas comosus* is becoming increasingly dangerous to the indigenous rock outcrop vegetation, due to its rapid vegetative spread (Fig. 10.1.4).

Fig. 10.1.3. Mats made up by the perennial Commelinaceae *Cyanotis archnoidea* are a common sight on rainforest inselbergs in the Côte d'Ivoire. (Photograph S. Porembski)

10.1.4.2 Vegetation of Rock Crevices

Vegetation of Horizontal and Vertical Crevices. Horizontal crevices usually result from exfoliation of large slabs of rock. They are similar to vertical crevices in their species composition. However, due to more shaded conditions, a higher percentage of shade-tolerant plants occurs, e.g., the ferns *Asplenium stuhlmannii*, *Pellaea doniana* and *Selaginella* spp. Vertical crevices, most of which are less than 2 cm in width (accumulating only small amounts of soil), are commonly colonized by short-lived plants. Cyperaceae (e.g., *Bulbostylis coleotricha*, *B. congolensis* and *Fimbristylis dichotoma*), Poaceae (e.g., *Anadelphia liebigiana* and *Sporobolus festivus*), Fabaceae (e.g., *Aeschynomene lateritia*, *Indigofera deightonii*, *Tephrosia mossiensis*) and ferns are widespread. The poikilohydric ferns *Actiniopteris radiata* and *A. semiflabellata*, which are typical elements on rock outcrops in northeastern and eastern tropical Africa, have been recorded from only a few localities in Nigeria and Cameroon (Kornas 1983).

Vegetation of Clefts. Shrubs and trees dominate. A high percentage of deciduous representatives such as *Bombax costatum*, *Hymenodictyon flori-*

Fig. 10.1.4. Invasive weeds are a potential danger to the indigenous vegetation of inselbergs. Occasionally the bromeliad *Ananas comosus* has become established on West African inselbergs. (Côte d'Ivoire, photograph N. Biedinger)

bundum, *Holarrhena floribunda*, and *Stereospermum acuminatissimum* is typical. Frequently, xeromorphic (e.g., *Ficus* spp., *Mimusops kummel*) and pachycaulous trees (e.g., *Hildegardia barteri*) occur. Occasionally, succulents can be found (e.g., *Sansevieria liberica* and *Euphorbia unispina*). Richards (1957) states in his study on the Nigerian Idanre Hills that the shrubs *Eugenia obanensis* and *Erythroxylum emarginatum* are found growing in deep crevices. The Nigerian Jos Plateau is particularly rich in succulent Euphorbias on rock outcrops.

According to Quézel (1965), rock crevices and clefts form the most suitable habitat for vascular plants on inselbergs in the western portion of the Sahara (former Spanish Sahara). A community largely consisting of small trees and shrubs which was decribed as association à *Maerua crassifolia* et *Ephedra rollandii* is characteristic. Further species that occur are *Periploca laevigata*, *Rhus tripartitum* and *Lycium intricatum*.

Vegetation Around Boulder Bases. Shading is the most important physical factor in this habitat type, which positively influences the moisture regime. Shade-tolerant ferns (e.g., *Asplenium* spp., *Pellaea* spp., *Selaginella* spp.) occur in particular. Among the ferns, the tiny *Selaginella tenerrima* is a frequent colonizer of boulders under overhanging rocks. Trees, which can be found as colonizers of clefts, also occur here.

Vegetation of Talus Slopes. Small trees, like *Ficus abutilifolia* and *Haematostaphis barteri*, or climbers, which typically are represented by *Cissus* spp. (prominent is the succulent *C. quadrangularis*), *Dioscorea* spp., and *Ipomoea* spp. are of frequent occurrence. Sometimes succulent Euphorbias and *Sansevieria liberica* are present. Under the shade of boulders ferns may also occur.

10.1.4.3 Vegetation of Depressions

Bornhardts, which consist almost only of precipitous flanks and which typically occur in rainforest regions (e.g., southern Nigeria, Equatorial Guinea) are comparatively rare in West Africa. More or less plain stretches of rock are therefore of widespread occurrence on West African inselbergs. The large number of man-made grinding holes which are very uniform in shape and in size, is remarkable (Fig. 10.1.5). Today, they are largely out of use and form temporarily water-filled pools.

Vegetation of Seasonal Rock Pools. Water-filled after rainfall, but rapidly dried out if not replenished by subsequent rain, they form temporal

Fig. 10.1.5. Despite the fact that grinding holes are of anthropogenic origin, they constitute ancient habitats. These rock pools are water-filled during the rainy season when they are colonized by a wide spectrum of species. Only a few of them, like the Scrophulariaceae *Dopatrium longidens* are rock pool specialists. (Bénin, photograph N. Biedinger)

habitats that may be thousands of years old. Their vegetational cover is usually relatively sparse and many pools are not colonized by vascular plants at all. However, cryptogams are always present and are responsible for a nearly complete coverage of the rock. Cyanobacteria occur frequently on the rocky bottom of the pools. Additionally, towards the pool edge several lichens (most of all *Peltula lingulata*) tend to develop. This zone is characterized by frequent fluctuations of the water level.

Among vascular plants, short-lived herbs predominate. Besides species which are otherwise widespread on marshy or swampy ground (e.g., *Ammania* spp., *Cyperus* spp., *Hygrophila* spp., *Ludwigia* spp.) specialists exist, which are more or less restricted to this habitat. Prominent examples are represented within the Scrophulariaceae, such as species belonging to the genus *Dopatrium* (e.g., *D. longidens*, *D. senegalense*) and within the Lythraceae (*Rotala* spp.). Characteristic, but frequently overlooked, are geophytic *Isoetes* species (e.g., *I. nigritiana*). Typical water plants like *Burnatia enneandra*, *Marsilea polycarpa*, *Nymphaea lotus*, *Najas* sp., and *Eichhornia natans* are restricted to larger rock pools. Occasionally, free-floating Lemnaceae can be observed (e.g., *Spirodela polyrrhiza*). Usually, a considerable amount of vagrants (i.e., opportunistic species) are encountered. Of major importance in this respect is *Cyanotis lanata*, which is one of the most frequently occurring species in West African rock pools. A prominent attribute of this habitat is the considerable spatial and temporal (i.e., year-to-year species turnover) variability in species composition (see Porembski et al. Chap. 12, this Vol.; Krieger et al., in press). Occasionally a successional stage towards communities in soil-filled depressions can be seen, with the grass *Acroceras amplectens* being a prominent species.

Vegetation of Permanently Water-Filled Rock Pools. Deeper (in part >1 m) and usually larger than seasonal rock pools, they contain at least a minimal amount of water in the dry season. They are only very occasionally colonized by vascular plants. Typical water plants occur here more frequently than in seasonal rock pools. Most notable are *Eleocharis acutangula*, *Ludwigia* spp., *Marsilea polycarpa*, *Sphenoclea zeylanica*, and floating species of *Utricularia* (e.g., *U. stellaris*). Rock pool specialists are largely absent.

Vegetation of Rock Debris. Soil cover is extremely thin (up to 2 cm in depth). Frequent colonizers of bare soil are mosses (e.g., *Bryum arachnoideum*), liverworts (*Riccia* spp.), and cyanobacteria (e.g., *Schizothrix* spp.). The thalli of the liverwort genus *Riccia* are a familiar sight in exposed positions over shallow soil. Most common on West African inselbergs are *R. congoana*, *R. discolor*, *R. lanceolata*, and *R. moenkemeyeri* (Perold 1995; Frahm and Porembski 1997). Vascular plants are represented by annuals, such as *Cyanotis lanata* and *Lindernia exilis*. Perennials occur only exceptionally, e.g., the poikilohydric *Lindernia yaundensis*, which is endemic to the rainforest region of Cameroon (Raynal 1966).

Herbaceous Vegetation of Soil-Filled Depressions. This habitat occurs where the rock surface is relatively flat, attaining a size of several m^2 (Fig. 10.1.6). Soils (5–15 cm in depth, increasing from the edge to the center) are coarsely textured and tend to dry out rapidly after rainfall. Vegetation cover is very sparse, with average percentage cover values for vascular plants usually below 40 %. A zonation of vegetation according to soil depth can frequently be observed. Relatively dense swards of cryptogams (mostly liverworts and cyanobacteria) dominate in the periphery of the depressions, occasionally accompanied by tiny therophytes (*Crepidorhopalon debilis*, *Lindernia exilis*). Towards the center, the number of vascular plant species increases. In particular in the savanna region this community is relatively rich in species. Characteristic elements recorded on Ivorian

Fig. 10.1.6. Shallow soil-filled depressions are mainly colonized by short-lived ephemerals. In West Africa *Cyanotis lanata* (Commelinaceae) is the dominant species in this habitat. (Côte d'Ivoire, photograph S. Porembski)

inselbergs include *Chamaecrista mimosoides*, *Cyanotis lanata*, *Heliotropium indicum*, *Indigofera astragalina*, *Microchloa indica*, *Tripogon minimus* (a poikilohydric grass) and the geophytes *Brachystelma simplex* and *Ophioglossum costatum* (for details see Porembski et al. 1995).

The species richness of this habitat decreases drastically in the drier Sudano-Sahelian region (i.e., north of 15° latitude). Interestingly certain Poaceae (e.g., *Diheteropogon hagerupii*, *Tripogon minimus*) and Cyperaceae (e.g., *Bulbostylis pusilla*) are still present on inselbergs in the deserts of Mauretania with an annual rainfall below 100 mm. Towards the rainforest region, the number of annuals also diminishes considerably. Instead, a relatively small number of longer-lived species tend to dominate. Of particular importance are *Mariscus dubius*, *Dissotis rotundifolia*, and *Clappertonia minor*, which typically occur in dense stands.

Woody Vegetation of Soil-Filled Depressions. Over deeper soil, small and usually deciduous trees and shrubs may establish. Even in the West African rainforest zone the inventory of this habitat is mainly recruited from species that are widespread in savanna areas. Therefore, a mainland/island type of distribution is typical for a large number of trees and shrubs which occur in only small populations on isolated outcrops in geographical separation. Anemochorous species are relatively frequent, for instance *Holarrhena floribunda* (Apocynaceae), the Bignoniaceae *Stereospermum acuminatissimum*, and *Markhamia tomentosa*. The number of inselberg specialists within this habitat is small. Examples of these are the trees *Elaeophorbia grandifolia* (Euphorbiaceae), *Haematostaphis barteri* (Anacardiaceae) and the Rubiaceae *Hymenodictyon floribundum*. Shrub-like rock outcrop specialists are, for instance, species of the genera *Dissotis* (e.g., *D. theifolia*), *Dolichos* (e.g., *D. tonkuiensis*), and *Virectaria* (*V. multiflora*).

In particular on inselbergs in rainforest regions, certain trees and shrubs are regular inhabitants of the ecotone between forest and open rock surface. Widespread in this zone, which is intermediate in regard to microclimate and soil conditions, are species which frequently grow epiphytically, like ferns (e.g., *Phymatosorus scolopendria*, *Nephrolepis biserrata*), angraecoid orchids, or Melastomataceae (e.g. *Calvoa pulcherrima*). Occasionally shrub-like species (e.g., *Clappertonia polyandra*) form dense thickets.

10.1.4.4 Ephemeral Flush Vegetation

The term ephemeral flush vegetation was introduced by Richards (1957), for a highly seasonal community occurring over shallow soils (0.5–12 cm), with lowest values at the margin towards the rock. A mosaic of different soil depths occurring close together (for details of abiotic and biotic features see Dörrstock et al. 1996a) is characteristic. Floristically, the ephemeral flush vegetation is relatively uniform and shows close affinities with communities on ferricretes. For example, Schnell (1952) described a Utricularieto-Eriocauletum pumili community from ferricretes in the Guinean part of the Nimba Mountains which is almost identical in regard to species composition. Phytosociologically, the ephemeral flush vegetation belongs to the class Eriocaulo-Utricularietea (Knapp 1966).

The ephemeral flush vegetation develops at the base of steep slopes, where water continuously seeps during the rainy season (Fig. 10.1.7). The vegetation cover is frequently interrupted by bare rock. Usually three to four intergrading zones can be distinguished according to soil depth: a transitional zone against the rock, a marginal zone, a central zone and an upper transitional zone (sometimes absent) towards habitats, like *Afrotrilepis pilosa* mats. The transitional zone against the lichen-covered rock is dominated by cyanobacteria. The marginal zone (soil depth 0.5–3 cm) is characterized by the scattered occurrence of tiny annuals (5–20 cm in

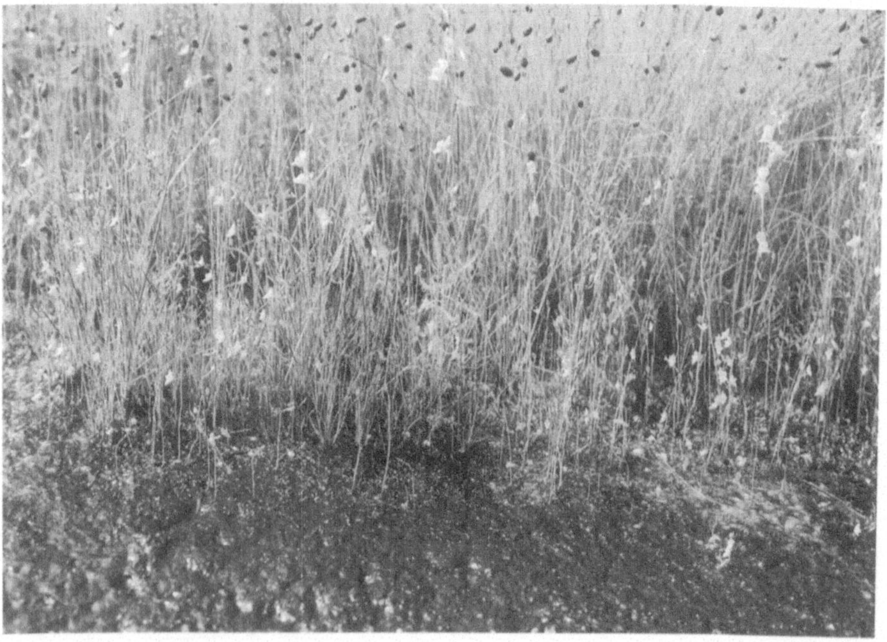

Fig. 10.1.7. Depending on seepage water, the ephemeral flush vegetation is best developed on the peak of the rainy season. The bulk of the vegetation consists of small herbs including a large percentage of carnivorous species, e.g., *Utricularia* and *Genlisea* spp. (Côte d'Ivoire, photograph S. Porembski)

height) with a high percentage of carnivorous species. Within the central zone (soil depth 3–10 cm) Cyperaceae and Poaceae (20–50 cm in height) dominate. Tall grasses are typical for the upper transitional zone. This zonation may become obscured due to the patchy distribution of different soil depths. Liverworts (*Riccia* spp.) and small thalli of the hornwort *Notothylas javanica* are widespread colonizers on open ground. Poaceae and Cyperaceae make up the largest part of the phytomass. Among the Poaceae, *Panicum griffonii*, *Panicum tenellum*, *Sporobolus pectinellus* and *Loudetiopsis capillipes* belong to the most important species. Frequently occurring Cyperaceae are *Scleria melanotricha*, *Ascolepis protea* and *Nemum spadiceum*. Most rich in species are tiny ephemerals, such as *Eriocaulon* spp. (e.g., *E. plumale*, *E. afzelianum*), *Xyris* spp. (most notably *X. straminea*), *Ophioglossum* spp. (e.g., *O. gomezianum*), *Burmannia madagascariensis* and carnivorous plants (Droseraceae and Lentibulariaceae: *Utricularia*, *Genlisea*). In contrast to the genus *Drosera*, which is usually represented by only a single species (*D. indica*), the genus *Utricularia* often occurs with four to six terrestrial species in a single locality. *Utricularia subulata* and *U. pubescens* are dominant, but the twining *U. spiralis* and *U. tortilis* are likewise frequent. Three species of the genus *Genlisea* (*G. barthlottii*, *G. hispidula* and *G. stapfii*) form characteristic elements of ephemeral flush communities on West African rock outcrops. Concerning life-forms, annuals clearly dominate, usually accounting for more than 70 % of the species present. In West Africa, this community is especially well developed (i.e., most rich in species) on inselbergs situated in savanna zones. On Ivorian inselbergs situated in the savanna zone, sites extremely rich in species were found with more than 30 species of vascular plants m^{-2}. In the rainforest zone, the ephemeral flush vegetation is either completely missing or forms small rims fringing monocotyledonous mats. In general the percentage of specialists is relatively high when compared with other habitats on inselbergs.

10.1.4.5 Mat Vegetation

The major component of monocotyledonous mats on West African inselbergs is the poikilohydric Cyperaceae *Afrotrilepis pilosa*, a species found only on rock outcrops (Schnell 1952; Richards 1957; Hambler 1961). Its distributional range is restricted to West Africa, extending from Senegal southwards to Gabon. *Afrotrilepis pilosa* is found in all major climatic zones of West Africa but declines in quantity towards the Sahel region, e.g., in southern Burkina Faso (Falaise de Banfora, ca. 11°N) and Mali (south of Bamako, ca. 13°N). The boundary of this species towards the Sahel region is obviously due to a decreasing amount of rainfall and an increased length of the dry period. On savanna zone inselbergs *Afrotrilepis pilosa* usually forms island-like mats surrounded by large expanses of open rock, whereas under a rainforest climate even steeply inclined slopes bear a nearly continuous cover of *Afrotrilepis pilosa* mats. Even under a rainforest climate these mats form a highly seasonal community, as is clearly demonstrated by the rapid change in leaf color which is grayish yellow in the dry state but turning to green shortly after rain.

Throughout West Africa there is almost no inselberg without *Afrotrilepis pilosa* mats, thus making this sedge the most characteristic element of this ecosystem over the whole area. Remarkably, however, *Afrotrilepis pilosa* is absent from small (i.e., less than 20 000 m^2 in size) and isolated rock outcrops. In particular on steeply inclined slopes on savanna zone inselbergs the *Afrotrilepis pilosa* mats are more or less circular in outline and attain a diameter of between 2 and 5 m. According to Hambler (1964), *Afrotrilepis pilosa* seeds germinate immediately (preferentially in capillary crevices) after shedding, provided moisture is available.

Afrotrilepis pilosa is a highly variable species, showing considerable morphological differences (e.g., growth-form, length, width and indumentum of leaves) between populations on different inselbergs (Fig. 10.1.8). Individual specimens may attain an age of more than 200 years (Bonardi 1966). The pseudostems (attaining a maximum height of 1.5 m) of *Afrotrilepis pilosa* mainly consist of a sheath made up by the remains of old roots and leaves, rendering the plants highly resistant to fire (Fig. 10.1.9). Due to the exposed position of inselbergs, natural fires caused by lightning have probably played an important role within the mat communities. Occasionally, individual mats struck by lightning could be found which were usually completely destroyed. Like most other mat-forming species on inselbergs, *Afrotrilepis pilosa* has developed the ability to spread horizontally through vegetative propagation by basal branching, resulting in the formation of clonal populations. On Nigerian inselbergs Richards (1957) and Hambler (1964) have reported the dislodgement of individual mats. The former location of dislodged mats is clearly visible for

Fig. 10.1.8. Mats formed by the Cyperaceae *Afrotrilepis pilosa* occur on nearly all West African inselbergs. Geographically isolated populations of *Afrotrilepis pilosa* show a large amount of morphological differentiation, e.g., prostrate versus upright pseudostems or glabrous versus densely indumented leaves. (Côte d'Ivoire, photograph S. Porembski)

Fig. 10.1.9. Old individuals of *Afrotrilepis pilosa* are effectively protected against fire by forming large clumps which consist of a dense sheath of roots and decaying leaves. (Côte d'Ivoire, photograph S. Porembski)

years by the bright color of the rock, which is only slowly recolonized by lichens and cyanobacteria.

Afrotrilepis pilosa mats are frequently a species poor (in cases monospecific, Porembski et al. 1996a) and floristically very uniform community which produce their own substrate (a product of decaying plant material). They are stabilized by an extensive root system which, furthermore, attaches the mat to bare rock. Most accompanying species are annuals, such as *Cyanotis lanata* and *Lindernia exilis*, which occur preferentially at the periphery of the mats. Several succulents (especially stem succulent *Euphorbia* spp.) and poikilohydric ferns (e.g., *Pellaea doniana*, *Asplenium stuhlmannii*) and geophytes (e.g., *Eriospermum abyssinicum*, *Raphionacme* spp.) are regular colonizers of *Afrotrilepis pilosa* mats. Certain mosses (e.g., *Brachymenium acuminatum*, *Campylopus* spp.) and lichens (e.g., *Cladonia* spec.) occur preferentially in unshaded peripheral parts of the mats or colonize the stems of *Afrotrilepis pilosa* (e.g., *Octoblepharum albidum*). Striking components of the supporting flora of the *Afrotrilepis pilosa* mats are several epiphytic orchids which are strictly limited to this community, for example *Polystachya microbambusa* (Guinea, Sierra Leone,

Liberia, Côte d'Ivoire) and *P. pseudo-disa* (Sierra Leone, Liberia, Côte d'Ivoire). Not restricted to *Afrotrilepis pilosa* mats, but frequently present, are the epiphytic orchids *Calyptrochilum christyanum* and *Tridactyle tridactylites*.

Afrotrilepis pilosa mats form a pioneer community which is fairly stable over long periods. Their typical appearance may be changed through the establishment of woody colonists, like *Hymenodictyon floribundum*, *Hildegardia barteri*, and *Dissotis* spp. which depress *Afrotrilepis pilosa* and initiate a succession towards woodland.

Apart from *Afrotrilepis pilosa*, only a few other monocotyledons occur as mat-formers on West African rock outcrops. The closely related Cyperaceae *Microdracoides squamosus* (disjunct in Guinea and Nigeria/Cameroon), *Afrotrilepis jaegeri* (Sierra Leone) and *Coleochloa abyssinica* (Nigeria/Cameroon extending to East Africa) are of only local importance. As already mentioned, the perennial *Cyanotis arachnoidea* occurs as a mat-former on rainforest inselbergs. The poikilohydric *Selaginella njam-njamensis* forms large patches on rock outcrops in drier parts of Nigeria (Richards 1957).

With the poikilohydric *Xerophyta schnitzleinia*, another mat-forming monocot occurs sympatrically with *Afrotrilepis pilosa* on inselbergs in northern Nigeria (Ayo-Owoseye and Sanford 1972). No information is, however, available about possible competitive interactions between these two species. Occasionally (in particular in the absence of *Afrotrilepis pilosa*), the usually lithophytic *Cyanotis arachnoidea* forms dense mats. Mats which are dominated by the grasses *Monocymbium ceresiiforme* and *Andropogon linearis* were reported from inselbergs in Nigeria (Hambler 1964) and Bénin (own unpubl. data). Today at several West African localities an accidental experiment can be observed concerning the competitive interactions between *Afrotrilepis pilosa* and the neophyte *Ananas comosus* where the latter has escaped from plantations. Our preliminary observations indicate an advantage of the relatively fast-growing bromeliad over *Afrotrilepis pilosa*. There is, for example, at least one inselberg (Mt. Brafouedi, Côte d'Ivoire) where the whole *Afrotrilepis pilosa* population has succumbed to *Ananas comosus*.

There are considerable differences between the typical habitats (i.e., exclusive forest vegetation) on inselbergs in regard to their spatial extent. Generally, the different vegetation types of rock surfaces, especially the cryptogamous crusts, are most important and often attain cover values of 70–80%. Monocotyledonous mats may cover between 10 and 15% of rock outcrops (even higher values in more humid areas). All other habitats usually do not reach values above 3%. Depending on geomorphological aspects (e.g., presence of smoothly inclined rocky slopes instead of steep flanks), this picture varies between individual inselbergs. In the latter case, ephemeral flush communities may extend over considerable areas of only slightly inclined slopes, provided sufficient seepage water is available during the rainy season. Moreover, there are variations due to climatic reasons. For example, under dry conditions (i.e., in a savanna climate) wet flush communities are restricted to narrow strips of wet rock, whereas the same community type is far more prominent on rainforest inselbergs.

10.1.5 Life-Forms and Ecological Characteristics

There is only scarce information available regarding life-forms among the vegetation on West African inselbergs (e.g., Adjanohoun 1964; Porembski and Brown 1995; Porembski and Barthlott 1997). Consequently, in the following account only a rough estimation can be provided, based on extensive data obtained in the Côte d'Ivoire. However, our own fieldwork on inselbergs in other West African countries (e.g., Bénin, Guinea) and data from other published sources (e.g., Hambler 1964; Richards 1957) confirmed that the observations made in the Côte d'Ivoire are largely compatible for the whole region.

In total, ca. 600 species of vascular plants have been recorded on Ivorian inselbergs. Amongst inselberg specialists therophytes are most rich in species, followed by hemicryptophytes, phanerophytes, and cryptophytes, whereas chamaephytes are only of minor importance. The predominance of annuals on inselbergs is less pronounced in the rainforest region. For example, on inselbergs in Equatorial Guinea which are surrounded by rainforest the number of therophytes is drastically reduced (own unpubl. data).

There is a tendency for annuals to increase in importance with diminishing rock outcrop size, while longer-lived species tend to behave in the opposite way (cf. Porembski et al., Chap. 12, this Vol.). The preponderance of annuals on small outcrops (i.e., covering only a few hundred m^2) is possibly due to their greater ability to survive under disadvantageous environmental conditions. Larger rock outcrops are better buffered against, e.g., climatic disturbances, and therefore provide safer growing sites for longer-lived k-selected species. Certain therophytes (e.g., *Lindernia exilis*) are characterized by having very short life cycles, i.e., only about 4 weeks are needed from germination to fruit set. In particular in ephemeral flush communities and in shallow depressions, the percentage of annual species may reach more than 80%. Families with high shares of annuals are Poaceae, Cyperaceae, and Fabaceae. Within the

Cyperaceae a particular broad range of life-forms is represented on inselbergs, with annuals (particularly *Cyperus* spp.), cryptophytes (e.g., *Mariscus dubius*) and poikilohydrics (e.g., *Afrotrilepis pilosa*). Most cryptophytes are monocots (e.g., *Albuca* spp., *Urginea* spp.); however, the widespread occurrence of ferns which possess either a bulbous rhizome (i.e., *Ophioglossum costatum*) or a corm-like subterranean axis (i.e., *Isoetes* spp.) is remarkable. Shrub-like species are found in particular within the genera *Dissotis* (Melastomataceae), *Solenostemon* (Lamiaceae), *Dolichos*, *Indigofera* and *Tephrosia* (all Fabaceae). Certain species of the genus *Solenostemon* (e.g., *S. graniticola*) possess succulent stems. At higher altitudes certain species (e.g., *Afrotrilepis jaegeri* which is endemic to the Loma Mountains of Sierra Leone) occur as cushion-forming plants. The number of pachycaulous trees on West African inselbergs is relatively low when compared to other regions. Characteristic representatives are *Hildegardia barteri* and several species of *Ficus*. Further drought-adapted trees on West African inselbergs include *Bombax costatum*, *Hymenodictyon floribundum*, and *Mimusops kummel*.

West African inselbergs are much less diverse in poikilohydrics and succulents than their counterparts in East Africa and Madagascar. The most prominent representatives of the poikilohydric vascular plants are the Cyperaceae *Afrotrilepis pilosa* and *Microdracoides squamosus* which possess pseudostems (attaining a height of more than 1 m) consisting mainly of adventitious roots. Similarly, the mat-forming *Afrotrilepis jaegeri* seems to be poikilohydric. Moreover other poikilohydric Poaceae (e.g., *Tripogon minimus*), ferns (e.g., *Asplenium stuhlmannii*, *Pellaea doniana*), and the Scrophulariaceae *Lindernia yaundensis* (Raynal 1966) are to be found. The latter species is closely related to other poikilohydric species of *Lindernia* that are mainly distributed in East Africa (Fischer 1992). Under phytogeographical considerations, the occurrence of the poikilohydric and mat-forming Velloziaceae *Xerophyta schnitzleinia* on rock outcrops in northern Nigeria is remarkable. The West African populations of this species, which is widespread in East Africa, form a phytogeographical outlier. Of considerable adaptive value concerning rapid water uptake is probably the presence of a velamen radicum in the adventitious roots of *Afrotrilepis pilosa* and *Microdracoides squamosus* (see Biedinger et al., Chap. 8, this Vol.). Like poikilohydrics, succulents are relatively poorly represented on West African inselbergs. In this region most succulents belong to the genus *Euphorbia*, which is represented by a broad range of growth forms, i.e., tree-like to geophytic, and which is particularly rich with regard to the number of species on the Nigerian Jos Plateau (Newton 1992). Furthermore, one encounters a small number of *Aloe* spp., *Sansevieria* spp., *Solenostemon* spp., and *Brachystelma* spp. The low number of succulents on West African inselbergs can be attributed to extinctions during major climatic changes that occurred in the Pleistocene (Newton 1989).

Over vast stretches of regularly burnt savanna, inselbergs provide shelter for fire-sensitive trees (i.e., species lacking an insulating bark, such as *Hildegardia barteri*) because of their large expanses of rocks. On the other hand, however, fire due to lightning stroke is a natural factor that certainly has influenced the vegetation cover of inselbergs. A widespread response of perennial plants to fire is vegetative insulation. Similarly, like other arborescent monocotyledons, the pseudostems of *Afrotrilepis pilosa* and *Microdracoides squamosus* are protected from fire by a dense coverage of adventitious roots and leaf remains. Under the recurrent influence of fire (today mainly due to human activities), the regular growth form of *Afrotrilepis pilosa* is modified towards bizarre clump-shaped forms (Fig. 10.1.9). Clump-shaped plants may have a diameter of more than 30 cm, thereby largely consisting of decayed organic material that serves to insulate the living meristems.

Vegetative reproduction, for instance by means of stolons, frequently occurs in some species (e.g., *Cyanotis lanata*). Data on the reproductive biology of inselberg plants are scarce. Judging from our own observations on Ivorian inselbergs, it can be stated that entomophilous species (mostly pollinated by hymenoptera) clearly dominate, followed by anemophilous taxa. Both chiropterophily and ornithophily are rare. Among the different habitat types, ephemeral flush vegetation shows the highest percentage of entomophilous species, whereas anemophily is most prominent in rock pools. On inselbergs situated in the savanna zone of the Côte d'Ivoire certain species are characterized by gregarious flowering (i.e., nearly all individuals of a population are in flower at the same time). In particular, several species of *Utricularia* demonstrated this phenomenon with all individuals (occasionally hundreds on a few m^2) being in flower over a relatively short time span. Remarkably, however, almost no visits of potential pollinators could be observed within these *Utricularia* fields. Among dispersal strategies endozoochory (more than 40 % of the species, including many Poaceae, Cyperaceae, and certain trees) and anemochory (ca. 25 %) are most important. The latter category comprises both annuals (e.g., *Utricularia* spp.) and trees (e.g., *Hymenodictyon floribundum*). If only the specific rock-outcrop colonizers are considered, the percentage of anemochorous taxa increases. Epizoochory is known from certain Lamiaceae (most of all *Aeollanthus* spp.). Several taxa do not possess effective dispersal modes that allow for a transport over longer distances. Prominent among these are certain succulent *Euphorbia* species (e.g., *E. kamerunica*, *E. letestui*) possessing seeds that simply fall to the ground and lack any obvious adaptations towards dispersal. These species have possibly to be considered when occurring on rainforest outcrops to represent relicts that date back to drier climatic periods in the Pleistocene where suitable growing sites existed in what is at present rainforest. Therefore the presence of drier vegetation types allowed better access to azonal sites even for plants with low dispersal efficiency. For shallow depressions on Ivorian inselbergs the existence of seed banks could be demonstrated (Porembski and Barthlott 1997). The seeds of certain annuals (e.g., *Cyanotis lanata, Aeollanthus pubescens*) maintain their germination abilities for more than 5 years (own unpubl. data).

10.1.6 Systematic and Ecological Affinities

10.1.6.1 Systematic Composition

No detailed account of species numbers of the whole West African inselberg vegetation exists. However, detailed information is available for the Côte d'Ivoire, which can be seen as representative because of the general uniformity of the West African inselberg vegetation. During extensive fieldwork over the past 10 years, more than 600 vascular plant species could be recorded which divide among more than 40 families. The number of species which occur on nearly all inselbergs is relatively small, with *Afrotrilepis pilosa, Cyanotis lanata, Ophioglossum costatum*, and *Xyris straminea* forming essential components. Table 10.1.1 summarizes the most typical species for West African inselbergs. It should be emphasized, however, that the species listed in Table 10.1.1 are most typical for inselbergs of the Upper Guinea Region. Only a few species are continuously distributed throughout the whole region from Senegal in the north up to Gabon in the south.

Largest families are Leguminosae (84 spp.), Poaceae (71 spp.), Cyperaceae (66 spp.), Scrophulariaceae (24 spp.), and Rubiaceae (23 spp.). The Poaceae and Cyperaceae are not only the most species-rich families but are also most important in regard to their biomass. Although relatively poor in species, Burmanniaceae, Eriocaulaceae, and Xyridaceae are typical inselberg families, which mainly occur within ephemeral flush communities. A comparison with the family spectrum of the Ivorian flora in general (according to Aké Assi 1984) reveals that families like Lentibulariaceae and Scrophulariaceae are overrepresented on inselbergs (Table 10.1.2). For species of these families, azonal habitats like rock outcrops and ferricretes are the most important growing sites. A comparison with data compiled by

Table 10.1.1. Characteristic plant species of West African inselbergs

Name	Family
Aeollanthus pubescens	Lamiaceae
Afrotrilepis pilosa	Cyperaceae
Asplenium stuhlmannii	Aspleniaceae
Bacopa floribunda	Scrophulariaceae
Bulbostylis coleotricha	Cyperaceae
Chamaecrista mimosoides	Leguminosae
Cyanotis lanata	Commelinaceae
Cyperus submicrolepis	Cyperaceae
Drosera indica	Droseraceae
Fimbristylis dichotoma	Cyperaceae
Hibiscus scotellii	Malvaceae
Hildegardia barteri	Sterculiaceae
Hymenodictyon floribundum	Rubiaceae
Indigofera deightonii	Leguminosae
Lindernia exilis	Scrophulariaceae
Mariscus dubius	Cyperaceae
Microchloa indica	Poaceae
Mimusops kummel	Sapotaceae
Nemum spadiceum	Cyperaceae
Neurotheca loeselioides	Gentianaceae
Ophioglossum costatum	Ophioglossaceae
Panicum griffonii	Poaceae
P. tenellum	Poaceae
Pellaea doniana	Pteridaceae
Rotala tenella	Lythraceae
Solenostemon latifolius	Lamiaceae
Sporobolus festivus	Poaceae
Tripogon minimus	Poaceae
Utricularia pubescens	Lentibulariaceae
U. subulata	Lentibulariaceae
Virectaria multiflora	Rubiaceae
Xyris straminea	Xyridaceae

Adjanohoun (1964) shows a considerable degree of similarity in regard to floristic composition. Particularly close affinities exist between ephemeral flush vegetation on inselbergs, and characteristic communities of ferricretes (i.e., association à *Dopatrium senegalense* et *Marsilea polycarpa*, Adjanohoun 1964). Remarkable for West African inselbergs is the almost complete absence of families like Asteraceae, Crassulaceae, and Gesneriaceae, which form important constituents of the inselberg flora of East Africa and Madagascar.

On the generic level *Cyperus* (18 spp.) is the most important genus on inselbergs in the Côte d'Ivoire, followed by *Indigofera* (17 spp.) and *Utri-*

Table 10.1.2. Floristic composition of the vegetation of the Côte d'Ivoire in comparison with the Ivorian inselberg vegetation. Given are the ten most speciose families (in order of importance)

Flora of Côte d'Ivoire	Species no.	Flora of Ivorian inselbergs	Species no.
Leguminosae	387	Leguminosae	84
Rubiaceae	306	Poaceae	75
Poaceae	248	Cyperaceae	68
Orchidaceae	222	Scrophulariaceae	24
Cyperaceae	178	Rubiaceae	23
Euphorbiaceae	150	Orchidaceae	19
Asteraceae	110	Lentibulariaceae	13
Acanthaceae	85	Commelinaceae	12
Apocynaceae	74	Euphorbiaceae	10
Annonaceae	71	Malvaceae	10

cularia (12 spp.). These three genera are completely distinguished from each other in regard to their ecological preferences. Almost all species of *Cyperus* are restricted to rock pools; the genus *Indigofera* shows strong affinities to rock crevices and fissures, whereas nearly all *Utricularia* species occur in ephemeral flush vegetation. The genus *Dissotis* (Melastomataceae) likewise occurs with a considerable number of species (more than a dozen) on West African rock outcrops. The species are either herbs (occasionally perennating with tubers, e.g., *D. antennina*) or shrubs (e.g., *D. leonensis*, *D. theifolia*), which frequently possess small distributional areas.

Similarly, as on the family level, the absence from West African inselbergs of certain genera which are widespread in East Africa is notable, for instance *Begonia*, *Caralluma* (except some inselbergs in Mauretania), and *Streptocarpus*.

10.1.6.2 Ecological Characteristics

Approximately 10% of the species recorded on Ivorian inselbergs are specialized to this ecosystem (i.e., found predominantly on inselbergs, like *Afrotrilepis pilosa*, *Hymenodictyon floribundum*, and *Tephrosia mossiensis*) but in contrast to other regions (e.g., southern Venezuela) only very few species are strictly limited to inselbergs. The largest proportion of the species (generalists) found occur in a broad range of habitats outside inselbergs. Floristically, the affinities of inselberg specialists are closest to those of other azonal sites. Of major importance are all types of rock outcrops (incl. ferricretes bearing a *Sporobolus pectinellus-Cyanotis lanata* community, Adjanohoun 1964), poor soils in grassland (in particular white sand savannas), and seasonally wet stands around ponds. There are considerable differences in the percentage of inselberg specialists between individual habitat types. Both *Afrotrilepis pilosa* mats (>50 %) and ephemeral flush vegetation (>40 %) comprise high percentages of inselberg specialists in contrast to all other habitats, with generalists being clearly dominant. The following species found in *Afrotrilepis pilosa* mats are almost completely restricted to inselbergs: *Asplenium stuhlmannii*, *Hibiscus scotellii*, the Gentianaceae *Oreonesion testui*, and the orchid *Polystachya microbambusa*. The latter species occurs only in mats of *Afrotrilepis pilosa*. Several carnivorous species, like *Genlisea stapfii* and *Utricularia juncea*, are restricted to ephemeral flush communities.

Among the generalists annuals, which are widespread in savannas or show weedy tendencies, predominate. The latter group contains mainly pantropical elements including neophytes (e.g., *Axonopus compressus* and *Panicum laxum*, both originating from South America). Inselbergs situated in the largely undisturbed rainforest of the Taï National Park (southwestern Côte d'Ivoire) have been successfully colonized by weeds. Obviously, this fact reflects the large dispersal potential of weeds, allowing them to reach patchily distributed sites and their ability to cope with unpredictable environmental fluctuations (e.g., droughts). Under the influence of disturbance, certain plant species indigenous to West African inselbergs also show a tendency towards weediness. *Cyanotis lanata* and *Microchloa indica*, for example, have been recorded as indicators of long-term overgrazing in the Sahel (Seghieri et al. 1997). Moreover, both species are widespread roadside weeds in the Côte d'Ivoire (own observ.).

Most plant species on Ivorian rock outcrops are anemochorous (e.g., *Afrotrilepis pilosa*, ferns); both epi- and endozoochory are less frequent. For a considerable number of species, the mode of dispersal is not known. However, it is suspected that exozoochorous transport by birds plays an important role, where seeds lying on the ground may stick to the feet of visiting birds and are thus dispersed between inselbergs.

The climatic conditions on Ivorian inselbergs have been measured in some detail (own unpubl. data). Investigations in the Comoé National Park (northeastern Côte d'Ivoire) have demonstrated that daytime temperatures on lichen-covered rock regularly approach 60 °C, whilst relative air humidity drops below 20 %. Moreover, there is a high degree of spatiotemporal year-to-year fluctuation in precipitation on a small local scale. There are indications that these environmental fluctuations influence the vegetation dynamics on rock outcrops and are important prerequisites for the coexistence of species-rich plant communities on inselbergs. For example, it could be demonstrated that stochastic climatic perturbations can cause local extinctions of 20 % of the species in certain plant communities on inselbergs in the Comoé National Park and may be an important factor for maintaining species-rich nonequilibrium plant assemblages. Local extinctions on inselbergs due to prolonged drought result in a temporal species turnover which was documented in the Côte d'Ivoire since 1990 (own unpubl. data; Krieger 1997). For a more detailed discussion of the relationships between species richness and temporal species turnover it is referred to Porembski et al., Chap. 12, this Volume.

Small rock outcrops (i.e., below 100 m^2) in the savanna zone are prone to the vagaries of unforeseeable climatic disturbances which renders them hostile for many perennials. While annuals like *Cyanotis lanata* regularly occur on small rock outcrops, chamaephytes such as *Afrotrilepis pilosa* or the trees *Hymenodictyon floribundum* and *Hildegardia barteri* do not occur below a certain minimum size. In contrast to this, small rock outcrops surrounded by rainforest contain a larger amount of perennials (e.g., *Dissotis rotundifolia, Mariscus dubius, Clappertonia minor*) due to lessening local climatic effects.

10.1.7 Phytogeography and Endemism

The West African inselberg vegetation is relatively uniform concerning its characteristic elements (i.e., species restricted to this ecosystem). With regard to the vegetation of inselbergs, West Africa is therefore probably the largest floristically homogeneous region in the tropics. Although certain local peculiarities exist, the inselberg vegetation exhibits remarkably little floristic variation over this vast region, when compared to other parts of the tropics where small-scale floristic variations are far more noticeable. Consequently, many characteristic vascular plants are widely distributed, like *Afrotrilepis pilosa*, which is found from Senegal south to Gabon, or occur on inselbergs throughout tropical Africa (e.g., *Cyanotis lanata, Hymenodictyon floribundum*). Several species belonging to the category of the widely distributed taxa are characterized by a large degree of morphological variability between populations from geographical distinct localities. Certain genera (e.g., *Aeollanthus, Dissotis, Solenostemon, Virectaria*) are represented on West African rock outcrops by a number of closely related vicariant species. In tropical Africa the genus *Dissotis* (Melastomataceae) comprises a large number (ca. 120) of mostly shrubby species which are often associated with azonal localities. Interestingly, the closely related neotropical genus *Tibouchina* (i.e., also belonging to the tribe Melastomeae, Renner 1993) includes shrubby species which form typical elements of the Brazilian rock outcrop vegetation. Distributional areas far beyond tropical Africa are typical of many rock outcrop cryptogams (i.e., in particular cyanobacteria) even comprising cosmopolitic species.

Floristic affinities of the West African inselberg vegetation are mainly Guineo-Congolian or Sudano-Zambezian. Inselbergs situated in rainforest are dominated by Guineo-Congolian elements, whereas in savanna regions the latter phytogeographical element is better represented. The Guineo-Congolian elements are frequently perennials, e.g., in the Côte d'Ivoire the Tiliaceae *Clappertonia minor*, in Cameroon/Gabon *Clappertonia polyandra*, and throughout West Africa the orchid *Habenaria procera* with general affinities to azonal sites, whereas therophytes are relatively rare. In general, the number of species (e.g., *Dissotis rotundifolia*, *Impatiens* spp., *Oreonesion testui*) which are restricted to rainforest inselbergs is low. Occasionally, orchids, otherwise mainly epiphytic (e.g., *Bulbophyllum* spp. and other angraecoid genera) and ferns (e.g., *Platycerium angolense*) can be observed as colonizers of rocky slopes on rainforest inselbergs. For reasons that are not entirely clear, there is a significant underrepresentation of certain ecological groups (e.g., carnivorous plants, succulents) and taxa (e.g., *Indigofera*) on rainforest inselbergs. The rare occurrence of stem succulent euphorbias on rainforest rock outcrops today might be a testimony to Pleistocene climatic fluctuations which caused a contraction and fragmentation of forests in favor of savanna formations. This certainly rendered easier access to inselbergs for succulent euphorbias. However, it is not only the isolating barrier effect of the rainforest that accounts for the underrepresentation of particular taxa and ecological groups. Moreover, due to the geomorphology (i.e., the mostly fairly steep slopes) of rainforest inselbergs, appropriate habitats for particular groups, like rock pools are simply not available.

There is a sharp floristic boundary between rainforest inselbergs with Guineo-Congolian elements dominating and savanna zone inselbergs with a preponderance of Sudano-Zambezian species which are sometimes separated by less than 100 km. Most Sudano-Zambezian species are not inselberg specialists but occur in a broad range of ecosystems. Principally on inselbergs in savanna areas, the percentage of widespread species (i.e., paleo- and pantropics), like *Burmannia madagascariensis*, *Cyperus compressus*, *Fimbristylis dichotoma*, and *Microchloa indica*, increases. Throughout West Africa savanna zone inselbergs are floristically more similar to each other than their counterparts in the rainforest zone. Consequently, there are only minor floristic differences concerning both cryptogams and vascular plants, for instance between inselbergs in northern Côte d'Ivoire and those situated more than 600 km to the east in the Dahomey Gap in Bénin (Frahm and Porembski 1997 and own unpubl. data). However, when comparing rainforest inselbergs in close neighborhood, a large amount of floristic divergence becomes apparent (Porembski et al. 1996b). Many characteristic elements of the West African savanna inselberg vegetation extend far beyond the above given limits of this area. This is particularly well expressed in the ephemeral flush communities, which occur almost identically in species composition in the Bamingui-Bangoran area (Central African Republic) ca. 400 km to the east of Cameroon.

On species level, the phytogeographical relationships with the Zambezian Region and with Madagascar are not very pronounced. The presence of *Coleochloa abyssinica* and *Xerophyta schnitzleinia* in isolated localities in northern Nigeria far away from their main distibutional areas in eastern Africa is, however, remarkable. For certain taxa and life-forms, West African inselbergs form only outliers of their general distribution pattern. Numerous genera show a marked decrease in species number on rock outcrops from East to West Africa (e.g., *Aloe*, *Brachystelma*, *Craterostigma*, *Encephalartos*, *Xerophyta*). Similarly, succulents, xerophytes, and poikilohydric vascular plants (incl. ferns) are poorly represented on West African rock outcrops in comparison (Kornas 1983; Newton 1989; Barthlott and Porembski 1996).

Certain species typical for rock outcrops have disjunct areas within West Africa. For instance, *Microdracoides squamosus* and *Asplenium jaundeense* occur on rock outcrops in Guinea/Sierra Leone and to the east of the Dahomey Gap in the border region between Nigeria and Cameroon. This is possibly a relictual distribution pattern, caused by climatic perturbations during the Pleistocene. The number of American-African disjuncts is low. Remarkably, most of them are to be found within ephemeral flush communities. *Utricularia juncea* is the only species of this community showing an American-West African disjunction and is, apart from its major center of distribution in the New World, known from just two inselbergs in the Côte d'Ivoire (Dörrstock et al. 1996b), which may indicate a rather recent arrival. The Gentianaceae *Neurotheca loeselioides* and the genus *Genlisea* provide further examples of American-African disjuncts which, however, are not restricted to West Africa. In this relation one should also mention the only Old World bromeliad, *Pitcairnia feliciana*, which occurs on sandstone outcrops in the Fouta Djalon highlands of Guinea. Comparative studies (Porembski and Barthlott 1999) have shown that the closest relatives of *Pitcairnia feliciana* probably occur on rock outcrops situated in the Brazilian Atlantic rainforest.

Apart from very few exceptions, most species found on West African inselbergs can be found in other sites, too. Species that are more or less restricted to inselbergs are considered to be characteristic for this ecosystem. Amongst this group the number of endemics (i.e., species restricted to West African inselbergs) is comparatively low. Prominent examples are *Afrotrilepis pilosa*, *Afrotrilepis jaegeri*, *Microdracoides squamosus*, *Djaloniella ypsilostyla*, *Genlisea barthlottii*, *Polystachya microbambusa*, *Lindernia yaundensis*, and several species of the genus *Dissotis* (see Table 10.1.3, see also Table 10.1.4 for inselbergs studied in the Côte d'Ivoire).

Climatic fluctuations during the Pleistocene possibly led to higher extinction rates, leading to both lower numbers of species and endemics in West Africa in comparison with other tropical regions. Higher extinction rates were probably a consequence of the contraction of the vegetation belts to small isolated remnants, since a southward shift towards the equator was restricted by the Atlantic coast (Newton 1989) in the Upper Guinea region.

Table 10.1.3. Endemic vascular plant species of West African inselbergs and their habitat preference

Name	Family	Distribution	Habitat[a]
Afrotrilepis pilosa	Cyperaceae	Senegal – Gabon	M
A. jaegeri	Cyperaceae	Sierra Leone	M
Dissotis glauca	Melastomataceae	Upper Guinea	M, C, SD
Djaloniella ypsilostyla	Gentianaceae	Upper Guinea	EFV, SD
Dolichos tonkouiensis	Leguminosae	Upper Guinea	C
Dopatrium longidens	Scrophulariaceae	Upper Guinea	RP
Euphorbia letestui	Euphorbiaceae	Cameroon, Gabon Equatorial Guinea	EC
Genlisea barthlottii	Lentibulariaceae	Guinea	EFV
Hibiscus scotellii	Malvaceae	Upper Guinea	M, C
Lindernia yaundensis	Scrophulariaceae	Cameroon	SD
Microdracoides squamosus	Cyperaceae	Guinea, Sierra Leone Nigeria, Cameroon	M, EFV
Ophioglossum thomasii	Ophioglossaceae	Upper Guinea	EFV, M
Oreonesion testui	Gentianaceae	Cameroon, Gabon Equatorial Guinea	M
Polystachya microbambusa	Orchidaceae	Upper Guinea	M
Raphionacme vignei	Asclepiadaceae	Upper Guinea	SD, M
Solenostemon graniticola	Lamiaceae	Upper Guinea	M, C
Tephrosia mossiensis	Leguminosae	Upper Guinea	C

[a] M, Monocotyledonous mats; C, crevices; SD, shallow depressions; EFV, ephemeral flush vegetation; RP, rock pools; EC, ecotone

It is likely that as a result of the ongoing destruction of West African rainforests a number of weeds have become established on inselbergs situated in formerly closed forest. These mostly neotropical neophytes (e.g., *Ananas comosus*, *Axonopus compressus*, *Panicum laxum*) may have reached rock outcrops by using roadsides and forest clearings as transport ways or, in the case of the latter, as stepping stones (Porembski et al. 1996b).

Table 10.1.4. Selected inselbergs investigated in the Côte d'Ivoire by the author. Detailed species lists for each inselberg are available from the author on request

No.	Name	Geographic location	Absolute height/area
IB 1	Mt. Niangbo	8°49'N/5°11'W	370 m/7 km^2
IB 2	Mt. Korhogo	9°27'N/5° 38'W	140 m/0.8 km^2
IB 3	Near Boundiali	9°32'N/6°29'W	175 m/1.2 km^2
IB 4	Near Boundiali	9°35'N/6°44'W	140 m/0.5 km^2
IB 5	Near Boundiali	9°32'N/6°30'W	260 m/1.5 km^2
IB 6	Near Séguéla	7°55'N/6°31'W	140 m/0.8 km^2
IB 7	Near Séguéla	7°55'N/6°31'W	120 m/0.8 km^2
IB 8	Near Séguéla	8°05'N/6°37'W	60 m/0.4 km^2
IB 9	Near Séguéla	8°05'N/6°25'W	70 m/0.5 km^2
IB 10	Near Man	7°30'N/7°35'W	310 m/1 km^2
IB 11	Near Danané	7°15'N/8°10'W	150 m/0.4 km^2
IB 12	Near Duékoué	6°45'N/7°21'W	60 m/0.3 km^2
IB 13	Near Duékoué	6°45'N/7°22'W	80 m/0.3 km^2
IB 14	Dent de Man	7°25'N/7°35'W	340 m/1 km^2
IB 15	Mt. Niénokoué	5°25'N/7°10'W	210 m/1.2 km^2
IB 16	Rocher d'Issia	6°29'N/6°35'W	70 m/0.4 km^2
IB 17	Near Dabakala	8°25'N/4°32'W	60 m/0.3 km^2
IB 18	Gbonkonou (near Nassian)	8°42'N/3°35'W	50 m/0.3 km^2
IB 19	Near Man	7°25'N/7°35'W	60 m/0.2 km^2
IB 20	Near Mt. Niénokoué	5°27'N/7°12'W	70 m/0.3 km^2
IB 21	Near Duékoué	6°45'N/7°22'W	60 m/0.2 km^2
IB 22	Sénéma	7°45'N/6°40'W	250 m/0.7 km^2
IB 23	Near Mankono	8°05'N/6°10'W	90 m/0.3 km^2
IB 24	Near Bouaké	7°45'N/5°10'W	70 m/0.3 km^2
IB 25	Near Boundiali	9°40'N/6°30'W	160 m/0.4 km^2
IB 26	Near Abengourou	6°40'N/3°40'W	50 m/0.2 km^2
IB 27	Near Abengourou	6°40'N/3°45'W	50 m/0.2 km^2
IB 28	Near Abengourou	6°40'N/3°45'W	40 m/0.2 km^2
IB 29	Near Abengourou	6°40'N/3°45'W	80 m/0.3 km^2
IB 30	Mt. Mafa	5°50'N/4°05'W	120 m/0.4 km^2
IB 31	Near Dabakala	8°30'N/4°35'W	30 m/0.2 km^2
IB 32	Near Dabakala	8°35'N/4°40'W	140 m/0.7 km^2
IB 33	Near Dabakala	8°35'N/4°40'W	250 m/2 km^2
IB 34	Near Toulépleu	6°30'N/8°10'W	70 m/0.2 km^2
IB 35	Comoé NP, Iringou/Bamago	8°55'N/3°38'W	40 m/0.3 km^2
IB 36	Rocher de Brafouédi	5°35'N/4°40'W	60 m/0.4 km^2

Up to now it is not clear whether these invasions may have severe consequences for the indigenous vegetation of West African rock outcrops. On a local scale, however, there are already several sites where exotic species are becoming dominant on account of the native flora. A striking example is represented by *Ananas comosus* which has replaced *Afrotrilepis pilosa* on Mt. Brafouédi, an inselberg which is situated in the rainforest zone of the Côte d'Ivoire. However, there also seems to exist an indigenous component among the West African weed flora which could have been originally restricted to small azonal habitat patches, such as rock outcrops. Most prominent examples are *Cyanotis lanata* and *Dissotis rotundifolia*, which today are widespread along roadsides in the rainforest region of the Côte d'Ivoire.

Acknowledgements. Research in West Africa was supported by the Deutsche Forschungsgemeinschaft (grant no. Ba 605/4-3 to W. Barthlott). Investigations in Equatorial Guinea were supported by CUREF (Conservación y Utilización Racional de los Ecosistemas Forestales de Guinea Ecuatorial, Bata). I am grateful to many colleagues for valuable comments, particularly N. Biedinger (Rostock), W. Barthlott, R. Seine, A. Krieger, and S. Dörrstock (all Bonn). L. Aké Assi (Abidjan), N. Nguema (Bata), M. Elad (Bata), and B. Sinsin (Cotonou) kindly cooperated in plant identification and fieldwork. I am much indebted to K. E. Linsenmair (Würzburg) for support during fieldwork in the Comoé National Park and to F. Stenmanns, J. E. Garcia (both Bata) and J. Lejoly (Brussels), who enabled studies in Equatorial Guinea. The authorities of Côte d'Ivoire, Equatorial Guinea, Guinea, and Bénin are thanked for the permission to conduct research.

References

Adjanohoun E (1964) Végétation des savanes et des rochers découverts en Côte d´Ivoire centrale. Mém ORSTOM 7:1–178

Aké Assi L (1984) Flore de la Côte d'Ivoire: étude descriptive et biogéographique, avec quelques notes ethnobotaniques. Thèse de doctorat, Université d'Abidjan

Anhuf D (1994) Zeitlicher Vegetations- und Klimawandel in Côte d'Ivoire. In: Lauer W (ed) Veränderungen der Vegetationsbedeckung in Côte d'Ivoire. Erdwissenschaftliche Forschung, Bd. XXX. Franz Steiner, Stuttgart, pp 7–299

Ayo-Owoseye J, Sanford WW (1972) An ecological study of *Vellozia schnitzleinia*, a drought-enduring plant of northern Nigeria. J Ecol 60:807–817

Barthlott W, Porembski S (1996) Biodiversity of arid islands in tropical Africa: the vegetation of inselbergs. In: van der Maesen LJG, van der Burgt XM, van Medenbach de Rooy JM (eds) The biodiversity of African plants. Proc XIVth AETFAT Congress. Kluwer, Dordrecht, pp 49–57

Barthlott W, Lauer W, Placke A (1996) Global distribution of species diversity in vascular plants: towards a world map of phytodiversity. Erdkunde 50:317–327

Bonardi D (1966) Contribution à l'étude botanique des inselbergs de Côte d'Ivoire forestière. Diplome d´études supérieures de sciences biologiques, Université d'Abidjan

Bornhardt W (1900) Zur Oberflächengestaltung und Geologie Deutsch-Ostafrikas. Reimer, Berlin
Büdel B, Becker U, Porembski S, Barthlott W (1997) Cyanobacteria and cyanobacterial lichens from inselbergs in the Ivory Coast, Africa. Bot Acta 110:458–465
Dörrstock S, Porembski S, Barthlott W (1996a) Ephemeral flush vegetation on inselbergs in the Ivory Coast (West Africa). Candollea 51:407–419
Dörrstock S, Seine R, Porembski S, Barthlott W (1996b) First record of *Utricularia juncea* (Lentibulariaceae) for tropical Africa. Kew Bull 51:579–583
Eldin M (1971) Le climat. In: Le milieu naturel de la Côte d'Ivoire. Mém ORSTOM 50:73–108
Falconer JD (1911) The geology and geography of northern Nigeria. Macmillan, London
Fischer E (1992) Systematik der afrikanischen Linderniae (Scrophulariaceae). Akad Wiss Lit, Mainz. Franz Steiner, Stuttgart
Frahm JP, Porembski S (1994) Moose von Inselbergen aus Westafrika. Trop Bryol 9:59–68
Frahm JP, Porembski S (1997) Moose von Inselbergen in Benin. Trop Bryol 14:3–9
Griffiths JF (1972) Climates of Africa. In: Landsberg HE (ed) World survey of climatology, vol 10. Elsevier, Amsterdam
Hambler DJ (1961) A poikilohydrous, poikilochlorophyllous angiosperm from Africa. Nature 191:1415–1416
Hambler DJ (1964) The vegetation of granitic outcrops in Western Nigeria. J Ecol 52: 573–594
Hutchinson J, Dalziel JM (1954–1972) Flora of West Tropical Africa. Revised edn. Crown Agents, London
Keay RJW, Hepper FN (1954–72) Flora of West Tropical Africa. Revised edn. Crown Agents, London
Isichei AO, Longe PA (1984) Seasonal succession in a small isolated rock dome plant community in western Nigeria. Oikos 43:17–22
Kershaw KA (1968) A survey of the vegetation in Zaria Province, N. Nigeria. Vegetatio 15:244–268
Knapp R (1966) Höhere Vegetations-Einheiten von West Afrika unter besonderer Berücksichtigung von Nigeria und Kamerun. Geobotanische Mitteilungen 34:1–31
Kornas J (1983) Pteridophyta collected in Northern Nigeria and Northern Cameroon. Acta Soc Bot Pol 52:321–335
Krieger A (1997) Vegetationsdynamik und Species Turnover in saisonalen Felsgewässern auf Inselbergen der Côte d'Ivoire (Westafrika). MSc Thesis, University of Bonn, Germany
Krieger A, Porembski S, Barthlott W (in press) Vegetation of seasonal rock pools on inselbergs in the savanna zone of the Ivory Coast (West Africa). Flora
Leneuf N (1959) L'altération des granites calco-alcalins et des granodiorites en Côte d'Ivoire forestière et des sols qui en sont dérivés. Thèse de doctorat, Paris
Létouzey R (1968) Étude phytogéographique du Cameroun. Lechevalier, Paris
Létouzey R (1985) Notice de la carte phytogéographique du Cameroon au 1 : 500 000. Institut de la Recherche Agronomique, Yaoundé, Institut de la Carte Internationale de la Végétation, Toulouse
Maley J (1989) Late Quaternary climatic changes in the African rain forest: the question of forest refuges and the major role of sea-surface temperature variations. In: Leinen M, Sarnthein M (eds) Paleoclimatology and paleometeorology: modern and past patterns of global atmospheric transport. NATO Adv Sc Inst, Ser C, Math Phys Sci. Kluwer, Dordrecht, pp 585–616
Miége J (1955) Savanes et forêts claires de la Côte d'Ivoire. Études Éburnéennes. IFAN 4:62–83

Mildbraed J (1922) Wissenschaftliche Ergebnisse der zweiten deutschen Zentral-Afrika-Expedition 1910–1911. Band II: Botanik. Klinkhardt & Biermann, Leipzig

Newton LE (1989) West Tropical Africa as an outlier in succulent plant distribution. Excelsa 14:11–13

Newton LE (1992) An annotated and illustrated checklist of the succulent euphorbias of West Tropical Africa. Euphorbia Journal 8:113–123

Perold SM (1995) A survey of the Ricciaceae of tropical Africa. Fragm Florist Geobot 40:53–91

Porembski S (1996) Notes on the vegetation of inselbergs in Malawi. Flora 191:1–8

Porembski S, Barthlott W (1992) Struktur und Diversität der Vegetation westafrikanischer Inselberge. Geobot Kolloq 8:69–80

Porembski S, Barthlott W (1993) Ökogeographische Differenzierung und Diversität der Vegetation von Inselbergen in der Elfenbeinküste. In: Barthlott W, Naumann CM, Schmidt-Loske K, Schuchmann K-L (eds) Animal-plant interactions in tropical environments. Results Annu Meet of the German Society for Tropical Ecology, Bonn 1992, pp 149–158

Porembski S, Barthlott W (1997) Seasonal dynamics of plant diversity on inselbergs in the Ivory Coast (West Africa). Bot Acta 110:466–472

Porembski S, Barthlott W (1999) *Pitcairnia feliciana*: the only indigenous African bromeliad. Harvard Papers in Botany 4:175–184

Porembski S, Brown G (1995) The vegetation of inselbergs in the Comoé National Park (Ivory Coast). Candollea 50:351–365

Porembski S, Barthlott W, Dörrstock S, Biedinger N (1994) Vegetation of rock outcrops in Guinea: granite inselbergs, sandstone table mountains and ferricretes – remarks on species numbers and endemism. Flora 189:315–326

Porembski S, Brown G, Barthlott W (1995) An inverted latitudinal gradient of plant diversity in shallow depressions on Ivorian inselbergs. Vegetatio 117:151–163

Porembski S, Brown G, Barthlott W (1996a) A species-poor tropical sedge community: *Afrotrilepis pilosa* mats on inselbergs in West Africa. Nord J Bot 16:239–245

Porembski S, Szarzynski J, Mund J-P, Barthlott W (1996b) Biodiversity and vegetation of small-sized inselbergs in a West African rain forest (Taï, Ivory Coast). J Biogeogr 23:47–55

Quézel P (1965) La végétation du Sahara, du Tchad à la Mauritanie. Gustav Fischer, Stuttgart

Raynal A (1966) Une scrophulariacée camerounaise peu connue: *Ilysanthes yaundensis* S. Moore. Adansonia NS 6:281–287

Reitsma JM, Louis AM, Floret JJ (1992) Flore et végétation des inselbergs et dalles rocheuses: première étude au Gabon. Bull Mus Nat Hist Nat B Adansonia 14:73–97

Renner SS (1993) Phylogeny and classification of the Melastomataceae and Memecylaceae. Nord J Bot 13:519–540

Richards PW (1957) Ecological notes on West African vegetation I. The plant communities of the Idanre Hills, Nigeria. J Ecol 45:563–577

Richards PW (1973) Africa, the "odd man out". In: Meggers BJ, Ayensu ES, Duckworth WD (eds) Tropical forest ecosystems in Africa and South America: a comparative review. Smithsonian Institute, Washington, pp 21–26

Schnell R (1952) Contribution à une étude phytogéographique de l'Afrique occidentale: les groupements et les unités géobotaniques de la Région Guinéenne. Mém IFAN 18:45–234

Seghieri J, Galle S, Rajot JL, Ehrmann M (1997) Relationships between soil moisture and growth of herbaceous plants in a natural vegetation mosaic in Niger. J Arid Environ 36:87–102

Sillans R (1958) Les savanes de l'Afrique centrale. Lechavalier, Paris
Villiers JF (1981) Formations climaciques et relictuelles d´un inselberg inclus dans la forêt dense camerounaise.Thèse de doctorat, Paris
Walter H, Lieth H (1967) Klimadiagramm - Weltatlas. Gustav Fischer, Jena
White F (1983) The vegetation of Africa. A descriptive memoir to accompany the UNESCO/AETFAT/UNSO vegetation map of Africa. Nat Resour Res 20. UNESCO, Paris
White F (1993) The AETFAT chorological classification of Africa: history, methods and applications. Bull Jard Bot Nat Belg 62:225–281
Wilson AF (1968) Geological report on granites in West Africa. In: Proceedings of the symposium on the granites of West Africa. UNESCO, Paris, pp 123–147
Zehnder A (1953) Beitrag zur Kenntnis von Mikroklima und Algenvegetation des nackten Gesteins in den Tropen. Ber Schweiz Bot Ges 63:5–26

10.2 East and Southeast Africa

R. Seine and U. Becker

10.2.1 Geography and Geology

East and southeast Africa are here considered as the territories covered by the flora of tropical East Africa and the Flora Zambesiaca (i.e., Botswana, Kenya, Malawi, Mozambique, Tanzania, Zambia, and Zimbabwe). Inselbergs of the Imatong Mountains (southern Sudan), eastern Congo, Rwanda, and Somalia are additionally included because of their floristic affinity.

The African shield, which underlies most of East and Southeast Africa, has been exposed to erosion over the past 450 million years. Through uplifts, downwarps, and punctuations, the relief has been formed. The highest East African plateaux are remnants of the Jurassic surface of Gondwana. The lower plateaux have been shaped in later pediplanation cycles (King 1951). Inselbergs are characteristic African landscape elements that have been generated by weathering, scarp retreat, and incision (Twidale 1982; Petters 1991).

In East and southeast Africa granitic and gneissic inselbergs are scattered in an extensive stretch from South Africa to Somalia and Sudan (Fig. 10.2.1). The indicated area encompasses the occurrence of inselbergs. However, punctuations of other minerals and large-scale deposits partly cover the basement complex and inselbergs are, therefore, not present everywhere within that area.

10.2.2 Climate

East and southeast Africa comprise various types of climate ranging from desert and steppe climate (BW, BS) to equatorial winter dry and monsoon climates (Aw, Am) and warm temperate climate (Cw) of highlands (Griffiths 1970). Mean annual precipitation lies between 155 mm in Lodwar,

Ecological Studies, Vol. 146
S. Porembski and W. Barthlott (eds.) Inselbergs

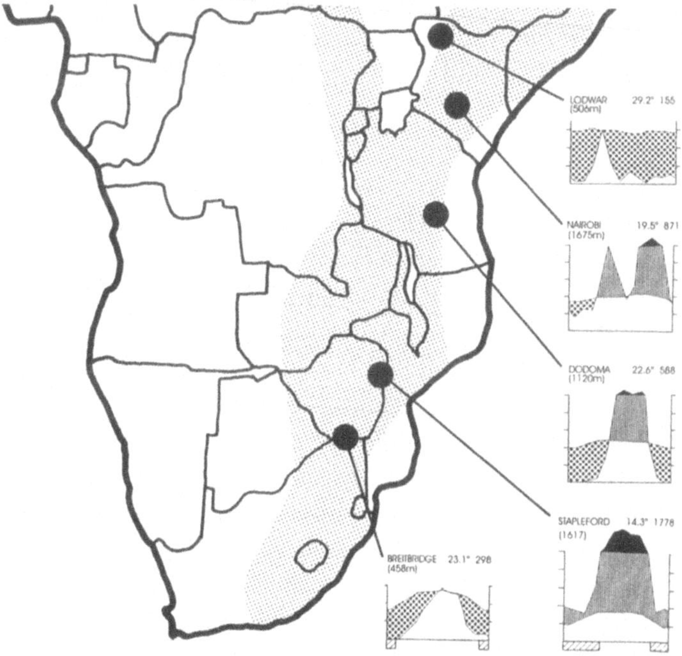

Fig. 10.2.1. Climate and occurrence of inselbergs in east and southeast Africa. *Shaded area* outline of inselberg distribution. (Climate diagrams after Walther and Lieth 1960)

Kenya, and 1778 mm in Stapleford, Zimbabwe (Walther and Lieth 1960). Orogenic precipitation on windward mountain slopes may largely exceed these figures; Plowes and Drummond (1990) report more than 5000 mm for Pungwe Gorge on the east side of Mt. Inyangani, Zimbabwe. One or two rainy seasons occur, lasting for 4 to 9 months, again modulated by exposition and elevation. Rainfall reliability decreases to the north and south of the equator (Torrance 1972). In adjacent central Africa moist tropical climate gains importance. In high montane regions of east and southeast Africa and in the extreme south of our area annual frosts are recorded (Knapp 1973).

10.2.3 Vegetation

East Africa comprises a number of vegetation types ranging from rainforests to dry savannas and afroalpine paramos. Dry woodland and savanna are the most widespread types of vegetation (Knapp 1973; White

1983). Phytogeographically, the territory here considered belongs to the Somali-Masai, the Zambezian, and the Afromontane archipelago-like regional center of endemism, and the Zanzibar-Inhambane regional mosaic (White 1983).

Low-Altitude Rainforest. A continuous stretch of low-altitude rainforests once covered the coast of the Indian Ocean (Knapp 1973) from Mozambique to Kenya. These forests have been reduced to small isolated patches. Outliers of this type of rainforest are found on the east-facing slopes of mountains (e.g., Usambara, Mt. Selinda, Chirinda, Pungwe Gorge, Nyika Plateau), where orographic rains amount to nearly 2000 mm per year.

Afromontane Forest. On the high mountains afromontane forests are found at altitudes between 1500 m and 2800 m. They have been reduced by deforestation. In Kenya, trees in afromontane forests reach a height of 40 m while they usually do not exceed 25 m in Malawi, Zambia, and Zimbabwe (Knapp 1973). Afromontane forests are either speciose stands or single species-dominated (White 1983). Speciose forests are found on the moister slopes of mountains.

Woodland. Dry deciduous woodland dominates the vegetation over vast areas. Four types of woodland are most prominent: *Acacia* woodlands at low altitude on valley floors, miombo woodlands which are spread over most of the south-central African plateaus above 1000 m asl, *Baikiaea* woodlands over Kalahari sands, mopane woodlands in the southern lowlands of the region (Menaut 1983).

Acacia woodlands are more or less restricted to low altitudes with a hot and dry climate.

In miombo woodlands *Brachystegia* spp. and *Julbernardia globiflora* are the dominant tree species. The trees form a loose canopy approximately 15–20 m tall. The ground is usually covered by grasses of Andropogonae.

Baikiaea plurijuga is the dominant species of *Baikiaea* woodlands. In addition, several species also native to miombo woodland are found, e.g., *Guibourtia coleosperma*, *Pterocarpus angolensis*, and *Marquesia macroura*.

Mopane woodland is a single species dominated formation usually consisting of more than 90 % *Colophospermum mopane* (Knapp 1973). The grass cover is made up of Paniceae.

Savanna. Vegetation types with a dominant grass stratum and interspersed woody plants, i. e., savannas in a broad sense, are widespread throughout the area. In Southeast Africa patches of savanna are interspersed into the dominant woodland whereas they are the dominant vegetation type over much of East Africa. Savanna grasslands with very few trees are relatively rare while savanna parklands and shrub savannas cover whole landscapes (Cole 1986). Extensive savanna grasslands are found in the Zambian Kafue Plains and Busango basin. Main grass genera in these grasslands are *Setaria*, *Themeda*, *Loudetia*, and *Hyparrhenia*. Savanna parklands and shrub savannas occupy large areas in the drier parts of east and southeast Africa, notably at lower altitudes associated with the rift valley. In the southern part of the region, savanna parklands and shrub savannas *Acacia*, *Piliostigma*, and *Terminalia* are the most conspicuous tree genera. *Eragrostis* is the dominant genus of Poaceae, *Aristida* and *Anthephora* spp. are also regularly present. To the north, *Acacia* is most important in savanna parklands. Shrub savannas in the north of the area are often formed by speciose *Commiphora* thickets. The grass stratum consists of *Aristida*, *Eragrostis*, *Chloris*, *Cenchrus*, and *Enteropogon* spp.

Outcropping serpentine on the Great Dyke in Zimbabwe results in a poor edaphic grassland vegetation of mainly *Loudetia simplex* and *Andropogon gayanus* with endemic species of *Aloe*, *Euphorbia*, *Barleria* and other genera (Cole 1986). Savanna formations gain increasing importance due to overgrazing and anthropogenic fires. Savanna parklands remain after excessive burning of miombo woodlands. *Hyparrhenia cymbaria*, *Andropogon amplectus*, *Themeda triandra*, and various *Loudetia* species form the grass stratum of these savannas. Overgrazed shrub savannas are replaced by annual grass communities with scarce ground cover. These are dominated by *Aristida* spp. with *Brachiaria*, *Eleusine*, and *Tetrapogon* spp. as codominants (Knapp 1973).

10.2.4 Inselberg Vegetation

Studies on inselberg vegetation in east and southeast Africa have been conducted in Zimbabwe (Seine 1996; Seine et al. 2000), Malawi (Porembski 1996), Tanzania (Bjørnstad 1976), Kenya (Bally 1968; R. Seine, unpubl.), the Imatong Mountains (I. Friis and K.B.Vollesen, unpubl.), eastern Kongo (Taton 1949; Lisowski 1992; Porembski et al. 1997), Rwanda (Porembski et al. 1997), Somalia (W. Lobin, pers. comm.), and in various localities over the entire area by Knapp (1973).

Inselbergs in East and Southeast Africa harbor at least 900 species of vascular plants, calculated from existing studies. Most probably, the complete inventory will amount to some 1800 to 2000 species; however, a survey of some completeness is available only for Zimbabwe (Seine 1996), reporting 549 species of vascular plants. The species composition of habitats given below includes many species that are not found throughout the entire area covered in this report. However, species from all countries of the area have been included to cover the diversity of inselberg vegetation in east and southeast Africa.

10.2.4.1 Cryptogamic Vegetation of Rock Surface

The gray, brown or colorful (orange, yellow) appearance of inselbergs results from a dense cover of cryptogams. From the lowland to most of the higher ground of the plateaus, the cryptogamic vegetation of the rock surface is dominated by lichens with cyanobacteria as the phytobiont. They form a low carpet a few millimeters high. *Peltula* is the most important genus, with approximately 20 species in the area, usually accompanied by Lichinaceae. Lichens with chlorophytic algae as phytobiont are restricted to small areas along drainage channels, rock pools, or of elevated microrelief. In these areas associations of *Buellia*, *Caloplaca*, *Acarospora*, *Dermatiscum*, *Parmelia*, *Parmotrema*, *Xanthoparmelia*, *Protoparmelia*, and *Rinodina* prevail.

Inselbergs on the highest parts of the plateaus, in the Eastern Highlands of Zimbabwe, and occasionally on the summits of other inselbergs, are covered with chlorophytic lichens and a few cyanophytic lichens. The former are represented by *Usnea*, *Bulbotrix*, *Heterodermia*, *Dimelaeana*, *Diploschistes*, *Lecanora*, *Xanthoparmelia*, and the genera mentioned above. *Usnea* species, e.g., *U. pulvinata*, and *U. welwitschiana* sometimes form an almost continuous “lawn” some 5 cm high (Fig. 10.2.2). *Dimelaeana oreina* may reach nearly 100 % surface cover to the effect that the summits appear entirely green.

10.2.4.2 Cryptogamic Vegetation of Boulders

Boulders are present on many inselbergs. The cryptogamic vegetation of boulders is dominated by chlorophytic lichens which may densely cover the rock surface. Many species on boulders are also found on the main rock surface (*Dermatiscum thunbergii*, *Acarospora* sp., *Buellia* sp., *Caloplaca*, *Diploschistes* sp., *Parmotrema*, *Pertusaria*, and *Xanthoparmelia*). Species that are restricted to boulders are found in the genera *Pertusaria*, *Heterodermia*, and *Phyrrospora*. Only one species of *Peltula* (*P. euploca*), the dominating genus on the rock surface, is present on boulders.

Fig. 10.2.2. *Usnea* "lawn" with *Usnea pulvinata* and *U. welwitschiana* on Mt. Dombo, Zimbabwe. (Photograph R. Seine)

10.2.4.3 Vegetation of Drainage Channels

Drainage channels are present on most larger inselbergs (Fig. 10.2.3). They are absent on flat outcrops where water runs off over the entire surface. Drainage channels are almost exclusively colonized by *Peltula* species, sometimes attaining 90% cover. *Peltula cylindrica*, *P. lingulata*, and *P. obscurans* grow in shaded parts of the channels. The remainder of the channels harbors *Peltula impressa*, *P. umbilicata*, *P. marginata*, *P. placodizans*, *P. umbilicata*, and *P. zahlbruckneri*, i.e., species that are also found on the rock surface. Among these, *P. umbilicata* may attain almost 100% ground cover. The raised margins of drainage channels are inhabited by chlorophytic lichens like other areas of elevated microrelief.

Mosses colonize drainage channels occasionally and among them *Bryum arachnoideum* is relatively common. *Rhacocarpus purpurascens* was found growing in vertical drainage channels on Mt. Dombo, Zimbabwe. Vascular plants are rare in this habitat; however, they may establish in patches of moss.

Fig. 10.2.3. Drainage channel, almost completely covered with *Peltula obscurans*, *P. placodizans*, *P. umbilicata*, and *P. zahlbruckneri*. Near Banket, Zimbabwe. (Photograph R. Seine)

10.2.4.4 Wet Flush Vegetation

Wet flush is a rarely encountered habitat. Two localities have been recorded in the Rhodes Matopos National Parc (Zimbabwe) and on Marendé on the road from Bukavu to Hombo (E. Congo). Two species of *Utricularia*, *U. gibba* and *U. striatula*, were found, each in one location. They were growing in dense carpets of cyanobacteria and algae.

10.2.4.5 Vascular Lithophytic Vegetation (Lithophytes)

Lithophytes are occasionally found in shaded positions on inselbergs. Of the few species recorded, the majority are orchids such as *Tridactyle tridactylites*, and *Angraecum* sp., *Ceterach cordatum*, a poikilohydric fern from Aspleniaceae, was recorded as a lithophyte in Zimbabwe. Succulents such as *Cissus quadrangularis*, *Aloe lateritica*, *A. dawei*, and other *Aloe* spp. may also grow as lithophytes.

10.2.4.6 Vegetation of Horizontal and Vertical Crevices

Exfoliation of rock produces horizontal fissures on most inselbergs. Only one vascular plant species, the succulent *Anacampseros rhodesica*, colonizes this habitat perpetuously (Fig. 10.2.4). The perennial Portulacaceae is more or less restricted to this habitat. Although there is an abundance of horizontal crevices, *Anacampseros rhodesica* is recorded on only relatively few inselbergs in Zimbabwe.

Vertical crevices are present on almost every inselberg (Fig. 10.2.5). Poaceae, Fabaceae, and Cyperaceae dominate the vegetation of vertical crevices. Grass species are commonly also found in the surroundings of inselbergs. They are savanna and open woodland species such as *Andropogon gayanus*, *Heteropogon contortus*, *Loudetia simplex*, and *Melinis*. *Indigofera* (e.g., *I. emarginelloides*, *I. setiflora*, *I. tanganyikensis*, *I. viscidissima*), and *Tephrosia* (e.g., *T. decora*, *T. longipes*, *T. reptans*) are the characteristic Fabaceae found in crevices. The parasitic *Striga gesnerioides* (Scrophulariaceae) often grows on them. The Cyperaceae are present with *Bulbostylis* spp. (*B. burchelllii*, *B. coleotricha*) and a number of *Cyperus* spp. (e.g., *C. holostigma*, *C. rupestris*). Poikilohydric ferns from the genera

Fig. 10.2.4. Horizontal crevice colonized by *Anacampseros rhodesica*. Mambo Hills, Zimbabwe. (Photograph R. Seine)

Cheilanthes (*C. leachii*, *C. viridis*, *C. multifida*) and *Pellaea* (*P. calomelanos*, *P. boivinii*) are regularly found in horizontal crevices. Succulent species such as *Adenium obesum*, *Aeollanthus densiflorus*, *Caralluma* spp., *Sansevieria* spp., *Sarcostemma viminale*, and *Senecio coccineus* are often found among rock crevice vegetation. Other characteristic vascular plant species are: *Anthospermum whyteanum*, *Ceratotheca triloba*, *Commelina africana*, *Cyanotis lanata*, *Gomphocarpus tenuifolius*, *Hibiscus* spp., *Oldenlandia herbacea*, and *Stemodiopsis* spp. A small tree, *Hymenodictyon floribundum*, is found in this habitat throughout the area.

10.2.4.7 Vegetation of Clefts and Around Boulder Bases

Clefts are found only on larger inselbergs. Tree species are able to colonize clefts and are the characteristic elements of their vegetation. The herbaceous species of horizontal crevices are usually also present in clefts. A number of tree species (e.g., *Brachystegia glaucescens*, *Bridelia mollis*, *Commiphora marlothii*, *Euphorbia matabelensis*, *Ficus glumosa*, and *Hymenodictyon floribundum*) show a strong preference for inselbergs.

Fig. 10.2.5. Vertical crevice with *Adenium obesum*. Samburu National Park, Kenya. (Photograph R. Seine)

Species commonly found on inselbergs are *Acacia* spp., *Croton gratissimus*, *Diplorhynchus condylecarpon*, *Ficus* spp., *Julbernardia globiflora*, *Maytenus* spp., *Ochna* spp., and *Peltophorum africanum*. Many other tree species from surrounding woodland or savanna vegetation are occasionally found on inselbergs.

Only large inselbergs feature boulders with vegetation around the bases, usually near the summit of the inselberg. The boulders shade off sunlight from accumulations of shallow soil around their base. Plants tolerant of shade such as *Arthropteris orientalis*, *Ceterach cordatum*, and *Selaginella* spp. may establish.

10.2.4.8 Vegetation of Talus Slopes

On talus slopes, woody vegetation with dense undergrowth may be found. The species encountered in this habitat are normally found in forest or savanna vegetation. Important constituents of the tree stratum are *Brachystegia* spp. (*B. allenii*, *B. manga*, *B. spiciformis*, *B. tamarindoides*), *Diospyros* spp., *Faurea saligna*, *Ficus* spp., *Monotes* spp. (*M. africanus*, *M. glaber*, *M. katangensis*), *Pterocarpus angolensis*, *Sclerocarya birrea*, and *Uapaca* spp. (*U. kirkiana*, *U. pilosa*). Shrubs, lianas, and a few herbs form the undergrowth. Important shrubs are *Grewia* spp., *Maytenus* spp. (e.g., *M. heterophylla*, *M. puberula*), *Poulzolzia mixta*, and *Barleria crassa*. Lianas found on talus slopes are *Cryptolepis cryptolepioides*, *Gloriosa superba*, *Clerodendron* spp., *Ipomoea* spp., *Strophanthus* spp., and *Tragia insuavis*. The scanty herb layer is composed of, e.g., *Kaempferia* sp., *Pavonia patens*, and grasses, e.g., *Heteropogon contortus* and *Loudetia simplex*.

10.2.4.9 Vegetation of Seasonal Rock Pools and Ponds

Rock pools are present on many inselbergs. Lichen vegetation at the rims of pools is similar to that described for drainage channels. While some rock pools support a dense vascular plant vegetation, others are not colonized at all. In Southeast Africa *Lindernia monroi* and *Lindernia conferta* (Fig. 10.2.6) are most important in rock pool vegetation. The former is a perennial poikilohydric species that regreens quickly after the onset of rains. *Lindernia conferta* is an annual species that regrows from seed. The two *Lindernia* species, *Aponogeton monostachyus* and *Rotala wildii* are specialists that are found only in rock pools. Other important species found in rock pools are *Dopatrium junceum*, *Isoetes* spp., *Ludwigia leptocarpa*, *Murdannia semiteres*, and *Eriocaulon* spp.

Fig. 10.2.6. Rock pool, widely covered with flowering *Lindernia conferta* and few *Lindernia monroi* in the front. Near Banket, Zimbabwe. (Photograph R. Seine)

Permanently water-filled rock ponds are hardly found on inselbergs. Two seemingly perennial ponds were found in Zimbabwe. Vegetation of these ponds comprised the perennial *Cyperus distans*, *Oryza barthii*, and *Eleocharis* cf. *acuta*. A floating bladderwort, *Utricularia stellaris*, was present in abundance. *Hygrophila auriculata* and *Nymphaea* sp. were also recorded in those ponds.

10.2.4.10 Vegetation of Rock Debris

Rock debris is present on almost every inselberg. Annuals, succulents, and poikilohydric species are most important in this habitat. The latter are represented by vascular plants, liverworts, and lichens. Among the vascular resurrection plants are: *Xerophyta humilis* and several Scrophulariaceae (i.e., *Craterostigma lanceolatum*, *Craterostigma plantagineum*, *Lindernia pulchella*, *Lindernia wilmsii*; cf. Seine et al. 1995). *Selaginella* spp. (*S. dregei*, *S. njam-njamensis*) are very common vascular colonizers of rock debris (Fig. 10.2.7). Several species of liverworts (i.e., *Exormotheca*, *Riccia*) may cover large portions of rock debris. Mosses include *Bryum arachnoideum*,

Fig. 10.2.7. *Selaginella dregei*, the most common plant on rock debris in Zimbabwe. Domboshava, Zimbabwe. (Photograph R. Seine)

Leucobryum madagassum, and *Tortula eubryum*. Lichens rarely grow on rock debris; however, cyanophytic and chlorophytic lichens have been recorded. Species such as *Peltula patellata* are also found on the rock surface while others have, on inselbergs, only been recorded in this habitat (*Catapyrenium lacinulatum* var. *lacinulatum* and *C. l.* var. *latisporum*).

Portulaca rhodesica, *Crassula schimperi*, and *Cyanotis lanata* are succulent species occurring in this habitat. *Cyanotis lanata* is widespread in tropical Africa and is regularly found on inselbergs. Annual species on rock debris are, e.g., *Chamaechrista mimosoides*, various Cyperaceae (*Bulbostylis burchellii*, *B. oligostachys*, *Cyperus holostigma*, *C. rupestris*), and *Microchloa indica*. Also, one *Eriospermum* sp. is regularly present.

10.2.4.11 Herbaceous and Woody Vegetation of Soil-Filled Depressions

Soil-filled depressions are found on most inselbergs. In East and Southeast Africa their vegetation is usually dominated by perennial herbaceous species (Fig. 10.2.8). Many of those species are common constituents of savanna and woodland vegetation. Vegetation is usually dense and domi-

Fig. 10.2.8. Herbaceous vegetation of soil-filled depression comprising several perennial Poaceae with edging *Lindernia pulchella*, typical for rock debris. Near Banket, Zimbabwe. (Photograph R. Seine).

nated by grasses (*Andropogon gayanus*, *A. schinzii*, *A. schirensis*, *Eragrostis nindensis*, *Heteropogon contortus*, *Loudetia flavida*, *L. simplex*, *Melinis repens*, *Themeda triandra*) and sedges (*Cyperus amauropus*, *C. angolensis*, *C. holostigma*, *C. rupestris*, *Kyllinga alba*, *K. merxmuelleri*). Fabaceae of the genera *Indigofera* (*I. astragalina*, *I. emarginelloides*, *I. setiflora*, *I. subcorymbosa*), and *Tephrosia* (*T. decora*, *T. malvina*, *T. rhodesica*), and Lamiaceae (*Aeollanthus buchnerianus*, *A. ukamensis*, *Leonotis nepaetifolia*, *Plectranthus masukuensis*) are regularly present. Some small perennial bushes are frequently encountered (*Corchorus kirkii*, *Melhania acuminata*, *Pavetta petraea*, and *Waltheria indica*). The depth of soil allows for various geophytes such as *Albuca melleri*, *Chlorophytum* spp., *Haemanthus multiflorus*, *Urginea bequaertii*, *Scilla kirkii*, *Gladiolus* spp., *Dipcadi viride*, and *Ornithoglossum vulgare*. *Caralluma prigonium*, *Crassula schlechteri*, *Aloe* spp., *Kalanchoe lanceolata*, *K. sexangularis*, and *Senecio coccineus* are among the succulents found in this habitat. Common therophytes are *Aneilema johnstonii*, *Ceratotheca triloba*, *Commelina africana*, *Crotalaria bequaertii*, *Oldenlandia herbacea*, *Hibiscus engleri*, and *Phyllanthus niruroides*. In the center of larger depressions, trees (especially *Euphorbia* spp. and *Ficus* spp.) may establish. The shallow edges of soil-filled depressions resemble rock debris and are therefore colonized by species typical for that habitat (Fig. 10.2.8).

In our area depressions dominated by woody vegetation are rare. Tree and shrub species found in depressions are largely the same as on talus slopes. A genus often found in soil-filled depressions and not recorded from talus slopes is *Dissotis*.

10.2.4.12 Mats

Coleochloa Mats. Coleochloa mats are found on all larger inselbergs. *Coleochloa setifera* is the dominant mat-forming species in the area. This Cyperaceae forms very dense vegetation (usually attaining ground cover of more than 70%) and may grow as tussocks up to 50 cm tall (Fig. 10.2.9). Common co-dominants in *Coleochloa*-mats are *Myrothamnus flabellifolia* and various species of *Xerophyta* (e.g., *X. equisetoides*, *X. spekei* , *X. splendens*, and *X. villosa*). All mat-forming species are poikilohydric. *Myrothamnus flabellifolia* often covers the shallow fringes of the mats

Fig. 10.2.9. Mat of *Coleochloa setifera* with *Aloe chabaudii* (*center*) and *Myrothamnus flabellifolia* (*right*). Near Mt. Darwin, Zimbabwe. (Photograph R. Seine)

completely while *Xerophyta* spp. are usually found interspersed in matrix of *Coleochloa setifera*. On Mt. Mulanje, *Xerophyta splendens* reaches heights of up to 4 m and thrives in altitudes of up to 2500 m asl. It is probably the Velloziaceae that grows at highest altitudes (S. Porembski, pers. comm.). Specialized orchids growing exclusively or preferentially on *Xerophyta*- or *Coleochloa*-stems are often found (e.g., *Polystachya johnstonii*, *P. mafingensis*, *P. modesta*, *P. songaniensis*). Poikilohydric ferns such as *Pellaea calomelanos* and *Cheilanthes viridis* are often found in mat vegetation. *Aloe* spp. are often present in *Coleochloa* mats and include: *Aloe cameronii*, *Aloe chabaudii*, and *Aloe labworana*. Usually, a number of annual plants is found in mats. These species, however, are not observed with any constancy.

Mats Formed by Other Species. As indicated above, *Coleochloa setifera* is the dominant mat-forming species and, therefore, mats formed by other species are quite rare. Monospecific mats are occasionally formed by *Myrothamnus flabellifolia* and *Selaginella dregei*.

10.2.4.13 Ephemeral Flush Vegetation

Ephemeral flush seems to be a relatively rare habitat in our region, being found only on few inselbergs. The most conspicuous plants in ephemeral flush vegetation are carnivorous species such as *Utricularia* spp., *Genlisea* spp., and *Drosera indica*. Burmanniaceae, Eriocaulaceae, Gentinaceae (*Sebaea*, *Swertia*), Xyridaceae, *Antherotoma*, *Ophioglossum*, and *Polygala* spp. are further prominent members of ephemeral flush vegetation. Cyperaceae (*Cyperus* spp., *Fimbristylis* spp., *Kyllinga* spp., and *Pycreus* spp.) are the dominant components of vegetation.

10.2.4.14 Woody Vegetation

Woody vegetation is usually present only on a few larger inselbergs. Small trees and bushes of woody vegetation are, e.g., *Azanza garckeana*, *Erythrina abyssinica*, *Maytenus* spp., *Ochna* spp., and *Grewia* spp. On deep soils, woodland species become dominant. *Isoberlinia angolensis*, *Julbernardia globiflora*, *Brachystegia* spp., *Combretum* spp., and *Terminalia sericea* were often found in this habitat. Other important species include *Sclerocarya birrea*, *Uapaca* spp., and *Vitex payos*.

10.2.5 Life-Forms

A complete survey of vascular plant life-forms on inselbergs is, in our region, available only from Zimbabwe (Seine et al. in press). The number and percentage of species per life-form is given in Table 10.2.1.

Therophytes are the most important life-form. Their short life cycle may enable them to reproduce in the short, favorable periods and to endure the dry season as seed. However, only few of these species have a preference for the inselberg habitat (Seine et al. 1998). Savanna and woodland species form the main portion among therophytes. Their presence on inselbergs may partly be due to mass effect (Auerbach and Shmida 1987) of the surrounding savanna. Phanerophytes constitute the second important group. Among them is a relatively high number of species that prefer inselbergs. The low fire frequency may be an important factor for these species. Chamaephytes, the third important life-form, harbor even a larger percentage of species with a preference for inselbergs. The reason is the multitude of poikilohydric and succulent plants. These are well adapted to the habitat and are generally weak competitors that will be outcompeted among dense vegetation.

10.2.6 Vegetative Adaptations

Inselbergs are extreme habitats due to dry microclimate and scarce soil cover. A number of plants show vegetative adaptations that may be advantageous in coping with these adverse conditions (succulence, poikilohydric plants, carnivorous plants). The possible advantage of caulescent rosette trees is not yet understood; however, this growth form will here be treated under this heading.

Succulents are well represented on inselbergs in the region. Some 80 species have been reported in the published accounts, but many more remain to be found. The number of species listed here are, therefore, somewhat preliminary. Plant families and genera present are Agavaceae (*Sansevieria* 3 spp.), Aloaceae (*Aloe* 10 spp.), Apocynaceae (*Adenium multiflorum*, *A. obesum*), Asclepiadaceae (*Caralluma* 6 spp., *Duvalia polita*, *Echidnopsis* 2 spp., *Gomphocarpus tenuifolius*, *Huernia* 2 spp., *Sarcostemma* 2 spp., *Stapelia* 2 spp.), Asteraceae (*Crassocephalum vitellinum*, *Senecio* 3 spp.), Cactaceae (*Rhipsalis baccifera*), Crassulaceae (*Crassula* 4 spp., *Kalanchoe* 9 spp.), Cucurbitaceae (*Gerrardanthus macrorhizus*), Euphorbiaceae (*Euphorbia* 7 spp., *Monadenium* 3 spp.), *Icacinaceae* (*Pyrena-*

Table 10.2.1. Life-forms represented on inselbergs in Zimbabwe (Data from Seine et al. 1998)

Life-form	Species no.	Species (%)
Phanerophytes	126	23.0
Chamaephytes	84	15.3
Lianas	15	2.7
Hemicryptophytes	73	13.3
Geophytes	41	7.5
Therophytes	190	34.6
Epiphytes	2	0.4
Lithophytes	2	0.4
Hydrophytes	3	0.6
No data	13	2.4
Total	549	100.2[a]

[a] Due to approximations

cantha 2 spp.), Lamiaceae (*Aeollanthus* 6 spp., *Plectranthus* 2 spp.), Moraceae (*Dorstenia* 2 spp.), Orchidaceae (*Ansellia africana*, *Bulbostylis* 2 spp., *Polystachya* 4 spp.), Passifloraceae (*Adenia globosa*), Portulacaceae (*Anacampseros rhodesica*, *Portulaca* 3 spp.), and Vitaceae (*Cissus* 2 spp.). A gradient in the distribution patterns of succulent Aloaceae and Asclepiadaceae is evident: to the south the number of *Aloe* species is larger; to the north, succulent Asclepiadaceae are increasingly speciose.

Resurrection plants are very numerous, especially in southeast Africa, which is considered a center of evolution for poikilohydric vascular plants (Seine et al. 1995). Inselbergs over the whole region harbor at least 30 species of resurrection plants. Almost half of these belong to the Pteridophyta: Adiantaceae (*Actiniopteris* 2 spp., *Cheilanthes* 5 spp., *Pellaea* 3 spp.), Aspleniaceae (*Ceterach cordatum*), Selaginellaceae (*Selaginella* 3 spp.). Spermatophyta include Cyperaceae (*Coleochloa* 2 spp.), Gesneriaceae (*Streptocarpus* spp.), Myrothamnaceae (*Myrothamnus flabellifolia*), Poaceae (*Eragrostis hispida*, *Microchloa kunthii*, *Sporobolus festivus*), Scrophulariaceae (*Craterostigma* 2 spp., *Lindernia* 4 spp.), and Velloziaceae (*Xerophyta* 4 spp.). Carnivorous plants are present with 17 species from Droseraceae (*Drosera* 2 spp.) and Lentibulariaceae (*Genlisea* 2 spp., *Utricularia* 13 spp.). Caulescent rosette trees are found in the genera *Aloe* (e.g., *A. arborescens*, *A. excelsa*), *Coleochloa* (e.g., *C. abyssinica*, *C. setifera*), and *Xerophyta* (e.g., *X. equisetoides*, *X. splendens*). Although these species have the same overall appearance, the morphology underlying the caulescent rosette habit in *Aloe* is completely different from that in *Coleochloa* and *Xerophyta*.

10.2.7 Reproductive Adaptations

The only data on pollination and dispersal syndromes of inselberg vegetation in our region are from Zimbabwe (Seine 1996). The most important pollination mode by far is entomophily (75 %), followed by anemophily (20 %). Ornithophily and chiropterophily occur in very few species.

Dispersal is often effected by barochory (50 %), anemochory (25 %), and vertebratochory (20 %). Other dispersal modes are not important. Ornithochory is most important among vertebratochorous species.

10.2.8 Systematic and Ecological Characteristics

The systematic composition of inselberg vegetation is difficult to assess, as complete inventories are lacking for most of east and southeast Africa. The published reports, however, support the assumption that inselberg florulae are relatively homogenous over the area.

The composition given here refers to Zimbabwe, lichen data were compiled by U. Becker, vascular plant data are from Seine (1996). It must be kept in mind that there is certainly some amount of variation in comparison to other countries within the region.

Approximately 250 species of lichens, from 66 genera and 28 families, have been recorded on Zimbabwean inselbergs. Vascular plants belong to 549 species, 308 genera, and 109 families. The ten most speciose families are given in Table 10.2.2, for lichens and Table 10.2.3 for vascular plants.

The ten most speciose lichen families comprise almost three quarters of the lichen flora recorded for inselbergs in Zimbabwe. They include some species-rich genera which are typical for the inselberg ecosystem but are not restricted to it: *Xanthoparmelia* (32 spp.), *Usnea* (16 spp.), *Buellia* (16 spp.), and *Rinodina* (16 spp.). The only family that is almost restricted to inselbergs is Peltulaceae.

The ten vascular plant families include half the species found. The ecological constraints of the inselberg environment have a dramatic effect on the systematic composition: three families (Adiantaceae, Lamiaceae, Scrophulariaceae) which are not among the ten most speciose in Zimbabwe (Gibbs Russel 1975) are important on inselbergs. Adiantaceae and Scrophulariaceae have a high number of poikilohydrous species while Lamiaceae species are partly succulent and can, therefore, tolerate dessication of the habitat. On the other hand, families that are important in the flora of Zimbabwe (Asclepiadaceae, Liliaceae, Orchidaceae) are not that

Table 10.2.2. The ten most speciose lichen families of Zimbabwean inselberg vegetation

Lichen family	Species on inselbergs	Species of inselberg flora (%)
Parmeliaceae	60	24.0
Physciaceae	51	20.4
Lecanoraceae	20	8.0
Peltulaceae	16	6.4
Teloschistaceae	8	3.2
Acarosporaceae	6	2.4
Verucariaceae	6	2.4
Pertusariaceae	5	2.0
Thelotremataceae	3	1.2
Trapeliaceae	3	1.2
Total	178	71.2

Table 10.2.3. The ten most speciose vascular plant families of Zimbabwean inselberg vegetation

Plant family	Species on inselbergs	Species of inselberg florula (%)
Fabaceae s. lat.	58	10.6
Poaceae	53	9.7
Cyperaceae	40	7.3
Asteraceae	33	6.0
Scrophulariaceae	22	4.0
Euphorbiaceae	18	3.3
Rubiaceae	18	3.3
Lamiaceae	14	2.6
Adiantaceae	12	2.2
Acanthaceae	10	1.8
Total	278	50.8[a]

[a] Due to approximations

speciose on inselbergs. The remaining families are approximately equally important in inselberg and national flora.

Physiognomically, open habitats with herbaceous or lichen cover are most important on inselbergs in east and southeast Africa. Woody vegetation covers only a small proportion of individual inselbergs, if any.

10.2.9 Biogeography, Endemism, and Speciation

Distribution patterns of vascular plant species found on inselbergs in the area ranges from cosmopolitic (e.g., *Pteridium aquilinum*) to local endemic (e.g., *Aloe inyangensis*). Importance of individual distribution patterns (in species numbers and percentages) for vascular plants from Zimbabwean inselbergs (Seine 1996) is given in Table 10.2.4.

Similar results have been reported by Taton (1949) for several habitats on inselbergs of Ituri (Kongo). White (1983) reports 54 % of species restricted to the Zambezian region for the complete Zambezian region. The figure found on Zimbabwean inselbergs is considerably smaller. The typical regional elements are hence underrepresented. Consequently, the importance of more widespread species is larger than would be expected from a random sample. This may be attributed to Quarternary climatic oscillations that facilitated migration of plants between the Zambezian, the Somali-Masaian, and the Sudanian floristic regions (Wild 1956). Especially dry periods must have aided the spread of xerotolerant species such as those found on inselbergs today. The high percentage of widespread species, most of which spread far into East Africa, together with the existing inventories, supports the assumption that the general picture of biogeography will be similar over most of East and Southeast Africa.

Table 10.2.4. Distribution patterns in Zimbabwean vascular inselberg flora

Distribution	No. of species	Species (%)
Endemic	15	3.7
Zambezian	150	37.2
Zambesi-Somali-Masaian	29	7.5
Sudano-Zambesi-Somali-Masaian	29	7.5
Sudano-Zambezian	14	3.5
Afromontane	6	1.5
Tropical-African	13	3.2
Tropical and South African	20	4.9
Palaeotropic	46	11.4
Pantropic	18	4.5
Cosmopolitic	1	0.2
Others (all further distributed than Zambezian)	54	13.4
Introduced	8	1.9
Total	403	100.1[a]

[a] Due to approximations

Strong floristic links are evident between Southeast Africa and Madagascar, which have many genera and even species in common (Seine 1996; Fischer and Theisen, Chap. 10.4, this Vol.). The same is true for Namibia (Jürgens and Burke, Chap. 10.3, this Vol.) and South Africa. Similarity to the more distant oceanic islands of the Seychelles is less obvious (Biedinger and Fleischmann, Chap. 10.5, this Vol.). Two distribution patterns show important connections between the inselberg floras of east, southeast, and west Africa: Sudano-Zambesi-Somali-Masaian, covering the complete area and Sudano-Zambezian disjunctions that support the theory of a western route of migration in the Quarternary proposed by Wild (1956). Similarly, species with Zambesi-Somali-Masaian distribution may not have spread westward. Several paleotropic species connect our area with Asia via the Arab Peninsula.

Only few introduced species have been able to colonize inselbergs. Among these are *Bryophyllum tubiflorum*, *Catharanthus roseus*, and *Opuntia* spp.

The number of endemic species on inselbergs in Zimbabwe is surprising as compared to total species number. In Zimbabwe, approximately 1.8 % of the flora are endemic (Brenan 1978), in the inselberg florula 3.7 % are endemic. The number of endemic species reported from inselbergs in the region (Taton 1949; Bjørnstad 1976; Lisowski 1992; Porembski 1996; Porembski et al. 1997) and the astonishing diversity of succulents in Somalia indicate that the same may be true for the complete area.

The importance of inselbergs in plant speciation may be illustrated by the many species that are restricted to this ecosystem. For some species that grow exclusively on inselbergs (e.g., *Kalanchoe wildii*, *K. luciae*, *Portulaca rhodesiana*, *P. hereroensis*), the closest relative is found in the surroundings (Seine et al. 1998). The adverse living conditions on inselbergs and climatic fluctuations in the Quaternary may well have led to speciation on inselbergs (Seine et al., Chap. 11, this Vol.).

Acknowledgements. We are indebted to the Deutsche Forschungsgemeinschaft for funding research in Zimbabwe. Studienstiftung des Deutschen Volkes and Graduiertenförderung des Landes Nordrhein-Westfalen kindly granted scholarships to R. Seine and U. Becker, respectively. Support during field work by National Herbarium and Botanic Gardens (Harare, Zimbabwe), the Research Council of Zimbabwe, and the German Embassy (Harare, Zimbabwe) is gratefully acknowledged. We would also like to thank S. Porembski (Rostock) for valuable comments on the manuscript.

References

Auerbach M, Shmida A (1987) Spatial scale and the determinants of plant species richness. Trends Ecol Evol 2:238–242

Bally PRO (1968) The Mutomo Hill Plant Sanctuary in Kenya. In: Hedberg I, Hedberg O (eds) Conservation of vegetation in Africa south of the Sahara. Acta Phytogeogr Suec 54, pp 164–166

Bjørnstad A (1976) The vegetation of Ruaha National Park, Tanzania. 1. Annotated check-list of the plant species. Serengeti Res Inst Publ 215:1–61

Bourliere F, Hadley M (1983) Present-day savannas: an overview. In: Bourliere F (ed) Tropical savannas. Ecosystems of the World 13. Elsevier, Amsterdam, pp 1–15

Brenan JPM (1978) Some aspects of the phytogeography of tropical Africa. Ann Mo Bot Gard 65:437–478

Cole M M (1986) The savannas. Academic Press, London

Fleischmann K, Porembski S, Biedinger N, Barthlott W (1996) Inselbergs in the sea: vegetation of granite outcrops on the islands of Mahé, Praslin and Silhouette (Seychelles). Bull Geobot Inst ETH 62:61–74

Frahm J-P, Porembski S, Seine R, Barthlott W (1996) Moose von Inselbergen aus Elfenbeinküste und Zimbabwe. Nova Hedwigia 62:177–189

Gibbs Russel GE (1975) Comparison of the size of various African floras. Kirkia 10:123–130

Griffiths JF (1970) General introduction. In: Griffiths, JF (ed) Climates of Africa. World survey of climatology, vol 10. Elsevier, Amsterdam

King LC (1951) South African scenery, 2nd edn. Oliver and Boyd, Edinburgh

Knapp R (1973) Vegetation Afrikas. Gustav Fischer, Stuttgart

Lisowski S (1992) Note floristique de l'Ituri (Haut-Zaire). Fragm Florist Geobot 37:215–220

Menaut J-C (1983) The vegetation of African savannas. In: Bourliere F (ed) Tropical savannas. Ecosystems of the world 13. Elsevier, Amsterdam, pp 109–149

Petters SW (1991) Regional geology of Africa. Lecture notes in earth science 40. Springer, Berlin Heidelberg New York

Plowes DCH, Drummond RB (1990) Wild flowers of Zimbabwe. Longman, Harare

Porembski S (1996) Notes on the vegetation of inselbergs in Malawi. Flora 191:1–8

Porembski S, Fischer E, Biedinger N (1997) Vegetation of inselbergs, quarzitic outcrops and ferricretes in Rwanda and eastern Zaire. Bull Jard Bot Nat Belg 66:81–99

Seine R (1996) Vegetation von Inselbergen in Zimbabwe. Archiv naturwissenschaftlicher Dissertationen, vol 2. Martina Galunder, Wiehl

Seine R, Fischer E, Barthlott W (1995) Notes on the Scrophulariaceae of Zimbabwean inselbergs - with the description of *Lindernia syncerus* spec. nov. Feddes Repert 106:7–12

Seine R, Becker U, Porembski S, Follmann G, Barthlott W (1998) Vegetation of inselbergs in Zimbabwe. Edinb J Bot 55:267–293

Seine R, Porembski S, Barthlott W (1999) Diversity and phytogeography of inselberg vegetation in the Zambesian region. In: Timberlake J, Kativu S (eds) African plants: biodiversity, taxonomy and uses. Royal Botanic Gardens, Kew, pp 153–164

Taton A (1949) La colonisation des roches granitiques de la région de Nioka. Vegetatio 1:317–332

Torrance JD (1972) Malawi, Rhodesia and Zambia. In: Griffiths JF (ed) Climates of Africa. World survey of climatology, vol 10. Elsevier, Amsterdam

Troupin G (1966) Étude phytocénologique du Parc National de l'Akagera et du Rwanda oriental. Inst Natl Rech Sci, Butare 2
Twidale CR (1982) Granite landforms. Elsevier, Amsterdam
Walter H, Lieth H (1960) Klimadiagramm-Weltatlas. Gustav Fischer, Jena
White F (1983) The vegetation of Africa. Natural resources research 20. Unesco, Paris
Wild H (1956) The principal phytogeographic elements of the southern Rhodesian flora. Proc Trans Rhod Sci Assoc 1956:53–62

10.3 The Arid Scenario: Inselbergs in the Namib Desert Are Rich Oases in a Poor Matrix (Namibia and South Africa)

N. JÜRGENS and A. BURKE

10.3.1 Geography and Geology

Topographically uniform, the Namib Desert consists of vast plains of low altitude, stretching from the coast some 60 to 120 km inland. Further inland, the desert is often fringed by a steep escarpment, leading up to a plateau attaining altitudes of 1000 to 1800 m. The desert is geologically characterized by very old rock formations and a predominance of erosional processes since the Cretaceous. These processes shaped the wide plains, tilted towards the Atlantic Ocean.

The homogeneous pattern is interrupted by the presence of several archipelagoes of inselbergs. These are mainly formed by basaltic pipes and dikes, and granitic plutons. Exceptionally, also quartzite (Hartmannsberge), basaltic plateaus (Damaraland), and metasediments (e.g., gneiss, schist and dolomite) (southern Namib) contribute to the formation of inselbergs, often in combination with old intrusions. The inselbergs are generally forming an important contribution to the structural diversity of the Namib Desert, but their occurrence is not evenly scattered. At least five important centers of inselbergs or archipelagoes of inselbergs can be distinguished in the Namib Desert region:

1. Northern Namib inselberg archipelago,
2. Central Namib inselberg archipelago,
3. Great Namib Dunefield inselberg archipelago,
4. Southern Namib inselberg archipelago, and
5. Namaqualand-Richtersveld Sandveld inselberg archipelago.

Other, less prominent or less isolated groups of inselbergs are found in Damaraland (mostly basaltic plateaus). An inland part of the Namib Desert, the East Gariep Namib in the southeast of Namibia (Jürgens 1991),

Ecological Studies, Vol. 146
S. Porembski and W. Barthlott (eds.) Inselbergs

as well as the adjacent arid Namaland contain a number of arid inselbergs (Tatasberg, Brukkaros, Karasberg). Similar to these, inselbergs are found in the adjacent Bushmanland of the South African Northern Cape Province. Using a somewhat wider definition of inselbergs, a number of isolated plateaus, e.g., in Kaokoland or the huge sandstone plateau of the Waterberg, should also be mentioned. However, these more humid regions are outside the focus of this chapter.

More detailed descriptions of the major inselberg archipelagoes are presented in the section on flora and vegetation, while the following section discusses some general information on the geographical and edaphic features of inselbergs.

10.3.2 Geomorphological Features

The shape of inselbergs varies according to their lithology and geological history. While phonolithic pipes and granitic plutones are generally more or less circular in extension, dolerite dikes can form linear inselbergs of a few hundred meters to about 100 km (!) length (Fig. 10.3.1).

Fig. 10.3.1 Dolerite ridges form linear inselbergs of several 10 km length in the central Namib desert plains

Very unusual is the horizontal extension of the Messum crater between the Brandberg and the coast, forming a crater-like ring of mountains, interpreted as a volcanic pipe. Another circular ring structure is represented by the Roter Kamm, a meteorite crater in the southern Namib Sperrgebiet.

Looking at the vertical structural patterns, the crystalline inselbergs of the Namib Desert region show the same rounded shapes as elsewhere. Depending on the degree of fragmentation and the length of their exposure to weathering agents, the structures can vary from huge granite domes with smooth surface (Fig. 10.3.2 und 10.3.3, Spitzkoppe) to deeply dissected arrangements of boulders (koppies). In contrast to other regions (cf. Porembski, Chap. 10.1, this Vol.), the differentiation between granite domes with steep slopes and flat shield inselbergs shows no correlation to climate, but seems to follow a regional pattern of geological units. However, extreme examples of shield inselbergs can be observed in those parts of the Namib which are exposed to high wind speed. In these regions, sand-blasting results in aerodynamic planation of granite domes (Fig. 10.3.4 and 10.3.5). Bimodal structures are found, where inselbergs in aerodynamic landscapes are high enough, to form dome-shaped tops in zones at higher altitudes above the impact of sand-blasting (e.g., Uri-Hauchab).

A very different type of inselbergs are plateau inselbergs, formed by sedimentary or basaltic rocks with constant-angle erosion. Their shape is controlled by the extension of sediment or basaltic layers and the erosive energy of the catchment areas of adjacent river systems, often controlled by older tectonic forms.

Again a different structure is provided by the linear inselbergs of the dolerite dikes of the central Namib Desert (Fig. 10.3.1). The hard dolerite filling of the dikes is most resistant to erosion, while the contact zone and surrounding crystalline material form a basement on either side of the dolerite ridges.

Locally, surface structures on inselbergs are strongly dependent on lithology. Rock pools are frequently found in crystalline rocks, while the formation of small channels, cut into the rock surface, seems to be dependent on the presence of cyanobacteria. Both structures are characteristic for the summer-rainfall part of the Namib Desert. In contrast, granite domes in the winter-rainfall climate of the southern Namib show green algae and lichens covering larger surface areas of the rock, although broader linear arrangement of these larger carpets can also be observed on steep slopes.

10.3.2

10.3.3

10.3.4

10.3.5

10.3.3 Soils

In humid regions, inselbergs differ edaphically from their surrounding landscapes, because they possess a high proportion of rocky surface without soil formation. However, in the desert environment, the soils of the surrounding zonal habitats also have to be classified as aridisols, inceptisols, and other substrate-dominated soil types of weak soil genesis.

Important large-scale soil gradients in the Namib are observed along coast-inland transects (Jürgens et al. 1997) where coastal predominance of NaCl is replaced by predominance of Ca_2SO_4 and, still further inland, $CaCO_3$ (Martin 1963; Scholz 1967). Along the coast, sand fields are dominated by white quartz sands, while increasing importance of iron coatings is observed with increasing distance to the coast, resulting in more reddish or brownish colors. Important dune fields are found in the southern central Namib (Great Namib Erg), the southern Namib (Diamond Area or Sperrgebiet), and the northern Namib (Skeleton Coast) (sequence reflecting decreasing importance). While soils are generally above pH 7 in the summer-rainfall parts, acid soils are somewhat more important in the winterrainfall part of the Namib Desert.

Owing to the concentration of runoff water in local catchment areas, rocky inselbergs harbor microhabitats of high and/or all-year-present soil humidity generating processes of intensive rock weathering and accumulation of organic matter. These humid microenvironments create – in an arid macroclimate – relatively better-developed and more diverse soil and vegetation types on inselbergs than in the surrounding zonal landscapes. In spite of the very high ecological importance of these wet pocket soils, no scientific data on the properties of these soils seem to exist.

Fig. 10.3.2. Granite dome of the Spitzkoppe, central Namib desert, Namibia. *Cleome elegantissima* fringing a soil-filled depression at its margin

Fig.10.3.3. Ephemeral rock pool on the Spitzkoppe inhabitated by the poikilohydric *Chamaegigas intrepidus*

Fig.10.3.4. Sand blasted aerodynamic granitic inselbergs of the Uri-Hauchab archipelago in the center of the dunefield of the southern Central Namib. Note presence of water in rock pools and runoff vegetation, fringing the inselberg on its right-hand side

Fig.10.3.5. Tracks of runoff on dunes at the margin of the inselberg shown in Fig. 10.3.4

10.3.4 Climate

10.3.4.1 Regional Climate

In the Namib Desert, climatic gradients can be very steep and are of extreme importance (Lancaster and Lancaster 1984; Lindesay and Tyson 1990; Jürgens et al. 1997). Mild temperatures and high air humidity along the coast (cold upwelling Benguela current) contrast with the high temperatures and generally lower air humidity further inland. Fog is very frequent along the coast and often driven by southwesterly winds to regions far inland (Olivier 1995). The salt content of the fog decreases with distance from the coast. The southwestern part of the Namib, southwest of a line running from Lüderitz to the Eastern Cape receives winter rains. These are mainly provided by cyclonic winter rains developing over the Atlantic Ocean and normally form a soft drizzle. Rare events of convective rains associated with thunderstorms provide the highest rains in the tropical summer-rainfall zone, northeast of the boundary mentioned above.

10.3.4.2 Local Climate

Inselbergs are subject to special climatic conditions. Local climates on inselbergs, beyond the general decrease of temperature with increasing altitude, can be observed in two scenarios:

1. Inselbergs along the coast catch a lot of precipitation from fog and can therefore be described as fog oases. These are particularly favorable habitats for lichens and succulents. Examples are the Laguneberg in the Northern Central Namib. Vogelfederberg inland of Walvisbay, several inselbergs north and south of Lüderitz, and, especially, the Boegoeberg Twins and Aughrabies Mountains south of Alexanderbay. Generally, the importance of fog and dewfall is increasing with increasing altitude. However, fog events are often localized and may be limited to the lower parts of high inselbergs!
2. Large inselbergs like the Brandberg create their own circulation system of thermal uplift, being strong enough to create local cloud formation and rainfall. However, no quantitative data are available.

10.3.4.3 Microclimate

The high proportion of bare rock and runoff on rocky inselbergs obviously creates desert-like, unfavorable conditions for the vegetation on most of the surface. Therefore, in humid regions, inselbergs are edaphically arid and biologically relatively species-poor ecosystems.

In a desert environment with generally low precipitation in the whole landscape, the scenario is very different. Here, the high proportion of (unfavorable) rocky surface area creates a process of redistribution and concentration of available water and organic material to a very small surface area. This provides a higher and more predictable moisture supply which in combination with greater habitat diversity results in higher biodiversity on inselbergs than in the surrounding zonal habitats, a contrasting trend to inselbergs in humid regions.

Comparing different seasonal types of identical mean annual precipitation, the effect of concentration of moisture on inselbergs discussed here is low in those regions where precipitation is subdivided into numerous events of very low quantity (e.g., fog precipitation), thus leading to small runoff quantities. Very rare events of very high precipitation, in turn, will create strong (and destructive) runoff, accompanied by very long intervals of high aridity.

Depending on the size of catchment areas, the storage volume of the focus area collecting the runoff, and the characteristics of the regional climate, precipitation events of medium frequency and quantity will generally be more favorable than the two extremes given above.

10.3.5 Flora, Vegetation, and Life-Forms

10.3.5.1 General

The climatic division of the Namib Desert into a winter-rainfall part and a summer-rainfall part controls the phytogeographical pattern: the Namib Desert is divided into two very different floras. The succulent Karoo floral region, a part of the Greater Cape flora, is prominent in the southwest of the above mentioned climatic boundary, while the Nama Karoo floral region, part of the paleotropical floral kingdom, dominates northeast of the boundary.

Similarly, the vegetation units are separated into these major entities. Further subdivision of vegetation is controlled by:

- the complex gradient between coast and inland,
- the latitudinal complex gradient,
- major lithologically and geomorphologically defined landscapes,
- edaphical factors,
- local hydrological differences, and
- land-use systems.

Details have been reviewed in Jürgens et al. (1997).

In general, the floristic inventories of the inselbergs form part of the respective floristic unit in the particular area to which they belong. However, in the summer-rainfall part of the southern Namib Desert and the Great Dunefield, inselbergs house a large proportion of taxa, forming part of the Succulent Karoo phytochorion, which largely occurs in the winter-rainfall region. These populations have been interpreted as relict populations and can be interpreted as signs for a movement of the winter-rainfall system in a southwestern direction. However, the shift from Nama Karoo flora to Succulent Karoo flora can also at least partly be explained as a response to the altitudinal gradient in the inselbergs. Due to the steep ecological gradients in the Namib Desert, it seems adequate to discuss the inselbergs of the various regions separately.

10.3.5.2 Northern Namib Inselberg Archipelago

Inselbergs in the northern part of the Namib Desert present a mixture of foothills of the Hartmanns Mountains and granitic outcrops extending into the dune fields of the northern Namib. Surrounded by mobile dunes at their western side and sandy plains to the east, a preliminary analysis of the floristic composition of some of these inselbergs indicated some correlation with the inselbergs of the Great Namib dune fields. Dune-related species such as *Acanthosicyos horridus* and *Centropodia glauca* are present on both inselberg archipelagoes, although the flora of the northern Namib inselbergs is otherwise largely composed of taxa of paleotropic origin characteristic of summer-rainfall conditions. Many of these, such as *Commiphora anacardiifolia* and *Turnera oculata*, are endemic to this region, but possibly not restricted to the inselberg habitats. An interesting record at one of the northwesternmost inselbergs is the occurrence of *Sclerocarya birrea* var. *birrea*, a subspecies of a widely cultivated indigenous fruit tree in north-central Namibia. It may have been distributed by hunter-gatherers in the past.

10.3.5.3 Central Namib Inselberg Archipelago

An important center of inselbergs is found in the central Namib Desert between the river beds of Kuiseb and Ugab. In this area a very clear zonation of the zonal flora (Giess 1981) and vegetation (Hachfeld 1996; Jürgens et al. 1997) can be observed between the coast and the escarpment.

Next to the coast, outliers of the succulent Karoo flora like *Brownanthus kuntzii* and *Zygophyllum clavatum*, etc. occur together with lichens and the two most important species of the Central Namib, *Arthraerua leubnitziae*, and *Zygophyllum stapffii*. Further inland, *B. kuntzii* and *Z. clavatum* disappear and, still further inland, only *Zygophyllum stapffii* forms the majority of the biomass in the minimum zone of vegetation. Again further inland, with increasing importance of summer rainfall, grassland, and thorn savanna form the transition zone to the savanna biome.

In this region various types of inselbergs occur:

- huge crystalline inselberg massivs like the Brandberg,
- medium- and small-sized inland crystalline inselbergs like the Spitzkoppe, the Roessing Mountain, Blutkuppe, Messum mountains, etc.,
- small coastal inselbergs like Vogelfederberg and Laguneberg, and
- a network of linear dike structures of basaltic material (e.g., dolerite ridges), dissected into numerous inselbergs. These harbor a quite different flora and vegetation.

10.3.5.4 Brandberg

The Brandberg (2579 m) is a unique structure. This huge crystalline massif is a mountain range by its size and diversity, but an inselberg with respect to isolation and due to its very special vegetation, which is supported by an annual precipitation of about 100 mm in the upper parts of the mountain (Breunig 1990). The plant cover of the mountain is much more dense than in the surrounding semidesert or thorn shrub savanna vegetation. Probably it has been so for at least the past 4000 years, as numerous rock engravings and archeological sites in the Brandberg emphasize (Lennsen-Erz 1997). The flora of the Brandberg has been analyzed in detail by Nordenstam (1974, 1982), although there is no detailed account on vegetation. Nordenstam reports a strong altitudinal zonation with a number of temperate taxa at higher altitudes.

The Brandberg provides a diversity and magnitude of landscapes and habitats which goes far beyond the spectrum of habitats and geomorphological structures which are usually linked to the term inselberg. Therefore, more detailed information or analysis of the Brandberg is outside the scope of this chapter.

10.3.5.5 Medium- and Small-Sized Inland Crystalline Inselbergs

In contrast to the Brandberg, a number of medium- and small-sized crystalline inselbergs of the central Namib Desert present perfect examples for isolated granitic inselbergs. The most famous example is found in the Spitzkoppe group (Fig. 10.3.2 and 10.3.3). The rock surfaces themselves are bare of vegetation of vascular plants and only cyanobacteria and lichens form blackish crusts and brown margins along smaller drainage channels.

More interesting is the vegetation of seasonal rock pools which house specialists like the poikilohydric *Chamaegigas intrepidus* (Fig. 10.3.3 and 10.3.6). Also soil-filled depressions can accumulate and store a lot of water after seasonal rainfalls. Depending on soil depth and quantity of accumulated water, these depressions can show either rich herbaceous vegetation, e.g., with dominance of *Cleome elegantissima* (Fig. 10.3.2), or even woody elements, e.g., including *Dichrostachys cinerea*.

Fig. 10.3.6. The poikilohydric *Chamaegigas intrepidus* in a rock pool of the Spitzkoppe inselberg

Rock crevices and larger clefts are more favorable for growth of tall-stem succulent growth forms like *Moringa ovalifolia* (Fig. 10.3.7) and *Commiphora* spp., although nonsucculent rare local or regional endemics like *Nicotiana africana* (Giess 1982) are also found in this habitat type. In contrast, the poikilohydric *Myrothamnus flabellifolia* and several Acanthaceae prefer shallow substrate covers over rock. With increasing annual precipitation towards the east, monocotlyedonous mats dominated by *Xerophyta viscosa* gain in importance.

The flora and vegetation of only few inselbergs of the central Namib has been thoroughly inventoried. One example is presented here: the vascular flora of the Roessing mountain (U. Tränkle and F. Hübner, unpubl.), which includes 66 species. However, a comparison of this species inventory with the surrounding regions (Table 10.3.1) shows that not a single species of this inselberg is restricted to the mountain. All inselberg species are known also on the surrounding plains, provided rocky habitats of the plains are included in the comparison analysis.

Other central Namib inselbergs in a more isolated position with few rocky habitats on the surrounding plains, such as the group of granite

Fig. 10.3.7. Stem-succulent *Moringa ovalifolia* on rocky outcrops close to the Brandberg in the central Namib desert

Table 10.3.1. Roessingberg flora and its presence in surrounding habitats

Roessingberg	Plains	Rocky outcrops in plains	Hauchab	Family
Acanthosicyos horridus	0	0	H	Curcurbitaceae
Aloe asperifolia	0			Aloaceae
Anacampseros papillosus	0			Portulacaceae
Arthraerua leubnitziae	0			Amaranthaceae
Asclepias buchenaviana	0			Asclepiadaceae
Asparagus denudatus	0			Asparagaceae
Barleria lancifolia	0	0		Acanthaceae
Blepharis grossa	0	0		Acanthaceae
Citrullus ecirrhosus	0			Cucurbitaceae
Commiphora saxicola	0			Burseraceae
Cotyledon arborea	0			Crassulaceae
Dyerophytum africanum	0		H	Plumbaginaceae
Enneapogon scaber	0			Poaceae
Euphorbia gariepina	0			Euphorbiaceae
E. lignosa	0		H	Euphorbiaceae
E. mauretanica	0		H	Euphorbiaceae
E. phylloclada	0			Euphorbiaceae
Forsskaolea hereroensis	0			Urticaceae
Galenia fruticosa	0		H	Aizoaceae
Gazania jurineifolia spp. *scabra*	0			Asteraceae
Helichrysum roseo-niveum	0		H	Asteraceae
Heliotropium tubulosum	0			Boraginaceae
Hoodia currori	0			Asclepiadaceae
H. pedicellata	0			Asclepiadaceae
Hydnora sp.	0			Hydnoraceae
Hypertelis salsoloides	0			Molluginaceae
Juncus rigidus	0			Juncaceae
Kleinia longiflora	0	0		Asteraceae
Kohautia virgata	0			Rubiaceae
Lavrania marlothii	0			
Lithops ruschiorum	0			Aizoaceae
Lycium cinereum	0			Solanaceae
Mesembryanthemum guerichianum	0			Aizoaceae
Monechma arenicola	0			Acanthaceae
M. cleomoides	0			Acanthaceae
Orthanthera albida	0		H	Asclepiadaceae
Petalidium canescens	0	0		Acanthaceae
Petalidium variabile	0	0		Acanthaceae
Polygala pallida	0			Polygalaceae
Psilocaulon salicornioides	0			Aizoaceae
Rhus marlothii	0			Anacardiaceae
Sesuvium sesuvioides	0			Aizoaceae
Salsola tuberculata	0			Chenopodiaceae
Salvia gariepensis	0			Lamiaceae

Table 10.3.1 (*continued*)

Roessingberg	Plains	Rocky outcrops in plains	Hauchab	Family
Sarcocaulon patersonii	0			Geraniaceae
S. salmoniflora	0			Geraniaceae
Sarcostemma viminale	0			Asclepiadeaceae
Sporobolus consimilis	0			Poaceae
Stipagrostis schaeferi	0			Poaceae
S. ciliata	0		H	Poaceae
S. sp.	0			Poaceae
Suaeda merxmuelleri	0			Chenopodiaceae
Sutera maxii	0		H	Scrophulariaceae
Tamarix usneoides	0			Tamaricaceae
Tephrosia dregeana	0		H	Fabaceae
Tetragonia reduplicata	0			Aizoaceae
Trichodesma africanum	0			Boraginaceae
T. angustifolium	0			Boraginaceae
Zygophyllum cylindrifolium	0			Zygophyllaceae
Z. simplex	0			Zygophyllaceae

inselbergs in the Ganab area, show a slightly different trend: although many species occur in both habitats, inselberg and plains, there are several, such as the tree *Cordia sinensis* and the shrubs *Croton gratissimus*, *Hibiscus elliottiae*, and *Helichrysum tomentosulum*, which are restricted to these inselbergs.

10.3.5.6 Small Coastal Inselbergs

Far more special and isolated in its surrounding landscapes is the Laguneberg, a series of hills and smaller mountains close to the coast at the same altitude as the Brandberg. The Laguneberg mountains are very special due to the strong catchment of coastal fog, resulting in a very dense cover by lichens and succulents. Further south, similar structures are provided by the Swartbank and Vogelfederberg. Interestingly, the Swartbank mountains, which are slightly lower than Vogelfederberg and composed of marble, show a dominance of lichen species, while lichens are nearly absent from the granitic Vogelfederberg. The underlying chemistry of the substrate and differences in exposure to fog may explain these patterns, which warrant further investigation.

10.3.5.7 Dolorite Dikes

Between all these structures, linear inselberg archipelagoes are formed by the dolerite dikes of the Damara orogene. Single dikes – with minor interruptions due to erosion processes – sometimes reach a length of many tens of kilometers. The vertical size of these structures ranges from some meters or tens of meters in altitude in the normal case to up to 300 m in exceptional cases.

The flora and vegetation of these dolerite inselbergs is different from the surrounding plains. Firstly, several species like, e.g., *Pelargonium otaviense* and *Euphorbia virosa*, occur only on the dolerite dikes and not in the surrounding plains. Secondly, the species richness of the dolerite inselberg ridges lies generally high above the species richness in the surrounding habitat types plains and washes.

In a detailed analysis of the vegetation of the central Namib Desert plains, transects from the foggy coast to the inland have shown that the graph representing species richness on dolerite dikes is very different from the corresponding graphs for the habitat types of the surrounding plains (Fig. 10.3.8; Hachfeld 1996): while plain habitats show the lowest species

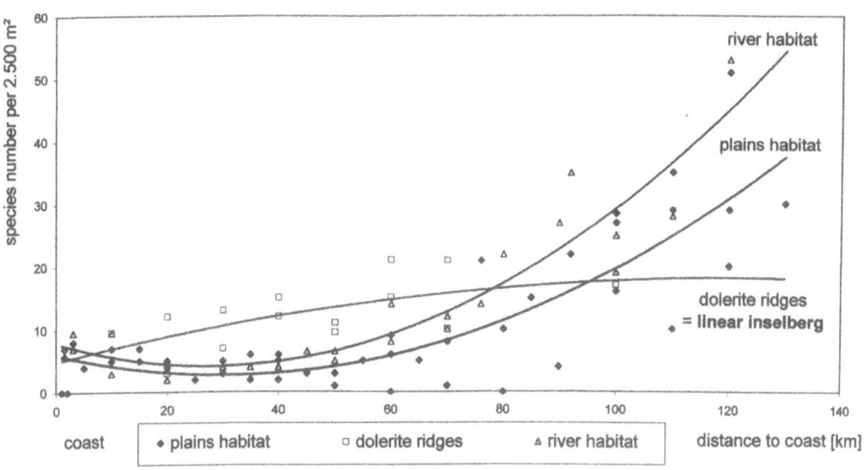

Fig. 10.3.8. Alpha-diversity along a transect cutting the central Namib Desert from coast (*left hand side*) to 120 km inland (*right-hand side*). Species richness of vascular plants on relevés of 2500 m² is shown for three habitat types, including linear inselbergs, formed by dolerite ridges. Biodiversity of the linear inselberg habitat shows steady increase from coast to inland and is higher than the biodiversity of plains and ephemeral river beds in the more arid parts of the Namib. With increasing precipitation further inland, species richness on the inselberg habitat is surpassed by plains and river bed habitats. (After Hachfeld 1996)

richness some 20 to 40 km inland from the coast, where neither coastal fog nor inland summer rain support higher vegetation, the graph for dolerite ridges shows more or less constant increase in species richness all along the transect from the coast to the inland escarpment. This curve may be explained by three factors: (1) the stronger fog-harvesting effect of the dolerite outcrops, combined, further inland, with (2) a relatively stronger response to the increasing, albeit low, amounts of summer rainfall, owing to the concentration effect by runoff from the rocky surfaces, and, still further inland, (3) the saturation of the limited water-storage capacity of rock fissures and crevices.

The graph also shows that species richness of the inselbergs is higher than the surrounding plains over the more arid western part of the transect. This relation is reversed at about 100 km inland: east of this position more species coexist in the plains than in the rocky dikes. This eastern scenario can perhaps be called the normal situation for tropical inselbergs, forming an arid island in a humid environment. The more arid 100 km of our Namib transect underlines that arid inselbergs, in contrast to the humid tropical scenario, possess higher biodiversity than the surrounding zonal habitats, due to a relatively higher moisture supply and higher habitat diversity.

10.3.5.8 Great Namib Dunefield Inselberg Archipelago

Another important archipelago of inselbergs is positioned in the Great Namib Dunefield with a maximum of density of inselbergs at the eastern margin of the dunefield close to the escarpment, but also including extremely isolated inselbergs like the Hauchab and the Blue Mountain. In the northeast, close to the Kuiseb Canyon, this sequence of inselbergs is nearly in contact with the central Namib inselberg archipelago. Again, isolated rocks in close vicinity to the coast like Sylvia Hill show very special environmental conditions and a specific flora. The occurrence of the locally endemic Mesemb *Jensenobotrya lossowiana* at the Spencer Bay hills forms the most famous example (Giess 1974; Robinson and Giess 1974).

All these inselbergs are very isolated with respect to the absolute interruption of rocky or stony habitat types between them: the matrix landscape of the whole region is formed by aeolian sandy deposits, mainly by dune sands without vegetation or with predominance of a few grass species, especially *Stipagrostis seelyae*, *Stipagrostis sabulicola*, and *Cladoraphis spinosa*; rarely a few more habitat specialists like *Monsonia ignorata* or *Trianthema hereroensis* are found. In striking contrast, the inselbergs show a very different composition. Already around the inselbergs a belt of *Stipagrostis lutescens* underlines the importance of the concentration effect of runoff from the rocky surfaces. On the mountains themselves, numerous species exist. As an example, the angiosperm flora of the Uri Hauchab archipelago in the center of the Great Namib Dunefield has been analyzed. The angiosperm flora of the Hauchab mountains is composed of 82 species with a large proportion of stem- and leaf-succulent taxa. Very interesting is the wide range of phytogeographical relationships of these species.

While the whole spectrum ranges from trees of the Sudano-Zambezian region of the Paleotropis to shrubs from the Greater Cape flora, the majority of species belongs to the Nama Karoo region of the Paleotropis (38 %) and the succulent Karoo region of the Greater Cape flora (19 %). For 42 % of the species, the Hauchab population forms an isolated locality in an extreme position in relation to the main area of distribution. Today, the Hauchab mountains are in the range of the summer rainfall climate, which should result in a Nama Karoo-related vegetation similar to the Roessing Mountain, shown above. However, as shown in the species list of the Roessing mountain (Tab. 10.3.1), very few of the Roessing species occur in the Hauchab as well. The high proportion of species with Greater Cape flora affinity could be attributed to a recent retreat of the succulent karoo flora after the last glaciation, but survival of many of these taxa in the inselberg situation owing to low immigration rate and resulting low competition.

10.3.5.9 Southern Namib Inselberg Archipelago

The region of highest density of inselbergs is found in the southern Namib Desert of Namibia in the vicinity of and between the two settlements Lüderitz and Oranjemund (Burke et al. 1998).

Overall, there is no doubt about the presence of numerous endemic species, like, e.g., *Microloma penicillatum*, *Pelargonium mirabile*, *Cynanchum meyeri*, and *Lessertia acanthorhachis*, all of which are associated with rocky habitats. Plant formations on the inselbergs show a strong contribution of leaf-succulent chamaephytes, while the surrounding plains are often dominated by grasses like *Stipagrostis ciliata*, *S. obtusa*, and *S. geminifolia*, or annuals like *Zygophyllum simplex*, in large parts also by the stem-succulent *Euphorbia gummifera*. A special element of those inselbergs which are within reach of the frequent sandstorms of the region are the psammophorous life-forms. Species like *Psammophora modesta* and *Psammophora nissenii* are well sheltered against the destructive force of sand-blasting by a layer of sand grains fixed to the surface of their leaves due to excreted sticky substances (Jürgens 1996).

The absence of detailed vegetation and floristic studies allows only a preliminary review and is not exhaustive. Complex geology and localized climatic influences call for a differentiation of these inselbergs into smaller units:

- low coastal outcrops,
- the Rechenberg-Tsaukhaib complex,
- Klinghardts Mountains, and
- eastern Sperrgebiet inselbergs.

10.3.5.10 Low Coastal Outcrops

Composed of metamorphic sediments such as gneiss, schist, and dolomite, low outcrops of less than 100 m above the surroundings occur all along the coast between Chameis and Lüderitz. They consist of isolated single outcrops and ridges and include spectacular rock features such as the Bogenfels Arch.

The flora of these coastal outcrops contains a large portion of regional endemic species – as high as 25 % (Burke 1997) – which could be attributed to their long isolation through mobile dune fields to the north and south from similar habitats along the coast. Many of these endemic species, such as *Eremothamnus marlothianus* and *Pelargonium cortusifolium*, are well adapted to thrive on low, but regular, moisture supply provided by fog and to endure strong, sand-blasting winds.

10.3.5.11 The Rechenberg-Tsaukhaib Complex

The inselbergs of the Rechenberg-Tsaukhaib complex are mostly composed of gneiss and granitic material and rise several hundred meters above their surroundings. A more southerly geographic position compared to the inselbergs of the Great Namib dunefield results in a larger component of species of the Cape Floristic region intermingled with regional endemics and a comparatively small (10–15 %) portion of species characteristic of summer rainfall conditions.

10.3.5.12 Klinghardts Mountains

The position of the Klinghardts Mountains in the southern Namib can be compared to the Brandberg Massif in the central Namib. Although not reaching the same impressive altitudes – the highest peak is 1114 m – its closeness to the coast (30 km), its large extent over approximately 20 × 20 km, and its isolated position justify the comparison. Geologically, the Klinghardts Mountains comprise the above-mentioned metasediments of the surrounding rocky habitats (e.g., like those along the coast) and volcanic intrusions mainly in the form of phonolite. This mosaic of different subtrates in combination with fog influence resulted in a unique flora composed mainly of elements from neighboring floristic zones (Williamson 1997). Approximately 150 species have been recorded so far, mostly taxa of the Cape Floristic region and some species endemic to this mountain complex, e.g., *Strumaria phonolithica* and *Tromotriche ruschiana*.

10.3.5.13 Eastern Sperrgebiet Inselbergs

Although the group of eastern Namib inselbergs lumped together here varies in substrate and is exposed to often very localized special climatic conditions, their importance as refuges for species with high conservation status warrants a brief, descriptive account of some of the most striking features. The relative proximity to the escarpment and other large inland mountain complexes, such as the Huns Mountains and further south to the Namaqualand Broken Veld (Acocks 1988), is clearly reflected in the floristic composition of these inselbergs.

A western extension of black limestone and sandstones of the Nama group makes up the Tsaus Mountains and surrounding inselbergs (Geological Survey 1980). The mountains form a flat plateau with steep slopes and are characterized by relatively low species richness with a larger portion of summer-rainfall taxa than the inselbergs north and south of Tsaus. Examples include several *Pteronia* species, *Kleinia longiflora*, and *Trianthema triquetra*. Evidently restricted to black limestone and dolomite is the regional endemic *Euclea asperrima*, which can also be found on dolomites of the Naukluft Mountains, some 300 km north of the Tsaus Mountains. The Aurus Mountains show the highest plant diversity and density of vegetation in the southern Namib (Williamson 1997). While the western slopes rise steeply from the surrounding plains, forming a barrier to catch and force fog clouds to ascent over the saddle, the interior and eastern part of the mountains descend gradually into the eastern plains. Protected by the southern and western ridges from the usually strong southwesterly winds and receiving additional moisture from fog, the interior of the mountains supports a diverse and dense vegetation, accompanied by many taller plants and moisture-loving taxa usually absent in this part of the Namib Desert. Examples include the trees *Aloe ramosissima*, *Maerua gilgii*, *Ozoroa dispar*, the desert orchid *Holothrix filicornis*, and several ferns and mosses. The southeastern corner of the southern Namib holds an array of inselbergs of various sizes mainly composed of quartzite, schist, and dolomite forming a transition to the foothills of the escarpment. The most prominent outposts are the Obib Mountains, where the presence of some taxa typical of the Richtersveld flora, such as *Aloe pearsonii*, indicates the transition to the Namaqualand-Richtersveld flora.

10.3.5.14 Northern Namaqualand-Richtersveld Sandfield Granite Inselbergs Archipelago

While in the mountains of Namaqualand (Northern Cape Province, RSA) numerous huge granite domes form parts of an uninterrupted mountain range and hence cannot be called inselbergs, similar, but isolated mountains are found in the sandfield of Coastal Namaqualand. This archipelago includes numerous more or less isolated mountains and hills from the Boegoe Twins south of Alexanderbay (van Jaarsveld 1987) or the Aughrabies Mountains near Port Nolloth to the Buffels River and further south. Due to the generally high air humidity and high incidence of fog in this coastal region, lichens and small leaf succulents reach extreme species numbers in these inselbergs. On the southwestern slope of the Boegoe mountains near Alexanderbay, lichens have formed pioneer vegetation which accumulated downslope on moving substrate. In later successional stages small leaf-succulent species including *Tylecodon schaeferanus*, *Conophytum saxetanum*, and *Senecio phonolithicus* established themselves on the microterraces thus formed. Through this process, sheer rock surfaces are transferred to species-rich mosaics of vegetation units of different soil depth and age (N. Jürgens, pers. observ.).

10.3.6 Human Impact and Conservation

As oases in an arid environment, the inselbergs of the Namib Desert are of high conservation value, which house a rich flora and many endemic taxa. Although the example of the Roessingberg in the central Namib (Tab. 10.3.1) has shown that very many species of the surrounding plains also exist on the inselbergs, examples of inselbergs in the southern Namib indicate the presence of endemic species restricted to these mountains. This is a general pattern due to the high diversity of microenvironments on inselbergs which normally includes a large part of the spectrum of habitat conditions of the surrounding plains. Therefore, inselbergs possess a very high regeneration potential for the flora of the zonal environments, a fact which can be of great importance when natural environmental oscillations or man-made degradation destroy the flora of the zonal habitats. Therefore, conservation of inselbergs should receive great attention.

Fortunately, in the Namib Desert, the actual protection status is relatively good due to the low intensity of human impact and the high proportion of inselbergs being included in conservation areas, consequently already falling under formal protection. Nevertheless, those inselbergs not included in conservation areas, such as the Brandberg and surrounding inselbergs, experience severe impacts due to increasing tourism and/or increasing grazing pressure and land use for small stock farming. This forms a potential threat to the flora of these inselbergs in the future.

In addition, the inselberg flora of the southern Namib contains many rare plant species sought for by succulent collectors. Although currently most of these inselbergs are difficult to access, the potential opening up of the restricted Diamond Area for tourism, may call for special measures and protection status for inselbergs in this area.

10.3.7 Research Needs

Although data on inselberg floras in many parts of the Namib are still lacking, the largely descriptive approach of the authors points clearly to a need for a comprehensive review and analysis of current data. Questions related to the origin and development of the Namibian flora, migration routes, and distribution patterns are some relevant research topics. Understanding ecological processes and environmental factors responsible for the distribution of inselberg floras will contribute invaluable information to establishing conservation needs, identifying potential threats, and evaluating inselberg floras in the light of regeneration potential of degraded areas.

Acknowledgements. We would like to thank B. Hachfeld, U. Traenkle, and F. Huebner for allowing the use of unpublished material from the central Namib Desert. Special thanks to Elke Erb for providing various data and valuable logistic support in the central Namib Desert.

References

Acocks JPH (1988) Veld types of South Africa. Mem Bot Surv S Afr 57:1–146

Breunig P (1990) Temperaturen und Niederschläge im Hohen Brandberg. J Namibia Sci Soc 42:7–24

Burke A (1997) Coastal vegetation between Chameis and Baker's Bay, Sperrgebiet. Report for NAMDEB, Oranjemund

Burke A, Jürgens N, Seely MK (1998) Floristic affinities of an inselberg archipelago in the southern Namib desert – relic of the past, centre of endemism or nothing special? J Biogeo 25:311–317

Geological Survey (1980) Geological map of South West Africa/Namibia. Geological Survey, Windhoek

Giess W (1974) Zwei Fahrten zur *Jensenobotrya lossowiana* Herre. Dinteria 10:3–12

Giess W (1981) Die in der zentralen Namib von Südwestafrika/Namibia festgestellen Pflanzenarten und ihre Biotope. Dinteria 15:14–71

Giess W (1982) Zur Verbreitung des Tabaks in Südwestafrika – *Nicotiana africana* Merxm. Dinteria 16:11–20

Hachfeld B (1996) Vegetationsökologische Transektanalyse in der nördlichen Zentralen Namib. Diplomarbeit, Univ Hamburg

Jürgens N (1991) A new approach to the Namib Region. Vegetatio 97:21–38

Jürgens N (1996) Psammophorous plants and other adaptations to desert ecosystems with high incidence of sandstorms. Feddes Repert 107:345–359

Jürgens N, Günster A, Seely MK, Jacobsen KM (1997) Desert. In: Cowling RM, Richardson DM (eds) Vegetation of Southern Africa. Cambridge University Press, Cambridge, pp 189–214

Lancaster J, Lancaster N (1984) Climate of the central Namib Desert. Madoqua 14:5–61

Lenssen-Erz T (1997) Metaphors of intactness of environment in Namibian rock paintings. In: Faulstich P (ed) Rock art as visual ecology. IRAC Proc, American Rock Art Research Association, Tucson, Arizona, 1:43–54

Lindesay JA, Tyson PD (1990) Climate and near-surface airflow over the central Namib. In: Seely MK (ed) Namib ecology – 25 years of Namib research. Transvaal Museum, Pretoria, pp 27–38

Martin H (1963) A suggested theory for the origin and a brief description of some gypsum deposits of South West Africa. Trans Geol Soc S Afr 66:345–351

Nordenstam B (1974) The flora of the Brandberg. Dinteria 11:3–67

Nordenstam B (1982) The Brandberg revisted. Dinteria 16:3–9

Olivier J (1995) Spatial distribution of fog in the Namib. J Arid Environ 29:129–138

Robinson ER, Giess W (1974) Report on the plants noted in the course of a trip from Lüderitz Bay to Spencer Bay, January 10–21. Dinteria 10:13–17

Scholz H (1967) Die Böden der Wüste Namib/Südwestafrika. Z Pflanzenernähr Bodenk 119:91–107

Tränkle U, Hübner F (1994) Vegetationskundlich-floristische Untersuchungen des Rössing-Berges östlich Swakopmund (Namibia). Unpublished report, Universität Stuttgart-Hohenheim

van Jaarsveld E (1978) The succulent riches of South Africa and Namibia. Aloe 24:45–92

Williamson G (1997) Preliminary account of the floristic zones of the Sperrgebiet (Protected Diamond Area) in southwest Namibia. Dinteria 25:1–68

10.4 Vegetation of Malagasy Inselbergs

E. Fischer and I. Theisen

10.4.1 Introduction

While the geomorphology of inselbergs in Madagascar is well known, their vegetation structure up to now has been less analyzed. Mostly, information from the surrounding savannas and forests is available. Until today only few publications concerning the vegetation of inselbergs are obtainable (Rauh 1973, 1992; Koechlin et al. 1974), and even in recent publications (e.g., Lowry et al. 1997), inselbergs are mentioned only marginally; new publications on Malagasy inselbergs are not available. However, due to the increasing habitat destruction on Madagascar, inselbergs play an important role in conservation. On the Central Plateau, they mostly represent the only vestiges of natural vegetation. The present chapter provides many new data and fills a gap in our knowledge of these inselbergs. A more detailed account will be published elsewhere. The data presented here are mainly based on field studies made by E. Fischer and I. Theisen from March to April in 1993.

10.4.2 Geography and Geology

With a total land area of 587 000 km^2, Madagascar, the fourth largest island in the world, is 1600 km long and 580 km wide. Two-thirds of this "micro-continent" consists of a Precambrian crystalline basement with granite and gneiss which had been extensively metamorphosed over the intervening period, covering an area of 400 000 km^2. A narrow coastal plain lies behind the seaboard, soon interrupted by a steep escarpment which rises abruptly from the warm lowlands to form a plateau varying in height from 800 to 1500 m. This upland region (Central Plateau), forms a vast area which gradually dips downwards towards the northern and southern lowlands. The Central Plateau is not a flat area, but consists of a complex blend of hills

Ecological Studies, Vol. 146
S. Porembski and W. Barthlott (eds.) Inselbergs

and valleys brought about by extensive reworking of the crystalline basement, by geological forces and by eons of erosion. This has given rise to numerous outcropping peaks made up of rocks such as quartz and granite or gneiss, which are more resistant to weathering (e.g., the granite peaks of Andringitra and the quartz of Itremo). The region between Fianarantsoa and Ambalavao is especially rich in inselbergs (Fig. 10.4.1). Here, isolated monoliths of smooth, rounded granite dominate the surrounding vegetation of grassy plains and scattered rice fields. These rocky islands protrude from a vast blanket of red lateritic clay which covers the main part of the Precambrian basement rocks, varying in depth from 10 to as much as 80 m. This characteristic red soil – responsible for the "great red island" – is easily eroded, when torrential rains sweep the lateritic soils into the rivers turning them red. Most of the central highlands are now covered with grassland, which is burned off annually. Many of the major massifs (e.g., Ankaratra, Itasy) on the Central Highland are not derived directly from the

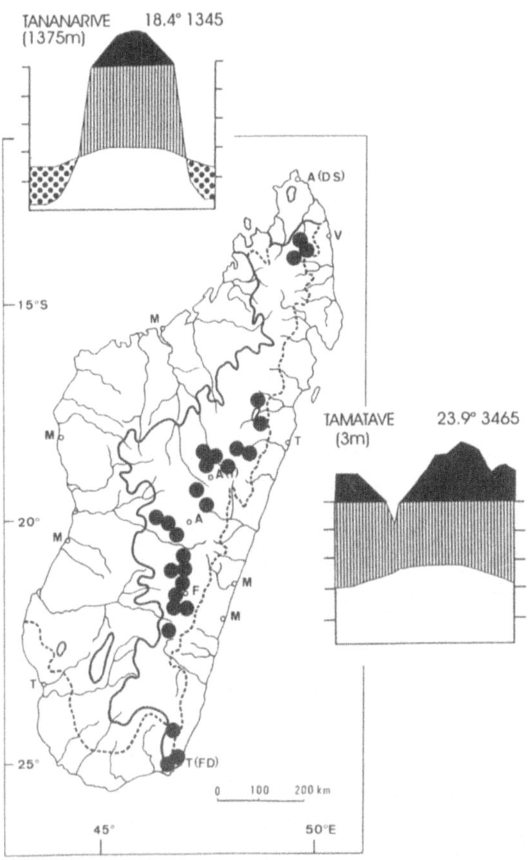

Fig. 10.4.1. Distribution of inselbergs in Madagascar

original Precambrian basement rocks, but are the results of more recent volcanic events. The western third of the island is generally of much lower relief, consisting of sedimentary deposits which exhibit a limited amount of folding. Ancient Permian deposits (280–225 Ma) are overlain by sediments from the Jurassic and Cretaceous periods, the latter being mainly of marine origin. These Mesozoic limestone and chalk forms the Plateau Calcaire, which drops down in three steps to the Mozambique Channel. In the Isalo mountains in the southwest the sedimentary sandstone rocks have been intensely eroded into a landscape of winding canyons. For a detailed survey of geology and soils see Battistini and Hoerner (1986).

10.4.3 Climate

Madagascar undergoes a high diversity of climates (Rauh 1973, 1992; Koechlin et al. 1974). Its geographic position between latitude 12° and 25° south under the tropic of Capricorn set the large island into the tropical zone. The prevailing influence on temperature and rainfall is the presence of the great eastern mountain ranges, which interrupt the moisture-bearing winds coming from the Indian Ocean (monsoon). They climb up the mountains, the air cooling down with rising altitude, and unload most of their moisture. Here, annual precipitation of 3500 to 5000 mm can be observed. From the eastern slopes, the winds blow westward across the Central Plateau, bringing with them only limited rain (up to 1000–1500 mm) with a winter dry season of 4 to 7 months. However, fog provides sufficient air humidity to enable rich growth of lichens (*Usnea* spp., *Cladonia* spp.). The dry west, southwest, and extreme north (Cap d'Ambre) have an annual rainfall between 500 and 900 mm, the dry season lasting between 7 and 9 months. The extreme southwest of Madagascar has a characteristic semidesert regime (precipitation about 350 mm), with a prolonged dry season broken by sparse and unpredictable rains. The temperature varies according to altitude and latitude, from subequatorial humid to subtropical dry weather (see Rauh 1992).

10.4.4 Vegetation

The main reason for the fascination of Madagascar is the species richness of its vegetation. The unique flora, of which about 80 % are endemic (Rauh 1973; Leroy 1978), consists of at least 12 000 species of flowering plants.

Madagascar is separated from Africa by the Mozambique Channel, a tract of water only 300 km wide at its narrowest point, yet of fundamental importance in having shaped the evolutionary destiny of Madagascar's plants and animals by allowing them to evolve in isolation and high endemism.

The vegetation of Madagascar can be differentiated into two main regions: eastern and western. Along the eastern coast, the richest formation of all vegetation types, the Eastern Lowland Evergreen Rainforest is situated. It is divided into the forest on the flat, sandy coastal plain – which has now been totally destroyed and replaced by grasslands or *Philippia* thickets – and the lowland evergreen rainforest of the eastern slope up to 800 m. The diversity of species in this type of forest is extraordinary: ca. 90 % of the constituent plants are endemic to Madagascar. Some examples are *Cycas thouarsii*, *Dypsis decaryi*, as well as 140 further palms of this genus, all endemic to Madagascar (Dransfield and Beentje 1995). In the swamps and *Sphagnum* bogs, the carnivorous *Nepenthes madagascariensis* and *N. masoalensis* are found. The evergreen montane forest, situated above 900 m on the mountain range of the Central Plateau, is especially rich in epiphytes. The palms of the lowland rainforest are replaced by tree ferns of the genus *Cyathea* (33 species, 31 of them endemic). The genus *Impatiens* occurs with ca. 120 species, nearly all of them endemic. Above 2500 m an ericaceous shrub forest similar to that occurring on East African high mountains can be observed. An alpine zone (paramo), however, is lacking. The main parts of the Central Plateau, nowadays devasted and covered with secondary grassland and rice fields, were inhabited by a sclerophyllous forest with *Uapaca bojeri* and several members of the endemic Sarcolaenaceae (*Sarcolaena* spp., *Leptolaena* spp.). Today, only scattered remnants can be found on the western slopes of the Central Plateau. Inselbergs are most frequent in this region and represent dry islands in a more or less humid climate, which have probably always been free from forest vegetation. The deciduous dry forest is widely distributed in western and northern Madagascar, occurring mainly on Mesozoic rocks. In northern Madagascar, *Euphorbia* species (e.g., *E. ankaranensis*, *E. neohumbertii*), *Pachypodium* species (e.g., *P. baroni*, *P. decaryi*, *P. ambongense*), and the baobab *Adansonia suarezensis* dominate. In the central and southern parts, pachycaul trees like *Delonix* spp., *Adansonia grandidieri*, *A. fony*, *A. madagascariensis*, and *A. za* can be found. The southwestern dry regions are covered by a xerophytic scrub with dominating Didiereaceae (*Didierea madagascariensis*, *D. trollii*, *Alluaudia procera* etc.) and arborescent *Euphorbia* species. For a more detailed account of vegetation see Koechlin et al. (1974) and Rauh (1973).

10.4.5 Inselberg Vegetation

Inselbergs occur in most of the vegetation types and climate zones of Madagascar. They are especially frequent on the Central Plateau (Fig. 10.4.1), but can also be found along the eastern coast south to Fort Dauphin (Tolianaro). In northwestern and southwestern Madagascar, granitic outcrops are replaced by limestone or sandstone rocks (see above).

In the following, a survey of characteristic habitat types observed on Malagasy inselbergs is presented.

10.4.5.1 Exposed Rock Surfaces

Cryptogamic Vegetation of Rock Surfaces. Sun exposed, bare rocks are usually covered by crustose lichens and cyanobacteria, the latter often restricted to wet parts. Thus cryptogamous settlement is responsible for the black and brown color of almost every inselberg. Dominating lichen species are *Acarospora* sp., *Buellia* sp., *Caloplaca* sp., *Peltula* cf. *obscurans*, *Protoparmelia* sp., and *Xanthoparmelia* sp.

Inselbergs in the Central Plateau are covered with chlorophytic lichens like *Heterodermia*, *Diploschistes*, *Verrucaria* and *Usnea*. Several *Usnea* species may form mat-like covers up to 5 cm high.

Cryptogamic Vegetation of Boulders. Isolated boulders on rock surfaces are covered by chlorophytic lichens, mainly species of *Caloplaca*, *Xanthoparmelia*, and *Tonninia*.

Vegetation of Drainage Channels. Drainage channels are cut into the rock surface by offrunning water. Phanerogamic vegetation is totally lacking and replaced by cyanobacteria and lichens. The whitish stripe known from other African inselbergs, which is a boundary zone of naked rock between cyanobacteria and lichens, has been observed on Malagasy inselbergs too.

Wet Flush Vegetation. Adjacent to the ephemeral flush vegetation, where water seeps continuously, communities which consist mainly of *Utricularia* species and other rainy-season ephemers occur on a carpet of cyanobacteria covering the rock. This habitat type is frequent on inselbergs and other plain rocky outcrops in the Central Highland of Madagascar. *Utricularia caerulea*, *U. firmula*, *U. subulata*, *Antherothoma naudinii*, *Emilia graminea*, and different Eriocaulaceae are characteristic elements for this vegetation type. All constituents of wet flush could also be observed in ephemeral flush vegetation.

Vascular Lithophytic Vegetation. Only few species are growing on free exposed rocks. Several succulents, especially *Aloe* spp., *Kalanchoe* spp., and taxa from the *Euphorbia milii* complex are found here. Also orchids are frequent as colonizers of bare rock surfaces, e.g., *Bulbophyllum oreodorum*, *Polystachya perrieri*, or *Angraecum sororium*. They all are closely related to epiphytic taxa. Occasionally, *Rhipsalis baccifera* subsp. *horrida* is observed.

One problem is the occurrence of neophytic weeds like *Furcraea foetida* (Agavaceae), which cover the rocks of some inselbergs nearly completely.

10.4.5.2 Vegetation of Rock Crevices

Vegetation of Horizontal Crevices, Vertical Crevices, and Clefts. Horizontal and vertical crevices bear a similar vegetation. Here ferns like the poikilohydric *Actiniopteris radiata*, *Pellaea calomelanos*, *P. viridis*, *P. tripinnata*, as well as some *Cheilanthes* and *Asplenium*, can be observed. Flowering plants found in this habitat are *Impatiens baroni* (under very humid conditions), *Senecio* spp., *Helichrysum* spp., *Tetradenia* spp., *Cyanotis nodiflorum*, and *Bulbostylis* spp.

Vegetation around Boulder Bases. Shady habitats are provided around boulder bases. Here, beside several ferns, *Impatiens baroni* and species of *Streptocarpus* are growing.

10.4.5.3 Vegetation of Rock Depressions

Vegetation of Seasonal Rock Pools. Seasonally water-filled rock pools are almost devoid of ephemerals. On the inselbergs studied rock pools seem to be of rare occurrence. While one of them presented no phanerogamic vegetation (inselberg near Zazafotsy), the other was covered by a mat of *Eriocaulon* spp. and *Rotala* spp. (inselberg near Andranovelona).

Other frequently observed elements are the ecologically widespread *Digitaria horizontalis*, *Lindernia rotundifolia*, *Eragrostis* sp. and the liverwort *Riccia trichocarpa*. Several species occur on fine debris (*Microchloa kunthii* and *Antherotoma naudinii*), in the ephemeral flush vegetation (*Utricularia caerulea*, *U. subulata*, *Bulbostylis* sp.) and in monocotyledonous mats (*Cyanotis nodiflorum*) as well. The endemic *Cyanotis nodiflorum* (Fig. 10.4.2) is a perennial species with close relationship to the East African perennials of this genus. While in western and central Africa, annual species are dominating, e.g., *Cyanotis lanata*, an undescribed perennial, was observed on inselbergs in eastern Zaire (Porembski et al. 1997).

Vegetation of Permanently Water-Filled Rock Ponds. Permanently water-filled rock ponds could not be observed.

Vegetation of Rock Debris. On gentle slopes or flattened parts of rock fine material of erosion has been accumulated. This fine debris is colonized by a pioneer vegetation type consisting mainly of the annuals *Perotis patens*, *Exacum* spp., and *Antherotoma naudini*. Accessory differentials are *Chamaecrista mimosoides*, *Emilia graminea*, and *Sporobolus pyramidalis*, which invaded from the *Loudetia* mats.

This plant community can be observed frequently adjacent to *Coleochloa* and *Loudetia* mats (see below).

10.4.5.4 Vegetation of Soil-Filled Depressions

Mats

As a most uniform characteristic plant community inhabiting rock surfaces monocotyledonous mats can colonize even steep slopes. In the Central Highland of Madagascar there are two plant communities,

Fig. 10.4.2. *Cyanotis nodiflorum*, Andranovelona

which form the monocotyledonous mats at lower and medium altitudes (ca. <2000 m). They are distinguished by their dominating species and several differentials. Most frequent are the *Coleochloa setifera* community as pioneer and the *Loudetia simplex* community on rocks with thicker humus layer. Above the altitude of 2000 m, a sharp impoverishment of the flora will be recognized, where *Xerophyta dasylirioides* is the dominant species while *Coleochloa* is greatly diminished or lacking.

Coleochloa Mats. The *Coleochloa setifera* community (Fig. 10.4.3) shows a lot of geographic differentiations, due to the occurrence of local endemic species in its range. Common species in the whole area are: *Coleochloa setifera*, *Myrothamnus moschata*, *Selaginella echinata*, *Sobennikoffia* spp., and *Xerophyta dasylirioides*. The following genera occur in nearly all of the vegetation plots, but represented with different species: *Aloe*, *Cynanchum*, *Euphorbia*, *Kalanchoe*, and *Senecio*.

Rocky outcrops in southern and central parts of the *Coleochloa* area bear either *Pachypodium densiflorum* or *P. horombense* and *P. rosulatum*; other geographical differentials (see below) are *Angraecum sororium*, *Euphorbia milii* agg., *E. duranii*, *E. didierioides*, *Gladiolus bojeri*, *Ischnolepis tuberosa*, *Lindernia andringitrae*, *Nematostylis anthophylla*, *Perrieriastrum oreophilum*, *Radamea montana*, *Xerophyta pinifolia*, and *X. eglandulosa*. Typical cryptogamic elements are *Cladonia pycnoclada*, *Leucobryum madagascariensis*, and *Pellaea boivinii*.

Fig. 10.4.3. *Coleochloa* mat near Fianarantsoa with *Angraecum sororium* (*left*) and *Kalanchoe synsepala* (*right*)

Mats Formed by Other Species. The floristically poor and very homogenous *Loudetia simplex* community is characterized by the dominance of Poaceae, especially *Loudetia simplex*, *Hyparrhenia rufa*, and *Aristida rufescens*. The following species have a scattered occurrence in this vegetation type: *Aloe macroclada*, *Antherotoma naudinii*, *Buchnera hispida*, *Exacum* spp., *Lasiosiphon madagascariensis*, *Rhynchelytrum roseum*, and *Tachiadenus longiflorus*. Koechlin et al. (1974) presumed that the *Loudetia* mats are the natural habitats for several today widespread grasses (*Aristida*, *Hyparrhenia*, *Loudetia*) and it is possible that they invaded the secondary grassland from here.

The *Xerophyta dasylirioides* community in the Central Highland on altitudes above 2000 m is characterized by *Xerophyta dasylirioides* and dominant cryptogams. *Cladonia*, *Leucobryum*, and *Polytrichum* are especially abundant and the stems of *Xerophyta* are covered by epiphytes such as *Usnea* spp., *Parmotrema* spp., and *Frullania* spp.

Ephemeral Flush Vegetation. Plant cover which is supplied with water during the rainy season often develops on thin layers of peaty soil, adjacent to monocotyledonous mats or larger rock pools on less steep slopes. It is the most diverse habitat type dominated by many Cyperaceae (*Bulbostylis*, *Cyperus*, *Scleria*), Poaceae, carnivorous plants (*Drosera*, *Utricularia*, *Genlisea*), Eriocaulaceae, and Xyridaceae.

Characteristic species of this vegetation type are: Cyperaceae (*Bulbostylis* spp., *Scleria* spp.), *Crepidorhopalon microcarpaeoides*, *Drosera madagascariensis*, *D. indica*, Eriocaulaceae, *Lindernia parviflora*, *L. viguieri*, *Lycopodiella affinis*, *Rhamphicarpa fistulosa*, *Schizaea pusilla*, *Micrageria filiformis*, *Utricularia* species (*U. caerulea*, *U. livida*, *U. subulata*, *U. firmula* and *U. prehensilis*), *Xyris* spp. and bryophytes (*Philonotis* spp., *Campylopus* spp.).

In the drier part of the Central Highland (mainly southwest of Amabalavao and around Ihosy) the typical seasonal ephemeral flush vegetation is developed. These are types influenced by dry-season lack of perennials. Here, annual species form the main differentials like *Bulbostylis* sp., *Lindernia parviflora*, *L. viguieri*, *Rhamphicarpa fistulosa*, *Utricularia subulata*, and *Drosera indica*. The latter replaces the perennial *Drosera madagascariensis* in the drier areas as vicarious species.

According to these observations, two communities can be distinguished: the more or less permanent humid *Scleria* community and the distinctly seasonal *Bulbostylis* community.

The ephemeral flush vegetation is frequent in the central Highlands where even permanent wet formations occur. Here, perennials such as the hydrophilous *Genlisea margaretae* can survive. Most characteristic for permanently humid formations are *Scleria* species, which are dominating elements within this vegetation type. With a high presence, *Drosera madagascariensis*, *Utricularia caerulea*, *Xyris anceps*, *X. humilis*, and *Paepalanthus* sp. occur. *Genlisea margaretae* has so far been recorded from only one station at Andranovelona (Klotz and Köck 1991).

As stated above, the ephemeral flush vegetation is the most species-rich habitat type occurring on Malagasy inselbergs; but it is a striking feature that endemic species are very rare or lacking. Thus, all species of Madagascan *Utricularia* occur in Africa or are even pantropical elements. *Drosera madagascariensis*, *D. indica*, and *Antherotoma naudinii* are widespread in Africa. Most of the Cyperaceae occur in tropical Africa as well and the putative endemic *Genlisea margaretae* (see below) is an eastern African element. Truly Malagasy endemics are: *Exacum* spp., *Cynorchis uniflora*, and *Crepidorhopalon microcarpeoides*.

Woody Vegetation. Belt forests normally surround inselbergs at the foot, where the offrunning water causes the growth of woody formations as an azonal vegetation type even in dry climates. Due to human influence, the belt forests of our study inselbergs have totally disappeared. Only vestiges may have survived in remote areas. The same is true for the summit forests in our research area. The only observation we made was a small relic forest consisting of *Uapaca bojeri*, *Sarcolaena grandiflora*, *Leptolaena* spp., *Xyloolaena perrieri*, and *Schizolaena microphylla* on the twin inselberg of Zazafotsy.

10.4.6 Systematic Characteristics

The granitic and gneissic inselbergs of Madagascar bear a rich flora, which comprises 423 species of ferns and flowering plants (compiled after our own records and herbarium studies). This species number is certainly not complete, as modern taxonomic treatments for several important families (e.g., Eriocaulaceae, Fabaceae, Cyperaceae, Poaceae, cf. Bosser 1969) are lacking and the data concerning these groups were obtained mainly by our own field observations. For other families, supplementary data from herbarium study and literature (e.g., Polhill 1982; Rauh 1995; Flore de Madagascar) could be included.

The most species-rich families on Madagascan inselbergs are Asteraceae (79 spp.), Poaceae (36 spp.), Orchidaceae (33 spp.), Asclepiadaceae (31 spp.), Cyperaceae (26 spp.), Euphorbiaceae (22 spp.), Aloaceae (21 spp.), Scrophulariaceae (20 spp.), Gentianaceae (20 spp.), and Fabaceae (19 spp.).

These 10 families comprise 307 species, i.e., 73% of the total inselberg flora. The surprisingly high number of Orchidaceae includes 21 lithophytic species from a mainly epiphytic relationship (*Angraecum*, *Bulbophyllum*, *Jumellea*, *Polystachya*, *Sobennikoffia*) and 12 terrestrial orchids from the genus *Cynorkis*.

The remaining 116 species belong to 25 families (see Table 10.4.1).

10.4.7 Life-Forms and Adaptations

An analysis of life-forms found in the Madagascan inselberg flora (definitions according to Seine 1996) reveals 29 phanerophytes (7%), 182 chamaephytes (43%), 38 hemicryptophytes (9%), 123 therophytes (29%), 30 geophytes (7%), 20 lithophytes (5%) and 1 liana (0.2%).

Succulents. A unique feature of the Madagascan inselbergs is the richness of succulents (Rauh 1995); 123 species (29%) have either succulent leaves and/or stems. They belong to the following genera: *Aloe* (21 spp., Reynolds 1966), *Angraecum* (7 spp.), *Bulbophyllum* (4 spp.), *Ceropegia* (8 spp.), *Cynanchum* (17 spp.), *Euphorbia* (12 spp.), *Jumellea* (4 spp.), *Kalanchoe* (9 spp.), *Pachypodium* (6 spp.), *Perrierastrum* (1 sp.), *Polystachya* (4 spp.), *Rhipsalis* (1 sp.), *Sarcostemma* (4 spp.), *Sedum* (1 sp.), *Senecio* (21 spp.), *Sobennikoffia* (2 spp.), and *Stapelianthus* (1 sp.).

Poikilohydrics. Twenty four species (5.6%) are resurrection plants and the majority of them (16 spp.) belongs to the ferns (*Actiniopteris*, *Cheilanthes*, *Pellaea*, *Selaginella*, see Alston 1932; Gaff 1971; Rauh and Hagemann 1991; Tardieu-Blot 1958). However, eight angiosperm species are poikilohydric as well. Within the monocotyledons, *Coleochloa setifera* (Cyperaceae) and the three *Xerophyta* species (*X. dasylirioides*, *X. eglandulosa*, *X. pinifolia*, Smith and Ayensu 1974) are such resurrection plants. Poikilohydric dicotyledons are *Myrothamnus moschata* (Myrothamnaceae), *Lindernia andringitrae*, *L. horombensis*, and *L. pygmaea* (Scrophulariaceae).

Carnivorous Plants. Typical elements of wet and ephemeral flush vegetation on inselbergs are carnivorous plants, which comprise 17 species (4%) in Madagascar. The cosmopolitan carnivorous genus *Drosera* occurs with five species (Keraudren-Aymonin 1982): *D. natalensis*, *D. burkeana*, *D. madagascariensis*, *D. humbertii*, which is confined to a small area in northern Madagascar (Marojejy massif), where it grows on granitic outcrops in the summit region above 1400 m, and *D. indica*.

The genus *Genlisea* is represented with one species, *G. margaretae* (*G. recurva*); it has been described from the surroundings of Tananarivo. (Bosser 1956, 1959) and was only known from this type locality. Rauh

Table 10.4.1. Species numbers and endemics of the Malagasy inselberg flora (based on our own records, herbarium specimens, and Flore de Madagascar et des Comores)

Family	No. of species on inselbergs	No. of endemic species	Endemics (%)	Endemic genera
Asteraceae	79	75	95	*Syncephalum*
Poaceae	36	15	42	*Isalus, Stenotaphris*
Orchidaceae	33	33	100	*Sobennikoffia*
Asclepiadaceae	31	30	97	*Stapelianthus, Ischnolepis*
Cyperaceae	26	0	0	
Euphorbiaceae	22	21	96	
Aloaceae	21	21	100	
Scrophulariaceae	20	12	60	*Pseudomelasma, Radamaea*
Gentianaceae	20	20	100	
Fabaceae	19	11	58	*Mundulea*
Adiantaceae	13	3		
Lentibulariaceae	12	0	0	
Crassulaceae	10	10	100	
Apocynaceae	9	9	100	
Melastomataceae	9	8	88	*Amphorocalyx, Dionychia, Grawesia*
Acanthaceae	8	6	75	*Achyrocalyx*
Rubiaceae	8	4	50	*Nematostylis*
Droseraceae	5	1		
Eriocaulaceae	5	1		
Lamiaceae	5	5	100	*Perrierastrum*
Sarcolaenaceae	5	5	100	*Leptolaena, Sarcolaena, Schizolaena, Xyloolaena*
Thymeleaceae	4	4	100	
Amaryllidaceae	3	0	0	
Commelinaceae	3	1		
Iridaceae	3	3	100	
Velloziaceae	3	3	100	
Balsaminaceae	2	2	100	
Selaginellaceae	2	2	100	
Burmanniaceae	1	0	0	
Cactaceae	1	0	0	
Dioscoreaceae	1	1	100	
Myrothamnaceae	1	1	100	
Ophioglossaceae	1	0	0	
Ranunculaceae	1	0	0	

(1973) stated that the locus classicus has been destroyed and so the species was thought to be extinct in Madagascar. Klotz and Köck (1991) discovered in 1988 a further locality near Andranovelona. The genus *Utricularia* also occurs in ephemeral flush vegetation, among them *U. arenaria*, *U. livida*, *U. subulata*, *U. scandens*, *U. prehensilis*, and *U. caerulea* (Taylor 1989).

10.4.8 Biogeography and Endemism

The flora of Madagascar shows a high degree of endemism which surpasses 80 % on species level and 20 % on the level of genus (Rauh 1973; Leroy 1978). Similar results can be obtained regarding the flora of inselbergs. From the 423 species of vascular plants (25 pteridophytes, 398 angiosperms), 304 (72 %) are endemic to Madagascar. On generic level, 20 genera of the 124 genera observed are endemic, which is 16 % (Tab. 10.4.1).

The degree of endemism varies considerably within the families. Many of the endemic species have close relatives in continental Africa. Examples are the poikilohydric species of Velloziaceae and Cyperaceae, which all show close relationship to East African taxa or even actually occur in eastern Africa (*Coleochloa*, Haines and Lye 1983). The same is true for *Myrothamnus moschata* (Myrothamnaceae), closely related to *M. flabellifolia*, and for the three poikilohydric *Lindernia* species (*L. andringitrae*, *L. horombensis*, *L. pygmaea*, Fischer 1995), which find their closest relatives in the African taxa *L. welwitschii*, *L. yaundensis*, *L. sudanica*, and *L. pulchella* (Fischer 1992). Koechlin et al. (1974) name another interesting member of the high mountain moncotyledonous mats, *Sedum madagascariense*, from the Andringitra massif, whose closest relative, *Sedum churchillianum*, occurs in the Ruwenzori mountains in central Africa. It is interesting to note that there are nearly no relations to the inselberg flora of the Seychelles or Southeast Asia (Biedinger and Fleischmann, this Vol.). In this respect, inselbergs sharply contrast with the surrounding vegetation, where Asian elements are quite well represented (see Leroy 1978), while they are almost lacking on Malagasy inselbergs.

The more widespread taxa usually were also recorded in eastern and southern Africa. Characteristic examples are *Actiniopteris australis*, *A. radiata*, *Pellaea boivinii*, *Coleochloa setifera*, *Drosera burkeana*, *D. madagascariensis*, *D. natalensis*, *Genlisea margaretae*, *Utricularia appendiculata*, *U. bisquamata*, *U. firmula*, *U. prehensilis*, and *U. welwitschii*. *Lindernia nummulariifolia* and *Utricularia arenaria* are widespread in tropical Africa and India. *Drosera indica*, *Utricularia foveolata*, and *U. scandens* are widespread paleotropic elements known from Africa to Australia. *Rhip-*

salis baccifera is known from South America, Africa, and Sri Lanka. It has developed a secondary evolution center on Madagascar. *Rhipsalis baccifera* was probably introduced by bird dispersal and beside the epiphytic subsp. *mauritiana*, a neotenic race with aberrant mesotonic branching and bristly stem, *R. baccifera* subsp. *horrida* (Barthlott and Taylor 1995) has evolved. *Utricularia livida* is recorded also from Mexico and Africa and *Utricularia subulata* is a pantropical element. *Sarcostemma australe* shows a Madagascan-Australian disjunction and is lacking in Africa. Another phytogeographically interesting species is *Utricularia caerulea*, which ranges from India to Japan, New South Wales, and north to central Madagascar and does not occur in Africa.

Mechanisms of speciation can be observed in the taxonomically most difficult *Euphorbia milii* aggregate which comprises at least 20 subspecies, varieties, and forms. On nearly each investigated inselberg, taxa from this compact succulent shruby group could be found. According to our field observation, however, it seems that not every inselberg is characterized by a special taxon, as has been stated (e.g., Rauh 1995). Neighboring inselbergs generally possess the same, mainly locally restricted, species or variety. However, different taxa can be observed in the Central Plateau inselbergs in larger neighboring areas, an observation which gave rise to the statement cited above. *Euphorbia milii* s.str. has been recorded from the region north to Antananarivo south to Ambrositra with different varieties. *E. fianarantsoae* occurs only around Fianarantsoa, while *E. duranii* is distributed from Fianarantsoa to Ambalavao. All these taxa preferably grow in *Coleochloa* mats. Another closely related species growing on gneissic or granitic outcrops is *E. didierioides* (Fig. 10.4.4) which has been observed on rock plateaus near Ihosy growing in fine debris.

One of the most striking features is a substrate specifity observed in most of the succulent genera. From the species of *Aloe*, 21 are confined to granite and gneiss, 13 taxa can be found only on quartz, 7 taxa are confined to limestone, and on triassic sandstone, mainly in the Isalo mountain, 4 endemic taxa can be found (Reynolds 1966). Like *Aloe*, the *Pachypodium* species show a strong connection to the rock type (Markgraf 1976). Only *P. baronii* var. *baronii* in North Madagascar; *P. densiflorum* (Fig. 10.4.5), *P. horombense* and *P. rosulatum* var. *rosulatum* on the Central Plateau and the recently discovered *P. inopinatum* (Lavranos 1996), known from only one inselberg in the Central Plateau west of Lac Alaotra, are typical inselberg taxa growing on granitic outcrops. *P. baronii* var. *windsori* and *P. decaryi* are found on limestone in northern Madagascar. *P. brevicaule* is confined to quartz in the Ibity and Itremo mountains. *P. rosulatum* var. *gracilius* prefers sandstone habitats in the Isalo mountains (Rauh 1992).

Fig. 10.4.4. *Euphorbia didierioides*, Ihosy

Fig. 10.4.5. *Pachypodium densiflorum*, Ihosy

Similar observations can be made in the genera *Kalanchoe*, *Senecio*, *Euphorbia*, and *Ceropegia*.

10.4.9 Considerations upon Diversity

Here, only a few remarks on diversity will be made. High alpha diversities were shown by ephemeral flush vegetation, where the Shannon index H varied between 1.93 and 2.14. The Shannon index of *Colechloa* mats was, however, considerably lower and ranged between 1.43 and 1.66. As rock pools have only been observed occasionally, no comparative data can be presented here. Beta diversity within *Coleochloa* mats varied from 0.375 to 0.761, the average being about 0.583. This can be explained on one hand by the occurrence of some regional differential species with only restricted distribution (especially from *Aloe*, *Pachypodium* and the *Euphorbia milii*-complex), which resulted in comparatively high beta diversity (low values), and on the other by the common basic structure of these mats, which bear generally the same characteristic species. In ephemeral flush vegetation, beta diversity was much higher (between 0.023 and 0.098) and only very few species are common to the inselbergs studied. This may be caused by the local distribution of several species (e.g., *Genlisea margaretae* and several *Utricularia* spp.) as well as by the different precipitation on the inselbergs studied. Thus, the different length of dry season seems to be one of the most important factors for this high diversity.

Acknowledgements. The authors are greatly indebted to the Deutsche Forschungs Gemeinschaft for financial support. Furthermore, we thank the Department d'Ecologie végétale, University of Antananarivo, for cooperation, and the Direction des Eaux et des Forêts (Antananarivo, Madagascar) for kindly giving permission to research and to export specimens for later identification. For valuable information and comments we wish to thank S. Porembski (Rostock) and R. Seine (Bonn). Thanks are also due to U. Becker (Köln) for determination of some lichens. Special thanks belong to D. Supthut, Director of the Succulent Collection (Zürich). Finally, we are indebted to F. Ditsch, W. Höller, and I. Meusel (all Bonn) without whose help the field work in Madagascar would not have been possible.

References

Alston AHG (1932) Selaginellaceae. In: Christensen A (ed) The pteridophyta of Madagascar. Dan Bot Ark 7:193–201

Barthlott W, Taylor NP (1995) Notes towards a monograph of Rhipsalideae (Cactaceae). Bradleya 13:43–79

Battistini R, Hoerner JM (1986) Géographie de Madagascar. Paris, pp 1–187

Bosser J (1956) Un nouveau genre malgache de Lentibulariacée. Naturaliste Malgache 8:27–30

Bosser J (1959) Sur deux nouvelles Lentibularicées de Madagascar. Naturaliste Malgache 10:21–29

Bosser J (1969) Graminées des pâturages et des cultures à Madagascar. Mém ORSTOM 35:1–440

Dransfield J, Beentje H (1995) The palms of Madagascar. Royal Botanic Gardens Kew and The International Palm Society, Kew

Fischer E (1992) Systematik der afrikanischen Lindernieae (Scrophulariaceae). Trop Subtrop Pflanzenwelt 81:1–365

Fischer E (1995) Revision of the Lindernieae (Scrophulariaceae) in Madagascar. 1. The genera *Lindernia* Allioni and *Crepidorhopalon* E. Fischer. Bull Mus Natl Hist Nat Paris, 4e Sér Sect B Adansonia 7:227–257

Gaff DF (1971) Desiccation-tolerant flowering plants in southern Africa. Science 174: 1033–1034

Haines RW, Lye KA (1983) The sedges and rushes of East Africa. East African Natural History Society, Nairobi, pp 1–404

Humbert H (ed) (1936ff) Flore de Madagascar et des Comores. Muséum National d'Histoire Naturelle, Laboratoire de Phanérogamie, Paris

Humbert H (1963) Composées III. In Humbert H (ed) Flore de Madagascar et des Comores, 189e fam. Muséum National d'Histoire Naturelle, Laboratoire de Phanérogamie, Paris, pp 623–911

Keraudren-Aymonin M (1982) Droseracées. In: Humbert H (ed) Flore de Madagascar et des Comores, 87e fam. Muséum National d'Histoire Naturelle, Laboratoire de Phanérogamie, Paris, pp 53–62

Klackenberg J (1990) Gentianacées. In: Humbert H (ed) Flore de Madagascar et des Comores, 168e fam. Muséum National d'Histoire Naturelle, Laboratoire de Phanérogamie, Paris, pp 5–167

Klotz S, Köck UV (1991) Neufund von *Genlisea recurva* (Lentibulariaceae) auf Madagaskar. Willdenowia 20:131–133

Koechlin J, Guillaumet JL, Morat P (1974) Flore et végétation de Madagascar. In: Tüxen R (ed) Flora et Vegetatio Mundi. J Cramer, Vaduz, pp 1–645

Lavranos JJ (1996) *Pachypodium inopinatum* (Apocynaceae), a new species from Madagascar. Cactus Succulent J 68:171–176

Leroy JF (1978) Composition, origin, and affinities of the Madagascan vascular flora. Ann Mo Bot Gard 65:535–589

Lowry PP, Schatz GE, Phillipson PB (1997) The classification of natural and anthropogenic vegetation in Madagascar. In: Goodman SM, Patterson BD (eds) Natural change and human impact in Madagascar. Smithsonian Institution Press, Washington, pp 93–123

Markgraf F (1976) Apocynacées. In: Humbert H (ed) Flore de Madagascar et des Comores, 169e fam. Muséum National d'Histoire Naturelle, Laboratoire de Phanérogamie, Paris, pp 1–318

Moldenke H (1955) Eriocaulacées. In: Humbert H (ed) Flore de Madagascar et des Comores, 36e fam. Muséum National d'Histoire Naturelle, Laboratoire de Phanérogamie, pp 1–41

Polhill RM (1982) *Crotalaria* in Africa and Madagascar. Balkema, Rotterdam, pp 1–389

Porembski S, Fischer E, Biedinger N (1997) Vegetation of inselbergs, quarzitic outcrops and ferricretes in Rwanda and eastern Zaïre (Kivu). Bull Jard Bot Nat Belg 66:81–99

Rauh W (1973) Über die Zonierung und Differenzierung der Vegetation Madagaskars. Trop Subtrop Pflanzenwelt 1:1–145

Rauh W (1992) Klima- und Vegetationszonierung Madagaskars. In: Bittner A (ed) Madagaskar – Mensch und Natur im Konflikt. Birkhäuser, Basel, pp 31–53

Rauh W (1995) Succulent and xerophytic plants of Madagascar 1, Strawberry Press, Mill Valley, pp 1–343

Rauh W, Hagemann W (1991) *Selaginella moratii*, spec. nova (Selaginellales), a remarkable new species from central Madagascar. Plant Syst Evol 176:205–219

Reynolds GW (1966) The aloes of tropical Africa and Madagascar. The trustees, the aloes. Book Fund, Mbabane, Swaziland, pp 1–537

Seine R (1996) Vegetation von Inselbergen in Zimbabwe. Archiv naturwissenschaftlicher Dissertationen, vol. 2. Martina Galunder, Wiehl

Smith LB, Ayensu ES (1974) Classification of Old World Velloziaceae. Kew Bull 29:181–205

Tardieu-Blot ML (1958) Polypodiacées I. In: Humbert H (ed) Flore de Madagascar et des Comores, 5e fam. Muséum National d'Histoire Naturelle, Laboratoire de Phanérogamie, Paris, pp 1–391

Taylor P (1989) The genus *Utricularia* – a taxonomic monograph. Kew Bull Add Ser 14:1–724

10.5 Seychelles

N. BIEDINGER and K. FLEISCHMANN

10.5.1 Introduction

A short survey on the vegetation of inselbergs of the Seychelles Islands is given. The Seychelles form a relatively intact and isolated archipelago, that offers many opportunities for comparative phytogeographical studies. Furthermore, the granitic islands of the Seychelles bear large numbers of inselbergs which are highly contrasted to their mainland counterparts elsewhere in being situated on oceanic islands.

10.5.2 Geography and Geology

Located in the western Indian Ocean, the geographical position of the Seychelles Islands is very isolated. The granitic islands are situated some 930 km to the north of Madagascar, approximately 1600 km from East Africa and Mauritius and more than 1700 km from India.

The whole archipelago counts approximately 115 islands; a part of them are coralline and the others granitic. The granitic islands are recognized as the oldest oceanic islands worldwide. They were separated as fragments from the original land surface of Gondwana during the Cretaceous period (Baker and Miller 1963).

During the last glaciation periods, the sea level had fallen considerably (approximately 100 m after Walsh 1984), so that the area of the whole Seychelles Bank above water was much larger. The islands of the Seychelles in the present state are only the very tips of large mountainous islands, which have their major part below the water surface today.

Mahé, which is the largest granitic island of the Seychelles, has an area of 154 km^2, Praslin, the second largest about 37 km^2 and Silhouette about 19 km^2. The highest elevation of the Seychelles is Morne Seychellois, reaching 914 m asl on Mahé. Inselbergs and rock outcrops occur all over

Ecological Studies, Vol. 146
S. Porembski and W. Barthlott (eds.) Inselbergs

the granitic islands and along the whole altitudinal gradient and over a broad spectrum of sizes. On the Seychelles inselbergs and rock outcrops are called glacis, which is a French word meaning steep, rocky slope. Most of them do not have the characteristic dome-shaped appearance of inselbergs but rather form shield-like rock outcrops where the vegetation cover is less dense and vegetation differs considerably in species composition from the surrounding flora. Some of the outcrops are extremely steep and are traversed by deep clefts (Fig. 10.5.1).

Precambrian alkali granite containing microperthite dominates on Mahé (Baker and Miller 1963). In western Mahé at Port Glaud and on the offshore islands of Thérèse and Conception, a porphyritic granodiorite occurs (Braithwaite 1984), whereas on the geologically much younger Silhouette island syenitic rock is prevailing.

10.5.3 Climate

The Seychelles have a tropical climate with an average annual rainfall of 2200 mm and only few differences from year to year. Mean monthly rainfall

Fig. 10.5.1. Dome-shaped outcrops which are surrounded by rainforest are a widespread view on Mahé. (Photograph N. Biedinger)

exceeds 100 mm except in June, July, and August. Mean annual temperatures at sea level are around 24 °C at Victoria (Mahé). Seasonal and diurnal temperature ranges are very small. Relative air humidity also varies little with season; in Victoria, monthly means of relative air humidity differ from 75 % in April to a maximum of 80 % in January at the height of the rainy season. For a detailed description of the climatic situation see Walsh (1984).

10.5.4 Vegetation of the Granitic Islands

Compared to other oceanic islands, the Seychelles possibly have had the longest time for vegetation to develop by purely natural immigration and evolutionary processes (Dalziel 1995). These islands must have been continuously available for plant colonization for millions of years before the first human settlers arrived in 1768 (Lionnet 1984). This rare situation and the geographically isolated position gave rise to the formation of relictual vegetation elements of considerable botanical interest and a high level of endemism (High 1982). Approximately 40 % of the native plant species are endemic, including many famous rarities like the palm species *Lodoicea maldivica* (coco de mer) and the systematically isolated *Medusagyne oppositifolia* (Medusagynaceae).

Several studies (e.g., Vesey-Fitzgerald 1940; Sauer 1967; Bailey 1971; Procter 1984a) give general descriptions of the vegetation of the granitic islands. Updated reports on their floristics are the works of Robertson (1989), Friedmann (1987, 1991), Fleischmann et al. (1996), and Fleischmann (1997).

There is a large variety of vegetation types with coastal formations including mangroves and different forest types. In the inner islands, particularly Mahé, the plant formations can be divided into four categories (Jeffrey 1962; Procter 1984b): vegetation of the coastal plateau, lowland forest (coastal plateau–300 m), intermediate forest (300–550 m), and mountain mist forest (550–914 m). Inselberg vegetation is the fifth type, the locally so-called glacis-type vegetation.

The flora of the Seychelles's granitic islands comprises approximately 240 species, including endemics. Phytogeographical affinities mainly can be found with Africa (especially East Africa), Madagascar and Indomalaysia. Generally, the indigenous vegetation is highly invaded by non-native plants: approximately 280 invader plant species grow in the Seychelles (Robertson 1989).

10.5.5 Inselberg Vegetation

In the Seychelles, inselbergs occur from the sea shore to the mountain tops. Extreme edaphic and climatic conditions (high degree of insolation combined with high evaporation rates and poor soil water storage) exert selective pressures, resulting in a vegetation that completely differs from that of the surroundings. It is typical for the inselbergs of the Seychelles that substrate which accumulates in pockets and fissures consists largely of coarse quartz sand with variable amounts of peaty organic matter. The thin soil and scarce vegetation cover of inselberg habitats can retain only a small percentage of the precipitation (Fleischmann et al. 1996). These limiting factors reduce plant growth and seedling establishment considerably.

Most of the habitat types that have been described for inselbergs in other regions (Barthlott et al. 1993) are also found on Seychelles inselbergs, such as exposed rock surfaces, drainage channels, rock crevices, rock pools and shallow depressions filled with debris. In total, 76 native (incl. endemics) and alien vascular plant species were found (Fleischmann et al. 1996; Fleischmann 1997; our own unpubl. data). In general it was found that inselbergs on Mahé are characterized by a relative uniform vegetation. Beta diversity, i.e., species turnover between the inselbergs studied on the granite islands, is low.

A short description of the most important habitat types on Seychelles inselbergs will be given in the following:

1. *Exposed Rock Surfaces.* Due to the humid climatic conditions, the rock is almost completely covered by cyanobacteria (*Gloeocapsa sabulosa*, *Schizothrix* cf. *epiphytica*, *Stigonema flexuosum*, *Stigonema minutum*), which is typical for inselbergs under these climatic conditions throughout the tropics. This type of inselberg has been described as cyanobacteria inselberg (Porembski and Barthlott 1992). Green algae lichens are restricted to sheltered places (i.e., often under overhanging rocks). In the vicinity of the sea cyanobacteria are completely lacking on the salt-water sprayed, approximately 3–5 m wide strip, which rises directly from the sea.

   Regarding cryptogamous plants, it is striking that the lichen genus *Peltula* does not occur on inselbergs on the Seychelles. This genus grows regularly on the exposed rock surfaces on inselbergs of the Old World, New World, and Australia (Büdel et al., Chap. 5, this Vol.). Possibly the geographically isolated position of the islands or climatical reasons is responsible for this distributional gap.

2. *Drainage Channels.* The erosive power of the runoff water has produced numerous picturesque drainage channels which cut deeply into the rock. They are covered by cyanobacteria, most probably by species different than those covering exposed rock surfaces. Often a small rim of bare rock of several centimeters width borders the drainage channels.
3. *Rock Crevices.* Dominating vascular plants – according to the dimensions of the crevices – are the following trees and shrubs: *Canthium bibracteatum, Dracaena reflexa, Erythroxylum seychellarum, Euphorbia pyrifolia, Memecylon elaeagni, Pandanus balfourii, Pandanus multispicatus, Pandanus seychellarum,* the herbaceous *Kyllinga* sp., and *Lophoschoenus hornei* (Fleischmann et al. 1996). Invaders such as *Cinnamomum verum* or *Chrysobalanus icaco* have so far rarely been found. Rock crevices vary considerably in their dimensions and occur very frequently on Seychellan inselbergs.
4. *Rock Pools.* No higher plants could be detected in rock pools, although adequate amounts of soil to allow for the establishment of vascular plants was present. This could be due to the lack of suitable habitat types, which consequently results in a relatively small species pool of plants, that are adapted to aquatic or semiaquatic conditions or seasonally water logged habitats. Moreover, rock pools occur only in small numbers on inselbergs of the Seychelles. The rock pools have relatively large sizes (up to approx. 1 m wide and up to 60 cm deep), which stands comparison with African rock pools, but most rock pool walls are steeper than in Africa.
5. *Shallow Depressions filled with Debris.* The vegetation of most shallow depressions is dominated by *Memecylon eleagni, Lophoschoenus hornei* (Fig. 10.5.2), *Pandanus* spp., and *Dianella ensifolia.* Annual Cyperaceae (e.g., *Bulbostylis barbata, Fimbristylis dichotoma*) were found only in small numbers. Several mosses were seen mainly in shallow depressions (e.g., *Brachymenium exile, Campylopus* spp., see O'Shea et al. 1996). Shallow depressions are numerous on inselbergs of the Seychelles but are frequently devoid of vegetation (Fig. 10.5.3).

The endemic palm species *Phoenicophorium borsigianum* and *Deckenia nobilis* grow in shallow depressions and in rock crevices on Seychellan inselbergs, which is one of the rare records for palms growing on inselbergs in the paleotropics.

Monocotyledonous carpet-like mats, that are typically formed by certain species of Cyperaceae on inselbergs in Africa and Madagascar (Porembski 1996) are lacking (Fleischmann et al. 1996). One exception is *Lophoschoenus hornei*, that sometimes forms almost monospecific, dense stands in small depressions and crevices; but in contrast to the definition

Fig. 10.5.2. The perennial Cyperaceae *Lophoschoenus hornei* is a frequent colonizer of clefts and fissures on exposed rocky slopes. (Photograph N. Biedinger)

Fig. 10.5.3. Shallow depressions are widespread on Seychellan inselbergs. Usually, they are devoid of vascular plants. (Photograph N. Biedinger)

of monocotyledonous mats on inselbergs (Barthlott et al. 1993), *Lophoschoenus* does not form carpet-like life-forms. Furthermore, ephemeral flush vegetation, which also represents a typical and species-rich habitat type on inselbergs in Africa and Madagascar, is missing. Wet flush vegetation, a rather rare habitat type on African and Madagascan inselbergs, is also not present. This is possibly due to the morphology of the Seychellan inselbergs, so that the usually favorable, flat regions, colonized by the above vegetation type, are lacking. Other habitats that potentially provide a species pool for wet flush vegetation are absent in the Seychelles.

10.5.6 Life-Forms and Adaptations

The flora of the inselbergs in the Seychelles mostly consists of nanophanerophytes (65 % of the species). This percentage is true for native and alien vascular plants. Typically, nanophanerophytes grow in shallow depressions and are characterized by their xeromorphic foliage. This adaptation to the harsh environmental conditions on inselbergs is also expressed by nanophanerophytes in other tropical regions (e.g., *Clusia* in the Neotropics) and seems to be a typical feature for small trees on Seychellan inselbergs. Only a few hemicryptophytes (15 %), mostly ferns (e.g., *Dicranopteris linearis*, *Nephrolepis biserrata*, *Schizaea confusa*, and *Tectaria pleiotoma*), were found among the indigenous flora of the inselbergs. Epiphytes and lianas make up 12 % of all vascular plants (e.g., the orchids *Angraecum eburneum*, *Malaxis seychellarum*, and the carnivorous *Nepenthes pervillei*, Fig. 10.5.4). This relatively high percentage is probably due to the low species number on the inselbergs of the Seychelles. A striking difference to the African inselberg vegetation is the relatively low percentage of annual species. This can be explained by the fact that generally the flora of the Seychelles does not comprise many annual plants. Moreover, in contrast to African inselbergs, the most species-rich habitat type in annuals, the ephemeral flush vegetation, does not exist in the Seychelles.

Only a few specific adaptations to the extreme environmental conditions were found: succulence could be observed for epiphytes and some xeromorphic lianas (i.e., the leafless orchid *Vanilla phalaenopsis* or the Asclepiadaceae *Sarcostemma viminale*), which are both adapted to the dry and hot climatic conditions on rock outcrops.

Contrary to inselbergs in Madagascar and Africa, which are closest to the Seychelles, succulents such as the widespread epiphytic *Rhipsalis baccifera* var. *mauritiana*, or *Sarcostemma viminale* occur only occasion-

Fig. 10.5.4. The pitcher plant *Nepenthes pervillei* is a prominent endemic which occurs on Seychellan rock outcrops. (Photograph N. Biedinger)

ally epilithically. On inselbergs in the southern part of Madagascar, *Rhipsalis baccifera* forms dense stands. Here the low-growing, neotenic *Rhipsalis baccifera* var. *horrida* (Barthlott and Taylor 1995) is strictly limited to rock outcrops.

A few fern species, e.g., *Dicranopteris linearis* and *Schizaea confusa*, were the only poikilohydric vascular plants recorded. The only carnivorous species found in this survey – a plant group which is quite numerous on inselbergs in Africa, Madagascar, and the neotropics (Seine et al. 1995) – was represented by *Nepenthes pervillei*, the only carnivorous species in the Seychelles. This species, together with the two Madagascan species, marks the western distributional boundary of the genus *Nepenthes* and the species is regarded as systematically relatively isolated. *Nepenthes pervillei* can be regarded as a link between the Madagascan and Asian species (Schmid-Hollinger 1979; Cheek 1997).

The relatively low number of species on Seychellan inselbergs is paralleled by a paucity of specific adaptive strategies among their vegetation. It can be assumed that this is due to the high degree of geographic isolation, the small size of the islands and the absence of other azonal sites. In other tropical regions a large number of additional azonal sites (i.e., mainly other types of rock outcrops) is considered to promote the species richness of the inselberg vegetation (Porembski 1996). The small variety of specific adaptations thus may simply be a consequence of the low number of species present. Certain groups of plants that dominate the vegetation of inselbergs in most parts of the world are nearly completely absent in the Seychelles: this is true for succulents, pachycaulous plants as well as for poikilohydric vascular plants. On the other hand, other groups are overrepresented, such as sclerophyllous species.

Interesting observations about the vegetational dynamics within certain habitats could be made by monitoring permanent plots. For example, shallow depressions which were studied in 1995 and 1997 showed only little floristic change. Interestingly, however, different results were obtained from sites which were characterized by a large percentage of invasive weeds. In 1997 it could be observed at several localities that invasive shrubs and trees (e.g., *Chrysobalanus icaco*, *Cinnamomum verum*) were severely damaged by preceding drought conditions. In contrast, the indigenous rock outcrop vegetation showed no signs of drought damage.

10.5.7 Systematic and Ecological Characteristics

According to Fleischmann et al. (1996) and our own unpublished data, 76 species of vascular plants out of 39 families were documented on Seychellan inselbergs. Some of the species, genera, and even families are very isolated regarding their systematic position. An outstanding example is *Medusagyne oppositifolia*, the only species of the monotypic Medusagynaceae, which is restricted to inselbergs at present. The species is highly endangered and only a few plants have remained. *Medusagyne* certainly cannot be classified as a typical inselberg specialist but the few remaining trees of this species found their only refuge on inselberg sites. Similarly, as in other tropical regions, the Seychellan inselbergs act as refuges for endangered and systematically isolated, relictual species. In this context, one should mention that all palm species that grow on the inselbergs of the Seychelles belong to endemic genera and represent a highly isolated section within the palm family (Uhl and Dransfield 1987).

The analysis of the pollination mechanism (by analyzing the syndrome) of Seychellan inselberg species shows that most of them are pollinated by insects. Melittophilous and psychophilous species dominate. Despite the fact that one plant-pollinating sunbird species, *Nectarinia dussumieri*, is native to the Seychelles, it is obvious that only very few bird-pollinated species (e.g., *Gastonia*) have been able to establish populations or even single individuals on the inselbergs of the Seychelles.

Among the dispersal strategies of the indigenous plants through analysis of the syndrome it was found that zoochory dominates (more than 60 % of the species). Endemic species show an even higher percentage of dependency an animal dispersal; more than 70 % are zoochorous. Dispersal by birds dominates, but some species are possibly propagated by bats (e.g., *Chrysobalanus icaco*, *Pandanus* spp.). Only one third of the species possess propagules which are wind-dispersed.

10.5.8 Biogeography and Endemism

On the granitic islands of the Seychelles, approximately 240 vascular plants are indigenous over an area of ca. 400 km^2. Compared to other islands or groups of islands, the species richness of the Seychelles cannot be regarded as exceptionally high. The same is true for the percentage of endemism: ca. 40 % of the flora is restricted to the archipelago.

Although biodiversity of the Seychellan inselbergs is low when compared with other tropical regions (Porembski et al., Chap. 12, this Vol.). The percentage of endemic species on inselbergs is relatively high: out of the 76 species found, 63 % of the 41 indigenous species were endemic, which represents nearly one third of all endemics in the Seychelles. The percentage of endemism of two inselbergs is even higher: Montagne Palmiste habors 89 % and Mont Sébert 85 % endemic species.

Analyzing the species composition of the local inselberg flora, the lack of plant groups that are typical of inselbergs in other regions becomes obvious. Carnivorous plants, succulents, annuals or ancient floristic elements which are most common on African, South American, or Madagascan inselbergs are almost completely absent. This is certainly due to the isolated position of the Seychelles.

Twenty-seven invasive plant species belong to typical invader groups of other regions, such as Africa or South America; for example, *Alstonia macrophylla*, *Ananas comosus*, *Catharanthus roseus*, and *Furcraea foetida*. Only *Cinnamomum verum* was never observed as an invader species on inselbergs elsewhere. This is due to the fact that this rather aggressive species has not been introduced to many other regions where studies on inselbergs have been conducted or did not reach the same dominance as on the Seychelles, where it takes the place of native species in many habitat types.

Looking at the biogeographical affinities of the inselberg flora, affinities to the Asian become obvious. Only the fern *Afropteris barklyae* has its closest relatives in West Africa (Tryon et al. 1990). About 33 % of the indigenous species found on inselbergs are distributed or have their closest relatives in Asia (e.g., *Diospyros seychellarum*, *Phoenicophorium borsigianum*, *Deckenia nobilis*, *Dillenia ferruginea*). All other species are distributed in – or derive from – relatives indigenous to the Africa-Asia (15%, e.g., *Memecylon elaeagni*), Africa-Madagascar (23%, e.g., *Mimusops seychellarum*, *Paragenipa lancifolia*), Asia-Madagascar (5%), or pantropical region (15%).

Taking a closer look at the endemic species on inselbergs, the same tendency is reflected: 32% of the species are related to Asian species (e.g., *Eugenia wrightii*). This might be due to the vegetation history of the archipelago, resulting from a relatively late separation of the Indian subcontinent from the rest of Gondwanaland.

Among the cryptogams the moss *Campylopus brevirameus*, which occurs in shallow depressions on Seychellan inselbergs, is worth mentioning. This species is endemic to rock outcrops of the Seychelles (O'Shea et al. 1996), which is unusual for mosses because their distribution patterns in general have a wide range: many mosses found on tropical inselbergs are distributed throughout the tropics, some are even cosmopolitic. Once more, this endemic moss species underlines the high degree and long duration of isolation of the Seychellan archipelago.

10.5.9 Conservation

In 1995 the ecology of inselbergs on the islands of Mahé and Silouette was quantitatively investigated for the first time by applying ecological status matrices (Fleischmann 1997). The method involves several parameters of investigation such as diversity, singularity, rejuvenation, frequency, and abundance of native plants to evaluate the conservation value of potentially important habitats. This investigation has shown that in terms of conservation all inselbergs investigated revealed outstanding floristic and ecological values (i.e., PTV = protection value). The study also showed that inselbergs are the only areas on the islands which do not show a decline in PTV with decreasing altitude. Strongly isolated inselbergs on Mahé scored a relative mean PTV of 41% while the surrounding habitats showed typically relative protection values of only 19%. The sudden and spatially abrupt change in ecological habitat qualities between inselbergs and the habitats in their immediate vicinity is striking. This is best seen by a comparison given in Table 10.5.1.

On comparing prominence values (in terms of abundance and frequency) of species between inselbergs and adjacent habitats (Fleischmann et al. 1996), it was shown that native, often rare, species are significantly

Table 10.5.1. Seychelles inselbergs show an outstanding ecological status compared with habitats in their immediate surroundings

	Inselbergs (strongly isolated)				Inselbergs (medium isolated)				Surrounding habitats			
	PV[a]	DI	SI	PTV	PV	DI	SI	PTV	PV	DI	SI	PTV
Relative mean	40.2	42.7	60.7	41.1	37.4	34.2	24.7	39.9	22.4	23.1	14.6	19.0

[a] PV, prominence value of natives; DI, diversity of natives; SI, singularity; PTV, protection value. The sum of the relative means of each criteria is 100 %.

related to inselbergs, while the presence of alien plant species and invaders is skewed towards other forest areas, located sometimes in the immediate vicinity of an inselberg. A high average degree of endemism per sampling unit has been found on Mont Sébert, Mahé, with 85 %, while in the surrounding habitats the average value was only 5 %.

A severe danger to inselbergs is the colonization of inselbergs by invader plants, in particular *Cinnamomum verum* and *Alstonia macrophylla* as well as the neotropic species *Ananas comosus* (Fig. 10.5.5a) and *Furcraea foetida* (Fig. 10.5.5b). It could be observed in other regions that the highly specialized inselberg flora is easily replaced by quick-growing competitors. The above results indicate strongly why Seychelles inselbergs must become a focal point in conservation policy and why they should be given high priority for community conservation planning in the future.

Acknowledgements. We thank Wilhelm Barthlott (Bonn) for valuable discussions on the manuscript. Furthermore, we owe thanks to John Collie, Department of Conservation and National Parks (Mahé, Seychelles) and Dr. Maureen Kirkpatrick, Seychelles Conservation Trust (Mahé, Seychelles).

References

Bailey JP (1971) Flowering plants and ferns of Seychelles. Government printer, Seychelles

Baker BH, Miller JA (1963) Geology and geochronology of the Seychelles islands and structure of the floor of the Arabian sea. Nature 199:346–348

Barthlott W, Taylor NP (1995) Notes towards a monograph of Rhipsalideae (Cactaceae). Bradleya 13:43–79

Barthlott W, Gröger A, Porembski S (1993) Some remarks on the vegetation of tropical inselbergs: diversity and ecological differentiation. Biogéographica 69:105–124

Fig. 10.5.5a,b. On certain Seychellan rock outcrops the neotropical invader *Ananas comosus* (**a**) and *Furcraea foetida* (**b**) have become established. Due to their vegetative propagation the species might become more widespread in the future. (Photographs N. Biedinger)

Braithwaite CJR (1984) Geology of the Seychelles. In: Stoddart DR (ed) Biogeography and ecology of the Seychelles islands. Dr W Junk, The Hague, pp 17–33
Cheek M (1997) A skeletal revision of *Nepenthes* (Nepenthaceae). Blumea 42:1–106
Dalziel IWD (1995) Earth before pangea. Sci Am 272:30–37
Fleischmann K (1997) Invasion of alien woody plants on the islands of Mahé and Silhouette, Seychelles. J Veg Sci 8:5–12
Fleischmann K, Porembski S, Biedinger N, Barthlott W (1996) Inselbergs in the sea: vegetation of granite outcrops on the islands of Mahé, Praslin and Silhouette, Seychelles. Bull Geobot Inst ETH 62:61–74
Friedmann F (1987) Flowers and trees of Seychelles. ORSTOM, Paris
Friedmann F (1991) The threatened species of the flora of the Seychelles and their conservation. ORSTOM, Paris
High J (1982) The natural history of the Seychelles. Phillips, London
Jeffrey C (1962) Report on the forests of the granitic islands of the Seychelles. Dept Techn Coop, London
Lionnet G (1984) Observations d'histoire naturelle faites aux Seychelles en 1768 par l'expédition Marion-Dufresne. Mauritius Inst Bull 10:15–73
O'Shea BJ, Frahm J-P, Porembski S (1996) Die Laubmoosflora der Seychellen. Trop Bryol 12:168–191
Porembski S (1996) Notes on the vegetation of inselbergs in Malawi. Flora 191:1–8
Porembski S, Barthlott W (1992) Struktur und Diversität der Vegetation Westafrikanischer Inselberge. Geobot Kolloq 8:69–80
Procter J (1984a) Floristics of the granitic islands of the Seychelles. In: Stoddart DR (ed) Biogeography and ecology of the Seychelles islands. Dr. W. Junk Publishers, Boston, pp 209–219
Procter J (1984b) Vegetation of the granitic islands of the Seychelles. In: Stoddart DR (ed) Biogeography and ecology of the Seychelles islands. Dr W Junk, Boston, pp 195–207
Robertson SA (1989) Flowering plants of Seychelles. Royal Botanic Gardens, Kew
Sauer JD (1967) Plants and man on the Seychelles coast. Madison, Wisconsin
Schmid-Hollinger R (1979) *Nepenthes*-Studien V. Die Kannenformen der westlichen *Nepenthes*-Arten. Bot Jahrb Syst 100:379–405
Seine R, Porembski S, Barthlott W (1995) A neglected habitat of carnivorous plants: inselbergs. Feddes Repert 106:555–562
Tryon RM, Tryon AF, Kramer KU (1990) Pteridaceae. In: Kubitzki K (ed) The families and genera of vascular plants, vol II. Springer, Berlin Heidelberg New York, pp 230–355
Uhl NW, Dransfield J (1987) Genera Palmarum. A classification of palms. Lawrence, Kansas
Vesey-Fitzgerald D (1940) On the vegetation of the Seychelles. J Ecol 28:465–483
Walsh R (1984) Climate of the Seychelles. In: Stoddart DR (ed) Biogeography and ecology of the Seychelles islands. Dr W Junk, Boston, pp 39–62

10.6 Flora and Vegetation of Inselbergs of Venezuelan Guayana

A. GRÖGER

10.6.1 Geography and Geology

The Guayana shield is one of the oldest land surfaces of the world, consisting of Precambrian rocks between 0.9 and 3.5 billion years old (Schubert et al. 1986). The shield is stratified with a granitic base and a huge overlaying cover of sandstone. The latter suffered intensive erosion, producing the masses of white quartzitic sand which are distributed all over the shield and its adjacent areas. Remnants of the sandstone layer are huge table mountains, the tepuis.

Part of the underlying granite socle are the inselbergs locally known as lajas, which occur especially along the periphery of the shield, from Tumuc-Humac (French Guiana) in the east to the Mitú area (southeastern Colombia) in the west. Fieldwork in the present study concentrated on the Venezuelan part of the Guayana shield. The study area ranges from Ciudad Guayana (Edo. Bolívar) to San Fernando de Atabapo (Edo. Amazonas), covering the northwestern margin of the shield (Fig. 10.6.1).

Owing to the specific structure of the granite and the climatic conditions, inselbergs of the Guayana shield show a characteristic erosion phenomenon: the pseudocarst (cf. Blancaneaux and Pouyllau 1977). It results in distinct channels, gullies, and depressions (Fig. 10.6.2). This richness in spatial substructures is reflected in the variety of vegetation types established on the inselbergs.

10.6.2 Climate

Altitudinally, the inselbergs examined in this study all lie within the macrothermic level, i.e., below 500 m asl and with a mean average annual

Ecological Studies, Vol. 146
S. Porembski and W. Barthlott (eds.) Inselbergs

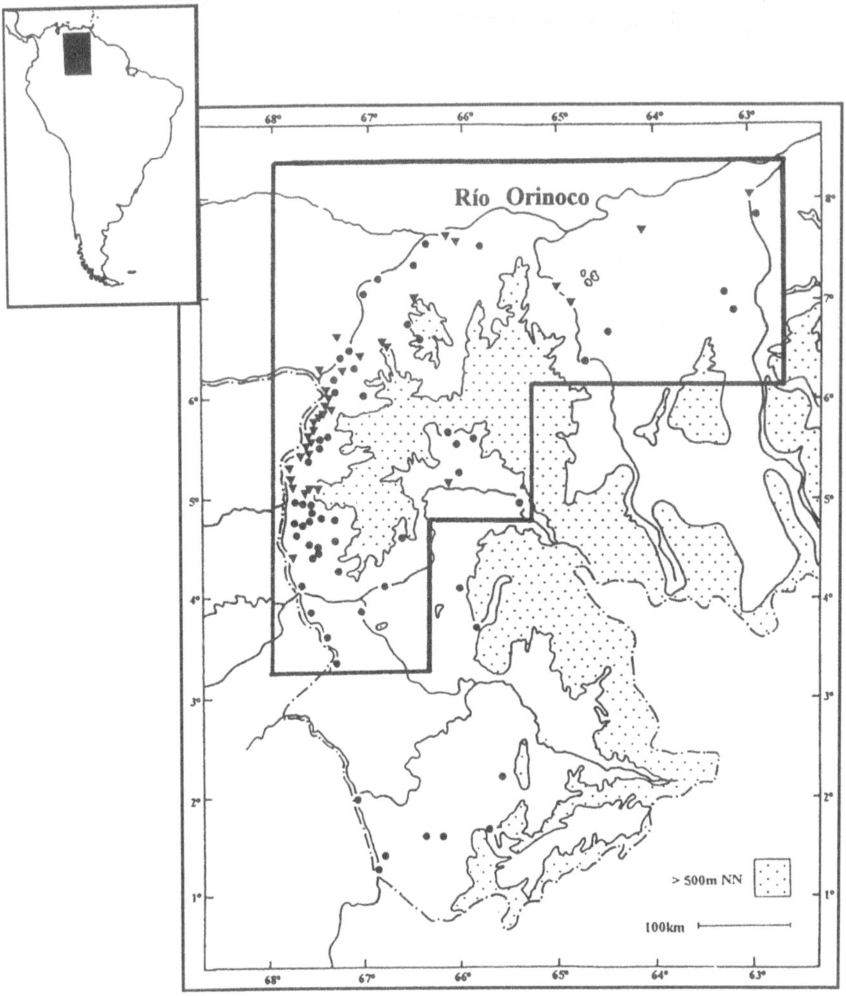

Fig. 10.6.1. Study area (*framed*) in southern Venezuela with locations of inselberg sites. *triangles* inselberg areas visited by the author; *points* inselberg areas visited by other botanists)

temperature of >24 °C. Latitudinally, the study area shows a macroclimatic gradient. Annual precipitations vary from 1000 mm around Ciudad Bolívar, to over 3000 mm south of San Fernando de Atabapo. From north (8°N) to south (4°N) the following climatic zones (according to Köppen 1932) can be distinguished: (1) Aw climate, savanna climate with dry winters, (2) Am climate, transitional climate between savanna and rainforest climate, and (3) Af climate, perhumid rainforest climate (Fig. 10.6.3).

Fig. 10.6.2. Results of pseudocarstic erosion on Venezuelan inselbergs: channels and depressions, filled in this case by *Pitcairnia* (*Pepinia*) *pruinosa*

Paleoclimatic events, especially those of the late Pleistocene and the early Holocene age, are of great importance for the present vegetation patterns in northern South America. During the last glaciation maximum (LGM), 18 000 years B.P., the decline in temperature and even more the diminution of precipitation affected the vegetation thoroughly. For northern South America, Manabe and Hahn (1977) calculated that, during the LGM, annual precipitation was 730–1825 mm below that of today. This drastic climatic change was caused by the southwards shift of the Caribbean trade-wind belt (Tricart 1985). Fossilized dunes with NE-SW orientation in the Venezuelan and Colombian Llanos bear witness to a drier climate during the late Pleistocene age and testify that the direction of the trade winds was the same as today (Schubert 1988).

10.6.3 Vegetation of Venezuela

The floristic richness of Venezuela – 14 000 vascular plant species are described – is partly due to its rich pattern of landscapes and ecosystems.

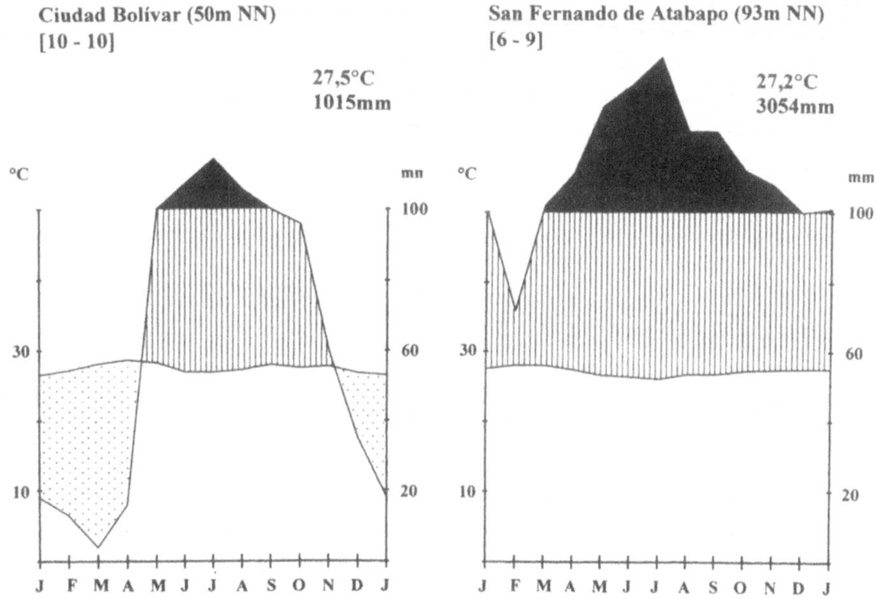

Fig. 10.6.3. Climatic diagrams (projected according to Walther and Lieth 1960) of Ciudad Bolívar (according to Baldridge et al. 1982) and San Fernando de Atabapo. (according to Huber 1982)

The Venezuelan vegetation contributes to three phytogeographical regions (Huber and Alarcón 1988): (1) the Andean region, formed by the extension of the Andes in the western part of the country; (2) the Caribbean region, including not only the coastal range but also the inland savanna plains (Llanos province); (3) the Guayana region, corresponding to the extension of the shield and some adjacent lowland areas. The latter region can be subdivided into the eastern, western, and central Guayana provinces, and the Pantepui province (Huber 1995).

The Venezuelan inselbergs are located exactly on the borderline between different phytogeographical units. In the northern part of the study area, from the lower Río Caroní to the mouth of the Río Sipapo, the inselbergs lie between the grass-dominated Llanos province and the central Guayana province. In the southern part, south of the mouth of the Río Sipapo, they lie in extensive forests, between the western and the central Guayana province. Therefore, the inselbergs are located along a macroclimatic gradient as well as within the transition zone between different phytogeographical units.

10.6.4 Inselberg Vegetation

Generally, inselberg vegetation represents a tesselated arrangement of different habitat types (Barthlott et al. 1993). Characteristic for the floristic mosaic found on Venezuelan granitic outcrops is the predominance of patches of low woody vegetation. In the following, each habitat type is presented with some of its typical species.

10.6.4.1 Vegetation of Rock Surface

In the northern drier region, lichens (mainly *Peltula tortuosa*) predominate in the cryptogamic crusts on bare rock (lichen type of inselbergs, cf. Porembski and Barthlott 1992). Southwards, they are increasingly being replaced by cynanobacteria, such as *Gloeocapsa sanguinea*, *Schizothrix telephoroides*, *Scytonema crassum*, and *Stigonema ocellatum*, and endolithic species of *Chroococcidiopsis* (cyanobacteria type of inselbergs). A detailed analysis of the cryptogamic flora of the Venezuelan inselbergs is provided by Büdel et al. (1994).

Also some vascular plant species may establish themselves on bare rock surfaces, e.g., *Vellozia tubiflora*, an extraordinarily widespread representative of the Velloziaceae. Most of the other vascular lithophytes belong to families which contribute significantly to the neotropical epiphytic flora: the Bromeliaceae (e.g., *Pitcairnia orchidifolia*, *Vriesea bibeatricis*, *Tillandsia flexuosa*), the Orchidaceae (e.g., *Encyclia leucantha*, *Pleurothallis* aff. *papillosa*), and the Piperaceae (e.g., *Peperomia maypurensis*, *P. magnoliaefolia*).

10.6.4.2 Wet Flush Vegetation

On flat rock surfaces where water flushes periodically, vascular species may take root in the cryptogamic layer. In arid regions, *Portulaca sedifolia* and even *Melocactus mazelianus* take root in layers of *Peltula*. In more humid regions, the gregarious rheophyte *Utricularia oliverana* grows in felts of cyanobacteria.

10.6.4.3 Vegetation of Drainage Channels and of Rock Depressions

Fast-running water in the drainage channels allows only cyanobacteria or lichens to settle. In the tub-like depressions water may be stagnant and eroded material may be deposited. Owing to varying water supply, depth of soil, and stage of succession, the depressions offer different ecological conditions, allowing different communities of vascular plants to establish themselves.

Seasonally water-filled rock pools with an ephemeral aquatic vegetation are quite scarce on Venezuelan inselbergs. They are colonized by widespread annuals, like *Bacopa callitrichoides*, *Eleocharis cellulosa*, *Ericaulon cinereum*, or *Utricularia subulata*. Depressions with shallow deposits of rock debris harbor different herb communities. In very shallow deposits with regular inundations, the stemmed rosettes of *Bulbostylis leucostachya* or the annual *Bulbostylis aturensis* can be found. Under drier conditions, geophytic *Portulaca pygmaea* hides its tiny bulbs in the shallow deposits.

Plant communities dominated by annuals, like *Acisanthera crassipes*, *Panicum arctum*, *Paepalanthus lamarckii*, and *Xyris stenostachya*, are normally found in small depressions with a soil layer only 1–2 cm deep. In hollows with deeper soil, these species are restricted to the periphery. Generally, the vegetation of depressions with deeper soil is characterized by concentric rings: ephemerals in the periphery, followed by perennial herbs, and shrubs and small trees in the center. This spatial sequence, determined by the thickness of the soil layer, partially reflects the successional sequence.

Perennial herbs occurring in rock depressions are mainly bromeliad species, which may also constitute extensive mat-like vegetation units (see below). As already mentioned, patches of low woody vegetation are another main characteristic of the Venezuelan inselbergs (Fig. 10.6.4). Medium-sized depressions are exceptionally rich in characteristic shrub species, such as *Acanthella sprucei*, *Diacidia galphimioides*, *Erythroxylum williamsii*, *Mandevilla lancifolia*, *Ouratea chaffanjonii*, and *Tabebuia orinocensis*.

10.6.4.4 Vegetation of Fissures

Beneath exfoliating granitic sheets, thin layers of organic substrate may accumulate. Depending on the amount of soil and the water supply, different plant communities establish themselves in these horizontal crevices. Around small scales with a very thin layer of soil, ephemerals such as *Borreria pygmaea*, *Sauvagesia ramosissima*, and *Portulaca insignis*, or succulents, such as *Melocactus* spp. and *Pitcairnia (Pepinia) armata*, can be found. Under better conditions shrubs, such as *Graffenrieda rotundifolia*, *Plumeria inodora*, and *Pseudobombax croizatii*, predominate. Typical for small crevices which split the rock vertically is *Ficus mollicula*. In broader

Fig. 10.6.4. Typical for Venezuelan inselbergs is woody vegetation even in shallow depressions: in the illustrated depression, shrubs of *Acanthella sprucei* and *Tabebuia orinocensis* are surrounded by a dense carpet of *Pitcairnia* (*Pepinia*) *armata*

vertical crevices, shrubs already mentioned for horizontal crevices and medium-sized depressions are rooted. In general, there is a fluent transition in species composition between the vegetation of depressions and that of fissures.

10.6.4.5 Mats

Monocotyledonous mats, composed of species of *Brewcaria brocchinioides, Pitcairnia* subgen. *Pepinia* (e.g., *P. armata, P. bulbosa, P. pruinosa*) or *Navia* (e.g., *N. arida*), are another characteristic of Venezuelan inselbergs. Only some geophytes, like *Caladium picturatum*, *Cleistes rosea*, or *Merremia maypurensis*, and deciduous orchids with large pseudobulbs, like *Catasetum bergoldianum* or *Cyrtopodium punctatum*, can coexist within these monospecific bromeliad mats. In some places *Selaginella* species (e.g., *S. revoluta, S. marginata*) form thin poikilohydric mats, accompanied by geophytes, such as *Echeandia bolivarensis, Habenaria dusenii*, and *Maranta linearis*.

10.6.4.6 Forests

Large depressions and talus slopes can be covered by low forests, with trees up to 6 m tall (Fig. 10.6.5). They are dominated either by palms (e.g., *Attalea racemosa, Syagrus orinocensis*), by Myrtaceae (e.g., *Myrcia floribunda*), or by Clusiaceae (e.g., *Clusia columnaris, Oedematopus obovatus*) and *Clusia*-like Rutaceae (*Decagonocarpus oppositifolius*). Even forests taller than 6 m can occur on Venezuelan inselbergs (Gröger 1994), but these were excluded from the present investigation.

Forest margins are another extremely species-rich habitat on Venezuelan inselbergs. They are composed of species which also occur in medium-sized depressions; but for a whole set of species, these margins represent the preferred habitat on inselbergs (e.g., *Byrsonima nitidissima, Eugenia emarginata, Mandevilla caurensis, Tabebuia pilosa, Tocoyena orinocensis, Zamia lecointei*).

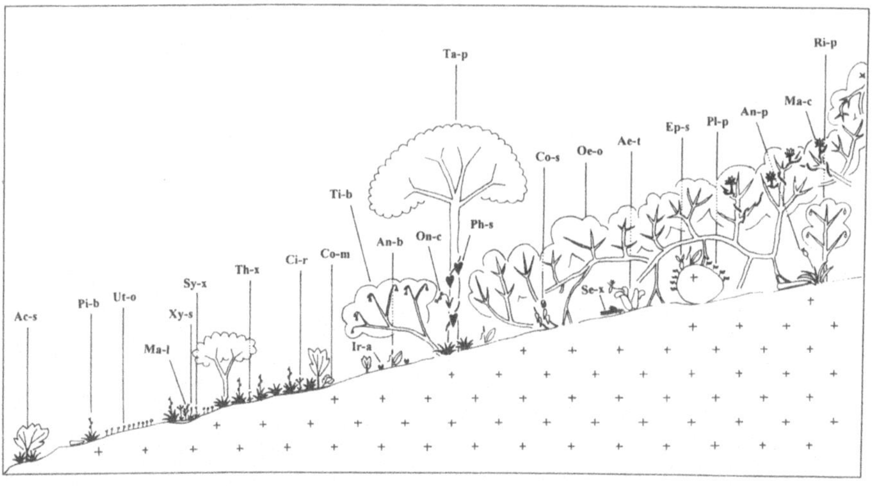

Fig. 10.6.5. Profile of a Clusiaceae forest and its margin on an inselberg northwest of Piedra Pintada (5°33'N 67°33'W, approx. 150 m asl): *Ac-s Acanthella sprucei; Ae-t Aechmea tocantina; An-b Anthurium bonplandii; An-p Ananas parguazensis; Ci-r Cipura rupicola; Co-m Comolia microphylla; Co-s Costus spiralis; Ep-s Epidendrum stanfordianum; Ir-a Irlbachia alata; Ma-c Mandevilla caurensis; Ma-l Mandevilla lancifolia; Oe-o Oedematopus obovatus; On-c Oncidium cebolleta; Ph-s Philodendron solimoesense; Pl-p Pleurothallis* sp. nov. (aff. *papillosa*); *Pi-b Pitcairnia* (*Pepinia*) *bulbosa; Ri-p Rinorea* cf. *pubiflora; Se-x Selaginella* sp.; *Sy-x Syngonanthus* sp.; *Ta-p Tabebuia pilosa; Ti-b Tibouchina bipenicillata; Th-x Thrasya* sp.; *Ut-o Utricularia oliverana; Xy-s Xyris stenostachya*

10.6.5 Life-Forms and Adaptations

10.6.5.1 Life-Forms

The 614 species of vascular plants recorded on Venezuelan inselbergs (see below) can be grouped by their life-forms, reflecting partially the structure of the vegetation cover. The diversity of the Venezuelan inselberg flora originates mainly from the variety of its woody species; 41 % of the species are phanerophytes, followed by 17 % therophytes, 13 % hemicryptophytes, 11 % chamaephytes, and 5 % geophytes (Fig. 10.6.6). Functional groups, such as lianas (11 %), epiphytes (4 %), and lithophytes (2 %) complete the spectrum of life-forms.

For Venezuelan savannas, Aristeguieta (1966) stated the following composition of life-forms (without separating the above-mentioned functional groups): 31 % hemicryptophytes, 29 % therophytes, 28 % phanerophytes, 7 % chamaephytes, and 5 % geophytes. There are no comparable data available for the forest surrounding inselbergs. Nevertheless, it can be concluded that the composition of life-forms on inselbergs lies between that of savannas and deciduous dry forests. This intermediate position reflects the forest margin character of inselberg habitats. Genera like *Cordia*, *Isertia*, or *Mandevilla* are indicators for forest margins, but also important elements of inselberg vegetation.

10.6.5.2 Vegetative Adaptations

The temporally and quantitatively restricted availability of water is one of the most important factors affecting the composition of vegetation on

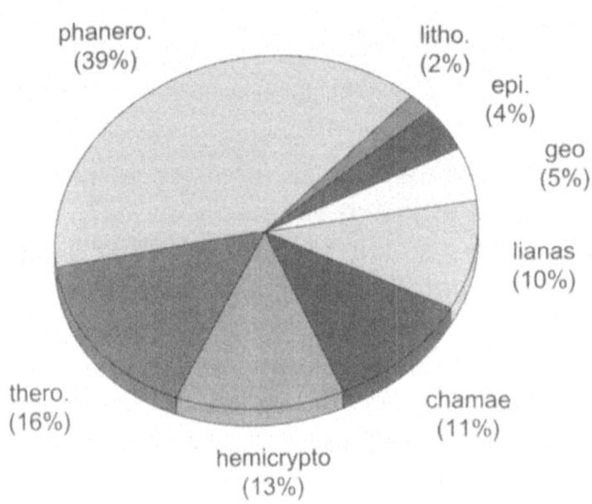

Fig. 10.6.6. Life-form spectrum

Fig. 10.6.7. Stages in the development of *Pseudobombax croizatii*: 1-year-old and 3-year-old juvenile plant; 3 m-tall shrub-like individual rooting beneath a granite lamina; 5 m-tall tree-like individual

granitic outcrops, also causing a variety of vegetative adaptations. Leaf deciduousity and leaf succulence are quite common within inselberg species. Geophytes and other taxa with subterraneous storage organs are relatively scarce, owing to the lack of thicker soil layers.

Species which situate their storage organs above ground are more frequent. Common syndromes are (1) creeping rhizomes (all mat-forming bromeliads, *Anthurium bonplandii*, etc.), (2) scarcely lignified stems with a large spongey medulla (orchids with pseudobulbs, *Begonia guaduensis*, *Chrysothemis dichroa*, *Guayania crassicaulis*, etc.), (3) classical stem succulents with thickened cortex (most of the cacti, *Mandevilla caurensis*, *Plumeria inodora*, etc.), and (4) pachycaulous stems. This latter phenomenon is typical for a whole number of deciduous shrubs and small trees, especially those endemic to inselberg sites of the region (e.g., *Acanthella sprucei*, *Ficus mollicula*, *Graffenrieda rotundifolia*, *Pseudobombax croizatii*,

Tabebuia orinocensis). Generally the bottle-shaped stems of these species are very pronounced during juvenile stages, as long as the root system cannot guarantee sufficient water supply (Fig. 10.6.7). Measurements of a bottle-shaped stem base of *Acanthella sprucei* at the end of the rainy season showed that more than 80 % of its weight was due to water storage (Gröger and Renner 1997).

Compared to inselbergs in the paleotropics, Venezuelan inselberg vegetation is poor in poikilohydric species. Only some ferns (e.g., *Anemia oblongifolia*) and fern relatives (*Selaginella* spp.) present this phenomenon.

Venezuelan examples for two other vegetative syndromes, which are common to inselberg vegetation in general, should be mentioned: carnivorous species (e.g., *Genlisea sanariapoana*, five species of *Utricularia*), and caulescent rosette trees (*Bulbostylis leucostachya*, *Vellozia tubiflora*).

10.6.5.3 Reproductive Adaptations

An analysis of dispersal syndromes (according to van der Pijl 1982 with some modifications) of the 614 recorded species reveals that wind (29 %) and birds (25 %) represent the most important dispersal agents of the inselberg flora. The actual percentage of anemochorous species may be even higher, because a considerable portion of those 27 % "without obvious syndromes" has small grain-like, probably wind-dispersed fruits and seeds.

López and Ramírez (1989) studied the dispersal syndromes of a shrubland community on sandstone, situated in the mesothermic level (1350 m asl) of Venezuelan Guayana. Ecological conditions, like exposition, lack of soil, and patchiness of the vegetation cover, are partially similar to those offered on inselbergs. Accordingly, the results of López and Ramírez considerably resemble those of the inselberg vegetation, with ornithochory and anemochory as dominating dispersal syndromes.

Generally, reproductive adaptations have to be considered with respect to the isolation of the populations. It will shown below that 85 species are restricted in their occurrence to granitic outcrops of the Guayana region. Populations of these inselberg-endemic species are extremely fragmented, imposing the question of consequences for their dispersal strategies. An analysis of dispersal syndromes of exclusively inselberg-endemic species shows the same general tendencies as the inselberg flora as a whole: anemochory and ornithochory are predominant; but some details differ. Within anemochorous syndromes, dust-like seeds are underrepresented in inselberg-endemic species, while dispersal units with appendages like hairs and little wings are overrepresented. This confirms the assumption of Rauh

Table 10.6.1. Dispersal syndromes in inselberg vegetation of Venezuela. The dispersal units of some of the species can be related to more than one dispersal syndrome. Therefore, the totals reached are greater than the total number of species

Dispersal syndrome	No. of sp. on inselbergs	Percent of 614	No. of insel-berg-end. sp.	Percent of 85
Zoochorous	**225**	**37**	**33**	**39**
Ornithochorous	154	25	18	21
Mammaliochorous	62	10	8	9
Chiropterochorous	24	4	2	2
Myrmecochorous	40	7	8	9
Epizoochorous	24	4	4	5
Anemochorous	**175**	**29**	**28**	**33**
Dust-like (<0.3 mm)	61	10	4	5
With hairs	43	7	11	13
With small wings (<0.5 cm)	30	5	8	9
With large wings	33	5	5	6
Autochorous	**32**	**5**	**3**	**4**
Without obvious syndromes	**164**	**27**	**24**	**28**
Without data	**6**	**1**	–	–
Total	**614**		**85**	

et al. (1975) that dust-like seeds evolved to reach very specific germination microsites (especially plant groups with saprophytic-mycotrophic or parasitic seedling stages) rather than to raise long-distance dispersal capability.

10.6.6 Systematic Characteristics

The species inventory of the Venezuelan inselbergs studied consists of 614 species of vascular plants from 107 different families (for a complete listing, see Gröger 1995). Genera with more than 5 representatives on inselbergs are *Rhynchospora* (17 spp.), *Mimosa* (9 spp.), *Erythroxylum* (8 spp.), *Miconia* (8 spp.), *Bulbostylis* (7 spp.), *Eugenia* (7 spp.), *Mandevilla* (7 spp.), *Portulaca* (7 spp.), *Selaginella* (7 spp.), *Chamaecrista* (6 spp.), *Clidemia* (6 spp.), *Pitcairnia* (including *Pepinia*, 6 spp.), *Tabebuia* (6 spp.), and *Xyris* (6 spp.).

The ten most diverse families are shown in Fig. 10.6.8. The representation of families on inselbergs differs remarkably from that in the Venezuelan flora as a whole. On an average, 4.3 % of the Venezuelan flora, which comprises described 14 000 species (R. Wingfield, in prep.), are represented

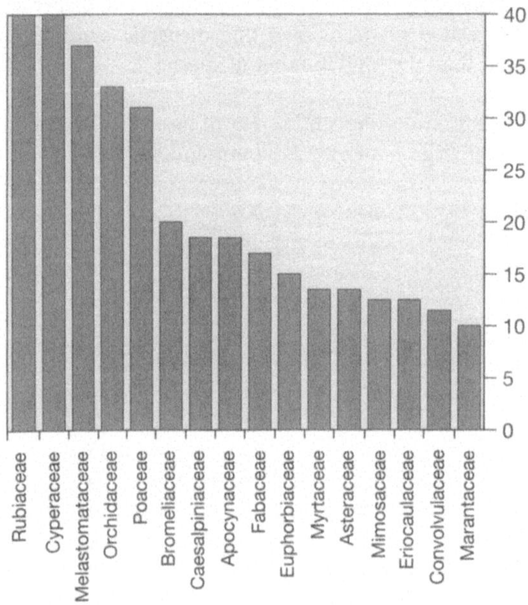

Fig. 10.6.8. Family spectrum

on the granitic outcrops, but the contribution of each single family varies. Some families are overrepresented, like the Cyperaceae, with 12% of all Venezuelan species found on inselbergs. On the other hand, some families are underrepresented, e.g., the Orchidaceae. In the Venezuelan flora they are by far the most diverse family. On the inselbergs only 3% of their species could be registered.

Generally, families with deciduous woody species, lianas, succulents, and geophytes, and families with ephemerals favoring open seepage habitats are overrepresented in number of species. These are, besides the already-mentioned Cyperaceae, the Apocynaceae, Cactaceae, Convolvulaceae, Eriocaulaceae, Erythroxylaceae, Lentibulariaceae, Marantaceae, Portulacaceae, Scrophulariaceae, and Sterculiaceae.

A comparison of the ten families richest in species number (1) on inselbergs in the study area, (2) in Venezuelan Guayana (Berry et al. 1995), and (3) in Venezuela as a whole (R. Wingfield, in prep.), shows Table 10.6.2.

At first glance, these lists look quite similar, but some differences are evident. Outstandingly rich in species number on inselbergs are Apocynaceae, a family dominated by lianas, shrubs, and small trees of forest margins, and Cyperaceae, a family typical for open, periodically inundated habitats. On the other hand, Araceae and Orchidaceae, both important epiphytic taxa, and Asteraceae, a family whose center of diversity lies in the Andean region, are comparatively underrepresented on Venezuelan inselbergs.

Tab. 10.6.2.

Inselbergs studied	Venezuelan Guayana	Total Venezuela
Cyperaceae (40)	Orchidaceae (698)	Orchidaceae (ca. 1250)
Rubiaceae (40)	Rubiaceae (530)	Asteraceae (725)
Melastomataceae (36)	Poaceae (420)	Rubiaceae (702)
Orchidaceae (33)	Melastomataceae (397)	Poaceae (645)
Poaceae (31)	Fabaceae (319)	Melastomataceae (584)
Bromeliaceae (20)	Bromeliaceae (273)	Euphorbiaceae (393)
Apocynaceae (18)	Asteraceae (257)	Fabaceae (391)
Caesalpiniaceae (18)	Cyperaceae (243)	Bromeliaceae (365)
Fabaceae (17)	Euphorbiaceae (237)	Cyperaceae (341)
Euphorbiaceae (15)	Caesalpiniaceae (203)	Araceae (286)

10.6.7 Biogeography and Endemism

10.6.7.1 Species Endemic to Granitic Outcrops of the Guayana Region

Generally, endemism in the Guayana region is proved to be considerably high. In the Pantepui province, endemism reaches 65 % (Berry et al. 1995), whereas in the Guayanan lowlands, Boom (1990) estimates a percentage of 10 % of endemic species. On the Venezuelan granitic outcrops, 144 species of the recorded inventory are endemic to the Guayana region. This comparatively high degree of endemism (24 %) for a Guayanan lowland ecosystem is mainly due to those 85 species which in their occurrence are obviously restricted to inselberg sites of the very same region. The majority of these inselberg-endemic species belongs to the generally diverse families on inselbergs (number of inselberg-endemic species in parantheses): Bromeliaceae (8), Apocynaceae (7), Melastomataceae (6), Cyperaceae (5), Rubiaceae (4), etc. Within this edaphic endemism, two gradations can be drawn:

1. Inselberg-endemic species sensu stricto: endemic to the Guayana region and exclusively occurring on granitic outcrops (57 species).
2. Inselberg-endemic species sensu lato: endemic to the Guayana region, mainly occurring on granitic outcrops with only very few reports from other ecologically similar sites, e.g., rock savannas or sandstone outcrops in lowland areas (28 species).

In conclusion, four families should be emphasized as characteristic of the Venezuelan inselberg flora: Apocynaceae, Cyperaceae, Erythroxylaceae, and Portulacaceae. Not only are these families disproportionately well represented in comparison to the total flora of Venezuela, but they also contain an extraordinarily high percentage of species endemic to granitic outcrops of the Guayana region.

10.6.7.2 Phytogeographical Subunits

Although the distributional data of the Venezuelan inselberg flora are far from complete, the relatively high number of inselberg-endemic species offers an ideal instrument for phytogeographical interpretations. Regions of high diversity in endemic species can be located and disparate phytogeographical subunits can be defined (Fig. 10.6.9).

By superimposing the distribution areas of the 85 inselberg-endemic species it becomes obvious that there is a conspicuous concentration of species in the Atures area, from approx. 60 km north to 50 km south of Puerto Ayacucho (Edo. Amazonas). Sixty one of the 85 species considered as endemic to granitic outcrops of the Guayana region are represented in this Atures center of endemism. North and south of this area the number of inselberg-endemic species drops significantly due to two circumstances:

Twenty species display a distribution area obviously restricted to the Atures center of endemism, e.g., *Comolia nummularioides*, *Cipura rupicola*, *Ernestia cordifolia*, and *Portulaca pygmaea*. It should be mentioned that some of these species are collected only at a few localities, and further field research may reveal a wider distribution. However, a second argument consolidates the existence of this center of endemism: the area forms an intersection zone of two different phytogeographical subunits of the inselberg flora, namely the northern and the southern inselberg subunit (Gröger and Barthlott 1998).

The northern inselberg subunit is defined by 24 species whose distribution area extends only northwards of the center of endemism [e.g., *Ficus mollicula*, *Mandevilla caurensis*, *Melocactus mazelianus*, *Pitcairnia* (*Pepinia*) *armata*]. This subunit, which is dominated by savanna climate, lies between the Llanos province and the central Guayana province.

The boundaries of the southern inselberg subunit are drawn by another 18 species which occur in the Atures center of endemism and/or only southwards of it [e.g., *Decagonocarpus oppositifolius*, *Diacidia galphimioides*, *Pitcairnia* (*Pepinia*) *bulbosa*, *Rhynchospora sanariapensis*]. The southern inselberg subunit lies on the borderline between the western and the central Guayana province, an area dominated by perhumid rainforest climate. Many species of this subunit show a southwestern extension of their distributional range to the Mitú area in Colombia (Dept. Vaupés).

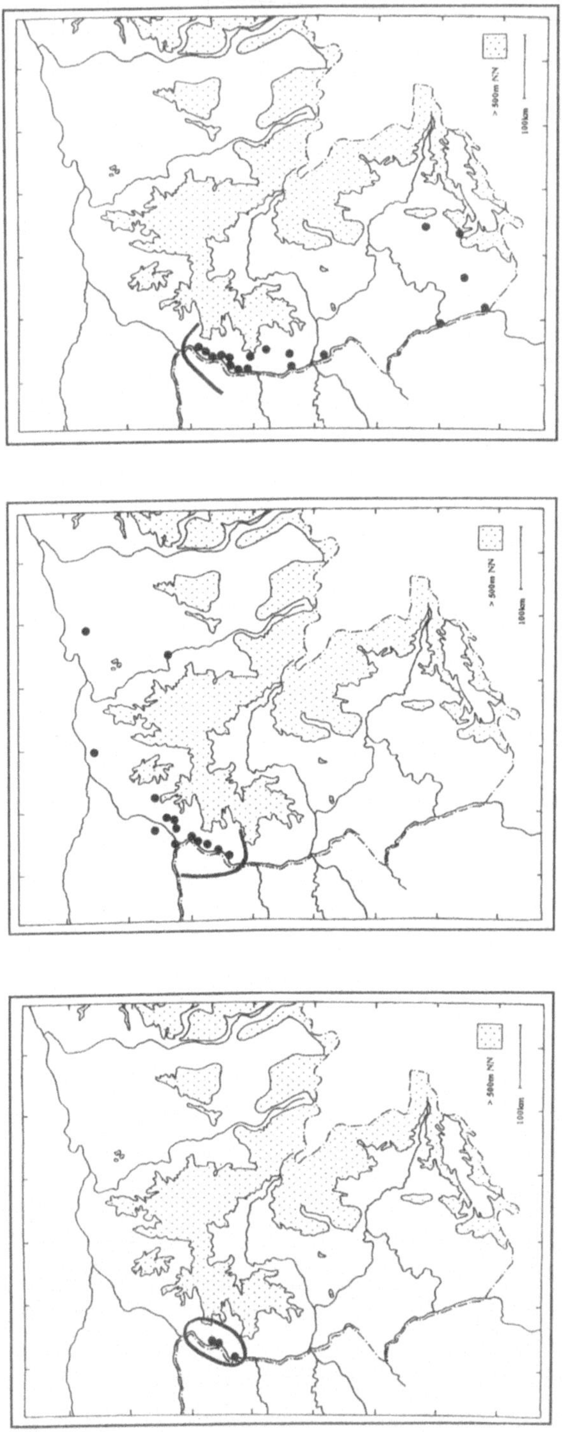

Fig. 10.6.9. Representative inselberg species for distinct phytogeographical subunits: (*left*) Atures center of endemism: *Ernestia cordifolia*; (*center*) northern inselberg subunit: *Mandevilla caurensis*; (*right*) southern inselberg subunit: *Pitcairnia* (*Pepinia*) *bulbosa*

What are the reasons for the concentration of endemic species in the Atures area? One argument may be that the Atures area represents a climatic and phytogeographical transition area. Climate changes from savanna to rainforest climate, and three floristic provinces border on each other. Therefore the Atures area may be interpreted as a melting pot of different floristic elements; but also another argument has to be considered. As shown above, inselberg-endemic species require very specific edaphic and climatic conditions. Their varying distributional ranges can be correlated with the recent macroclimate. This binding of the inselberg flora to specific environmental conditions involves an increased susceptibility to climatic changes. Clearly, the inselberg vegetation demonstrated macroclimatic preferences also in the late Pleistocene age. During these last periods of severe climatic changes, the macroclimatic zones in the study area shifted mainly to the south and southeast. Inselberg plants with specific climatic preferences had to follow the shift by moving their distributional area into their climatic optimum. The two fundamental geographical prerequisites for this island-hopping were (1) climatic zone shifts only over minor distances, and (2) the existence of sufficient stepping stones.

Both preconditions are fulfilled in the Atures center of endemism. The steep climatic gradient of the area, i.e., the dense grouping of different climatic belts, implies that climatic zones moved only a relatively small distance during late Pleistocene; also, granitic outcrops cover more than 15% of the land surface in the Atures area (Blancaneaux et al. 1977), facilitating the migration of inselberg species. Therefore, the Atures area offers ideal conditions for the differentiation and endurance of inselberg-endemic species, even in times of climatic changes. In other areas, species with such specific edaphic and climatic requirements were subject to a greater risk of extinction.

10.6.7.3 Floristic Affinities

Examining the floristic affinities of the Venezuelan inselberg flora, especially three groups provide illuminating hints: (1) inselberg-endemic species and their relatives, (2) species with disjunct distribution, and (3) species which reach the limits of their distributional ranges on the inselbergs. The results confirm the separation into two phytogeographical inselberg subunits by differing floristic affinities (Fig. 10.6.10).

The northern subunit is characterized by a pronounced Caribbean affinity, especially to the south Caribbean province. A considerable number of Caribbean species reach their southernmost limit of distribu-

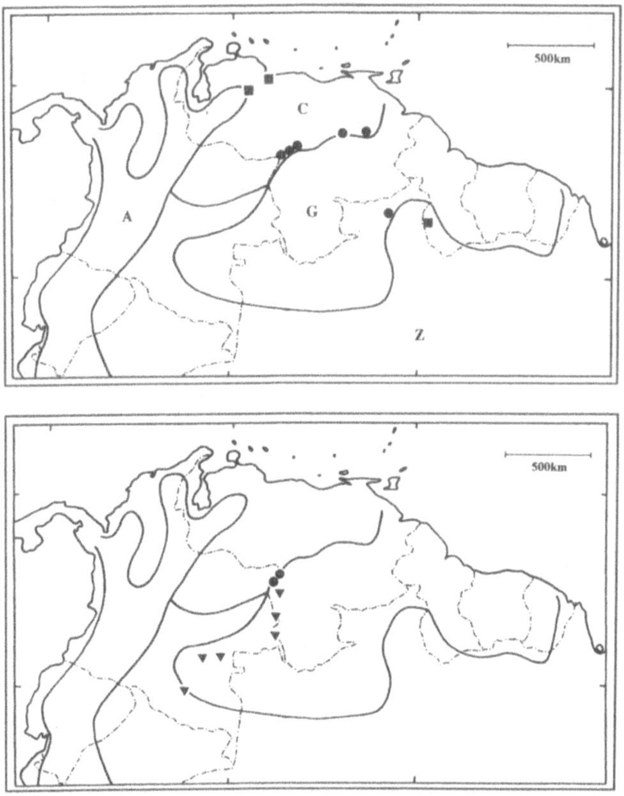

Fig. 10.6.10. Floristic relations of the inselberg flora: (*above*) to the Caribbean region and to the semiarid region in southern Guyana: *Cyrtocarpa velutinifolia* (*circles* inselberg localities; *squares* other localities; *A* Andean region; *C* Caribbean region; *G* Guayana region; *Z* Amazonas region; floristic regions according to Berry et al. 1995); (*below*) to the Guayana region: inselberg-endemic *Burmannia sanariapoana* (*circles*) and its next relative, *B. dasyantha* (*triangles*)

tion on the inselbergs of the northern inselberg subunit, such as, e.g., *Acanthocereus tetragonus*, *Calathea panamensis*, *Pereskia guamacho*, and *Tillandsia schiedeana*. By their occurrence on the northern inselbergs, some other taxa (e.g., *Bursera simaruba*, *Bursera tomentosa*, *Cyrtocarpa velutinifolia*, *Tabebuia ochracea* var. *heterotricha*) form connecting stepping stones between the xeric Caribbean coast and the semiarid areas in southern Guyana and adjacent Brazil.

The flora of the southern inselberg subunit and the Atures area shows a distinct affinity to the provinces of the Guayana region. Relations to the central Guayana province are shown by the next relatives (in parantheses) of inselberg-endemic species, like *Decagonocarpus oppositifolius*

(*D. cornutus*), *Graffenrieda rotundifolia* (*G. polymera*), *Kunhardtia radiata* (*K. rhodantha*), *Oyedaea wurdackii* (*O. tepuiana*), etc. Affinities to the western Guayana province are proved by the next relatives of some other inselberg-endemic species, like *Burmannia sanariapoana* (*B. dasyantha*), *Catasetum bergoldianum* (*C. ochraceum*), *Diacidia galphimioides* (*D. parvifolia*), etc.

Botanical studies on other inselberg sites of the Guayana shield, as in Suriname (J. van Donselaar and J.P. Schulz, unpubl.) and French Guiana (e.g., Sastre 1976; de Granville 1991; Sarthou 1992; Larpin 1993), provide inventories which are quite similar in their basic composition to that of the Venezuelan outcrops; but a closeup reveals that, although lying within the same phytogeographical region, there is hardly any common inselberg-endemic species occurring in the eastern as well the western part of the Guayana shield. Species endemic to outcrops of western Guayana (in parantheses) are partially replaced by relatives in eastern Guayana, e.g., *Ernestia granvillei* (*E. cordifolia*) and *Pitcairnia* (*Pepinia*) *geyskesii* (*P. armata, P. bulbosa, P. pruinosa*). Past climatological events, such as a Pleistocene dry belt crossing the Guayana shield diagonally, as postulated by Reinke (1962), may provide arguments for this distribution pattern.

In spite of the vast perhumid Amazonian lowland forests which separate the vegetation of semiarid areas north and south of the equator, some affinities exist between the Venezuelan inselberg vegetation and that of the Brazilian shield. One example is given by *Commiphora leptophloeos*, a characteristic Burseraceae species of the Brazilian caatinga and of Bolivian inselbergs (Prado and Gibbs 1993; Ibisch et al. 1995). North of the Amazon basin its occurrence is restricted to granitic outcrops of the Guayana region. A similar disjunct distribution pattern is displayed by *Helicteres heptandra* and some other inselberg taxa, which demonstrates that a floristic exchange between the semiarid sites of both of the shields took place.

10.6.8 Appendix

Enumeration of species endemic to the Guayana region and restricted to granitic outcrops. *Endemism*: categories of endemism (1a, inselberg-endemic sensu stricto, and known from at least three different localities; 1b, inselberg-endemic sensu stricto, and known from less than three different localities; 2, inselberg-endemic sensu lato). Geography: phytogeographical subunit to which the species belongs

Inselberg-endemic species	Family	Endemism	Geography[a]
Aspidosperma sp. nov.	Apocynaceae	1b	x
Mandevilla anceps Woodson	Apocynaceae	2	s
Mandevilla caurensis Markgraf	Apocynaceae	1a	n
Mandevilla filifolia Monach.	Apocynaceae	1a	n
Mandevilla lancifolia Woodson	Apocynaceae	1a	t
Mandevilla steyermarkii Woodson	Apocynaceae	1a	t
Plumeria inodora Jacq.	Apocynaceae	2	t
Caladium macrotites Schott	Araceae	2	a
Ditassa carnevalii (Morillo) Morillo	Asclepiadaceae	1a	n
Marsdenia guanchezii Morillo	Asclepiadaceae	1a	n
Mikania chaetoloba Pruski	Asteraceae	1b	t
Oyedaea wurdackii Pruski	Asteraceae	2	a
Tabebuia orinocensis (Sandw.) A. Gentry	Bignoniaceae	1a	n
Tabebuia pilosa A. Gentry	Bignoniaceae	1a	t
Pachira rupicola (Robyns) Alverson	Bombacaceae	2	n
Pseudobombax croizatii A. Robyns	Bombacaceae	1a	n
Cordia polystachya H.B.K.	Boraginaceae	2	s
Cordia stenostachya Killip ex Gaviria	Boraginaceae	1b	a
Brewcaria brocchinioides (L.B. Smith) Holst	Bromeliaceae	1b	s
Pitcairnia (*Pepinia*) *agavifolia* L. B. Smith	Bromeliaceae	1b	s
Pitcairnia (*Pepinia*) *armata* Maury	Bromeliaceae	2	n
Pitcairnia (*Pepinia*) *bulbosa* L. B. Smith	Bromeliaceae	1a	s
Pitcairnia (*Pepinia*) *nematophora* L. B. Smith & Read	Bromeliaceae	2	n
Pitcairnia (*Pepinia*) *pruinosa* H. B. K.	Bromeliaceae	1a	t
Vriesea bibeatricis Morillo	Bromeliaceae	1a	n
Vriesea melgueiroi Ramirez & Carnevali	Bromeliaceae	1b	s
Burmannia sanariapoana Steyerm.	Burmanniaceae	2	a
Melocactus mazelianus Riha	Cactaceae	2	n
Melocactus neryi K. Schumann	Cactaceae	1a	n
Merremia maypurensis H. Hallier	Convolvulaceae	2	a
Bulbostylis aturensis (Maury) C.B. Clarke	Cyperaceae	1a	t
Bulbostylis leucostachya (Kunth) C. B. Clarke	Cyperaceae	1a	t
Bulbostylis schomburgkiana (Steudel) M. T. Strong	Cyperaceae	1b	s
Rhynchospora agostiniana Koyama	Cyperaceae	1b	s
Rhynchospora sanariapensis Steyerm.	Cyperaceae	1a	s
Erythroxylum foetidum Plow.	Erythroxylaceae	2	n
Erythroxylum lindemanii Plow.	Erythroxylaceae	1a	t

Inselberg-endemic species	Family	Endemism	Geography[a]
Erythroxylum williamsii Standl. ex Plow.	Erythroxylaceae	1a	n
Croton romeroi Berry ined.	Euphorbiaceae	1b	s
Manihot tristis M.-Arg.	Euphorbiaceae	2	t
Desmodium orinocense (DC.) Cuello	Fabaceae	1a	t
Chrysothemis dichroa Lwbg.	Gesneriaceae	1a	n
Cipura rupicola Goldblatt & Henrich	Iridaceae	2	a
Hyptis guanchezii R. Harley	Lamiaceae	1b	a
Echeandia bolivarensis Cruden	Liliaceae	1b	n
Byrsonima nitidissima H. B. K.	Malpighiaceae	1a	t
Diacidia galphimioides Griseb.	Malpighiaceae	1a	s
Hiraea bifurcata Anderson	Malpighiaceae	1a	n
Maranta linearis Andersson	Marantaceae	2	t
Acanthella pulchra Gleason	Melastomataceae	1a	s
Acanthella sprucei Hook. f.	Melastomataceae	2	t
Comolia aff. *smithii* Wurdack	Melastomataceae	2	t
Comolia nummularioides (Bonpl.) Naud.	Melastomataceae	2	a
Ernestia cordifolia Berg ex Triana	Melastomataceae	1a	a
Graffenrieda rotundifolia (Bonpl.) DC.	Melastomataceae	1a	n
Votomita orinocensis Morley	Memecylaceae	1b	s
Mimosa brachycarpoides Barneby	Mimosaceae	1a	n
Ficus mollicula Pittier	Moraceae	1a	n
Eugenia callichroma Mc Vaugh	Myrtaceae	1b	n
Eugenia emarginata (H. B. K.) DC.	Myrtaceae	1b	x
Eugenia umbonata Mc Vaugh	Myrtaceae	1a	a
Guapira ayacuchae Steyerm.	Nyctaginaceae	2	a
Neea ignicola Steyerm.	Nyctaginaceae	1b	a
Ouratea chaffanjonii (Van Tieghem) Sastre	Ochnaceae	1a	t
Catasetum bergoldianum Foldats	Orchidaceae	1a	s
Pleurothallis aff. *papillosa* Lindl. (sp. nov.)	Orchidaceae	1a	n
Schomburgkia heidii Carnevali	Orchidaceae	1a	a
Peperomia maypurensis H. B. K.	Piperaceae	2	s
Panicum petrense Swallen	Poaceae	2	s
Thrasya setosa Swallen	Poaceae	2	a
Thrasya stricta Burman	Poaceae	1b	a
Polygala sanariapoana Steyerm.	Polygalaceae	1a	s
Portulaca pusilla H. B. K.	Portulacaceae	1a	t
Portulaca pygmaea Steyerm.	Portulacaceae	1a	a
Portulaca sedifolia N. E. Brown	Portulacaceae	2	t
Kunhardtia radiata Maguire & Steyerm.	Rapateaceae	1b	a
Gouania wurdackii Steyerm.	Rhamnaceae	2	a
Borreria pygmaea Spruce ex Schum.	Rubiaceae	2	n

Inselberg-endemic species	Family	Endemism	Geography[a]
Rudgea maypurensis Standl.	Rubiaceae	1a	t
Tocoyena brevifolia Steyerm.	Rubiaceae	1a	a
Tocoyena orinocensis Standl. & Steyerm.	Rubiaceae	1a	n
Decagonocarpus oppositifolius Spruce ex Engler	Rutaceae	1a	s
Ecclinusa parviflora Pennington	Sapotaceae	2	a
Stachytarpheta sprucei Moldenke	Verbenaceae	2	t
Xyris stenostachya Steyerm.	Xyridaceae	2	t

For a complete list of the species inventory: see Gröger (1995)

[a] a, Atures center of endemism; n, northern inselberg subunit; s, southern inselberg subunit; t, total area comprising n and s; x, undetermined

Acknowledgements. To my scientific mentors, W. Barthlott and O. Huber, I would like to express my sincere acknowledgements. Cordial thanks are extended to the "inselberg crew" of the Botanical Institute of Bonn and all specialists in various herbaria, who provided determinations and additional information. Much more than logistical support I received from the herbaria TFAV, VEN, and MO, and the Ministerio del Ambiente y Recursos Naturales Renovables (MARNR). The project received generous support from the Studienstiftung des Deutschen Volkes, the Deutscher Akademischer Austauschdienst (DAAD), and the Deutsche Forschungsgemeinschaft (DFG). Thanks also go to J. Wainwright-Klein and E. Blencowe for the revision of the manuscript.

References

Aristeguieta L (1966) Flórula de la Estación Biológica de los Llanos. Bol Soc Venez Cienc Nat 110:228-307

Baldrige M, Byrne JV, Courain ME (1982) World weather records 1961–1970. Vol. 3: West Indies, South and Central America. U S Department of Commerce, Washington, DC

Barthlott W, Gröger A, Porembski S (1993) Some remarks on the vegetation of tropical inselbergs: diversity and ecological differentiation. Biogéographica 69:105–124

Berry P, Huber O, Holst BK (1995) Floristic analysis and phytogeography. In: Steyermark JA, Berry P, Holst BK (eds) Flora of the Venezuelan Guayana. Missouri Botanical Garden, St Louis, Missouri, and Timber Press, Portland, Oregon, pp 161–191

Blancaneaux P, Pouyllau M (1977) Les relations géomorpho-pédologiques de la retombée nord-occidentale du massif guyanais (Vénézuela), 1. partie: Les concepts et définitions. Cah ORSTOM Sér Pédol 15:437–448

Blancaneaux P, Hernandez S, Araujo J (1977) Estudio edafológico preliminar del sector Puerto Ayacucho, Territorio Federal Amazonas, Venezuela. MARNR, Dir Suelos Veg Fauna, Caracas

Boom BM (1990) Flora and vegetation of the Guayana-Llanos ecotone in Estado Bolivar, Venezuela. Mem N Y Bot Gard 64:254–278

Büdel B, Lüttge U, Stelzer R, Huber O, Medina E (1994) Cyanobacteria of rocks and soils of the Orinoco Lowlands and the Guayana Uplands, Venezuela. Bot Acta 107:422–431

de Granville J-J (1991) Remarks on the montane flora and vegetation types of the Guianas. Willdenowia 21:201–213

Donselaar J van, Schulz JP (unpubl.) On the flora and vegetation of granite exposures in the Voltzberg region (Surinam). Utrecht, Paramaribo

Gröger A (1994) Análisis preliminar de la florula y vegetación del Monumento Natural Piedra La Tortuga, Estado Amazonas, Venezuela. Acta Bot Venez 17:128–153

Gröger A (1995) Die Vegetation der Granit-Inselberge Südvenezuelas: Ökologische und biogeographische Untersuchungen. PhD Thesis, Botanical Institute, University of Bonn

Gröger A, Barthlott W (1998) Biogeography and diversity of the inselberg (laja) vegetation of southern Venezuela. Biodivers Lett 3:165–179

Gröger A, Renner SS (1997) Leaf anatomy and ecology of the Guayana endemics *Acanthella sprucei* and *A. pulchra* (Melastomataceae). BioLlania Ed Espec 6:369–374

Huber O (1982) Significance of savanna vegetation in the Amazon territory of Venezuela. In: Prance GT (ed) Biological diversification in the tropics. Columbia University Press, New York, pp 221–244

Huber O (1995) Geographical and physical features. In: Steyermark JA, Berry P, Holst BK (eds) Flora of the Venezuelan Guayana. Missouri Botanical Garden, St Louis, Missouri, and Timber Press, Portland, Oregon, pp 1–61

Huber O, Alarcón C (1988) Mapa de vegetación de Venezuela. Ministerio del Ambiente y de los Recursos Naturales Renovables (MARNR), The Nature Conservancy, Fundación BIOMA, Caracas

Ibisch PL, Rauer, G, Rudolph D, Barthlott W (1995) Floristic, biogeographical, and vegetational aspects of Pre-Cambrian rock outcrops (inselbergs) in eastern Bolivia. Flora 190:299–314

Köppen W (1932) Die Klimate der Erde. De Gruyter, Berlin

Larpin D (1993) Les formations ligneuses sur un inselberg de Guyane Française. Thèse de doctorat, Université Pierre et Marie Curie, Paris

López M, Ramírez N (1989) Caracteristicas morfológicas de frutos y semillas y su relación con los síndromes de dispersión de una comunidad arbustiva en la Guayana Venezolana. Acta Cien Venez 40:354–371

Manabe S, Hahn DG (1977) Simulation of the tropical climate of an ice-age. J Geophys Res 82:3889–3911

Porembski S, Barthlott W (1992) Struktur und Diversität der Vegetation westafrikanischer Inselberge. Geobot Kolloq 8:69–80

Prado DE, Gibbs PE (1993) Patterns of species distributions in the dry seasonal forests of South America. Ann Mo Bot Gard 80:902–927

Rauh W, Barthlott W, Ehler N (1975) Morphologie und Funktion der Testa staubförmiger Flugsamen. Bot Jahrb Syst 96:353–374

Reinke R (1962) Das Klima Amazoniens. PhD Thesis, University of Tübingen

Sarthou C (1992) Dynamique de la végétation pionnière sur un inselberg en Guyane Francaise. Thèse de doctorat, Université Pierre et Marie Curie, Paris

Sastre C (1976) Quelques aspects de la phytogéographie des milieux ouverts guyanais. In: Descimon H (ed) Biogéographie et evolution en Amérique tropicale. Laboratoire de l'École Normale Supérieure, Paris, pp 67–74

Schubert C (1988) Climatic changes during the last glacial maximum in northern South America and the Caribbean: a review. Interciencia 13:128–137

Schubert C, Briceño HO, Fritz P (1986) Paleoenvironmental aspects of the Caroní-Paragua river basin (southeastern Venezuela). Interciencia 11:278–289

Tricart J (1985) Evidence of Upper Pleistocene dry climates in northern South America. In: Douglas I, Spencer T (eds) Environmental change and tropical geomorphology. British Geomorphological Research Group, London, pp 197–217

van der Pijl L (1982) Principals of dispersal in higher plants (3rd edn). Springer, Berlin Heidelberg New York

Walter H, Lieth H (1960). Klimadiagramm-Weltatlas. Gustav Fischer, Jena

Wingfield R (in prep.) Venezuela's native vascular plants: number and diversity of the taxa, with observations on endemicity, danger of extinction, and subjectivity of taxa. IUTAG, Coro

10.7 The Guianas (Guyana, Suriname, French Guiana)

U.P.D. RAGHOENANDAN

10.7.1 Introduction

This chapter is the result of data taken from publications concerning flora and vegetation of inselbergs in the Guianas. Notice should be taken that research on the vegetation of inselbergs in the Guianas, and especially the ecological aspect, has rarely been conducted. Extensive studies, however, were made by de Granville (1979, 1982, 1989, 1991), de Granville and Sastre (1974), Sastre and de Granville (1975) in the Tumuc Humac Mountains area, which lies in the area between French Guiana and Suriname at the southern frontier with Brazil, and Sarthou (1992), and Sarthou and Villiers (1998), who studied the pionier vegetation on inselbergs in French Guiana. J. van Donselaar and J.P. Schulz (unpubl.) made a study of the Voltzberg area in the center of Suriname, whereas Oldenburger et al. (1973) studied the Sipaliwini Savanna area in southwestern Suriname. As for Guyana, data were available from several botanical explorations in the Rupununi Savanna area in southern Guyana (Maas et al. 1988; ter Welle and Jansen-Jacobs 1991, 1995; ter Welle et al. 1987, 1989, 1993, 1994).

10.7.2 Geography and Geology

The Guianas consists of three countries (Guyana, Suriname, French Guiana) with a total area of approximately 500 000 km^2 and situated on the northeastern coast of South America to the north of Brazil (Boggan et al. 1997), between 1°15' and 8°33'N, and between 51°20' and 61°10'W (Fig. 10.7.1). Guyana is the westernmost country, between Venezuela and Suriname, and has a land area of approximately 231 800 km^2 (Boggan et al. 1997), while Suriname lies to the east of Guyana with a land area of ap-

Ecological Studies, Vol. 146
S. Porembski and W. Barthlott (eds.) Inselbergs

proximately 164 000 km² (Suriname Planatlas 1988). French Guiana is situated on the east of Suriname, and has a land area of approximately 90 000 km² (de Granville 1989).

The Guianas are mainly constituted on the Guayana Shield, which also includes much of southern Venezuela, a portion of southeastern Colombia and northern Brazil and is roughly bordered by the Amazon River, Rio Negro, and Orinoco River (Fig. 10.7.2). The Guayana Shield covers more than 80 % of the interior of the Guianas and consists of Precambrian igneous and metamorphic rocks, which originated during the period of Trans-Amazonian Orogenesis about 1900 Ma ago. It is a stable area where there are no longer any volcanic activities, earthquakes, or orogenesis (Leeflang et al. 1976). Where the crystalline Guayana Shield is covered by Roraima sandstones (Fig. 10.7.2) the table top mountains or tepuis are found. In the Guianas, tepuis exist only in western Guyana (e.g., Mt. Ayanganna, Mt. Roraima, etc.) and in central Suriname (Tafelberg, and, according to Jonker and Wessink 1960, in the region of the southern Yzermantop of the Emma Range). Besides Roraima sandstones, the crystalline shield is also covered by Tertiary and Quarternary unconsolidated sediments of the major river basins and coastal areas (Lindeman and Mori 1989).

Fig. 10.7.1. The Guianas [After: Map Utrecht Herbarium (U)]

According to Daniel (1984), the Suriname Planatlas (1988), and de Granville (1989), the following geomorphological regions are recognized in, respectively, Guyana, Suriname, and French Guiana: (1) The Coastal Plain, a narrow strip along the coast, further divided into Young Coastal Plain (0–4 m asl), and Old Coastal Plain (4–12 m asl). (2) The White Sand area, which begins halfway west of Guyana between the Coastal Plain and the crystalline rock outcrops further south, continues in Suriname, where it is named the Cover Landscape or Savanna Belt (elevation of 12–100 m asl), and ends in French Guiana where it occurs only in the northeast. (3) The Precambrian Plateau: in Guyana it is named the Lowland Region (general elevation up to 120 m asl, and peaks over 300 m asl in the north and 900 m in the south) and is mostly under tropical rainforest with the exception of the Rupununi Savanna Region; in Suriname it is named the Hilly/Mountainous area (up to 1230 m asl); in French Guiana there is a general elevation up to 500 m asl, and peaks up to 860 m asl in the central part and some places in the west. (4) Pakaraima Mountain Region found only in Guyana with elevated plateaus at heights up to 2772 m asl (Mt. Roraima).

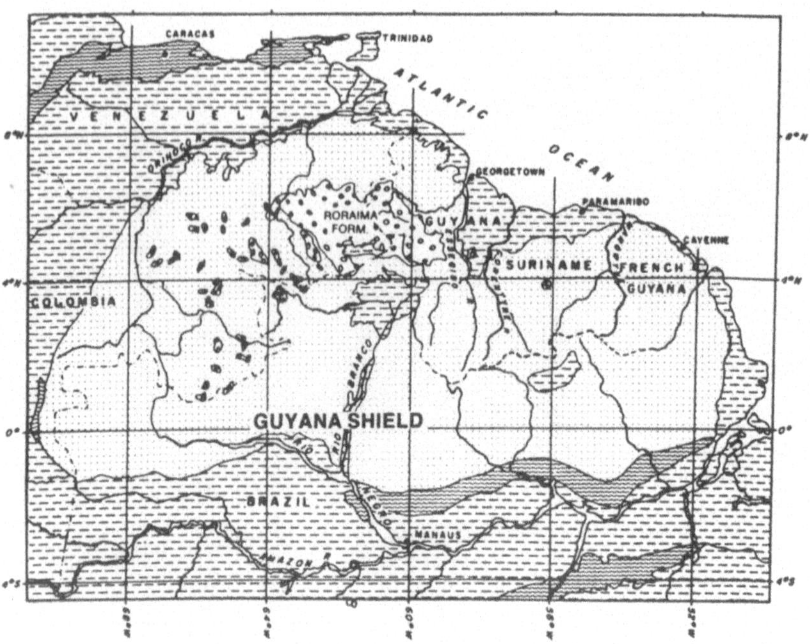

Fig 10.7.2. The Guayana Shield (*dotted*), covered by remnants of the Roraima sandstone formation (incl. Tafelberg) (*circled*), and unconsolidated sediments (*horizontal, interrupted lines*). (Map GMD, Geological Mining Service)

10.7.3 Climate

The climate of the Guianas (Fig. 10.7.3) is tropical humid, and is characterized by high rainfall and a high relative humidity (on average 80–86 %), and a relatively high and constant yearly mean temperature with a narrow range (Leeflang et al. 1976; Daniel 1984). The mean annual temperature on the coast is approximately 27 °C with a narrow range of approximately 2 °C. The average daily temperature varies from 26 to 32 °C (Leeflang et al. 1976; Daniel 1984).

Rainfall is mainly influenced by the northeasterly tradewinds, the convectional currents, and the orographic lifting. In Guyana the average annual rainfall distribution varies from 1400 mm in the southern savanna area (Rupununi) to 4400 mm in the northwest district (Daniel 1984); in Suriname from 1450 mm at Coronie in the northwestern coastal area to 3000 mm in the center at the Tafelberg (Suriname Planatlas 1988); in French Guiana the rainfall distribution varies from less than 2200 mm in the southwest in the Tumuc Humac area to over 4000 mm at Kaw Mts. in the northeast (de Granville 1989).

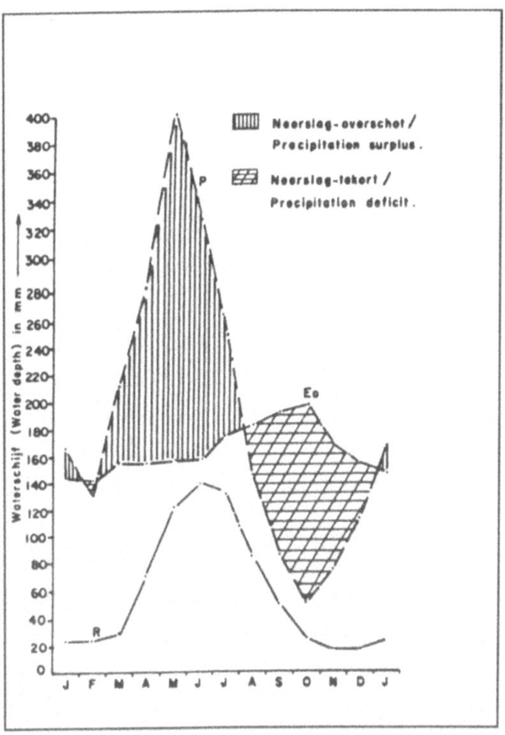

Fig. 10.7.3. Average monthly precipitation (P), potential evaporation (Eo), and runoff (R) in the Corantijn River Basin upstream of Mataway (1965–1980). (Suriname Planatlas 1988)

Throughout the Guianas, four annual seasons occur: a short dry season from February to April, a long wet from May to mid-August, a long dry one from mid-August to November, and a short wet one from December to January. This seasonality is due to the northerly and southerly movements of the intertropical convergence zone (ITCZ) (Daniel 1984; Suriname Plan-atlas 1988). According to Lindeman and Mori (1989), due to variability of the annual rainfall, the short dry or short wet season may be less distinct or may not occur at all, as in the Rupununi Savanna of Guyana.

10.7.4 Vegetation of the Guianas

The vegetation of the Guianas is determinded by the geology, geomorphology, soils, topography, altitude, and the availability, amount, and quality of water (de Granville 1989; Lindeman and Mori 1989; Boggan et al. 1997). It is estimated that the forest cover of the total land area varies from 86% in Guyana through 95% in Suriname to 97.7% in French Guiana. Apart from certain areas where shifting cultivation is practiced by local inhabitants and logging activities take place, the majority of the interior of the Guianas is still untouched. Mainly the area along the coast has been subject to cultivation.

According to Lindeman and Mori (1989), most of the forests of the Guianas are classified as seasonal evergreen forest (or seasonal wet forest). Besides the seasonal evergreen forest, they mention eight more vegetation types depending on soil conditions or temperatures decreasing with altitude viz: the mangrove swamps of the coast in brackish water, the salt-sprayed strand vegetation (beach vegetation), the periodically inundated marsh forests, the permanently moist freshwater swamp forests, herbaceous swamps, savannas, montane vegetation, and the vegetation of granitic outcrops (inselbergs). However, this list of vegetation types, according to Boggan et al. (1997), focuses mainly on Suriname and French Guiana and omits several forest types, which are recognized by Fanshawe (1952), who listed 28 vegetation types for Guyana. A combination of the two lists gives a broader, more complete range of vegetation types found in the Guianas.

The Guianas are estimated to possess over 9200 plant species, of which only 3% are introduced and naturalized, with approximately 6500 from Guyana, 5100 from Suriname, and 5400 from French Guiana; the largest families are Leguminosae with more than 800 species, Orchidaceae with about 700 species, Rubiaceae and Poaceae both with more than 400 species (Boggan et al. 1997).

10.7.5 Inselberg Vegetation

Inselbergs are found throughout much of the interior of the Guianas on the Precambrian basement rock in seasonal evergreen forest as well as in the drier savanna areas like the Rupununi Savanna area in southern Guyana and the Sipaliwini Savanna area in southwestern Suriname. They may occur as whalebacks which are rather easy to climb or on mountain ranges with bare summits which sometimes can be very steep (e.g., Kassikassima Mts. in southeastern Suriname). Some few inselbergs exceed 500 m, mostly occurring, with a few exceptions, in the most southern part of the Guianas (de Granville 1991). However, there are many inselbergs with low altitudes (below 500 m) occurring in, for instance, the Rupununi Savanna area, the central and southeastern part of Suriname, also the Sipaliwini Savanna area, and the eastern and southern part of French Guiana. Also many rock plateaus (shield inselbergs) or boulders are found in forest or savanna with the characteristic rock vegetation, e.g., *Cyrtopodium* (Orchidaceae), *Melocactus* (Cactaceae), *Cnidoscolus* (Euphorbiaceae), and/or *Furcraea* (Agavaceae). In general, the inselbergs higher than 500 m are more easy to indicate. Those are properly named and are either geologically or botanically explored.

For Guyana, from de Granville (1991) and several botanical explorations in the northern and southern Rupununi Savanna area, and the Kanuku Mts. area (Maas et al. 1988; ter Welle and Jansen-Jacobs 1991, 1995; ter Welle et al. 1987, 1989, 1993, 1994), the following inselbergs are mentioned: Mt. Nappi (960 m), Acarai Mts. (906 m), Mt. Makarapan (above 900 m), Wokrama Mts. (710 m), Mt. Kusad (up to 650 m), Mt. Shiriri (ca. 600 m), Mt. Toucan (360 m), Shea Rock (300 m, Fig. 10.7.4), Mt. Tawatawun (up to 200 m). In the southern Rupununi Savanna many small and large rocky outcrops with altitudes varying from a few meters to over 200 m above the ground level occur (ter Welle and Jansen-Jacobs 1995).

For Suriname the following inselbergs are mentioned (Oldenburger et al. 1973; Teunissen 1978; Suriname Planatlas 1988; de Granville 1991; Reichart 1993) viz: Wilhelmina Mts. with the Juliana Top (1230 m), Emma Range with the Hendrik Top (1030 m), southern part of the Bakhuis Mts. (1027 m), Acarai Mts. (906 m), Eilerts de Haan Mts. (900 m), Franssen Herderscheepiek (822 m), Apalagadi (771 m), Oranje Mts. (728 m), Asch van Wijck Mts. with the Ebba Top (721 m), Kassikassima (718 m), Vier Gebroeders Mts. (554 m), Rooseveltpiek (514 m), Bemau Top (501 m), Knopoiamoi (490 m), Van der Wijck Top (453 m), Magneetrots (376 m), Tebu Top (374 m), Voltzberg area (incl. Van Stockumberg 362 m, Voltzberg 209–245 m, Granmisibergi 210 m, and Granmisiweri), and Potihill near Palumeu (203 m).

Fig. 10.7.4. Shea Rock, an inselberg located in the south Rupununi savanna (Guyana). (Photograph F. Scheplitz)

For French Guiana the following inselbergs are indicated by de Granville (1991): Tumuc Humac area with Mitaraka as highest point (700 m), situated at the southern border between Suriname and French Guiana, Bakra Mts. with Pic Condreau (700 m), Mt. Saint-Marcel (635 m), Massif des Emerillons (570 m), and many rocky outcrops with an altitude lower than 500 m concentrated in the mid-eastern, eastern, and southern part of the country (de Granville 1979, 1982).

10.7.5.1 Vegetation Types

The vegetation on inselbergs varies depending on the thickness of the soil and the steepness of the slopes. On steeper slopes the soil is more seriously influenced by superficial runoff and erosion, and as the thickness of the soil decreases, the height of the forest decreases. Consequently, on places where soil is absent, bare rocks come to the surface. Several authors (Oldenburger et al. 1973; Holthuijzen and de Jong 1977; Moonen 1984; ter Welle et al. 1987, 1989, 1993, 1994; Maas et al. 1988; Lindeman and Mori 1989; de Granville 1991; ter Welle and Jansen-Jacobs 1991) distinguished more or less three vegetation types on inselbergs in the Guianas:

1. *Seasonal Evergreen Forests on Crystalline Basic Rock.* In the Guianas this vegetation type is mainly found on the larger inselbergs with altitudes overtopping 500 m, and is quite the same as forests growing on low altitudes in the interior. Tall trees are found in valleys, on slopes, and on ridges, where it presents a drier facies. Often, large granitic boulders are found in the understory, sometimes harboring a few epilithic plants growing in half shade. In Guyana also cloud forest was observed by ter Welle et al. (1987) at the top of a few inselbergs in the Kanuku Mountains area only at the highest altitudes (above 800 m). In the cloud forest many epiphytes, most of which are mosses, are locally abundant.
2. *Low, Dry Forests on Crystalline Basic Rock.* Some authors (Oldenburger et al. 1973; Holthuijzen and de Jong 1977) describe this type of vegetation as shrub-woodland, mountain savanna forest, or rock savanna forest, occurring mainly on very thin soil on the summit of the outcrops, in transition zones, and/or in belts between the base of the bare slopes and the seasonal evergreen forest. Treelets and shrubs rarely overtop 10–25 m, for example: *Tapirira* (Anacardiaceae); *Terminalia* (Combretaceae); *Pterocarpus* (Fabaceae); *Guarea, Trichilia* (Meliaceae); *Inga* (Mimosaceae); *Calyptranthes, Eugenia* (Myrtaceae); *Guapira* (Nyctaginaceae); *Ecclinusa, Chrysophyllum* (Sapotaceae), etc. (de Granville 1991; Holthuijzen and de Jong 1977). Herbs and xerophytic epiphytes are common in the understory. According to Holthuijzen and de Jong (1997), these epiphytes are scarce in the rock savanna forest on the up to 250 m high Voltzberg (Suriname).
3. *Xerophytic Vegetation on Exposed Granitic Rock.* This type of vegetation is also designated rock savanna or rock pavement vegetation (Holthuijzen and de Jong 1977), savanes-roches (de Granville 1991), and occurs on nearly bare granite rock around the summit and on slopes exposed to the sun. It is discontinuous with lithophytes, annual herbs, succulents, sclerophylls, etc. dominated by Bromeliaceae, Cactaceae, Orchidaceae, Poaceae, and Schizaeaceae. Up to 4 m tall, often deciduous, shrubs (Clusiaceae, Melastomataceae, Myrtaceae) and herbs are growing in patches. These bushes are scattered on the summit and the slopes of inselbergs at places where enough weathering residua and organic matter can accumulate in depressions and crevices. Some species may be abundant locally, such as *Clusia* species.

10.7.5.2 Habitat Types

Of most interest is the xerophytic vegetation growing on the exposed granite rocks on which different habitat types like the above-mentioned can be distinguished. The terminology of the different habitat types is according to Barthlott et al. (1993) and Porembski et al. (this Vol.). The following survey is mainly based on information from the Voltzberg area with altitudes up to 360 m (central Suriname) and the Sipaliwini Savanna area with altitudes up to ca. 550 m (southwestern Suriname) extracted from the National Herbarium of Suriname (BBS). In addition, information from several sources, as indicated, are mentioned.

Vegetation on Rock Surfaces. The seemingly bare rock surface is covered by cryptogams like cyanobacteria and lichens. The cyanobacteria give the rocks their dark color: purplish brown if they are dried up or brownish green when they come alive during the rainy season, making the rocks slippery (Teunissen 1978). Sarthou et al. (1995) mentioned several cyanobacteria occurring on the Nouragues inselberg in French Guiana, e.g., *Gloeocapsa*, *Schizothrix*, *Scytonema*, and *Stigonema*. Recently, many lichens have been collected on Shea Rock in the Rupununi Savanna (Flora of the Guianas Project), but detailed information will be available only after closer examination of the specimens. Among the lichens, *Peltula* was found on Shea Rock.

Drainage channels usually are scarcely inhabited by vascular plants because there is a quick runoff of water, but cyanobacteria (e.g., *Doliocatella formosa*) and green algae (e.g., *Oedogonium guyanense*) are frequently found (Sarthou et al. 1995). Where there is sand and humus *Melochia melissifolia* (Sterculiaceae) is found.

The vascular lithophytic vegetation on the rock surface comprises (Fig. 10.7.5, 10.7.6), for example, *Furcraea foetida* (Agavaceae), *Melocactus neryi*, *M. smithii* (Cactaceae), and *Cyrtopodium glutiniferum* (Orchidaceae).

Vegetation of Rock Crevices. On the edges of broken granite exfoliated sheets (horizontal crevices), where some erosion material accumulates, Cactaceae are found, e.g., *Melocactus* and *Cereus* species (Teunissen 1978), *Melocactus* also occurs frequently on the so-called dragon's teeth (Oldenburger et al. 1973). These are enormous granitic rock blocks weathered very typically, looking like huge teeth rising up to 5 m from the soil surface, and occur in the Sipaliwini Savanna area (southwestern Suriname) and the Voltzberg area (central Suriname) often near the edge of the forest (M.C.M. Werkhoven, pers. comm.).

Along and in vertical crevices *Bulbostylis* species (Cyperaceae), *Thrasya granitica* (Poaceae), and many herbs like *Schwenkia* (Solanaceae), *Waltheria indica* (Sterculiaceae), *Turnera ulmifolia* (Turneraceae), ferns, e.g., *Anemia* species (Schizaeaceae), *Doryopteris collina* (Adiantaceae), and

Fig. 10.7.5. Inselbergs in the Guianas provide natural growth sites of the Agavaceae *Furcraea foetida*. (Photograph F. Scheplitz)

Fig. 10.7.6. Typical colonizers over shallow soil are *Pitcairnia* (*Pepinia*) *nuda* (Bromeliaceae) and *Melocactus neryi* (Cactaceae). (Photograph F. Scheplitz)

woody elements like *Stylosanthes guianensis* (Fabaceae), *Byrsonima spicata* (Malpighiaceae) are found.

Boulder bases are inhabited by species like *Cerathosanthes palmata* (Cucurbitaceae), *Cyperus laxus* (Cyperaceae), and *Phyllanthus orbiculatus* (Euphorbiaceae).

In crevices on steep slopes there is *Monstera adansonii* var. *laniata* (Araceae), and ferns like *Hemionitis rufa* (Adiantaceae).

Vegetation of Depressions. Depressions may occur on the summit and also on slopes of the exposed granite inselbergs. They harbor herbaceous plants and/or shrubs up to 4 m high, depending on the amount of soil, available.

In seasonally water-filled rock pools of more or less 30 cm diameter, we find no vascular plants, but cryptogams (cyanobacteria). On the Voltzberg, according to J. Reichart (pers. comm.), birds use the rock pools to build their nests. Typical water plants do not occur on the inselbergs in the Guianas.

Open depressions with shallow soil on granite are mainly inhabited by herbs; *Operculina alata* (Convolvulaceae) grows in full sun together with *Pitcairnia (Pepinia) nuda* (Bromeliaceae) and a woody herb *Cnidoscolus urens* (Euphorbiaceae). *Panicum micranthum* (Poaceae) and a shrub *Stylosanthes guianensis* (Fabaceae) also occur. In such depressions occurring on slopes, *Caladium bicolor* f. *robustum* (Araceae), *Phyllanthus stipulatus* (Euphorbiaceae), *Sinningia incarnata* (Gesneriaceae), and *Axonopus ramosus* (Poaceae) are found. A purplish to red (sometimes greenish), flat, succulent herb, *Portulaca sedifolia* (Portulacaceae), with tufts of white hairs at the nodes may occur in small depressions. Mosses like *Campylopus savannarum*, *Octoblepharum cylindricum*, and *Syrrhopodon gaudichaudii* are found between roots or as small mats on the open soil.

In depressions with deep soil, bushes may establish, harboring many woody plants, e.g., *Clusia* (Clusiaceae), *Ernestia*, *Miconia* (Melastomataceae), *Mimosa* (Mimosaceae), *Eugenia* (Myrtaceae), and also herbaceous plants and epiphytes, e.g., *Encyclia granitica*, *Trigonidium acuminatum*, *Vanilla grandiflora* (Orchidaceae), *Tillandsia flexuosa*, *Vriesea procera*, *Ananas nanus* (Bromeliaceae), and ferns, e.g., *Hemionitis palmata* (Adiantaceae), *Aneilema umbrosum* (Commelinaceae), *Croton stahelianus* (Euphorbiaceae). The edges of the bushes are preferred by *Catasetum planiceps* (Orchidaceae), *Lasiacis anomala* (Poaceae) and many lianas of the families Apocynaceae, Marcgraviaceae, and Convolvulaceae.

According to Teunissen (1978), Poaceae, Cyperaceae, and Melastomataceae, and many other herbaceous flowering plants are the pioneers in depressions, but soon shrubs, like *Clusia* species (Clusiaceae), take over. In the shade of this scrub the pioneers disappear and are replaced by shade species, e.g., *Encyclia granitica* (Orchidaceae). When the shrubs become too large for the thin soil layer on which they grow, they fall and die off. The pioneer species can therefore appear again in full sun, starting a new succession.

Mat Vegetation. Monocotyledonous mats cover large areas of the bare rock and are dominated by *Pitcairnia (Pepinia)* species, e.g., *P. geyskesii*, *P. nuda*, and *P. sastrei* (Bromeliaceae). Also *Trilepis kanukuensis* (Cyperaceae), until now been mentioned only for the Kanuku Mts. and the Tumuc Humac Mts., is a perennial herb forming dense mats on the rocks (M.J. Jansen-Jacobs pers. comm.; Maguire et al. 1965). According to de Granville (1991) *Ischaemum guianense* and other Poaceae form dense patches in the moister places, while *Pitcairnia (Pepinia) geyskesii* (Bromeliaceae) is characteristic of the more arid zones of the slopes of the Tumuc Humac Mountains. This also occurs on the inselbergs in and around the Rupununi Savanna in Guyana. Even steep slopes can be covered by these mats.

In between the mats several vascular plants are found, e.g., *Begonia guaduensis* (Begoniaceae), *Cereus hexagonus* (Cactaceae), *Dichorisandra hexandra* (Commelinaceae), *Evolvulus alsinoides* (Convolvulaceae), *Bulbostylis capillaris* var. *major*, *Cyperus miliifolius* (Cyperaceae), *Manihot brachyloba*, *M. tristis* ssp. *saxicola* (Euphorbiaceae), *Irlbachia alata* ssp. *alata* (Gentianaceae), *Cyrtopodium glutiniferum*, *Encyclia calamaria* (Orchidaceae), *Panicum polycomum* (Poaceae), *Borreria latifolia* var. *latifolia*, *Spermacoce tenuior* (Rubiaceae), and *Cissus erosa* (Vitaceae).

The edges of bare granite plateaus are occupied by species like *Furcraea foetida* (Agavaceae), *Philodendron muricatum* (Araceae), *Epiphyllum phyllanthus* (Cactaceae), *Mimosa microcephala* var. *cataractae* (Mimosaceae), the more or less shady places by *Bomarea edulis* (Alstroemeriaceae), and *Ananas* (Bromeliaceae) with *Maranta gibba* (Marantaceae).

Ephemeral Flush Vegetation. This habitat usually occurs where water seeps from the mats or at the feet of steep slopes. Mostly species of *Paepalanthus*, e.g., *P. fasciculatus*, *Syngonanthus*, e.g., *S. gracilis* (Eriocaulaceae), *Utricularia*, e.g., *U. hispida* (Lentibulariaceae), and *Xyris*, e.g., *X. jupicai* (Xyridaceae) occur in this habitat. Also *Mesosetum cayennense* (Poaceae) and sometimes species of Iridaceae are found in the moist places. At the feet of steep slopes in moist well-watered zones *Polygala adenophora* (Polygalaceae) and *Costus* species (Costaceae) occur.

10.7.6 Life-Forms and Adaptations

As mentioned earlier, on exposed granitic outcrops a typical xerophytic vegetation occurs of scattered herbs and shrubs in an environment with extremely varying temperatures and hydrological conditions. Temperatures can be extremely high and water is only temporarily available. According to Reichart (1993), the relative humidity on the rock plateau may decrease to 40%. During rainfall, water is kept by the soil only in depressions for several days, whereas on the greater part of the outcrops the water is lost immediately by superficial runoff and rapid evaporation. Many of the plants show specific adaptations to the extreme habitat in both life-forms and growth patterns. According to Oldenburger et al. (1973) and Teunissen (1978), the following features occur:

1. Disproportional rootsystems are developed to preserve access to the water supply: *Philodendron acutatum* (Araceae).
2. To capture and store water supply as reserve for times of severe drought succulence occurs: Teunissen (1978) distinguishes stem succulence, e.g., Cactaceae and leaf succulence, e.g., *Furcraea foetida* (Agavaceae), Bromeliaceae, *Ernestia* sp. (Melastomataceae), *Portulaca sedifolia* (Portulacaceae).
3. Tubers are formed in: *Mandevilla tenuifolia* (Apocynaceae), *Manihot* (Euphorbiaceae).
4. Distinctive pseudobulbs are formed in: *Cyrtopodium* (Orchidaceae).
5. Thorns and hairiness are developed: *Mimosa microcephala* var. *cataractae*, *M. brachycarpoides* (Mimosaceae), Cactaceae, *Cnidoscolus urens* (Euphorbiaceae), several lianas.
6. To reduce transpiration micro- and sclerophylly is observed in, respectively, *Mimosa microcephala* var. *cataractae* (Mimosaceae) and *Clusia* species (Clusiaceae).
7. To meet the challenge of limited spatial conditions and, when established, to keep control of their environment, formation of facies and hummock is observed as adaptation in growth patterns: *Axonopus ramosus* (Poaceae).
8. Poikilohydry occurs in some plants: *Trilepis kanukuensis* (Cyperaceae), and several species of Schizaeaceae, mosses.
9. Annuals pass the unfavorable season as embryos in the seed: species of Poaceae and Cyperaceae.

10.7.7 Systematic and Ecological Characteristics

According to de Granville (1991), who studied granite outcrops in the Tumuc Humac area, the more or less xeric vegetation types, with the most characteristic flora on granitic outcrops and other crystalline rugged reliefs, are related to the driest flora of Roraima sandstone formation, e.g., *Clusia annularis* (Clusiaceae), *Sauvagesia tafelbergensis* (Ochnaceae), *Sipanea wilson-brownei* (Rubiaceae), *Asplenium zamiifolium* (Aspleniaceae), and *Doryopteris sagittifolia* (Adiantaceae), and to the coastal as well as the sandy savannas in the far interior of the Guianas, even as far as south of the Amazon basin. Teunissen (1978) mentions affinities also with epiphytes and lianas from the canopy of the seasonal evergreen forest and the flora of brackish coastal sand ridges, which also suffer physiological drought. Detailed information on ecological characteristics has not as yet been gathered. The paragraph on habitat types, however, gives a survey of some plants which choose certain specific habitats to grow.

De Granville (1991) also noticed that each mountain or group of mountains has its own characteristic plants, like *Epidendrum nocturnum* var. *tumuc-humaciense*, which is restricted to the Tumuc Humac area and the Kassikassima in southeastern Suriname. *Aechmea lanjouwii* (Bromeliaceae) and *Thrasya granitica* (Poaceae), on the other hand, have been reported only from the Voltzberg. *Sporobolus temomairemensis* (Poaceae) is at present only found in the Tumuc Humac region and the Sipaliwini Savanna area.

In Table 10.7.1, an attempt is made to give the floristic composition of some granitic outcrops in Suriname, i.e., the Voltzberg area and the Sipaliwini Savanna area. However, the list is far from complete. A total amount of 157 species of flowering plants are recorded, of which 115 were found in the Voltzberg area and 83 in the Sipaliwini Savanna area. The families which are well represented in these areas are as follows: Poaceae (23 spp.); Cyperaceae (15 spp.); Orchidaceae (14 spp.); Bromeliaceae, Rubiaceae (10 spp.); Euphorbiaceae (8 spp.); Melastomataceae, Cactaceae (7 spp.); Eriocaulaceae, Araceae (5 spp.).

The number of species found in both areas consists of 26 % (41 spp.); for the Voltzberg area 47 % (74 spp.) were recorded, and for the Sipaliwini Savanna area 27 % (42 spp.). Of all 157 species of flowering plants 19 % (30 spp.) are typical of granitic rock outcrops.

Table 10.7.1. Species list of granitic rock outcrops in the Voltzberg area (central Suriname) and the Sipaliwini Savanna area (southwestern Suriname). Data are extracted from labels of herbarium vouchers from the National Herbarium of Suriname (BBS), from Pulle (1932–1984), Oldenburger et al. (1973), Holthuijzen and de Jong (1977), Görts-van Rijn (1985–1997), Werkhoven (1986, pers. comm.), Boggan et al. (1997), van Donselaar and Schulz (unpubl.)

Species	Vo[a]	Si[b]	R[c]	D[d]
Agavaceae				
Furcraea foetida	+	+	+	N
Apocynaceae				
Himatanthus articulatus		+		
Mandevilla scabra var. *intermedia*	+	+		
Mandevilla surinamensis	+		+	GuV
Mandevilla tenuifolia	+	+	?	SV
Araceae				
Anthurium gracile	+			
Caladium bicolor f. *robustum*	+	+		tAS
Monstera adansonii var. *laniata*	+			
Philodendron acutatum	+	+		
Philodendron muricatum	+			
Asclepiadaceae				
Gonolobus ligustrinus		+		
Asteraceae				
Calea solidaginea ssp. *deltophylla*	+			
Unxia camphorata	+	+		
Begoniaceae				
Begonia guaduensis	+		+	@
Boraginaceae				
Cordia polycephala	+			
Cordia tomentosa	+		+	SF
Bromeliaceae				
Aechmea aquilega	+			
Aechmea lanjouwii	+		+	S
Aechmea tocantina		+	+	Z
Ananas ananassoides	+		+	Z
Ananas nanus	+			
Pitcairnia (Pepinia) geyskesii	+	+	+	Gu
Pitcairnia (Pepinia) nuda	+		+	SG
Tillandsia fasciculata		+		
Tillandsia flexuosa	+			
Vriesea procera	+			
Cactaceae				
Cereus hexagonus	+	+		
Epiphyllum phyllanthus var. *phyllanthus*	+			
Hylocereus scandens	+			SG
Melocactus neryi	+	+	+	@
Pereskia aculeata	+			
Rhipsalis baccifera	+			

Table 10.7.1 (*continued*)

Species	Vo[a]	Si[b]	R[c]	D[d]
Caesalpiniaceae				
Chamaecrista fagonoides		+		
Chrysobalanaceae				
Hirtella racemosa		+		
Clusiaceae				
Clusia fockeana	+			Gu
Clusia leprantha	+	+		Gu
Clusia nemorosa	+			
Clusia panapanari	+	+		Gu
Commelinaceae				
Aneilema umbrosum ssp. *ovato-oblongum*	+			
Dichorisandra hexandra	+			
Convolvulaceae				
Evolvulus alsinoides	+	+		
Evolvulus filipes		+		
Operculina hamiltonii	+			
Costaceae				
Costus arabicus	+			
Cucurbitaceae				
Ceratosanthes palmata	+	+		
Cyperaceae				
Bulbostylis capillaris var. *major*	+	+	+	S
Bulbostylis junciformis		+		
Bulbostylis surinamensis		+		Gu
Bulbostylis vestita		+		
Cyperus aggregatus		+		
Cyperus capillifolius		+		
Cyperus laxus	+			
Cyperus miliifolius	+			
Rhynchospora armerioides		+		
Rhynchospora barbata	+	+		
Rhynchospora cephalotes		+		
Rhynchospora fallax	+	+	+	Gu
Rhynchospora tenella var. *tenella*		+		
Scleria setacea	+			
Scleria staheliana	+		+	SF
Eriocaulaceae				
Paepalanthus fasciculatus	+	+		
Paepalanthus lamarckii	+			
Paepalanthus subtilis		+		
Syngonanthus simplex		+		
Syngonanthus caulescens		+		
Euphorbiaceae				
Cnidoscolus urens	+	+	+	N
Croton stahelianus	+		+	S
Manihot brachyloba	+		?	
Manihot tristis ssp. *saxicola*	+		+	Gu

Table 10.7.1 (*continued*)

Species	Vo[a]	Si[b]	R[c]	D[d]
Euphorbiaceae				
Omphalea diandra	+			
Phyllanthus orbiculatus	+			
Phyllanthus stipulatus	+			
Plukenetia verrucosa	+			
Fabaceae				
Aeschynomene brasiliana		+		
Stylosanthes guianensis	+	+		
Gentianaceae				
Irlbachia alata ssp. *alata*	+			
Neurotheca loeselioides ssp. *loeselioides*		+		
Gesneriaceae				
Sinningia incarnata	+	+		
Lentibulariaceae				
Utricularia hispida	+			
Malpighiaceae				
Byrsonima spicata	+	+		
Malvaceae				
Sida viarum	+	+		
Briquetia spicata	+			
Marantaceae				
Calathea mansonis		+		
Maranta gibba	+		?	
Marcgraviaceae				
Norantea guianensis	+			
Souroubea guianensis	+			
Melastomataceae				
Comolia villosa	+			
Ernestia blackii		+	+	SFB
Ernestia pullei	+		+	@
Ernestia rubra	+	+	+	SFB
Miconia alata		+		
Miconia rufescens	+	+		
Miconia serialis	+			
Mimosaceae				
Mimosa microcephala var. *cataractae*	+	+		
Moraceae				
Ficus amazonica				
Myrtaceae				
Campomanesia aromatica	+			
Eugenia monticola		+		S
Eugenia ramiflora	+	+		
Psidium sp.		+		
Ochnaceae				
Sauvagesia pulchella	+	+		
Orchidaceae				
Catasetum planiceps	+			

Table 10.7.1 (*continued*)

Species	Vo[a]	Si[b]	R[c]	D[d]
Orchidaceae				
Cyrtopodium glutiniferum	+	+	+	@
Encyclia aemula	+			
Encyclia calamaria	+			
Encyclia granitica	+		+	N
Epidendrum microphyllum	+			
Epidendrum nocturnum	+			
Liparis nervosa	+			
Oncidium baueri	+			
Psygmorchis pumilio	+			
Psygmorchis pusilla	+			
Schomburgkia crispa		+		
Trigonidium acuminatum	+			
Vanilla grandiflora	+			
Poaceae				
Aristida capillacea	+	+		
Aristida riparia		+		
Aristida setifolia var. *genuina*		+		
Axonopus ramosus	+	+	+	SF
Eragrostis maypurensis		+		
Gymnopogon foliosus		+		
Ichnanthus tenuis	+		+	N
Ischaemum guianense	+			
Lasiacis anomala	+			
Mesosetum cayennense	+			
Mesosetum rottboellioides	+		+	@
Panicum arctum	+		+	SF
Panicum hirtum	+			
Panicum micranthum	+			
Panicum polycomum	+			
Panicum pyrularium	+	+		
Paspalum convexum		+		
Paspalum melanospermum		+		
Paspalum multicaule	+	+		
Paspalum parviflorum	+	+		
Schizachyrium sanguineum		+		
Sporobolus temomairemensis		+	+	S
Thrasya granitica	+		+	S
Portulaceae				
Portulaca sedifolia	+	+	+	GuV
Rubiaceae				
Alibertia myrciifolia var. *myrciifolia*		+		
Borreria hispida		+		
Borreria latifolia var. *latifolia*	+	+		
Faramea crassifolia		+		
Guettarda spruceana		+		
Richardia scabra		+		

Table 10.7.1 (*continued*)

Species	Vo[a]	Si[b]	R[c]	D[d]
Rubiaceae				
Sabicea romboutsii		+	+	S
Sipanea pratensis		+		
Spermacoce tenuior	+			
Tocoyena surinamensis	+	+	+	S
Solanaceae				
Schwenkia americana	+	+		
Schwenkia guianensis	+	+		
Sterculiaceae				
Melochia melissifolia	+			
Waltheria indica	+	+		
Turneraceae				
Turnera caerulea var. *surinamensis*	+	+		
Turnera odorata	+	+		
Vitaceae				
Cissus erosa	+			
Xyridaceae				
Xyris jupicai		+		
Xyris paraensis var. *paraensis*		+		
Ferns + fern allies				
Adiantaceae				
Doryopteris collina		+		
Hemionitis rufa	+			
Hemionitis palmata	+			
Schizaeaceae				
Anemia ferruginea var. *ahenobarba*	+	+		
Anemia oblongifolia		+		
Anemia villosa		+		
Selaginellaceae				
Selaginella densifolia	+	+		
Selaginella kochii		+		
Mosses				
Calymperaceae				
Syrrhopodon gaudichaudii	+			
Dicranaceae				
Campylopus savannarum	+			
ssp. *bartlettii*	+			
Campylopus surinamensis	+			
Octoblepharum cylindricum	+			

[a] Vo, Voltzberg area; [b] Si, Sipaliwini Savanna area; [c] R, typical granitic rock outcrop species; [d] D: distribution (for convenience only recorded for the rock and endemic species); @, restricted to the Guayana Shield; Gu, the Guianas; G, Guyana; S, Suriname; F, French Guiana; V, Venezuela; Z, South America; tA, tropical America; N, Neotropics; B, Brazil; combinations of areas are possible, e.g., SV means restricted to Suriname and Venezuela. Note: This list is far from complete.

10.7.8 Biogeography and Endemism

Of the 157 species of flowering plants of inselbergs in Table 10.7.1 14 % (22 spp.) are endemic to the Guianas, among which 9 species are restricted to Suriname only viz: *Caladium bicolor* f. *robustum* (Araceae), *Aechmea lanjouwii** (Bromeliaceae), *Bulbostylis capillaris* var. *major** (Cyperaceae), *Croton stahelianus** (Euphorbiaceae), *Eugenia monticola* (Myrtaceae), *Sporobolus temomairemensis**, *Thrasya granitica** (Poaceae), *Sabicea romboutsii**, *Tocoyena surinamensis** (Rubiaceae). The species with an asterisk are typical granitic rock outcrop species.

As is reported above for the Voltzberg area and the Sipaliwini Savanna area, 19 % of the species are typical to rock outcrops. Of these, 23 % (7 spp.) are endemic to Suriname, 13 % (4 spp.) endemic to Suriname-French Guiana. 3 % (1 sp.) endemic to Suriname-Guyana, 7 % (2 spp.) are restricted to Suriname-French Guiana-Brazil, 7 % (2 spp.) to Guyana-Venezuela, 10 % (3 spp.) are restricted to the three Guianas, 17 % (5 spp.) to the Guayana Shield, and the rest is widely distributed with 13 % (4 spp.) neotropic and 7 % (2 spp.) South American. In his study on inselbergs in the Tumuc Humac area, de Granville (1991) recorded 184 species growing on bare, granitic slopes and in dry, shrubby low forests. Of these species he found 42 % to be widely distributed in South America, 27 % were also found in the Amazonian Basin on inland savannas and boulders along streams at lower altitudes, and 22 % were recorded as endemic to the Guianas. Among the endemic species 36 % are typical montane plants growing on granitic mountains most often above 500 m like *Clusia kanukuana* (Clusiaceae), *Trilepis kanukuensis* (Cyperaceae), *Mandevilla surinamensis* (Apocynaceae), *Pitcairnia* (*Pepinia*) *geyskesii*, *P. sastrei* (Bromeliaceae), *Epidendrum nocturnum* var. *tumuc-humaciense* (Orchidaceae), and *Croton stahelianus* (Euphorbiaceae).

The typical inselberg Cactaceae, the globular *Melocactus neryi*, is restricted to Venezuela, Brazilian Roraima, Suriname, and southwest Guyana, while *Melocactus smithii* is found only in Brazilian Roraima and southwest Guyana (Leuenberger 1997).

Furcraea foetida (Agavaceae) is a New World species which is found in the eastern coastal area of Central and tropical America from Costa Rica to about the 25th south parallel, southeast Antilles, the Guianas, and Colombia. It occurs naturally on rocks (on mountains, in rivers), but is also cultivated on sandy savannas and in gardens, and now naturalized in other tropical and subtropical areas in warm, mostly arid regions (Berry 1995; Boggan et al. 1997; Pulle 1932–1947).

The phytogeography of some plants, e.g., *Mandevilla surinamensis* (Apocynaceae), *Begonia prieurei* (Begoniaceae), *Pitcairnia* (*Pepinia*) *geys-*

kesii (Bromeliaceae), *Ernestia confertiflora*, *E. pullei* (Melastomataceae), *Sauvagesia tafelbergensis* (Ochnaceae), and *Portulaca sedifolia* (Portulacaceae), growing on inselbergs at the border of French Guiana with Suriname and Brazil, are discussed by Sastre and de Granville (1975).

Some species have phytogeographical links to the African continent, e.g., *Neurotheca loeselioides* (Gentianaceae) occurs on sandy and rock savannas in Venezuela, Guyana, Suriname, and Brazil, and is also an element of African inselbergs; *Schwenkia americana* (Solanaceae) is a common weed found widespread in Central and South America on sandy and rock savannas, and on roadsides, but without preference for a specific habitat. In tropical Africa it is a widespread weed on roadsides (S. Porembski and M.J. Jansen-Jacobs pers. comm.).

For certain species, e.g., species of *Trilepis*, *Doryopteris*, *Anemia*, *Mandevilla*, and *Cnidoscolus*, there are phytogeographical links to inselbergs in the Atlantic rainforest of Brazil (Porembski et al. 1998). De Granville (1991) also made note of *Banisteriopsis gardneriana* (Malpighiaceae) having affinities with southeastern Brazil.

Up till now Velloziaceae have not been reported as occurring on inselbergs in the Guianas, but according to M.J. Jansen-Jacobs (pers. comm.), *Vellozia tubiflora* is found on the Pakaraima Mts. in Guyana on sandstone or rocks. Porembski et al. (1998) state that *Vellozia tubiflora* is the only representative of the Velloziaceae in the Guayana region (in southern Venezuela), in Brazil and adjacent Bolivia, while the species number increases towards southeastern Brazil, where the family has its center of diversity on rock outcrops and sand localities.

In the National Herbarium of Suriname (BBS) there is an unidentified Velloziaceae voucher specimen (ONS 562), probably *Vellozia tubiflora*, which is collected on the Kantani Mountain at an altitude of ca. 550 m. This inselberg is situated very near the frontier of Suriname with Brazil in the Paru Savanna area, which is an extension of the Sipaliwini Savanna in southwestern Suriname. Could Velloziaceae possibly occur in Suriname?

10.7.9 Conclusion

There are no significant differences between the flora of inselbergs over short distances, and the difference on inselbergs over far distances, for example, inselbergs from Guyana compared to those of French Guiana, comprises only some species and mostly of the same genus, e.g., *Pitcairnia (Pepinia) geyskesii* in Guyana and *P. sastrei* in French Guiana.

As we already saw, the flora of the rock savannas has affinities with that of the sandy savannas. From the somewhat isolated appearance of the xeric flora on inselbergs, it can be assumed that because of the climatic changes which took place in the last glacial period (Würm), the inselbergs act as important savanna refugia for this flora during today's tropical rainforest climate.

From calculations of Table 10.7.1 it is found that 30 spp. (19 %) out of 157 are typical rock species. This high percentage of endemism on rock outcrops is probably due to lack of recent information about most of the recorded plants (information on ecology and distribution is extracted only from Pulle 1932–1984 and Görts-van Rijn 1985–1997).

It is obvious that more and extensive research needs to be done on inselbergs in the Guianas. However, this is a difficult task to accomplish because the interior of the countries, especially Suriname, is for the greater part inaccessible. To obtain a better and complete list of species living in their special habitats, it is important to visit the inselbergs during different seasons of the year. Most inselbergs has been inventoried during dry seasons.

Acknowledgements. I am very thankful to Dr. Stefan Porembski (Rostock), Drs. Marga Werkhoven from the National Herbarium of Suriname (BBS), Drs. Marion Jansen-Jacobs and Ing. Ben ter Welle both from the Utrecht Herbarium (U) for providing publications and information concerning the vegetation on inselbergs of the Guianas, and for all the useful comments and professional advice on the manuscript. I am also grateful to Dorothy Traag, documentation assistant of BBS, for bringing some useful, additional publications to my notice, and Judi Reichart from the Surinaams Museum, and Jan Vermeer, photographer, for their photographs of inselbergs. Extra thanks go to Marion Jansen-Jacobs for supervising throughout.

References

Barthlott W, Gröger A, Porembski S (1993) Some remarks on the vegetation of tropical inselbergs: diversity and ecological differentation. Biogéographica 69:105–124

Berry PE (1995) Agavaceae. In: Flora of the Venezuelan Guayana, vol 2. Timber Press, Portland, p 374

Boggan J, Funk V, Kelloff C, Hoff M, Cremers G, Feuillet C (1997) Checklist of the plants of the Guianas (Guyana, Surinam, French Guiana). Biological diversity of the Guianas program. Smithsonian Institution, Washington, DC

Daniel JRK (1984) Geomorphology of Guyana. An integrated study of natural environments. Occasional paper no. 6. Department of Geography, University of Guyana, Georgetown

de Granville J-J (1979) Atlas des départments d'Outre-Mer. La Guyane – Planches 12–13. Office de la Recherche Scientifique et Technique Outre-Mer, Paris de Granville J-J (1982) Rain forest and xeric flora refuges in French Guiana. In: Prance GT (ed) Biological diversification in the tropics. Colombia University Press, New York, pp 159–181

de Granville J-J (1989) Priority conservation areas in French Guiana. Office de la Recherche Scientifique et Technique Outre-Mer, Paris

de Granville J-J (1991) Remarks on the montane flora and vegetation types of the Guianas. Willdenowia 21:201–213

de Granville J-J, Sastre C (1974) Aperçu sur le végétation des inselbergs du sud-ouest de la Guyane Française. C R Soc Biogéogr 439:54–58

Fanshawe DB (1952) The vegetation of British Guiana. A preliminary review. Institute Paper 29. Imperial Forestry Institute, Oxford

Görts-van Rijn ARA (ed) (1985–1997) Flora of the Guianas. Koeltz Scientific Books, Koenigstein; Royal Botanic Gardens, Kew

Holthuijzen AMA, de Jong BHJ (1977) Vegetatiekundig onderzoek in het Natuurreservaat Raleighvallen/Voltzberg. CELOS Rapport 122

Jonker EP, Wessink JJ (1960) De Natuurwetenschappelijke expeditie naar de Emmaketen in Suriname, Juli–Oktober 1959. K Ned Aardrijkskundig Genootschap 77(2):145–161

Leeflang EC, Kolader JH, Kroonenberg SB (1976) Suriname in geografisch perspektief. Bolivar Editions, Paramaribo

Leuenberger BE (1997) Cactaceae. In: Görts-van Rijn ARA, Jansen-Jacobs MJ (eds) Flora of the Guianas. Royal Botanic Gardens, Kew

Lindeman JC, Mori SA (1989) The Guianas. In: Campbell DG, Hammond HD (eds) Floristic inventory of tropical countries. New York Botanical Garden, New York, pp 376–390

Maas PJM, Koek-Noorman J, Lall H, ter Welle BJH, Westra LYT (1988) Botanical exploration in the northern part of the Rupununi Savanna and the Mabura Hill Area (Guyana). Internal Report, Institute of Systematic Botany, Utrecht University

Maguire B, Wurdack JJ et al. (1965) The botany of the Guayana Highland – Part VI. Mem N Y Bot Gard 12(3):1–285

Moonen J (1984) Kassikassima. Report of an expedition untertaken from April 13–23, 1983. Suralco Mag 16(2):19–25

Oldenburger FHF, Norde R, Riezebos HT (1973) Ecological investigations on the vegetation of the Sipaliwini Savanna area (southern Suriname). Internal Report, University of Utrecht, Laboratory Physical Geography, Utrecht

Porembski S, Martinelli G, Ohlemüller R, Barthlott W (1998) Diversity and ecology of saxicolous vegetation mats on inselbergs in the Brazilian Atlantic rainforest. Divers Distrib 4:107–119

Pulle AA and subsequent editors (1932–1984) Flora of Suriname 1–6. JH de Bussy, Amsterdam; EJ Brill, Leiden

Reichart HA (1993) Raleighvallen/Voltzberg Natuurreservaat. Beheersplan 1993–1997. WWF, the Netherlands and USA, and Forestry Service Suriname, Paramaribo

Sarthou C (1992) Dynamique de la végétation pionnière sur un inselberg en Guyane Française. Thèse de doctorat, Université Pierre et Marie Curie, Paris 6

Sarthou C, Villiers J-F (1998) Epilithic plant communities on inselbergs in French Guiana. J Veg Sci 9:847–860

Sarthou C, Thérézien Y, Couté A (1995) Cyanophycées de l'inselberg des Nouragues (Guyane Française). Nova Hedwigia 61:85–109

Sastre C, de Granville JJ (1975) Observations phytogéographiques sur les inselbergs du Bassin Supérieur du Maroni. C R Soc Biogéogr 444:7–15

Suriname Planatlas (1988) Stichting Planbureau Suriname, Department of regional development, O. A. S., Washington, DC ter Welle BJH, Jansen-Jacobs MJ, Chanderbali A, Raghoenandan U (1994) Botanical exploration in Guyana. Eastern Kanuku Mountains/Crabwood Creek. Internal Report, Herbarium Division, Utrecht University

ter Welle BJH, Jansen-Jacobs MJ (1991) Botanical exploration in Guyana. Rupununi District. Internal Report, Herbarium Division, Utrecht University

ter Welle BJH, Jansen-Jacobs MJ (1995) Botanical exploration in Guyana North and South Rupununi Savanna and Kanuku Mountains. Internal Report, Herbarium Division, Utrecht University

ter Welle BJH, Jansen-Jacobs MJ, Görts-van Rijn ARA, Ek R (1987) Botanical exploration in the northern part of western Kanuku Mountains (Guyana). Internal Report, Institute of Systematic Botany, Utrecht University

ter Welle BJH, Jansen-Jacobs MJ, Nic Lughada EM (1989) Botanical exploration in the Wai-Wai area of southern Guyana (Guyana). Internal Report, Institute of Systematic Botany, Utrecht University

ter Welle BJH, Jansen-Jacobs MJ, Sipman HJM (1993) Botanical exploration in Guyana. Rupununi District and Kuyuwini River. Internal Report, Herbarium Division, Utrecht University

Teunissen PA (1978) Savanne-ecosystemen: rotssavannes. Instituut voor de Opleiding van Leraren, Paramaribo, Suriname

van Donselaar J, Schulz JP (no date indicated) On the flora and vegetation of granite exposures in the Voltzberg region, Suriname, with notes on some similar areas in Suriname (unpublished)

van Donselaar J, Schulz JP (no date indicated) Lijst van rots planten aangetroffen in Suriname (unpublished)

Werkhoven, MCM (1986) Orchideeën van Suriname/Orchids of Suriname. Vaco, Paramaribo

10.8 Southeast Brazil

H.D. Safford and G. Martinelli

10.8.1 Introduction

Inselbergs *sensu lato* are found throughout eastern Brazil, from the semi-arid, thorn scrub (caatinga) dominated northeastern interior, to the cool, misty, subtropical highlands of Rio Grande do Sul, more than 3000 km to the south. Although these often isolated granitic sentinels are a familiar part of the landscape for perhaps 75 % of the Brazilian population, they are poorly studied, and the systematic, biogeographical, and ecological details of their flora and fauna are largely unknown. The south and the southeast from Minas Gerais and the Bahian "panhandle" to the Uruguayan border are the only parts of Brazil for which even a modicum of quantitative information on inselberg vegetation exists.

In this chapter, we focus on the inselbergs of coastal southeastern Brazil, largely coincident with the state of Rio de Janeiro, but including much of Espírito Santo, the southeastern extremities of Minas Gerais, and eastern São Paulo state. This is not only because this is the part of Brazil with which we are botanically and ecologically most familiar, but, more importantly, because it is in this small region that the full variability of Brazilian inselbergs is best expressed. Within a circle of 150 km radius centered at Rio de Janeiro, one may encounter examples of all of the major vegetation types of eastern Brazil, from superhumid montane rainforest to semiarid thorn scrub, from coastal mangroves and restinga dune forest to páramo-like grasslands nearly 3000 m high. This variegated mosaic of vegetation is the progeny of a rich biological heritage, a dynamic climate and a fabulously rugged landscape. This fortuitous marriage of biotic and physical factors has also resulted in perhaps the most diverse and heterogeneous inselberg flora on earth.

In this chapter we consider an altitudinal sequence of inselbergs in the state of Rio de Janeiro, comparing the vegetation of granitic/migmatitic outcrops at two different altitudinal levels – 0 to 398 m, and 1700 to 2263 m – in the context of its systematics, biogeography, and ecology. Coastal,

Ecological Studies, Vol. 146
S. Porembski and W. Barthlott (eds.) Inselbergs

lowland inselbergs are represented by the famous Pão de Açúcar (Sugarloaf) in Rio de Janeiro city; the highland inselberg group is comprised of summit outcrops just north of Rio, in the Serra de Araras and the Serra dos Órgãos, subranges of the Serra do Mar (Fig. 10.8.2). Although the brevity of this chapter allows us to merely touch on many of the important issues that such a comparison raises, we hope our treatment will prove informative enough to be a catalyst for further, more intensive study of the inselbergs of Brazil.

10.8.2 Geography

Geographically, coastal southeastern Brazil presents the most rugged topography in all of extra-Andean South America. Ab'Sáber (1966) called this region the *mares de morros* (seas of hills), underlining the almost complete absence of extensive plains outside the littoral, and the overwhelming importance of inselberg-type landforms to the regional landscape. Three main geomorphological compartments define the coastal southeast, all of them closely following ENE-trending megastructures in the Precambrian basement. From south to north, these are: (1) the Serra do Mar, (2) the Valley of the Rio Paraíba do Sul, and (3) the Serra da Mantiqueira (Fig. 10.8.1). Inselbergs and inselberg-type formations are found in all three of these geomorphological compartments, as well as along the littoral.

The Serra do Mar stretches nearly 1500 km along the Brazilian Coast, from Rio Grande do Sul at ca. 30°S, to central Espírito Santo, at nearly 20°N. Between the cities of Santos and Rio de Janeiro, this mountain chain represents one of the steepest, most abrupt coastal escarpments on earth, rising 800 to 1000 m directly out of the Atlantic Ocean. In the vicinity of Rio de Janeiro, the coast bends temporarily to the east, while the main chain of the Serra do Mar continues east-northeastward, first rising to nearly 2300 m in the Serra dos Órgãos subrange (29°30'S, 43°W) (Fig. 10.8.1), then losing altitude as it disaggregates into a series of isolated serrinhas and inselbergs. The ramparts of the Serra do Mar are replete with barren to semibarren inselbergs: domes, spires, ridges, and rock plains are exceptionally common, especially above 1500 m in the Serra dos Órgãos.

From Rio de Janeiro eastward, a series of bays and sand-barred lagoons and estuaries punctuates the coast, with swarms of often spectacular inselbergs rising above the coastal plain. Most famous is the so-called Serra da Carioca, within the city of Rio de Janeiro proper, where postcard peaks like Pão de Açúcar (Sugarloaf), Corcovado, Gávea, and Tijuca rise from 300 to more than 1000 m above the city and the Baía de Guanabara.

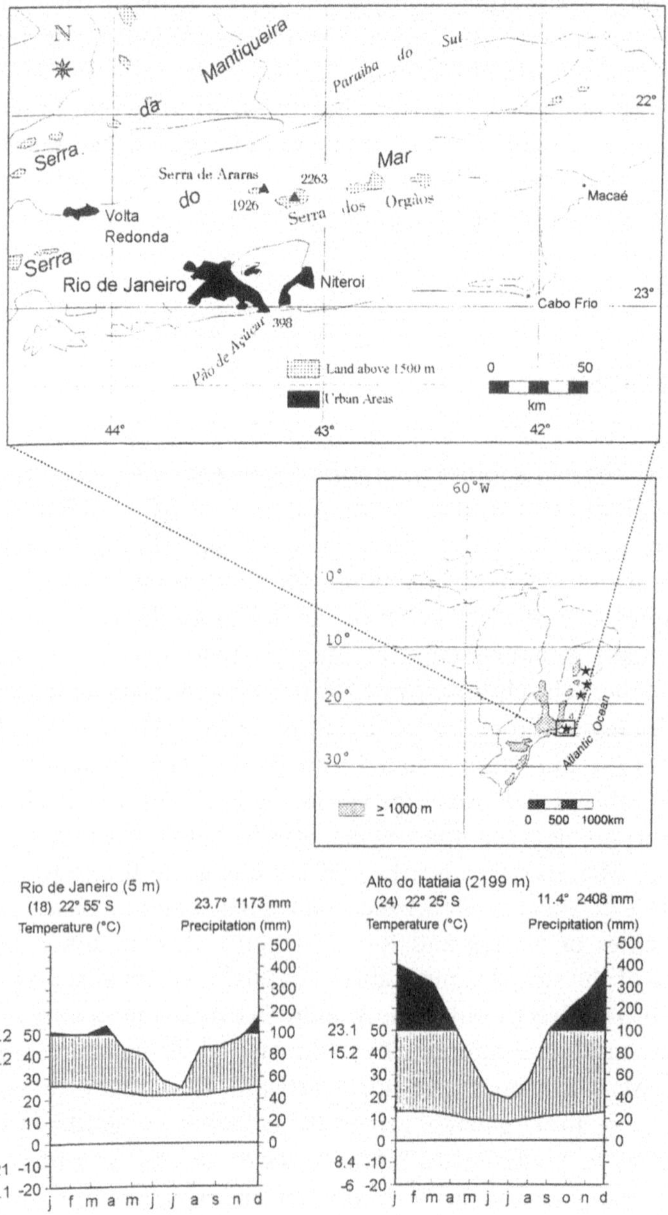

Fig. 10.8.1. Major physiographic and hydrographic features of coastal, southeastern Brazil, and Walter-type climate diagrams for Rio de Janeiro (Pão de Açúcar) and Alto do Itatiaia (RJ; for Serra do Mar summits). *Stars* on reference map mark dense concentrations of inselberg formations. Rio de Janeiro from AEERJ (1993/1994), data from 1973–1990; Alto do Itatiaia from Segadas-Vianna and Dau (1965), data from 1916–1940

Fig. 10.8.2. Steep-sided inselbergs are scattered throughout southeastern Brazil. (Photograph S. Porembski)

Entrenched along the northern flanks of the Serra do Mar, the Rio Paraíba do Sul is the only major river in coastal southeastern Brazil. First flowing southwest from its source in the Serra da Bocaina (Serra do Mar), the Paraíba do Sul makes an abrupt turn just beyond the Tropic of Capricorn, entering a deep, ENE-trending graben defined by the horsts of the Serras do Mar and da Mantiqueira. Though never a canyon, the Paraíba do Sul Valley is nonetheless an impressive tectonic depression: in some places, valley-bottom to mountain top relief may exceed 2300 m. For much of its length, the floor of the Paraíba do Sul Valley is a strange maze of mamelonized hills and sometimes rocky hummocks; only in the state of São Paulo, and near Volta Redonda, Rio de Janeiro (Fig. 10.8.1), does the river course a true valley bottom. At about 21°20'S, the Paraíba do Sul finally bends to the east, finding an easier path to the sea through the disjunct remnants of the northern Serra do Mar.

Paralleling the left bank of the Rio Paraíba do Sul, the Serra da Mantiqueira is the highest continuous mountain chain in eastern South America, a tall cordilheira of forested mountains that runs from northern São Paulo state (ca. 23°S), through the states of Rio de Janeiro and Minas Gerais until its merger with the Serra do Espinhaço near 20°S (Fig. 10.8.1). Representing the upturned eastern edge of the Brazilian Plateau, this great escarpment averages nearly twice the altitude of the Serra do Mar, with elevations reaching 2787 m at Pico das Agulhas Negras (Serra do Itatiaia). Most of the major rivers of both the coastal and interior southeast emanate from the humid heights of this Serra, including the Rio Doce, the Rio Grande, and many major tributaries of the Rios São Francisco and Paraíba do Sul. The upper heights of the Serra da Mantiqueira are often plateau-like, and locally characterized by large expanses of bare rock, cliffs, rocky peaks, and tors, particularly in the Itatiaia and Passa-Quatro Massifs (22°23', 44°35'–55'). These alkaline granite and syenite outcrops are among the highest inselberg-type landforms in eastern South America.

10.8.3 Geology and Paleoclimatology

Lithologically, the inselbergs of southeastern Brazil (as well as the rest of eastern Brazil) are almost entirely the product of Late Proterozoic intrusive suites of granites and associated migmatites emplaced into both the Brazilian and Congo cratons and the intervening (meta)sedimentary wedge during and after tectonic collisional events associated with the formation of the ancient southern continent of Gondwana (the so-called Brasiliano-Pan African Orogeny) (Trompette 1994). A limited number of Late Cretaceous alkaline intrusions also support inselberg landscapes, for example on the Ilha de São Sebastião (23°50', 45°25'), the Serra da Madureira (22°50', 43°30'), just east of the Baía de Guanabara and the Serra da Carioca, and most importantly, in the Passa-Quatro and Itatiaia Massifs of the Serra da Mantiqueira. Weathering and exhumation of these originally subterranean pods of resistant rock were mediated through subsequent interactions of climatic cycling and tectonism, particularly in the Neogene.

Beginning in the Latest Jurassic, the contiguous metamorphic terranes of central Gondwana – and their accompanying granitic plutons – were sundered by intercontinental rifting, giving birth to the African and South American Plates (the inselbergs of eastern Brazil and western Africa, presently widely separate, have a common origin). Following final separation of Africa from South America in the Late Cretaceous, southeastern Brazil was subjected to a long period of severe denudation, lasting at least until the Late Oligocene (Schobbenhaus et al. 1984; Bigarella 1991). The resultant pediplain represents the ancestral surface from which most of eastern Brazil's landscape has been subsequently carved (King 1967). Many of the inselberg formations in eastern Brazil clearly originated during this long erosional interval; some of them may thus have geomorphological ages exceeding 50 million years.

As in all of the tropics, the very existence of inselbergs in eastern Brazil is first and foremost a product of past climatic change (Thomas 1974). Our understanding of Quaternary climatic oscillations and their geomorphological results gives us insight into the types of morphogenetic shifts that must have occurred throughout the Tertiary. Pleistocene glacial-interglacial cycles appear to have been manifested in Brazil by semiarid-pluvial climatic shifts (Damuth and Fairbridge 1970; Ab'Sáber 1977), resulting in a distinct (albeit hypothetical) temporal rotation of geomorphological processes (Bigarella et al. 1965; Clapperton 1993). Interglacial periods, such as today, were generally characterized by warmer, more humid climatic conditions. Profound chemical weathering formed deep soils, and forests advanced, stabilizing steeper slopes; fluvial erosive processes were restricted to vertical entrenchment and dissection of the highland landscape. With the onset of glacial conditions in temperate and polar latitudes, climates are thought to have become relatively drier and cooler in eastern Brazil, forest retreated to be replaced by evergreen/semideciduous formations, stream networks were largely reduced to ephemeral channels, and high-energy, mechanical erosive processes dominated. Relict Tertiary and Quaternary-aged pediments and pediplains throughout southeastern Brazil are testimony to the past importance of semiarid morphogenetic processes in presently humid, forest-covered landscapes.

Due to episodic tectonism throughout the Tertiary, the remnants of these once regional pediplains – and the inselbergs embedded within them – are now found arrayed along a distinct altitudinal gradient. Following creation of the Oligocene pediplain, other major surfaces were carved in the Early and Late Pliocene, with additional minor pediment systems originating in the Miocene and the Pleistocene (Modenesi 1988; Bigarella 1991; Valeton et al. 1991). Brief periods of epeirogenic uplift punctuated each of these erosional events, beginning in the Late Oligocene-Early Eocene, and continuing at least until the Late Pleistocene (Almeida 1976; Riccomini 1989). Total relative uplift along the Serra da Mantiqueira range front has probably exceeded 2300 m. Each period of tectonism disconnected the Serras and their "passenger" inselbergs from previously contiguous erosional surfaces, subjecting older, higher surfaces to progressively more severe erosional segmentation. Today, it is possible to roughly correlate the oldest erosional remnants of the Serra do Mar and the Serra da Mantiqueira – now restricted to the highest peaks above timberline – with more extensive surfaces in adjacent lowlands, thousands of meters below (Ab'Sáber and Bernardes 1958). Put more succinctly, the inselbergs of the Brazilian littoral are lowland contemporaries of the domes, rock outcrops, and peaks of the southeastern highlands. Their original formation was largely coeval and conterminous (indeed, all three sites considered in this chapter were carved from the Serra dos Órgãos Intrusive Suite); the great gap in elevation that presently separates them is a geologically recent phenomenon.

The relative importance of climatic and tectonic geomorphological processes in exhuming and molding the modern landscape of eastern Brazil follows a distinct latitudinal gradient (Ab'Sáber 1977; Moreira 1977; Bigarella 1991). In the semiarid northeast, where Neogene folding and faulting has been of a very limited magnitude, dryland morphogenetic processes dominate even today; the extensive, thorn scrub-covered pediplains of the Bahia interior are host to hundreds of small to moderately sized, sugarloaf-like inselbergs. In the presently (mostly) superhumid southeast, fluvial downcutting and dissection of the ancient pediplain surfaces was enhanced by both higher average rainfall and the much greater magnitude of epeirogenic uplift. Exhumation of resistant rock structures was also accelerated by the more pronounced alternation of chemical and mechanical weathering processes from wet period to dry. As a result, inselbergs in the southeast, often perched on trophy-like bases of stacked pediment remnants, are generally much larger than their northeastern cousins – some rock formations are truly colossal, such as Pedra da Gávea in Rio de Janeiro, or Pedra Azul in Espírito Santo, both of which soar 800 to 900 m into the air (Fig. 10.8.2).

In summary, inselbergs in coastal southeastern Brazil can trace their origin to the climatic shifts of the Cenozoic, beginning with the onset of extreme aridity in the Late Paleogene. Their present geographic situation – with interior inselbergs raised hundreds and even thousands of meters above their lowland kin – is the more recent result of ongoing Neogene tectonism. Tectonism and climate change have thus worked together to create the highly dissected southeastern landscape, and the abounding inselbergs that characterize it.

10.8.4 Modern Climate

Despite its tropical position, climate and weather in southeastern Brazil are largely dominated by subtropical and temperate influences. Throughout the year, northward-moving polar fronts pass over the region, interacting with one of two different airmasses: the cool, stable South Atlantic anticyclone in the winter, and the warm and unstable tropical continental mass in the summer. Precipitation amounts are mostly high (1200 to >2500 mm) – although the coast east of Rio de Janeiro may receive as little as 750 mm in some locations – with the bulk of rain falling between November and April. A short dry season of one to two (three) months occurs in most of the region (Fig. 10.8.1). Orographic effects lead to enhanced rainfall amounts in the middle and upper reaches of both the Serra do Mar and Serra da Mantiqueira: the highest precipitation amounts in all of Brazil are registered in the Serra do Mar of São Paulo, where yearly rainfall maxima may exceed 5000 mm. Regional coefficients of variation for annual precipitation hover between 15% and 25% (Nimer 1977). Monthly average free-air humidities range from 77–83% at the coast, to 69–76% in the Paraíba do Sul Valley (440 m as1), to 81–86% in montane rainforest in the Serra da Mantiqueira (816 m asl), to 67–87% in the campos de altitude (mountaintop grasslands) of Itatiaia (2199 m) (Segadas-Vianna and Dau 1965; AEERJ 1993/1994).

Mean annual temperatures in the coastal southeast range from 23 to 24 °C at the coast to ca. 8 °C at the highest summits (2600 to 2800 m) (Ratisbona 1976; Safford 1999); mean maxima and minima are 27.2 °C and 21 °C, respectively, in Rio de Janeiro, 28 and 17 °C at 440 m in the Paraíba do Sul Valley, and 15.2 and 8.4 °C at 2199 m in the Serra da Mantiqueira (Itatiaia) (Fig. 10.8.1). As determined by measurements of soil temperature at 80 cm depth, mean annual temperature on the summit of Pedra do Sino (2263 m), one of the highland sites employed in the comparison, is approximately 10.5 °C (H.D. Safford, unpubl. data). Absolute maxima/minima are ca. 38/11 °C at sea level, ca. 40/1 °C in the Paraíba do Sul Valley, and ca. 24/–10 °C at 2199 m (Segadas-Vianna and Dau 1965; AEERJ 1993/1994; H.D. Safford, unpubl. data). Frost is possible (though rare) both in inland basins above about 500 m elevation and on mountain slopes above ca. 1300 m. In the highest summits of the Serra da Mantiqueira and Serra do Mar frost may occur 30 to 50 times a year (Nimer 1977; Safford 1999).

Formal microclimatological measurements have apparently not been made on any lowland inselbergs in southeastern Brazil, although Carauta and Oliveira (1982) report a rock-surface temperature from Pão de Açúar of 61.5 °C on a particularly hot November day. In general, it can be assumed that microclimatic patterns already described for, e.g., western Africa (Porembski et al. 1996; see also Phillips 1982) are valid for coastal southeastern Brazil as well. Rock outcrops in the uplifted inselbergs of the Serras do Mar and da Mantiqueira do not reach such high diurnal temperatures as their counterparts nearer sea level – on sunny days rock surface temperatures appear to regularly reach the high-40's and low-50's – but nighttime temperatures are much colder, and rock surface temperature may dip into the teens on winter days (H.D. Safford, pers. observ.).

Figure 10.8.1 includes Walter-climate diagrams from Rio de Janeiro (for Pão de Açúcar) and Alto do Itatiaia, a highland site in the Serra da Mantiqueira (for the Serra do Mar sites). Comparing these diagrams, two features are most salient: (1) mean, mean maximum, and mean minimum temperatures in the Serra do Mar highlands are 12 to 13 °C colder than their counterparts at sea level, and absolute maxima and absolute minima are much lower as well, with the latter dipping well below 0 °C on highland inselbergs (to –10 °C in the Serra do Mar); (2) rainfall in the highlands is more than twice as heavy as in Rio de Janeiro city [although two things should be noted here: (a) Rio is somewhat drier than most of the southeast coast, and (b) the Serra do Mar sites probably receive 20 to 25 % more precipitation than the Serra da Mantiqueira].

10.8.5 Vegetation of Southeastern Brazil

The Mata Atlântica – the Atlantic Forest – was one of the world's great forests. Occupied by Portuguese colonists beginning in the first decade of the 16th century, this great center of endemism and diversity, originally 3500 km long, and over 1 000 000 km^2 in area, had been reduced to 5 % of its former extent by the late 1980's (Fig. 10.8.3) (Por 1992; Dean 1995). As Dean (1995) lamented, for much of eastern Brazil "it is hard to say whether it is appropriate to refer to the Atlantic Forest in the present tense." Remarkably, considering its extremely high population density, southeastern Brazil is one place where the present tense *is* still appropriate. Between Rio de Janeiro and Curitiba, the maritime escarpment of the Serra do Mar retains much of its original forest cover, and a few large tracts remain in the Serra da Mantiqueira (Fig. 10.8.3). A series of national and state parks and reserves also blankets the area, although in many cases protection may be more illusory than real.

Isolated for millions of years from the rest of wet-tropical South America by the dry diagonal of the Chaco-Cerrado-Caatinga, the Mata Atlântica probably contains the most "aberrant" biota in the neotropics (Fonseca 1985; Por 1992). This is particularly true for southern Bahia and Rio de Janeiro. For example, Mori el al. (1981) found 53.5 % endemism among tree species in south Bahia, and 77.4 % endemism among nonarborescent species, including epiphytes. In addition, Calderón and Soderstrom (1980) documented 41 % endemism among bamboo genera, ferns are approximately 45 % endemic (Tryon 1972), southeast Brazil has the highest number of endemic species of lycopods (18) in the neotropics (Öllgard 1993), and Luteyn (1989) showed that 97 % of species of Ericaceae found in southeast Brazil are restricted to that region. The endemic nature of the fauna is even more pronounced (Por 1992). Typically, endemism in southeast Brazil is

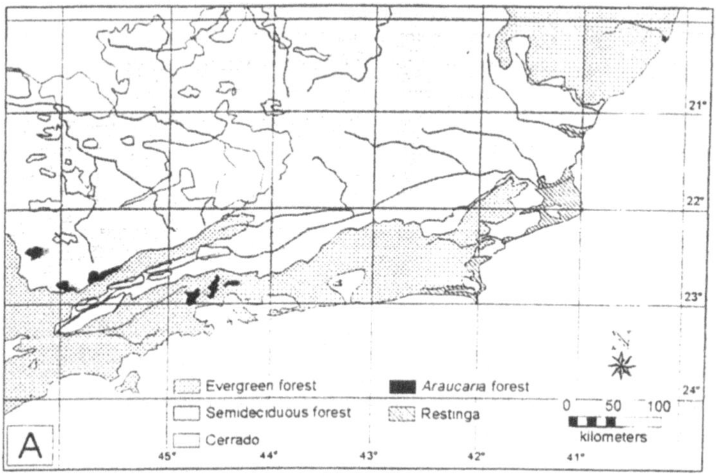

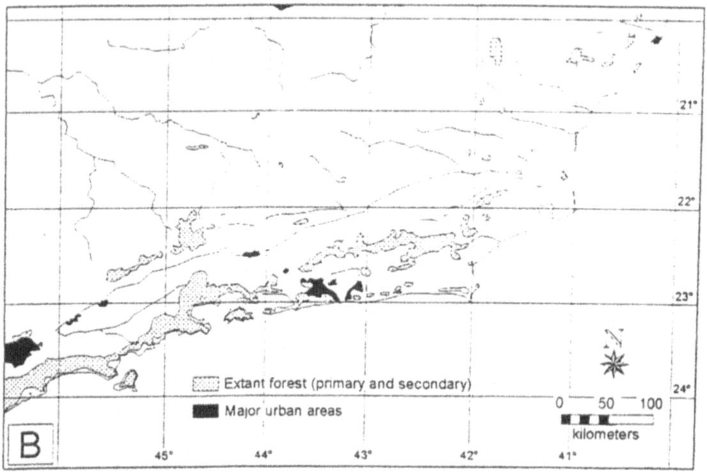

Fig. 10.8.3. **A** Original (pre-1500) extent of natural vegetation in southeastern Brazil, with approximate distributions of major vegetation types (After Ururahy et al. 1983). **B** Approximate extent of primary and secondary vegetation in 1994. [After Ururahy et al. 1983; Consórcio Mata Atlântica (no date); IEF 1994; Dean 1995]

highest on the generic level, while the high diversity and endemism of the wet-equatorial Andes is most pronounced at the species level. As Lynch (1979) and Por (1992) point out, this is probably the result of the much greater age of the Mata Atlântica and its flora and fauna, the Andes having risen and acquired forest cover only as recently as the Late Miocene.

Figure 10.8.3 outlines the vegetation of Rio de Janeiro and adjoining parts of Espírito Santo, Minas Gerais, and São Paulo. Because of its topographic and climatic complexity, coastal southeastern Brazil is host to a variety of major vegetation types, including Mata Atlântica (evergreen, semideciduous, and *Araucaria* forests, and campos de altitude), cerrado, restinga, and even a disjunct occurrence of caatinga near Cabo Frio. Brief treatments of each of these types of vegetation are given below.

10.8.5.1 Evergreen Forest

Evergreen forest – Hueck's (1966) *Brazilian Coastal Rainforest* and Rizzini's (1979) *Atlantic Rainforest* – provides the primary vegetational matrix within which are embedded the great majority of inselbergs in the southeast. Most authors have divided the evergreen forest into a number of altitudinal belts, although the actual widths of the belts, as well as the floristic or physiognomic justification upon which they have been identified have varied considerably. For most authors, lowland evergreen forest is found between the coast and altitudes of 50 to 60 m. Extensive tracts of this forest type in the southeast are (were) found only on the coastal plain and low bluffs east and north of Rio de Janeiro (today, an area largely devoid of natural vegetation). Canopy trees in the lowland evergreen forest – where it still grows – typically reach 25 to 30 m in height, but there is a marked paucity of the tall emergents that characterize other lowland rainforests; buttressed trees are also notably rare.

Montane forests have been typically split into upper and lower members (e.g., Hueck 1966; Rizzini 1979), often with an intermediate belt included (e.g., Ururahy et al. 1983; Guedes-Bruni and Lima 1996). The submontane and montane belts – canopy 20 to 30 m high, with 40 m emergents – are the centers of Atlantic Forest floral diversity: for example, some remnant submontane stands in south Bahia contain over 200 tree species (>10 cm dbh) per hectare. Work at the Macaé de Cima Reserve in Rio de Janeiro (22°25'S, 42°30'N) has thus far identified 1100 species of vascular plants on 7000 ha, the same number found on 10 000 ha at Reserva Ducke, near Manaus in the Amazon Basin, and only about 15 % less than the number found on Barro Colorado Island, Panamá, which is more than twice the size of the Macaé Reserve (Lima and Guedes-Bruni 1997b). The most diverse families in the Macaé montane forests are, in order: Orchidaceae, Melastomataceae, Rubiaceae, Fabaceae, Myrtaceae, Bromeliaceae, Lauraceae, Asteraceae, Solanaceae, and Piperaceae; woody plants are dominated by Melastomataceae, Lauraceae, Myrtaceae, Fabaceae, and Rubiaceae (Lima and Guedes-Bruni 1997c). Above 1400 to 1700 m, the heavy vascular-epiphytic load of the montane forests gives way to the wispy beards of foliose and filamentous lichens, and straight, tall, tree boles become the twisted, xeromorphic physiognomies of the elfin cloud forest. Physiognomy is not all that changes: while floristic composition of the lower montane forests is distinctly "Atlantic" (Rizzini 1979), upper montane forests are dominated by genera found nearly universally in such cool-moist habitats: *Clethra* (Clethraceae), *Drimys* (Winteraceae), *Ilex* (Aquifoliaceae), *Rapanea* (Myrsinaceae), *Roupala* (Proteaceae), *Weinmannia* (Cunoniaceae), etc.

10.8.5.2 Campos de Altitude

High altitude grasslands are found above 1700 to 2000 m in the Serra do Mar and above 2100 to 2400 m in the Serra da Mantiqueira, although Massenerhebung effects lead to the existence of similar formations on the summits of much lower mountains as well. As with the *Araucaria* forests, the campos de altitude are to a great extent relictual formations, mountain-top remnants of a vegetation type more extensive during, e.g., Pleistocene glacial maxima. Why treeline is so low in southeastern Brazil is an unresolved issue, although heavy frost, dry season length, and periodic fire probably all play a role (Safford 1999, and in prep.). The vegetation of the campos de altitude is dominated by bunch grasses (e.g., *Cortaderia* and *Calamagrostis*) and montane bamboo (*Chusquea*), with dwarf trees and sclerophyllous shrubs in varying admixtures (Martinelli and Bandeira 1989). The overall physiognomic appearance is reminiscent of lower Andean páramo (if somewhat rockier) and, indeed, many taxa are disjunctly shared with the southern and/or equatorial Andes (Brade 1956). The most speciose families within the campos de altitude are the Asteraceae, Melastomataceae, Orchidaceae, Poaceae, Lamiaceae, Lycopodiaceae, Ericaceae, Cyperaceae, Rubiaceae, and Eriocaulaceae (Brade 1956; Martinelli et al., unpubl. data). The highland inselbergs considered in this chapter are found primarily within this vegetation type.

10.8.5.3 Semideciduous Forest

This extensive formation is essentially found in areas of more moderate rainfall (ca. 1200–1500 m), where more than 2 months of profound drought occur per year, and/or where arenitic substrates significantly reduce the water retention capacity of the soil. On the average, about 20 to 50 % of the trees (not species) in this formation lose their leaves on a seasonal basis (Rizzini 1979; Ururahy et al. 1983); important deciduous taxa are *Erythrina*, *Schizolobium* (Fabaceae), *Tabebuia* (Bignoniaceae), and the Bombacaceae. The most important woody families in montane semideciduous forests are Myrtaceae, Melastomataceae, Rubiaceae, Fabaceae, and Lauraceae (Hueck 1966; Rizzini 1979; Oliveira-Filho and Machado 1993), simply a reordering of the important families in the montane evergreen forests. Disjunct stands of semideciduous forest often cap inselbergs, even in the wettest parts of the coastal southeast.

10.8.5.4 *Araucaria* Forest

This relictual formation is characterized by the presence of the conifer *Araucaria angustifolia*, sometimes in association with other conifers of the genus *Podocarpus*. *Araucaria* forest is usually found in small basins above about 800 m in the Serra do Mar (particularly in the Serra da Bocaina), and above 1200 m in the Serra da Mantiqueira, principally near Campos do Jordão, São Paulo. Smaller copses of *Araucaria* (not mapped) are also found in the Itatiaia Massif, and as far north as the Serra do Caparão at 20°S (Gollte 1993).

10.8.5.5 Cerrado

The presence of cerrado (savanna) in the coastal southeast is (was) limited to the rain-shadowed floor of the upper Paraíba do Sul Valley and the São Paulo Basin, although much more extensive stretches occupy the Brazilian Plateau beyond the Serra da Mantiqueira (Fig. 10.8.3). There are usually two major strata: more or less openly spaced trees, 2 to 6 m tall (sometimes taller), with pronounced xeromorphic characteristics, such as tortuous branching, thick corky bark, and sclerophylly; and a lower stratum of bunch grasses and subshrubs. Important woody taxa include members of the families Fabaceae, Vochysiaceae, Velloziaceae, and Annonaceae, and the genera *Caryocar* (Caryocaraceae), *Kielmeyera* (Clusiaceae), *Curatella* (Dilleniaceae), and *Byrsonima* (Malpighiaceae). Rainfall in the cerrado zone tends to vary between 1300 and 1700 mm, with a profound dry spell of 1 to perhaps 5 months (Hueck 1966; Rizzini 1979; Eiten 1982).

10.8.5.6 Restinga

Found on littoral sands and inland dunes, restinga encompasses a shore to backdune successional sequence that ranges from low herbs in the spray zone, to 10- to 15-m-tall fixed-sand forests inland. Although restinga is only mapped as occurring in eastern Rio de Janeiro and Espírito Santo (Fig. 10.8.3), much of the western coast maintains localized occurrences of this vegetation type as well. Sclerophylly is widespread, and succulent plants (ground bromeliads, Cactaceae) locally abound. The most typical woody species come from the families Myrtaceae and Ericaceae, with a variety of other families contributing single genera, such as *Protium* (Proteaceae), *Coccoloba* (Polygonaceae), and *Aristolochia* (Aristolochiaceae) (Hueck 1966; Rizzini 1979; Ururahy et al. 1983). Extensive restinga formations formerly abutted Pão de Açúcar on the west, but urban expansion has all but eliminated this vegetation type from the greater Rio de Janeiro metropole.

10.8.5.7 Caatinga (Not Mapped)

Characteristic of the semiarid northeast, caatinga is only found in one disjunct location in coastal southeastern Brazil, near Cabo Frio, Rio de Janeiro (23°S, 42°W; Fig. 10.8.1). Caatinga is a classic thorn-scrub vegetation, composed primarily of woody, xerophyllic, spiny, deciduous taxa. Soils in caatinga tend to be thin, argillic, and compact, derived in hot, dry climates from Precambrian basement. Characteristic taxa are *Mimosa* and *Caesalpinia* (Fabaceae), *Croton* (Euphorbiaceae), and various members of the Cactaceae and Myrtaceae (Rizzini 1979; Ururahy et al. 1983).

10.8.6 Inselberg Vegetation of Southeastern Brazil

Sadly, for the most part we know as yet very little about the vegetation of Brazilian inselbergs. What follows is a descriptive summary of each of the seven basic forms of inselberg vegetation defined by Barthlott et al. (1993), as they occur in coastal southeastern Brazil. Each section is divided into two parts, the first treating the vegetation of Pão de Açúcar (and other coastal inselbergs, where appropriate), the second treating that of the highland inselbergs of the Serra do Mar (specifically, the Serra dos Orgãos and Serra de Araras subranges). As will be obvious, we presently know more about the vegetation of highland inselbergs than we do about rock outcrops in downtown Rio de Janeiro!

10.8.6.1 Cryptogamic Crusts on Open Rock

Coastal. Coastal rock outcrops are typically of the "cyanobacterial" type, with exposed surfaces often stained darkly by a thin cover of photosynthetic bacteria; the most common taxa include *Stigonema*, *Scytonema*, and *Chroococcidiopsis* spp. (M. Weber and S. Porembski, pers. comm.). Lichens are also common, particularly where rock surfaces are somewhat drier; on Pão de Açúcar the genera *Caloplaca* and *Cladonia* appear to be especially well represented (H.D. Safford, pers. observ.). Although we have no data on the diversity of lichens on Pão de Açúcar, single granite domes of similar height and substrate petrology in the coastal ranges of Rio Grande do Sul may sustain many species (Fleig 1990).

Highland. High altitude rock surfaces in the Serra do Mar of Rio de Janeiro are inhabited by surface-hugging rupicolous (growing directly on rock) lichens like *Lecanora* (Lecanoraceae) and *Peltula* (Peltulaceae). Rock balds with sufficient year-round humidity also support localized turfaceous colonies of *Usnea* (Usneaceae). These lichen mats often catch wind-blown soil and seeds of ruderal vascular plants, providing safe haven for successional pioneers. Gray-green bush lichens of the family Cladoniaceae are common as well, particularly where rock surfaces are rough, and where wind speeds are diminished by aspect or microtopography. Although tropical rock outcrops are often darkly tinted by a patina of cyanobacteria (Barthlott et al. 1993), black-rock surfaces in the southeastern highlands appear primarily to be the result of lichen cover, with cyanobacterial skins mostly restricted to humid slopes and microsites.

10.8.6.2 Drainage Channels

Coastal. Drainage channels are usually solely inhabited by cyanobacteria, although thin strips of cyanobacterial-free rock are often found along both walls of the channel. Porembski and Barthlott (1993) hypothesized that such strips may be the result of bacterial allelopathy.

Highland. Ephemeral drainage channels are widespread in the Serra do Mar, although they are not as common or as strongly expressed as in the highly fluted syenites of the Serra do Itatiaia; sometimes these channels connect systems of small rock pools (see below). Lichens inhabiting the edges of these ephemeral pools and channels produce organic acids that may be essential for the further disaggregation of the host rock (Martinelli and Bandeira 1989). Where water runs often through such channels, lichens are rare, and cyanobacterial (and algal) slimes coat the rock.

10.8.6.3 Seasonally Water-Filled Rock Pools

Coastal. Rock pools are not particularly common on coastal or even lower montane inselbergs, as the most likely site for such pools – the summit – is often occupied by semideciduous forest or other vegetation. Neither Carauta and Oliveira (1982) nor Miranda and Oliveira (1983) mention the existence of seasonal pools on Pão de Açúcar, although we have seen water-filled basins on nearby outcrops. Typically, most ephemeral pools are devoid of vegetation, although, given a modicum of substrate, semiaquatic taxa like *Cyperus*, *Utricularia*, and *Commelina* may colonize.

Highland. Seasonal pools are much more common on the bald summits and rock plains of the Serra do Mar. The great majority of pools (with depths ≤10 cm) are mostly dry during the winter and devoid of vascular plants, although a ring of lichens may accompany waterline. Even those pools that maintain water throughout the winter dry season are usually devoid of obvious life. From May to September, temperatures as low as −10 °C periodically freeze the shallowest of these water bodies to their bottom, making survival difficult even for the hardiest of plants. Given a few centimeters of bottom sediment, pools with permanent water may support a variety of aquatic vascular plants, including *Ranunculus montevidensis*, *Hydrocotyle* cf. *ranunculoides* and *Lilaeopsis* sp., *Viola* sp., *Utricularia* spp., *Isoetes* spp., and a variety of Cyperaceae. Permanent pools are infrequently encountered in the Serra do Mar, being much more common in higher parts of the Serras da Mantiqueira and do Caparaó.

Over time, unbreached, deeper pools may sufficiently accumulate algal remains, wind-transported soil, etc. to support bryophytes, followed by rupicolous or saxicolous (requiring at least a modicum of soil) sedges (*Trilepis* or *Rhynchospora* spp.). Thereafter, succession will generally follow one of two paths: (1) the bryophyte-sedge mats will expand and collect sufficient soil to begin to support grasses, forbs and then shrubs (see following section, 10.8.6.4) or (2) rupicolous, spiny, tankless bromeliads – *Tillandsia* spp., or sometimes *Pitcairnia* spp. – will colonize and completely fill the pools, to the exclusion of anything beyond sparse bryophytes and sedges (in the higher, more interior ranges, other bromeliads often fill this role, including *Fernseea itatiaiae*, endemic to Itatiaia, or *Dyckia* spp., in Minas Gerais and southern Brazil). The cactus *Schlumbergera obtusangula* may also dominate former ephemeral pools, but then only in mesic conditions provided by rock shelter or the shade of dense shrubs or dwarfed trees.

10.8.6.4 Shallow Depressions

Coastal. Shallow, soil-filled depressions occupy large expanses of Pão de Açúcar (and neighboring inselbergs), especially on the east and northeast faces, and elsewhere where slopes are less than ca. 70°. Typically vegetation islands filling these depressions grade inward from bryophytes (often Dicranaceae and Bryaceae) and mat-forming monocots (e.g., Bromeliaceae, Orchidaceae, Velloziaceae), to cacti (*Coleocephalocereus fluminensis*, *Cereus* sp.), *Anthurium* spp., grasses (mostly invasive, like *Melinis minutiflora* and *Panicum maximum*), and shrubs (including *Tibouchina* spp., *Sinningia* sp., *Cassia patellaria*, *Croton compressus*, *Lantana* spp., etc.). Saxicolous orchids like *Prescottia plantaginea* and *Zygopetalum mackayi* often root in the shallow soil fringing these islands. Widespread, weedy species like *Emilia sagittata*, *Laportea aestuans*, *Phenax sonneratii*, and the therophytes *Apium leptophyllum* and *Lepidium virginicum* frequently colonize open soil in these depressions as well. If depressions are small and pocket-like, thick bromeliaceous clumps may come to completely dominate the locality; on Pão de Açúcar common gregarious species are found in the genera *Nidularium*, *Tillandsia*, and *Vriesea*.

Soil chemistry data from a shallow depression on Pão de Açúcar are provided in Table 10.8.1. Note the extremely low pH, the high Al toxicity, and the very low values of Ca and P.

Table 10.8.1. Comparison of soil chemistry of shallow, soil filled depressions on lowland and highland inselbergs in the vicinity of Rio de Janeiro city

	Pão de Açúcar (Carauta and Oliveira 1982)	Morro do Cuca (Martinelli and Bandeira 1989)	Pedra do Sino (H.D. Safford, pers.data)
Mean elevation of sampled outcrop	199 m	1780 m	2200 m
pH	3.8	5.2	3.46
Al (mEq/100 ml)	2.1	5.2	–
Ca (mEq/100 ml)	0.7	1.0	0.5
Mg (mEq/100 ml)	–	0.2	0.7
P ppm	2	3	81
K ppm	53	58	60
N (%)	0.50	–	1.41

Highland. Shallow, soil-filled depressions are ubiquitous on the Serra do Mar summits, particularly on the Serra dos Órgãos Plateau, where they form veritable archipelagos of vegetated islands in a sea of crystalline plains and domes. On the highest summits of the Serra dos Órgãos (2100 to 2263 m), closed depressions with some degree of protection from the direct wind are first colonized by lichens, then bryophytes and/or rupicolous monocots, followed by small sedges (*Trilepis* spp., *Rhynchospora* spp., and *Carex* sp.). The orchid *Zygopetalum mackayi* will often colonize these edge habitats as well, its beautiful purple and yellow-veined flowers forming a remarkable frame around many of these vegetation islands. Once 1 to 5 cm of soil have accumulated, small monocots like *Sisyrinchium* spp. and pioneering dicots like *Eryngium* spp., *Baccharis stylosa*, and *B. organensis* may colonize the depression; the *Baccharis* species root toward the center of the depression, sending their shoots decumbently to the edge of the island. On the downhill edge of the depression, where conditions are more mesic – with respect both to wind and the variability in water supply – and soils are deeper, *Xyris* spp., *Lycopodium* spp., and *Senecio* cf. *cuneifolius* may accompany the species noted above.

With 5 to 12 cm of soil accumulated, bunch grasses (*Agrostis*, *Calamagrostis*) become the dominant life-form in shallow depressions, sharing their space with a variety of rosettes (*Eryngium*, *Hypochoeris gardneri*), small shrubs (*Croton migrans*, *Hyptis* sp., *Esterhazya splendida*) and dwarf bamboo (especially *Chusquea pinifolia*); the average height of this sere is generally between 0.5 and 1 m. If the underlying rock surface is very irregular, depressions may support a wide variety of species on only 10 to 20 cm of soil. If the surface topography allows for the development of soils much deeper than 20 cm, however, the huge bunchgrass *Cortaderia modesta* will invade, and quickly drive the diversity curve in the opposite direction; there are only a few phanerophytes which can survive in *Cortaderia* grassland, including the tall sedge *Cladium ficticium*, which grows in very wet locations, the pachycaulous *Blechnum imperiale*, which may reach 1.5 m in height, and *Chusquea pinifolia*. Shade-tolerating chamaephytes and hemicryptophytes (e.g., *Polygala* spp., *Coccocypselum* spp., *Relbunium* spp.) are found on the pseudotrunks of *Cortaderia* or in "open" areas on the dark ground below. Shrubs that may be found growing out of the middle of such tall, dense clumps either predated the *Cortaderia* or are perched on or around broken blocks or clefts in the rock substrate; the most common saxicolous shrubs are *Myrceugenia alpigena*, *Symplocos* cf. *itatiaiae*, *Hyptis sp.*, and *Grazielia serrata*.

In "open" depressions (i.e., concave slopes, poorly defined nooks), *Barbacenia seubertiana* (or sometimes *Tillandsia* spp.) will often form incipient monotypic mats. These mats are secondarily colonized by bryophytes, sedges and rupicolous orchids, then, after the accumulation of significant organic and mineral sediment, by many of the same species listed above. Given sufficient slope, these communities may remain well-drained and shallow-soiled, and thus free for the most part from domination by *Cortaderia* or *Calamagrostis*: species diversity in these types of soil depressions is often greater (per unit area) than in the deeper and edaphically more hospitable (and hence, competitively more inhospitable) depressions described above.

Soil chemistry data for shallow, soil-filled depressions on the Morro do Cuca (Serra de Araras) and Pedra do Sino (Serra dos Órgãos) are provided in Table 10.8.1. Chemically speaking, conditions in general appear to be somewhat less hostile than on the lowland inselbergs, although levels of phosphorus are still critically low.

10.8.6.5 Monocotyledonous Mats

Coastal. The composition of monocot mats on Pão de Açúcar appears to depend heavily on the aspect and inclination of the slope in question (Carauta and Oliveira 1982; Miranda and Oliveira 1983). For example, the hot, dry north face of the rock is populated almost exclusively by highly dispersed mats of rupicolous species like *Brassavola tuberculata* (Orchidaceae) and *Vriesea brassicoides*; the cactus *Rhipsalis* sp. also covers large expanses of the north face, especially the lower 70 m. The shady and humid south face, which directly faces prevailing maritime winds, is covered almost entirely by an exuberant vegetation inhabiting monocot mats formed by *Barbacenia* spp. or *Vellozia* spp., orchids like *Brassavola tuberculata*, *Epidendrum robustum*, and *Maxillaria acicularis*, and the bromeliads *Tillandsia araujei* and *Vriesea* spp. Slope per se appears to play an important role in the distribution of bromeliads on Pão de Açúcar. For example, *Vriesea brassicoides* is found exclusively on the vertical and nearly vertical upper walls of the inselberg, whereas *V. procera* var. *rubra* and *V. regina* inhabit lower "runout" slopes, ledges, and the broken ramparts of the rock's eastern face (Carauta and Oliveira 1982). Lastly, although the sedge *Trilepis ihotzkiana* is not listed by Carauta and Oliveira (1982) as occurring on Pão de Açúcar, it exists nearby; *T. ihotzkiana* is generally one of the most important mat formers on both coastal and lower montane inselbergs throughout eastern Brazil.

Highland. In the Serra do Mar, monocotyledonous mats may be formed by a variety of species, depending on the environmental circumstances. The most obvious mats are those formed by species of *Tillandsia*, *Vriesea*, or *Barbacenia* on often extremely hostile ground, e.g., otherwise unvegetated, exposed, and often (very) steep rock surfaces. Where conditions are slightly less rigorous, particularly with respect to wind and cold exposure (and where the original colonization site is somewhat more than a microscopic holdfast on bare rock), sedges (*Trilepis* spp., *Rhynchospora* spp.) may form clonal mats as well. Miniature *Paepalanthus* spp. may form carpet-like associations in some cases, but these are found over at least a modicum of preexisting soil. Compared to coastal outcrops, orchid mats in the southeastern highlands are conspicuously rare.

10.8.6.6 Ephemeral and Wet Flush Vegetation

Coastal. We have no first- or secondhand data on flush vegetation on Pão de Açúcar, although such formations probably exist. In general, lowland flush communities are found at the feet of runout slopes. They are primarily cryptogamic, with local occurrences of Poaceae, Cyperaceae, Lentibulariaceae (*Utricularia*, *Genlisea*), and Xyridaceae, among other phanerogams.

Highland. Water seeps, whether seasonal or year-round features, are common in all of the campos de altitude. Ephemeral flush vegetation is principally composed of cyanobacterial and algal slimes on open rock, with bryophytic mats (*Pohlia* spp., *Atractylocarpus* spp., *Campylopus* spp.) along wet soil edges. More permanent water seeps (that may dry for a few weeks during the dry season) are extremely diverse, and usually include thick bryophytic mats (often including red or orange *Sphagnum* spp., and sometimes >0.5 m deep), foliose lichens (Parmeliaceae, Peltigeraceae), various marginal sedges and grasses (*Rhynchospora*, *Fimbristylis*, *Cladium*; *Calamagrostis*, *Briza*), tall members of the genus *Xyris*, various *Lycopodium* and *Selaginella* species, *Polygala* spp., *Drosera villosa*, *Paepallanthus* and *Eriocaulon* spp., *Plantago* spp., *Burmannia aprica*, *Utricularia* spp., and a variety of herbaceous composites.

10.8.7 Life-Forms and Adaptations

Rock outcrops in tropical latitudes are subject to extremely high surface temperatures and to drought-like conditions, even in rainforest climates. Phillips (1982) appropriately called the rock outcrop environment a microenvironmental desert. Metabolically active tissue in vascular plants can be irreparably damaged with prolonged exposure to temperatures of around 60 °C (Osmond et al. 1987), a figure that is probably exceeded with some regularity on lowland inselbergs of the coastal southeast. Ecologically, highland inselbergs clearly differ most from their lowland counterparts in the reduced importance of potentially lethal high temperatures and the concomitantly increased likelihood of killing frost; the absolute magnitude of physiological drought is significantly less in the higher sites as well. Although the inselberg flora in southeast Brazil exhibits many of the same tolerance or avoidance mechanisms common to desert and rock outcrop plants throughout the world [including succulence, crassulacean acid metabolism (CAM), poikilohydry, and – to a remarkably minimal extent – deciduousness and the adoption of an annual life cycle], the relative importance of each of these adaptations varies as one moves from low to high altitude. In addition, the high-altitude inselberg flora of southeastern Brazil exhibits a growth-form spectrum that has as much in common with tropical-alpine sites like the equatorial Andes as with the lowland inselbergs of the southeastern littoral (Safford 1999).

Succulent plants on southeastern lowland inselbergs are found primarily in the cactus family (*Coleocephalocereus*, *Cereus*, *Opuntia*, *Melocactus*, etc.); others include *Anthurium* (Araceae), a leaf succulent, the portulaca *Talinum paniculata*, *Kalanchoe brasiliensis* (Crassulaceae), and the rupicolous orchids, whose pseudobulbs are essentially succulent stems as well. Above 1700 m, succulence is much less common, but still found in the genera *Schlumbergera* (Cactaceae), *Peperomia* (Piperaceae), and *Prepusa* (Gentianaceae), not to mention the various orchid species. *Aloe*-like pseudosucculent pitcairnioid bromeliads are found in some southeastern mountains, like *Fernseea itatiaiae* in Itatiaia, and *Dyckia bracteata* in Caparaó. Of course, strict succulence in many bromeliads is obviated by their capacity to store water externally in rosette tanks; some huge species of *Vriesea* (subgenus *Alcantarea*) may store up to 20 l of liquid (Leme and Marigo 1993). Other bromeliad adaptations to xeric conditions include water- and nutrient-absorbing leaf trichomes, and (sometimes) deciduous leaves (Smith and Downs 1974). CAM carbon dioxide fixation occurs in some bromeliads as well, with nighttime temperature optima in some highland species running as low as 5 °C (Coutinho and Schrage 1970). Among the southeastern inselberg flora, *Kalanchoe*, some species of *Peperomia*, and a number of xerophyllic orchids are also CAM plants.

Poikilohydric plants in the southeast are found mostly in genera common to both lowland and highland inselbergs. These include *Barbacenia*, *Doryopteris*, *Anemia*, *Trilepis*, and some species of *Selaginella*;

Vellozia is found as high as 1600 to 1700 m in the Serra do Mar, but is generally absent on exposed summit outcrops. Poikilohydric shriveling not only protects these taxa from extreme drought, but the coincidence of the dry season with the cold season in the southeastern highlands also means that they are better able to protect metabolically active tissues from killing frosts.

Deciduous taxa are not common on southeastern inselbergs, perhaps because the profundity of the dry season is simply not of sufficient magnitude, but probably for biogeographic reasons as well. Trees inhabiting summit forests are more likely to be deciduous than are truly rupicolous or saxicolous plants on slope outcrops. For example, *Tabebuia* (Bignoniaceae) and various genera of the family Bombacaceae (e.g, *Chorisia*, *Pseudobombax*) are commonly encountered atop inselbergs throughout Brazil, to about 1600 m elevation; above this elevation, summit phanerophytes are entirely evergreen. Note that deciduousness is significantly more common among the inselberg flora of the interior northeast, where rainfall may drop below 700 mm, and the dry season extends for many months.

The cool, wet climate of the Brazilian highlands leads to the predominance of a series of growth forms in the campos de altitude of the Serra do Mar summits that are broadly characteristic of tropical alpine (and subantarctic maritime) environments (Cuatrecasas 1968; Rauh 1978; Safford 1999). These include tall-stemmed bunchgrasses (e.g, *Cortaderia*, *Calamagrostis*), dwarfed, sclerophyllous shrubs with often pilose leaves (e.g., *Baccharis*, *Lychnophora*, Ericaceae, and Myrtaceae), acaulescent rosettes (*Eryngium*, *Paepalanthus*, *Plantago*, *Prepusa*), and pachycaulous plants (e.g., *Blechnum imperiale*), all of which play important roles in the vegetation of highland rock outcrops. Although xeromorphic in some respects, these growth forms are widely thought to principally represent adaptations to the freezing temperatures common on dry-season nights in the strongly diurnal tropical-alpine climate (Rundel et al. 1994).

With respect to life-forms, clearly the most remarkable thing about southeastern Brazilian inselbergs is their extreme poverty in annual plants (therophytes). Desert climates were referred to by Raunkiaer (1934) as "therophyte climates", because of the relative dominance of annual plants in the desert flora. Investigations of rock outcrop and inselberg floras the world over have sustantiated the dominance of therophytes in inselberg and rock outcrop vegetation as well (Phillips 1982; Baskin and Baskin 1985; Barthlott et al. 1993; Porembski 1996; Porembski et al. 1996), but neither Pão de Açúcar nor the Serra do Mar highlands fit this pattern (Table 10.8.2). Data from elsewhere in southeastern Brazil confirm the generality of this finding (Porembski et al. 1998; G. Martinelli et al., unpubl. data;

Safford 1999). Note that the extremely low values for therophytes in the Brazilian inselbergs would doubtless be much lower were it nor for man: all of the (identified) annual species found on Pão de Açúcar are widespread ruderals, and at least three of them (*Apium leptophyllum*, *Lepidium virginicum*, *Petroselinum crispum*) are introduced species. The depauperate cast of therophytes on southeastern Brazilian inselbergs cannot be purely due to biogeographic circumstances: of 13 annual species listed by Porembski et al. (1996) from rainforest outcrops in the Ivory Coast, 6 come from genera with representatives in eastern Brazil. The same can be said of Mt. Mulanje in Malawi, where 12 of 26 therophytic species also have infrageneric cousins in eastern Brazil (Porembski 1996). Why most of these genera (and others) are not also found on southeastern Brazil outcrops remains a mystery, although we believe that the final answer will necessarily involve not only biogeography, but considerations of ecological competition, modern micro- and mesoclimatology, and past macroclimatological change. Much more work is necessary before we can begin to tease apart the complex ecological and evolutionary issues that govern this regional departure from worldwide norms.

Even though there are very few annual plants on coastal inselbergs in southeastern Brazil, the remainder of the life-form spectrum differs dramatically from that of the surrounding forest; this is obviously due to the steep environmental gradients that characterize the forest-outcrop ecocline. In general, species in neotropical rainforests are around 70% ground-rooted phanerophytes, with a further 15–25% epiphytes and 10–15% chamaephytes. The percentages of hemicryptophytes, geophytes, and especially therophytes are vanishingly small in a mature, primary forest (Vareschi 1992; Richards 1996). From Table 10.8.2, we can see that even on Pão de Açúcar, which is dominated by stress-tolerating phanerophytes and chamaephytes, the remainder of the life-form spectrum still makes up nearly 20% of the flora.

The taxa inhabiting the Serra do Mar outcrops are very different from those growing on Pão de Açúcar, both in terms of floristics (see below) and characteristic life-forms (Table 10.8.2). The increased importance of hemicryptophytes and geophytes in the flora is probably due to the enhanced seasonality of the high montane climate, and the overwhelming importance of frost avoidance on these cold, windy summits; the existence of somewhat deeper soils on highland inselbergs is fundamental to the survival of these seasonally subterranean plants. In stark contrast to the lowland situation, the inselberg flora of the Serra do Mar is remarkably similar to the surrounding campos de altitude in terms of its component life-forms. A representative spectrum for the campos de altitude can be derived from the Serra do Itatiaia, where botanical exploration is further

Table 10.8.2. Comparison of Raunkiaer life-forms spectra for Brazilian inselbergs with spectra of desert/dryland and rock outcrop floras

Location[a]	Phanero-phytes	Chamae-phytes	Hemicrypto-phytes	Geo-phytes	Thero-phytes
(1) S. Appalachians (USA)	23 %	4 %	26 %	15 %	32 %
(2) Death Valley (USA)	26	7	17	7	43
(3) "Tropical Dry" (S. Am.)	9	14	19	8	50
(4) Ivory Coast (Africa)	18.3	21.7	6.7	8.3	45
(5) Pão de Açúcar	38.9	42.1	8.4	5.3	5.3
(6) Serra do Mar	33.8	38.1	16.9	10.6	0.6

[a] (1) Phillips (1982); (2) Raunkiaer (1934); (3) Rizzini (1979); (4) Porembski et al. (1996); (5) See Table 10.8.3; (6) See Table 10.8.4

advanced than in the Serra dos Órgãos (Brade 1956; Safford 1999): phanerophytes, 39.4 %; chamaephytes, 38 %; hemicryptophytes, 14.1 %, geophytes 7.8 %, therophytes 0.7 % (compare with Serra do Mar in Table 10.8.2). In essence, the environment of the campos de altitude in general – which includes large expanses of open rock and often thin-soiled veneers over otherwise rocky plains – is not extremely different from that of the inselbergs embedded within it.

Along with floristic data (see next section, 10.8.8), Tables 10.8.3 and 10.8.4 provide information on pollination and seed dispersal syndromes in the southeast Brazilian inselbergs. At both altitudes, about 70 % of the listed plants use insects (either obligately or facultatively) to disperse their pollen, but anemophily is more important in the highland sites (ca. 14 % of the highland flora versus 9.5 % lowland), and ornithophily in the lowland sites (35 versus 23 % highland). Altitudinal differences are somewhat more pronounced with regard to seed dispersal: although both sites show a predominance of anemochory, the importance of wind dispersal is exagerated in the Serra do Mar (51 % of the highland flora versus 42 % lowland). Endozoochory and anemochory are more important in the lowland flora (22 versus 15 % highland, and 30 versus 25 % highland, respectively).

10.8.8 Systematics

Table 10.8.3 outlines the floristic composition of rock outcrop vegetation on Pão de Açúcar (and neighboring coastal outcrops); Table 10.8.4 outlines the composition of summit outcrop vegetation in the Serra de Araras and the Serra dos Órgãos, neighboring mountain groups in the Serra do Mar of Rio de Janeiro. Although these lists are not complete, we believe they represent the vast majority of species that commonly occur on outcrops in these two sites. Note that as these are inventories of multiple community types, Tables 10.8.3 and 10.8.4 do not bespeak the often great intrainselberg variation in floristic composition that characterizes, e.g., the monocot mat associations (Porembski et al. 1998). Both tables combine floristic data from relatively wide altitudinal belts: Since there are a variety of environmental gradients associated with these changes in altitude [e.g., the temperature lapse rate (ca. 0.55 °/100 m in southeast Brazil), wind exposure, changes in airborne salinity (at Pão de Açúcar), etc.], Tables 10.8.3 and 10.8.4 are inherently unable to account for species turnover along these clines. Lastly, since Table 10.8.4 combines data from two separate (although adjacent) mountain groups, it masks differences in the inselberg flora of the component mountains.

In Table 10.8.3, 98 species are listed for Pão de Açúcar, representing 86 genera from 49 families. In terms of species numbers, the most important families are the Bromeliaceae (15 species), the Orchidaceae (11 species), and Cactaceae (4 species); seven families are represented by three species each. The Orchidaceae, Bromeliaceae, and Poaceae probably contribute similar amounts to total biomass. The most diverse genera are *Vriesea* (including subgenus *Alcantarea*; 5 species) and *Tillandsia* (3 species) (both Bromeliaceae), and *Tibouchina* (Melastomataceae; 3 species).

Tables 10.8.4 lists 153 species from the Serra do Mar, representing 101 genera from 51 families. The most diverse families on the highland inselbergs are the Asteraceae (17 species), the Bromeliaceae (12 species), the Orchidaceae (11 species), the Cyperaceae, and Melastomataceae (10 species each), and the Poaceae (8 species); as in the lowlands, many families contribute one to three species each. In terms of actual biomass, the order shifts: (1) Poaceae, (2) Cyperaceae, (3) Asteraceae, with Apiaceae (*Eryngium*), Xyridaceae, Eriocaulaceae, and Bromeliaceae also entering the top three depending on local ecological conditions (Safford 1999). The most diverse highland genera are *Tibouchina* (Melastomataceae; 7 species), *Tillandsia* (Bromeliaceae; 6 species), *Mandevilla* (Apocynaceae; 5 species, but rare above 2000 m), and *Baccharis* (Asteraceae; 5 species).

Pão de Açúcar and the Serra do Mar inselbergs share more than two-thirds of the families listed in Tables 10.8.3 and 10.8.4. In the case of the former, 33 of 49 families (67 %) are shared with the highland sites, in the case of the latter, 38 of 51 families (75 %) are shared with Pão de Açúcar. Of the families from the lowland site not found in the highland sites, four are primarily warm-tropical in distribution (Cecropiaceae, Malpighiaceae,

Table 10.8.3. Summary of taxa found on rock outcrops on and near Pão de Açúcar (0 to 398 m), Serra da Carioca, Rio de Janeiro

Family	Name	Raunkiaer Life-form	Generic Chorology[b]	Rup/ Sax	Succ./ Poik	Poll[b]	Disp[b]	End
Adiantaceae	***Doryopteris*** sp.	Ch	W	RS	Po	W	A	
	Pteris longifolia L.	Ch	C			W	A	
Amaranthaceae	*Amaranthus spinosus* L.	T	C	S				
Amaryllidaceae	***Hippeastrum*** sp.	G	N	S		Av	AP	
Apiaceae	*Apium leptophyllum* (Pers.) F. Müller var. *leptophyllum*	T	C	S		E	AP	
	Petroselinum crispum Nyman	T	H	S		E	AP	
Apocynaceae	***Mandevilla*** *crassinoda* Gardner	L	N			AvE	AP	
Araceae	***Anthurium coriaceum*** G. Don	Ch	N	RS	*Su*	E	Zen	
	A. solitarium Schott	Ch	–	RS	*Su*	E	Zen	
Aspleniaceae	*Dryopteris setigera* (Bl.) O. Ktze.	Ch	C		W		A	
Asteraceae	***Baccharis*** *serrulata* (Lam.) Pers.	Ph	N			E	A	
	Emilia sagitatta (Vahl) DC	H	W	S		E	A	
	Vernonia *scorpioides* Pers.	Ph	W	S		E	A	
Blechnaceae	***Blechnum*** *unilaterale* Sw.	Ch	C	S		W	A	
Boraginaceae	*Cordia corymbosa* (L.) G. Don	Ph	W			AvE	AP	
Brassicaceae	*Lepidium virginicum* L.	T	C	S		E	AP	
Bromeliaceae	*Billbergia pyramidalis* (Sls.) Lindl.	Ch	N	S		Av	Zen	
	Cryptanthus bromelioides Mez	Ch	B	S		Av	Zen	
	Nidularium sp.	Ch	B	S		Av	Zen	
	Orthophytum sp.	Ch	B	RS		Av	Zen	
	Pitcairnia *albiflos*[a] Herb.	Ch	N	RS		AvE	A	Guana
	P. flammea Lindl.	Ch	–	S		AvE	A	
	Streptocalyx floribundus (Mart. ex Sch.) Mez	Ch	N			Av	Zen	
	Tillandsia *araujei* Mez	Ch	N	R		AvE	A	

Table 10.8.3 (*continued*)

Family	Name	Raunkiaer Life-form	Generic Chorology[b]	Rup/ Sax	Succ./ Poik	Poll[b]	Disp[b]	End
Bromeliaceae	***T. brachyphylla*** Baker	Ch	–	R		AvE	A	Cst/ SdMRJ
	T. stricta Sol.	Ch	–	R		AvE	A	
	Vriesea brassicoides[a] (Baker) Mez	Ch	N	R		Av	A	Carioca
	V. procera (Mart. ex Schult.) Wittm.	Ch	–	S		Av	A	
	V. geniculata (Wawra) Wawra	Ch	–	RS		Ch	A	
	V. goniorachis[a] (Baker) Mez	Ch	–	R		Av	A	Carioca
	V. regina[a] (Vell.) Beer	Ch	–	RS		Ch	A	WRJ
Cactaceae	*Coleocephalocereus fluminensis* (Miq.) Backeb.	Ph	N	S	Su	AvE	Zen	
	Cereus sp.	Ph	N	S	Su	ChE	Zen	
	Opuntia brasiliensis (Willd.) Haw.	Ph	N	S	Su	AvE	Zen	
	Rhipsalis sp.	Ph/L	W	R	Su	AvE	Zen	
Capparidaceae	*Capparis* sp.	Ph	W	S		AvE	Zen	
	Cleome sp.	H/T	W	S		AvE	AP	
Cecropiaceae	*Cecropia glazioui* Snethlage	Ph	N	S		E	Zen	
	Coussapoa microcarpa (Schott) Rizz.	Ph	N	S		E	Zen	
Celastraceae	***Maytenus*** sp.	Ph	W	S		E	A	
Clusiaceae	***Clusia*** sp.	Ph	N	S		E	Zen	
Convolvulaceae	*Ipomoea* sp.	L/Ch	W	S		AvE	AP	
Crassulaceae	*Kalanchoe brasiliensis* Camb.	Ch	W	RS	Su	E	AP	
Cyperaceae	*Fimbristylis* sp.	T/H	T			A	AP	
	***Trilepis** ihotzkiana* Nees ex Arnott	Ch	N	R	Po	A	AP	
Dilleniaceae	*Davilla* sp.	Ph	N			E	AP	
Dioscoreaceae	***Dioscorea*** sp.	G	W	S		E	A	

Table 10.8.3 (*continued*)

Family	Name	Raunkiaer Life-form	Generic Chorology[b]	Rup/ Sax	Succ./ Poik	Poll[b]	Disp[b]	End
Euphorbiaceae	***Croton** compressus* Lam.	Ph	W	S		E	Auto	
	Manihot sp.	Ph	N	S		E	Auto	
	***Phyllanthus** corcovadensis* Mart.	Ch/Ph	W	RS		E	Auto	
Fabaceae	***Cassia** patellaria* DC	Ph	W	S		E	AP	
	Canavalia parviflora Benth.	Ph/Ch	W			E	AP	
	Stylosanthes viscosa Sw.	Ph	W			E	AP	
Gesneriaceae	***Sinningia*** sp.	L/H	N	S		Av	AP	
Gleicheniaceae	*Gleichenia bifida* (W.) Spr.	Ch	A	S		W	A	
Lentibulariaceae	***Utricularia*** sp.	Ch	C	S		E	WZep	
Malpighiaceae	*Heteropteris chrysophylla* (Lam.) HBK	Ph	N	S		E	A	
Melastomataceae	***Tibouchina** bulbosa*	Ph	N			AvE	AP	
	T. grandiflora Cogn.	Ph	–	S		AvE	AP	
	T. granulosa (Desv.) Cogn.	Ph	–	S		AvE	AP	
Moraceae	*Ficus enormis* (Mart. ex Miq.) Miq.	Ph	W	S		E	Zen	
Myrtaceae	*Eugenia* spp. (2)	Ph	W			AvE	Zen	
	Psidium littorale Raddi	Ph	N	S		AvE	Zen	
Ochnaceae	***Ouratea*** sp.	Ph	W	S		E	Zen	
Orchidaceae	*Bifrenaria harrisoniae* (Hook.) Reichb.	Ch	N	S		E	A	
	Brassavola tuberculata Hook.	Ch	N	R		E	A	
	***Cyrtopodium** andersonii* R. Br.	Ch	N	RS		E	A	
	***Epidendrum** denticulatum* Barb.	Ch	N	R		E	A	
	E. robustum Cogn.	Ch	–	RS		E	A	
	Laelia lobata (Lindl.) Veitch	Ch	N	R		E	A	

Table 10.8.3 (*continued*)

Family	Name	Raunkiaer Life-form	Generic Chorology[b]	Rup/ Sax	Succ./ Poik	Poll[b]	Disp[b]	End
Orchidaceae	***Maxillaria** acicularis* Herb.ex Lindl.	Ch	N	R		E	A	
	Polystachya estrellensis Reichb.	Ch	W	R		E	A	
	Prescottia plantaginea Lindl.	G	N	S		E	A	
	Sarcoglottis biflora (Vell.) Schltr.	Ch	N	R		E	A	
	Zygopetalum mackayi Hook.	Ch	N	S		E	A	
Oxalidaceae	***Oxalis** barrelieri* L.	G	C	S		E	Auto	
Passifloraceae	*Passiflora* sp.	L	N	S		AvE	Zen	
Plantaginaceae	***Plantago*** sp.	H	C	S		A	AP	
Poaceae	*Melinis minutiflora* P. Beauv.	Ch	W	S		A	A	
	***Panicum** maximum* Jacq.	Ph	W	S		A	A	
	Rhynchelytrum repens (Willd.) C.E. Hubb.	H	W	S		A	A	
Polygalaceae	***Polygala paniculata*** L.	H	C	S		E	Zen	
Portulacaceae	*Talinum paniculatum* (Jacq.) Gärtn.	Ph	W	S	Su	E	AP	
Rubiaceae	***Borreria verticillata*** (L.) Meyer	H	W	S		E	AP	
Sapindaceae	*Serjania* sp.	L	N	S		E	AP	
Sapotaceae	*Pouteria caimito* (Ruiz & Pav.) Radlk.	Ph	W	S		AvE	Zen	
Schizaeaceae	***Anemia** phyllitidis* (L.) Sw.	Ch	W	S	Po	W	A	
Selaginellaceae	***Selaginella*** sp.	Ch	W	RS	Po	W	A	
Solanaceae	*Capsicum* sp.	Ph	N	S		AvE	Zen	
Ulmaceae	*Trema micrantha* (L.) Blume	Ph	W	S		E	Zen	
Urticaceae	*Laportea aestuans* (L.) Chew	H	W	S		A	AP	
	Phenax sonneratii (Poir.) Weddel	Ph	N	S		A	AP	
	Urera sp.	Ph	W	S		A	Zen	

Table 10.8.3 (*continued*)

Family	Name	Raunkiaer Life-form	Generic Chorology[b]	Rup/ Sax	Succ./ Poik	Poll[b]	Disp[b]	End
Velloziaceae	***Barbacenia purpurea***[a] Hook.	Ch	N	R	Po	AvE	AP	Carioca
	Vellozia candida[a] Mikan	Ph	N	R	Po	Av	AP	CstRJ
Verbenaceae	***Lantana*** sp.	Ph	W	S		E	AP	
	Stachytarpheta sp.	H	W	S		E	AP	

Taxa in boldface type are common to Tables 10.8.3 and 10.8.4

[a] Taxa endemic within the study area; location noted in *End* column (see key).

[b] Data in these columns were figured completely (in the case of chorology), or primarily (in the case of pollination and dispersal) at the level of genus; thus, in the case of pollination and dispersal, combinations are not only possible for real biological reasons, but also because the possible vectors for the genus in question could not be differentiated at the level of species. Dashes (–) indicate that the datum in question does not have meaning at the species level; see the first species in the appropriate genus.

Key: Life-form: Ch = chamaephyte, E = epiphyte, G = geophyte, H = hemicryptophyte, L = liana/vine, Ph = phanerophyte, T = therophyte. Chorology (phytogeographic groups from Cleef 1979; see text): A = australantarctic, B = (extraAmazonian) Brazil; C = cosmopolitan; H = holarctic; N = neotropical; T = widespread temperate; W = widespread tropical. Rup/Sax: R = rupicolous; S = saxicolous. Succ./Poik (italics = somewhat): Su = succulent; Po = poikilohydric. Poll: A = anemophily; Av = ornithophily; Ch = chiropterophily; E = entomophily; W = aqueous pollination (ferns). Disp: A =anemochory; AP = passive anemochory; Auto = autochory; Zen = endozoochory; Zep = epizoochory. End: Araras = Serra de Araras; Carioca = Serra da Carioca s.l.; CstRJ = Rio de Janeiro coast; Guana = environs of Guanabara Bay; Órgãos = Serra dos Órgãos; WRJ = western half of the state of Rio de Janeiro; SdMRJ = Serra do Mar of Rio de Janeiro.

Data: Carauta and Oliveira (1982), Miranda and Oliveira (1983); H.D. Safford (pers. observ.); pollination and dispersal from Mabberley (1996), Zomlefer (1994), Martinelli (1997), monographs and pers. observ.; endemics from Smith and Ayensu (1976), and Martinelli and Vaz (1988).

Table 10.8.4. Summary of taxa found on summit-level rocks (1700 to 2263 m) in the Serra do Mar, north of Rio de Janeiro

Family	Name	Raunkiaer Life-form	Generic Chorology[b]	Rup/ Sax	Succ./ Poik	Poll[b]	Disp[b]	End
Adiantaceae	***Doryopteris*** spp.	Ch	W	RS	Po	W	A	
Amaryllidaceae	***Hippeastrum aulicum*** Herb.	G	N	S		Av	AP	
	Worsleya[a] *rayneri* (J.D. Hooker) Traub & Moldenke	G	B	R	*Su*	Av	AP	Araras
Apiaceae	*Eryngium* spp. (2)	Ch	C	S		E	AP	
Apocynaceae	***Mandevilla*** spp. (5)	L/Ph	N			E	A	
Araceae	***Anthurium coriaceum*** G. Don	Ch	N	RS	*Su*	E	Zen	
	A. spp. (2)	Ch	–	RS	*Su*	E	Zen	
Asclepiadaceae	*Oxypetalum* sp.	L/Ch	N	S		E	A	
Aspleniaceae	*Elaphoglossum* spp.	Ch	W	S		W	A	
Asteraceae	*Achyrocline satureoides* DC	Ch	W	S		E	A	
	Baccharis *calvescens* DC	Ph	N			E	A	
	B. cf. *myriocephala* DC	Ph	–	S		E	A	
	B. organensis Baker	Ph	–	S		E	A	
	B. platypoda DC	Ph	–	S		E	A	
	B. stylosa Gardner	Ph	–	S		E	A	
	Chaptalia sp. nov. (?)	Ch	N			E	A	
	Dasyphyllum cryptocephalum (Baker) Cabrera	Ph	N	S		E	A	
	D. leptocantus Gard. & Cabrera	Ph	–	S		E	A	
	Erigeron maximus DC	H	C	S		E	A	
	Grazielia serrata (Sprengel) R. King & H. Robins.	Ph	B	S		E	A	
	Hypochoeris cf. *gardneri* Baker	H	C	S		E	A	
	Koanophyllon baccharifolium (Gardner) R. King & H. Robins.	Ph	N	S		E	A	

Table 10.8.4 (*continued*)

Family	Name	Raunkiaer Life-form	Generic Chorology[b]	Rup/ Sax	Succ./ Poik	Poll[b]	Disp[b]	End
Adiantaceae	*Lychnophora* sp.	Ch	B	S		E	A	
	Senecio cuneifolius Gardner	H	C	S		E	A	
	S. cf. *itatiaiae* Dusén	H	–	S		E	A	
	***Vernonia** glazioviana* Baker	Ph	W	S		E	A	
Blechnaceae	***Blechnum** imperiale* (Fee & Glaz.) Christ.	Ph	C			W	A	
Bromeliaceae	***Pitcairnia flammea*** Lindl.	Ch	N	S		AvE	AP	
	P. glaziovii[a] Baker	Ch	–	S		AvE	A	Araras
	Quesnelia lateralis Wawra	Ch	B	RS		Av	Zen	
	Tillandsia brachyphylla Baker	Ch	N	R		AvE	A	Cst/ SdMRJ
	T. carminea Till	Ch	–	R		AvE	A	
	T. gardneri Lindl.	Ch	–	RS		AvE	A	
	T. grazielae[a] Sucre & Braga	Ch	–	R		AvE	A	Araras
	T. reclinata[a] Pereira & Martinelli	Ch	–	R		AvE	A	Araras
	T. stricta Sol.	Ch	–	RS		AvE	A	
	***Vriesea** atra*[a] Mez	Ch	N	RS		Ch	A	SdMRJ
	V. imperialis[a] Carrière	Ch	–	RS		Ch	A	SdMRJ
	V. itatiaiae Wawra	Ch	–	S		A	A	
Burmanniaceae	*Burmannia aprica* (Malme) Jonker	T	W			E	A	
Cactaceae	*Schlumbergera obtusangula* (Loefgr.) Hunt	Ch	B	RS	Su	Av	Zen	
Celastraceae	***Maytenus*** sp.	Ph	W	S		E	A	
Clusiaceae	***Clusia*** sp.	Ph	N	S		E	Zen	
	Hypericum brasiliensis Choisy	H	T			E	AP	
Cunoniaceae	*Weinmannia organensis* Gardner	Ph	A			E	A	

Table 10.8.4 (*continued*)

Family	Name	Raunkiaer Life-form	Generic Chorology[b]	Rup/ Sax	Succ./ Poik	Poll[b]	Disp[b]	End
Cyperaceae	*Bulbostylis* sp.	G	W	S		A	A	
	Carex cf. *fuscula* D'Urb.	G	T	S		A	A	
	Cladium ficticium Hernst.	Ph	C			A	A	
	Lagenocarpus polyphyllus (Nees) O. Ktze.	G	N			A	A	
	Rhynchospora coriifolia Boeck.	G	C	S	*Po*	A	A	
	R. splendens Lindm.	G	–	S		A	A	
	Trilepis *microstachya* (Clark) H. Pfeiff.	Ch	N	RS	Po	A	A	
	T. spp. (3)	Ch	–	S	Po	A	A	
Dioscoreaceae	***Dioscorea*** spp.	G	W	S		E	A	
Droseraceae	*Drosera villosa* St. Hil.	Ch	C			E	AP	
Ericaceae	*Gaultheria ferruginea* Cham. & Schlecht.	Ch	A	S		E	AP	
	G. organensis Meissn.	Ch	–	S		E	AP	
	Gaylussacia fasciculata Gardner	Ch	N	S		E	Zen	
Eriocaulaceae	*Paepalanthus* cf. *glabrifolius* Ruhl.	H	N	S		A	AP	
Escalloniaceae	*Escallonia organensis* Gardner	Ph	A			E	AP	
Euphorbiaceae	***Croton*** *migrans* Casar.	Ph	W	S		E	Auto	
	Phyllanthus sp.	Ch	W	S		E	Auto	
Fabaceae	***Cassia*** *organensis* Glaz. ex Harms	Ph	W	S		E	AP	
	Crotalaria sp.	Ph	W	S		E	AP	
	Mimosa sp.	Ph	W	S		E	AP	
Flacourtiaceae	*Abatia americana* Gardner	Ph	N	S		Av	Zen	
Gentianaceae	*Hockinia montana* Gardner	H	B	S		E	AP	
	Prepusa connata[a] Gardner	Ch	B	S	Su	E	AP	Araras

Table 10.8.4 (*continued*)

Family	Name	Raunkiaer Life-form	Generic Chorology[b]	Rup/ Sax	Succ./ Poik	Poll[b]	Disp[b]	End
Gentianaceae	*P. hookeriana*[a] Gardner	Ch	–	S	Su	E	AP	Órgãos
Gesneriaceae	***Sinningia*** spp. (4)	Ch/H	N	S		AvE	AP	
	Vanhouffea leptopus O. Ktze.	Ch	B	RS		AvE	AP	
Grammitidaceae	*Grammitis* sp.	Ch	W	S		W	A	
Iridaceae	*Sisyrinchium alatum* Hooker	G	A	S		E	AP	
	S. incurvatum Gardner	G	–	S		E	AP	
Lamiaceae	*Hyptis* spp. (2)	Ph	N	S		AvE	AP	
Lentibulariaceae	***Utricularia*** spp. (3)	C	C	S		E	WZep	
Lycopodiaceae	*Lycopodium* sp.	Ch	C			W	A	
Melastomataceae	*Benevidesia organensis*[a] Sald. & Cogn.	Ph	B	S		AvE	Zen	SdMRJ
	Lavoisiera glazioviana Cogn.	Ph	B	S		AVE	AP	
	Leandra scabra DC	Ph	N	S		AvE	Zen	
	Tibouchina *alba* Cogn.	Ph	N	S		AVE	AP	
	T. gardneriana (Triana) Cogn.	Ph	–	S		AvE	A	
	T. grandifolia Cogn.	Ph	–	S		AvE	A	
	T. martusiana Cogn.	Ph	–	S		AvE	A	
	T. spp. (3)	Ph	–	S		AvE	A	
Myrsinaceae	*Rapanea gardneriana* (DC) Mez	Ph	W	S		E	Zen	
Myrtaceae	*Gomidesia kunthiana* Berg	Ph	B	S		AvE	Zen	
	Myrceugenia alpigena (DC) Landrum	Ph	A	S		AvE	Zen	
Ochnaceae	*Luxemburgia glaziouviana* Baker	Ph	N	S		E	A	
	Ouratea sp.	Ph	W	S		E	Zen	
Orchidaceae	***Cyrtopodium*** sp.	Ch	N	RS		E	A	
	Epidendrum sp.	Ch	N	S		E	A	
	Habenaria spp.	G	W	S		E	A	

Table 10.8.4 (*continued*)

Family	Name	Raunkiaer Life-form	Generic Chorology[b]	Rup/ Sax	Succ./ Poik	Poll[b]	Disp[b]	End
Orchidaceae	***Maxillaria** gracilis* Ruiz & Pavon	Ch	N	RS		E	A	
	M. sp.	Ch	–	S		E	A	
	Oncidium blanchetii Reichb.	Ch	N	RS		E	A	
	Pleurothallis spp.	Ch	N	S		E	A	
	Polystachya estrellensis Reichb.	Ch	W	RS		E	A	
	Pseudolaelia vellozicola (Hoehne) Porto & Brade	E/Ch	B			E	A	
	Stenorrhynchos sp.	G	N	S		E	A	
	Zygopetalum mackayi Hooker	Ch	N	S		E	A	
Oxalidaceae	***Oxalis*** spp. (3)	G	C	S		E	Auto	
Piperaceae	*Peperomia galioides* HBK	H	W	RS	Su	AE	Zep	
	P. spp. (2)	H	–	RS	Su	AE	Zep	
Plantaginaceae	***Plantago*** cf. *australis* L.	H	T	S		AE	AP	
Poaceae	*Axonopus* sp.	H	N	S		A	A	
	Briza sp.	H	T	S		A	A	
	Calamagrostis sp.	Ch	T	S		A	A	
	Chusquea pinifolia Nees	Ph	N	S		A	A	
	Cortaderia modesta (Doell.) Hack.	Ph	A			A	A	
	Ichnanthus tenuis (Presl.) Hitchc. & Chase	H	N			A	A	
	Glaziophyton[a] *mirabile* Franchet	Ph	B	S		A	A	Araras
	Panicum sp.	H/Ch	C	S		A	A	
Polygalaceae	***Polygala paniculata*** L.	H	C	S		E	Zen	
	P. spp. (4)	H	–	S		E	Zen	
Proteaceae	*Roupala lucens* Meissn.	Ph	N	S		Av	A	
Rubiaceae	***Borreria verticillata*** (L.) Meyer	H	C	S		E	AP	

Table 10.8.4 (*continued*)

Family	Name	Raunkiaer Life-form	Generic Chorology[b]	Rup/ Sax	Succ./ Poik	Poll[b]	Disp[b]	End
Rubiaceae	*Manettia sarcophylla* Rizzini	L/Ch	N	S		Av	Zen	
	Relbunium spp. (2)	L/H	N	S		E	Zen	
Schizaeaceae	***Anemia*** sp.	Ch	W	S	Po	W	A	
Scrophulariaceae	*Esterhazya splendida* Mik.	Ph	N	S		AvE	AP	
Selaginellaceae	***Selaginella*** sp.	Ch	W	RS	Po	W	A	
Smilacaceae	*Smilax campestris* Griseb.	L/Ph	C	S		E	Zen	
Symplocaceae	*Symplocos itatiaiae* Wawra	Ph	W			E	Zen	
Velloziaceae	***Barbacenia*** *seubertiana*[a] Goeth. & Henr.	Ch	N	RS	Po	AvE	AP	RJ
	Vellozia *variegata* Goeth. & Henr.	Ph	N	RS	Po	Av	AP	
Verbenaceae	***Lantana*** spp. (2)	Ph	W			E	Zen	
Xyridaceae	*Xyris* spp. (2)	H	W	S		A	AP	

Taxa in boldface type are common to Tables 10.8.3 and 10.8.4.
[a] Taxa endemic within the study area; location noted in *End* column (see Table 10.8.3 key).
[b] See Table 10.8.3 for explanation.

Data: Martinelli and Bandeira (1989), Martinelli et al. (unpubl. data); Safford (unpubl. data); for pollination/dispersal and endemics see Table 10.8.3.

Moraceae, Ulmaceae), five families (Boraginaceae, Convolvulaceae, Dilleniaceae, Gleicheniaceae, Sapindaceae) would have been included in Table 10.8.4 had its range been extended 100 m downslope, and a sixth (Solanaceae), occurs throughout the campos de altitude, but it was simply not sampled. The remaining seven families (Amaranthaceae, Boraginaceae, Brassicaceae, Capparidaceae, Crassulaceae, Portulacaeae, Urticaceae) all have neotropical representatives in the High Andes, but they are not found in the campos de altitude. Of the 13 families restricted to the highland inselbergs, 9 of them are typical of the neotropical montane and alpine flora (Ericaceae, Gentianaceae, Grammitidaceae, Iridaceae, Lamiaceae, Lycopodiaceae, Scrophulariaceae, Symplocaceae, Xyridaceae), as are the genera *Rapanea* (Myrsinaceae), and *Peperomia* (Piperaceae). The remaining two families (Asclepiadaceae, Smilacaceae) are also found throughout the lowland and submontane southeast, and both contain rupicolous and saxicolous species; their absence from Pão de Açúcar may well be an artifact of sampling.

Of the families held in common, only three can be said to be composed largely of rupicolous or saxicolous species (whether facultatively epiphytic or not). These are the Bromeliaceae, Orchidaceae, and – especially – Velloziaceae. The Cyperaceae and Poaceae are also common members of the inselberg/outcrop flora, but many sedges and grasses are ecological generalists, and their occurrence on southeast inselbergs is due in most cases to ecological tolerance rather than to habitat specificity (*Trilepis* being an exception).

Genera and species shared between Pão de Açúcar and the Serra do Mar inselbergs are listed in boldface type in Tables 10.8.3 and 10.8.4. Out of a total of 152 genera at both sites, 36 are held in common. A further ten (*Coleocephalocereus*, *Cordia*, *Davilla*, *Eugenia*, *Ficus*, *Ipomoea*, *Passiflora*, *Psidium*, *Rhipsalis*, *Serjania*) occur on Pão de Açúcar and on mountain outcrops just below the treeline in the Serra do Mar (up to ca. 1700 m in the Serra das Araras, up to ca. 2000 m in the Serra dos Órgãos); presumably these taxa have not evolved local species that can withstand the rigorous growing conditions of the open campos de altitude.

Of 243 total species in Tables 10.8.3 and 10.8.4, only eight are found on both the lowland and highland inselbergs (Jaccard similarity = 0.033, Sørensen = 0.064), even though the two are separated by less than 60 km; this is testimony to the great environmental differences between the two sites. Considering that inhabitants of both sites must be genetically prepared to deal with temperatures ranging from +60 to −10 °C it is actually somewhat remarkable that any tropical species can survive both environments. Note that six of the eight shared species are bromeliads, orchids, or aroids, families typified by facultatively rupicolous and epiphytic taxa.

Although we can environmentally explain the great taxonomic differences between the inselberg floras of Pão de Açúcar and the Serra do Mar, it is more difficult to account for the high species turnover that we find in the southeast between nearby inselbergs within the same altitudinal belt. This is particularly the case with regard to monocot mat communities (Porembski et al. 1998). Apparently, a large regional pool of rupicolous species with mat-forming capabilities, combined perhaps with somewhat stochastic patterns of colonization, leads to very high beta-diversity both within and among inselbergs. From the Rup/Sax column in Tables 10.8.3 and 10.8.4, there are at least 27 rupicolous species (28.4% of the listed flora) found on Pão de Açúcar, and 25 in the Serra do Mar (19.7% of the listed flora). As was noted above, the primary components of rupicolous mats on Pão de Açúcar change abruptly simply with altered aspect or slope of the rock surface. Observations (Safford) on summits in the Serra do Mar suggest that similar differences prevail in the south-eastern high-lands, although the magnitude of intrainselberg variation may be somewhat muted vis à vis the exaggerated turnover found in some lowland sites. Porembski et al. (1998) note that floristic affinities tend to be better developed between inselbergs of more northern parts of the east coast. We suggest that the more dynamic tectonic and paleoclimatic history of the coastal southeast has played an important role in producing this regional disparity (see sect. 10.8.3), not only by producing a more variable landscape, but by providing multiple opportunities over geological time for local adaptation to and from barren rock substrates (see Axelrod 1972).

Studies of inselberg vegetation often find little or nothing in common between the outcrop flora and the flora of the surrounding vegetational matrix. Following this paradigm, the floristic differences between inselberg and forest vegetation in southeastern Brazil are large, but compared to results reported from, e.g., Africa (Porembski et al. 1994, 1996; Porembski 1996), the great environmental gaps separating these two habitats seem to have been bridged by an uncommonly high number of taxa. For example, of the taxa listed for Pão de Açúcar in Table 10.8.3, 44 genera and 11 species (11.2% of the flora) are also found in the mid-montane forests of Macaé de Cima. Nine of the 11 shared species are facultative epiphytes, 8 of them from the families Araceae, Bromeliaceae, or Orchidaceae, families rich with examples of species that have adapted to physiological drought on both rock and treelimb. In contrast to the situation in the lowlands, environmental differences between highland outcrops and the surrounding campos de altitude are relatively slight, a fact reflected not only in highly similar life-form spectra (see above), but in taxonomic composition as well: of the 153 species listed for the Serra do

Mar inselbergs in Table 10.8.4, less than 95 do not also occur in other habitat types subsumed within the campos.

Although we are as yet limited in eastern Brazil by the absolute scarcity of basic floristic and biogeographic information (and happily this is beginning to change), we may compare the floristic data presented in this chapter with species lists from other locations outside of the coastal southeast to begin to enumerate those genera – and, where possible, species – that most characterize the inselberg/rock outcrop flora of humid, eastern Brazil (Table 10.8.5). It would be perhaps more satisfying to construct a list of inselberg species, but species widely distributed even among coastal southeastern inselbergs are few and far between, and those that are are almost invariably weeds, or species of relatively wide ecological amplitude that occur in various – albeit related – habitat types (e.g., inselbergs, restinga, campos rupestres, etc.) throughout eastern Brazil and often Latin America. Examples of widely distributed species of this type include *Achyrocline satureoides*, *Baccharis serrulata*, *B. trimera*, *Gamochaeta americana* (Asteraceae), *Tillandsia gardneri*, *T. geminiflora*, *T. stricta*, *Vriesea philippocoburgi*, *V. procera* (Bromeliaceae; all five species show great plasticity in habit as well, adopting terrestrial, epiphytic, rupicolous, or saxicolous life styles depending on the location and environmental conditions in question), *Zygopetalum mackayi* (Orchidaceae; terrestrial, epiphytic or saxicolous), *Borreria verticillata* (Rubiaceae), and *Anemia phyllitidis* (Schizaeaceae). Compared to sympatric congeners (and even, in many cases, of confamilials) of more restricted geographic distribution, these widespread taxa are often of greater local abundance on southeastern inselbergs, an affirmation of the "nonindependence" of abundance and range size recognized by many authors for many types of biota (Brown 1995). Note that even the "widespread" species listed above and/or in Table 10.8.5 are never found in all potentially colonizable sites on a given inselberg, nor even on all inselbergs in a given region.

Although the data are clearly incomplete, the Brazilian inselberg flora thus far appears to be significantly more diverse than that of almost any other region on earth (cf. Ibisch et al. 1995; Porembski et al. 1998). We can think of at least four reasons why the species pool of available rock outcrop colonizers may be larger in Brazil than elsewhere: (1) the predominant families on Brazilian inselbergs – the Bromeliaceae and Orchidaceae (and, to a lesser extent, the Araceae and Cactaceae) – are largely composed of taxa able to switch facultatively from epiphytic to rupicolous and saxicolous habitats; for genetically predisposed taxa, an inselberg's granite surface may simply represent an extension of the forest canopy. (2) The absolute number of barren rock slopes and inselbergs is probably greater in eastern Brazil than in any other part of the tropics, providing a

Table 10.8.5. Rupicolous and saxicolous taxa characteristically found on inselbergs of eastern Brazil (from southern Bahia to São Paulo, including eastern Minas Gerais)

Family	Genus	Widespread species (species distribution)
Adiantaceae	***Doryopteris***	
Apocynaceae	*Mandevilla*	
Araceae	***Anthurium* sect. *pachyneurium***	*A. coriaceum, A. solitarium* (SEBr)
Asteraceae	*Achyrocline, Baccharis*	*A. satureoides* (Ven-Arg), *B. serrulata* (BA-SP)
Bromeliaceae	*Billbergia,* ***Dyckia***[a], *Encholirium*[a], *Orthophytum, Pitcairnia,* ***Tillandsia, Vriesea***	*P. flammea* (SEBr), *T. gardneri, T. stricta* (NWSAm-E/SE/SBr) *V. philippocoburgi* (RJ-RS)
Cactaceae	***Cereus*** sensu lato	*Coleocephalocereus fluminensis* (SEBr)
Cyperaceae	*Bulbostylis, Rhynchospora,* ***Trilepis***	*T. ihotzkiana* (EBr)
Eriocaulaceae	*Eriocaulon*[a], *Paepalanthus, Syngonanthus*[a]	
Euphorbiaceae	*Croton, Phyllanthus*	*P. corcovadensis* (RJ-SP)
Fabaceae	*Cassia, Mimosa, Stylosanthes*	
Gesneriaceae	*Sinningia* sensu lato	
Lentibulariaceae	*Utricularia*	
Melastomataceae	*Tibouchina*	various
Ochnaceae	*Ouratea*	
Orchidaceae	*Cyrtopodium, Epidendrum Maxillaria, Oncidium,* ***Zygopetalum***	*C. andersonii* (Ven-SEBr), *M. acicularis* (BA-SC) *O. blanchetii* (SEBr), *Z. mackayi* (ES/MG-RS)
Piperaceae	***Peperomia***	
Poaceae	*Panicum*	
Polygalaceae	*Polygala*	*P. paniculata* (ES/MG-SC)
Rubiaceae	*Borreria*	*B. verticillata* (Ven-Uru)
Schizaeaceae	***Anemia***	*A. phyllitidis* (Mex/Car-Arg/Br)

Table 10.8.5 (*continued*)

Family	Genus	Widespread species (species distribution)
Selaginellaceae	*Selaginella*	*S. sellowii* (Bol-EBr)
Velloziaceae	***Barbacenia, Vellozia***	*V. variegata* (ES, SdMRJ, MG)
Verbenaceae	*Verbena*	
Xyridaceae	*Xyris*	

Genera in bold face type are "core" genera of the Brazilian inselberg flora, i.e., they are nearly always present.
[a] Taxa not found in Tables 10.8.3 or 10.8.4

Abbreviations: Arg = Argentina; BA = Bahia; Bol = Bolivia; Br = Brazil; Car = Caribbean; ES = Espírito Santo; Mex = Mexico; MG = Minas Gerais; RJ = Rio de Janeiro; RS = Rio Grande do Sul; SAm = South America; SC = Santa Catarina; SdMRJ = Serra do Mar of Rio de Janeiro; SP = São Paulo; Uru = Uruguay; Ven = Venezuela.

Data from: Angely (1970), Barroso (1973), Carauta and Oliveira (1982), Giulietti et al. (1987), Grandi et al. (1988), Hoehne (1953), Leoni (1997), Lima and Guedes-Bruni (1997a), Martinelli and Bandeira (1989), Martinelli and Vaz (1988), Mickel (1962), Miranda and Oliveira (1983), Pirani et al. (1994), Rambo (1956), Ruschi (1950), Smith and Ayensu (1976), and Safford (pers. observ.).

multiplicity of sites for colonization and local evolution, but also for metapopulation and source-sink dynamics that may "save" collapsing populations in periods of environmental stress. (3) Southeastern Brazil has experienced more tectonic uplift than perhaps any other inselberg-rich region, leading to altitudinally driven environmental differences between otherwise nearby inselbergs and subsequent genetic divergence within species. (4) Regional dynamics in Late Tertiary and Quaternary climates – particularly in the southeast – may have acted as a sort of evolutionary pump, alternately joining and dividing inselbergs and their rupicolous/saxicolous flora as forests contracted and expanded.

10.8.9 Biogeography and Endemism

Tables 10.8.3 and 10.8.4 provide data on the biogeographic distributions (chorology) of the genera found on Pão de Açúcar and the Serra do Mar summits. These data are summarized and compared to phytogeographic spectra from other eastern Brazilian vegetation types in Table 10.8.6. The phytogeographic groups employed in Table 10.8.6 were borrowed from Cleef (1979), and distributions determined using Mabberley (1996), various monographs, and herbarium collections at the Jardim Botânico do Rio de Janeiro and the Museu Nacional (Rio de Janeiro). In Table 10.8.6 columns 1 and 2 and columns 3 and 4 represent couplets of, respectively, the outcrop flora of Pão de Açúcar and its surrounding forest matrix (roughly approximated by the well-known midmontane forest at Macaé de Cima), and the Serra do Mar inselberg flora and its surrounding matrix of campos de altitude. Columns 5 and 6 represent vegetation types with marked environmental similarities to inselbergs: campos rupestres – montane grasslands and outcrop formations on quartzite ridges in the cerrado, and restinga – coastal beach and sand terrace vegetation.

Comparing the phytogeographic spectra for Pão de Açúcar and the Serra do Mar inselbergs in Table 10.8.6, the most obvious similarities are with respect to the neotropical group, the most obvious differences with respect to the remaining tropical groups and the total temperate component. The greater influence of the widespread tropical group in the lowland inselberg flora is due primarily to the presence of heliophilic weeds (e.g., *Emilia sagitatta*, *Vernonia scorpioides*, the urticaceous genera) common throughout the warm, lowland tropics. Perhaps most interesting is the much greater importance of eastern Brazilian genera in the highland sites, an indication of the largely montane nature of endemism in this corner of the continent. The proportions of the total tropical and total temperate

Table 10.8.6. Comparison of generic phytogeographic spectra for Brazilian inselbergs with other vegetation types of southeast Brazil

Phytogeographic Group	Vegetation type, and percentage of flora belonging to given phytogeographic group					
	(1) Lowland inselberg (Pão de Açúcar)	(2) Midmontane rainforest (Macaé de Cima)	(3) Highland inselberg (Serra do Mar)	(4) Campos de Altitude (Serra do Mar)	(5) Campo Rupestre (Cerrado Ambrosio)	(6) Restinga (Ilha do Cardoso)
Neotropical	41.9 %	55.6	37.6	32.7	53.1	45.9
Widespread tropical	38.4	23.2	23.8	19.6	31.3	38.7
Extra-Amazonian Brazil	3.5	9.2	12.9	10.1	9.4	3.6
Total tropical	83.8	88	74.3	62.4	93.8	88.2
Widespread temperate	1.2	1.1	5	10.1	0	0.5
Holarctic	1.2	0.5	0	4	0	0.5
Australantarctic	1.2	4.0	5.9	7	0	2.1
Total temperate	3.6	5.6	10.9	21.1	0	3.1
Cosmopolitan	12.8	6.3	14.9	16.6	6.3	8.8

Phytogeographic groups from Cleef (1979); see text for details.
Data: (1) This chapter (see Table 10.8.3); (2) Lima and Guedes-Bruni (1997a); (3) This chapter (see Table 10.8.4); (4) Rizzini (1954), Martinelli et al. (unpubl. data), Safford (unpubl. data); (5) Pirani et al. (1994); (6) De Grande and Lopes (1981).

components in the highland inselberg flora differ more from those of the campos de altitude than the same proportions at Pão de Açúcar differ from midmontane forest. This would seem somewhat ironic, given the much greater environmental similarity between the highland sites and surrounding campos, but none of the attendant temperate genera are rupicolous, so appropriate sites for colonization are almost completely lacking. The exceedingly warm climate of the southeastern lowland is clearly unsuitable for most taxa of temperate extraction. Of campos rupestres and restinga, the latter shows the closest biogeographic affinity to the inselberg sites, especially vis à vis the lowland spectrum from Pão de Açúcar, to which it shows a remarkable similarity. This broad congruity in phytogeographic composition underlines the important historic role that the restinga corridor has putatively played as both source and refuge for a significant portion of the inselberg flora in times of climatic change.

Going one step further than Table 10.8.6, and including a few other taxa typically found on southeastern inselbergs but not listed in Tables 10.8.3 and 10.8.4, we may divide the inselberg flora into subgroups based on biogeographic distributions within Brazil and Latin America. Biogeographic connections between the inselberg flora and restinga vegetation are very strong, with a variety of genera and even species inhabiting both coastal sands and lowland and/or highland rock outcrops. Here we may list: *Achyrocline satureoides*, *Anthurium*, *Baccharis serrulata*, *B. trimera*, *Borreria verticillata*, *Canavalia*, *Cereus*, *Clusia*, *Daphnopsis* (Thymelaeaceae), *Eriocaulon*, *Eugenia*, *Gamochaeta americana*, *Gaylussacia*, *Gomidesia*, *Hydrocotyle*, *Mimosa*, *Opuntia*, *Oxypetalum*, *Pouteria*, *Psidium*, *Rapanea ferruginea*, *Roupala lucens*, Mez, *Stylosanthes*, *Tillandsia geminiflora*, *T. stricta*, *Utricularia*, *Vriesea philippocoburgi*, *Weinmannia organensis*, and *Xyris*, among others.

Inselberg taxa shared with the campos rupestres of the cerrado include *Baccharis* (various species; see Barroso 1973), *Barbacenia*, *Eriocaulon*, *Lychnophora*, *Mimosa*, *Paepalanthus*, *Vellozia*, and *Xyris*. Biogeographic connections with campos rupestres are best expressed in the highlands of the Serras do Mar and da Mantiqueira, for both geographic and environmental reasons.

A number of genera – primarily from lowland inselbergs – are shared with the semiarid caatinga of northeastern Brazil (and the more nearby Cabo Frio area), as well as with the *cardonales* and *espinares* of the northern Venezuelan and Colombian coasts (Hueck 1966; Vareschi 1992). This group includes the genera *Capparis*, *Canavalia*, *Cassia*, *Coleocephalocereus*, *Cereus*, *Cleome*, *Croton*, *Ipomoea*, *Manihot*, *Melocactus*, *Opuntia*, *Tillandsia*, and the Bombacaceae, which are characteristic members of inselberg summit forests up to about 1600 m.

Two groups of taxa are found primarily or exclusively on highland inselbergs and in the surrounding campos de altitude and high montane forests; these are generally taxa adapted to cool, moist climatic conditions, and rocky or sandy substrates. The Andean group is comprised of genera disjunctly shared with the southern, maritime Andes and/or the high Andes of the equatorial tropics. Component inselberg taxa include *Abatia*, *Agarista* (primarily Brazilian), *Blechnum*, *Carex*, *Chionolaena* (Asteraceae), *Chusquea*, *Cortaderia*, *Gaultheria*, *Escallonia*, *Rapanea*, *Sisyrinchium*, *Symplocos*, and *Weinmannia*; many bryophytes show the same distribution pattern (Frahm 1991). Inselberg taxa primarily or entirely restricted to the southeastern highlands belong to the Brazilian Montane group, which includes the genera *Fernseea* (Bromeliaceae; restricted to the Rio de Janeiro-São Paulo border), *Hockinia*, *Microlicia* (Melastomataceae), *Prepusa*, *Schlumbergera*, and *Worsleya*, and, among others, the species *Abatia americana*, *Baccharis platypoda*, *B. stylosa*, *Burmannia aprica*, *Drosera villosa*, *Gomidesia kunthiana*, *Myrceugenia alpigena*, *Rapanea gardneriana*, *Vellozia variegata*, two species of *Chionolaena*, four species of *Agarista*, at least seven species of *Barbacenia* (the latter all restricted to single ranges or mountaintops), and ten of the twelve local and subregional (i.e., at scales much smaller than the state of Rio de Janeiro) endemics noted in Table 10.8.4.

As noted above (see sect. 10.8.6), southeastern Brazil is a hotbed of floristic endemism. Inselbergs of both the lowland and highland southeast appear to make significant contributions to both the character and magnitude of this endemism. Seven of 98 (7.1 %) of the species listed for Pão de Açúcar, and 12 of 153 (7.8 %) of the Serra do Mar species are local or subregional endemics, and most of the remaining species are restricted to Rio de Janeiro, the coastal southeast, or the Mata Atlântica. Barren, rocky sites are often centers of endemic populations (Major 1988), and indeed endemism among the southeastern inselberg flora not only seems to closely correlate with the rupicolous life style, but it is largely restricted to those few monocotyledonous neotropical families that have generally succeeded as rock outcrop colonizers, i.e., the Bromeliaceae, Orchidaceae, and Velloziaceae, and to a lesser extent, the Araceae. Five of the seven locally or subregionally endemic species on Pão de Açúcar are bromeliads, two are Velloziaceae (Table 10.8.3). In addition, the two *Anthurium* species (both from section *pachyneurium*, the only section of Araceae specifically adapted to xeric growing conditions; Croat 1991) are both basically restricted to the coastal southeast (*A. solitarium* has an outlying population in Goiás). Although there are no locally endemic orchids on Pão de Açúcar – interestingly, orchids are rarely local endemics – almost all of them are restricted either to the southeast, or to some slightly larger portion of the

Mata Atlântica. The endemic species in the Serra do Mar are less tied to barren rock substrates, but even so, all of the endemic taxa are either rupicolous (seven) or saxicolous (five); six of these species are bromeliads, one is a Vellozia. Significantly, of the remaining five endemics, two are members of monotypic genera restricted to a few mountaintops in the tiny Serra de Araras (*Worsleya rayneri* and *Glaziophyton mirabile*), and two others are from the succulent-leaved gentian *Prepusa*, another genus basically restricted to the Serra do Mar of Rio de Janeiro, with each of its five species limited to single or neighboring mountain tops.

10.8.10 Summary

1. High gamma (regional) diversity, the overall pool of species adapted to rock outcrops in southeastern Brazil, is comprised of both high alpha (local) diversity, and high intra- and interinselberg turnover in species composition (beta diversity); this is particularly the case for mat-forming species.
2. Reasons for high gamma diversity in the southeastern Brazilian inselberg flora are probably both taxonomic and environmental, and may include the predominance of facultatively rupicolous-epiphytic taxa, the absolute areal extent of barren rock, and the region-specific interworkings of Cenozoic tectonics and paleoclimates.
3. Typical rock outcrop adaptations like poikilohydry, succulence, and crassulacean acid metabolism are common in the inselberg flora of southeastern Brazil, but annual plants are remarkably rare. The near-complete absence of therophytes on southeastern inselbergs begs a sound explanation.
4. Floristic similarity between lowland and highland inselbergs in south-eastern Brazil is low, with shared taxa largely restricted to rupicolous/epiphytic families like Bromeliaceae, Orchidaceae, and Araceae. Interestingly, local endemisms are primarily within these families as well, as well as within the rock specialist Velloziaceae.
5. Important biogeographic connections exist between the Brazilian inselberg flora and various other vegetation types found on rocky or sandy substrates throughout South America, including the restinga of the Brazilian coast, the espinares and cardonales of Venezuela and Colombia, and the campos rupestres of the Brazilian Plateau. Highland inselbergs also show close affinities with the Andean Cordillera.

10.8.11 Postscript

The highly diverse and endemic inselberg flora of eastern Brazil is gravely threatened by the same short-sighted species that so deeply admires it. Anthropogenic disturbances long ago reached catastrophic proportions in most of the Mata Atlântica, but it is only recently that forest buffers in the topographically difficult terrain surrounding many southeastern inselbergs have begun to fall victim to roadbuilding, squatting, fire, grazing, and illegal extraction, the protected status of many of these sites notwithstanding. With the accelerated loss of these *tampões florestais*, cosmopolitan and pantropical weeds have begun to invade inselberg outcrops, cattle and goats graze their shoulders, and dry-season fire sweeps even their steepest brow. Between Niteroi/Rio de Janeiro and the southern suburbs of São Paulo, a distance of about 400 km, live approximately 35 million people, twice as many as in the similarly sized Philadelphia-NewYork-Boston metropole. The ironic – and grievously unfortunate – juxtaposition of the last stronghold of one of the world's great forests with what may be the fastest growing urban corridor on earth is now staging the final act in a tragedy of Hadean proportion.

Acknowledgements. The authors would like to thank S. Porembski (Universität Rostock), and M. Rejmánek (University of California-Davis) for critically perusing early drafts.

References

Ab'Sáber AN (1966) O domínio morfoclimático dos mares de morros no Brasil. Geomorfologia/USP 2:1–9

Ab'Sáber AN (1977) Espaços ocupados pela expansão dos climas secos na America do Sul, por ocasião dos periodos glaciais quaternários. Paleoclimas/USP 3:1–19

Ab'Sáber AN, Bernardes, N (1958) Vale do Paraíba, Serra da Mantiqueira e arredores de São Paulo. Guia de Excursão, 18° Congr Int de Geografia. Conselho Nacional de Geografia, Rio de Janeiro

AEERJ (Anuário Estatístico do Estado do Rio de Janeiro). 1993/1994. Centro de Informações de Dados do Rio de Janeiro, Governo do Estado do Rio de Janeiro

Almeida FFM (1976) The system of continental rifts bordering the Santos Basin, Brazil. An Adad Brasil Ciênc 48:15–26

Angely J (1970) Flora analítica e fitogeográfica do Estado do São Paulo. Edições Phyton, São Paulo

Axelrod I (1972) Edaphic aridity as a factor in angiosperm evolution. Am Nat 106: 311–320

Barroso GM (1973) Compositae – Subtribo Baccharidinae Hoffman. Estudo das espécies occorentes no Brasil. PhD Thesis, Universidade Estadual de Campinas, São Paulo

Barthlott W, Gröger A, Porembski S (1993) Some remarks on the vegetation of tropical inselbergs: diversity and ecological differentiation. Biogéographica 69:105–124

Baskin JM, Baskin CC (1985) Life cycle ecology of annual plant species of cedar glades of southeastern United States. In: Lieth H (ed) The population structure of vegetation. Handbook of vegetation science. Part III. Dr W Junk, Dordrecht, pp 371–398

Bigarella JJ (1991) Aspectos fisicos da paisagem. In: Cartesão J et al. (eds) Mata Atlântica. Editora Index, Fundação SOS Mata Atlântica, São Paulo, pp 63–93

Bigarella JJ, Mousinho MR, da Silva JX (1965) Processes and environments of the Brazilian Quaternary. Universidade de Paraná, Curitiba

Brade AC (1956) A flora do Parque Nacional do Itatiaia. Bol Parq Nac Itatiaia 5:1–112

Brown JH (1995) Macroecology. The University of Chicago Press, Chicago

Brown KS Jr, Ab'Sáber AN (1979) Ice-age forest refuges and evolution in the Neotropics: correlation of paleoclimatological, geomorphological and pedological data with modern biological endemisms. Paleoclimas 5:1–30

Calderón C, Soderstrom T (1980) The genera of Bambusoideae (Poaceae) of the American continent: keys and comments. Smithsonian Contrib Bot 44:1–27

Carauta JPP, Oliveira RR (1982) Fitogeografia das encostas do Pão de Açúcar, Rio de Janeiro. Flora Alguns Estados 2:9–31

Clapperton C (1993) Quaternary geology and geomorphology of South America. Elsevier, Amsterdam

Cleef AM (1979) The phytogeographical position of the neotropical vascular páramo flora with special reference to the Colombian Cordillera Oriental. In: Larsen K, Holm-Nielsen LB (eds) Tropical botany. Academic Press, London, pp 175–184

Consórcio Mata Atlântica (no date) Working map of forest remnants and management zones of the Atlantic Rainforest Biosphere Reserve.

Coutinho LM, Schrage CAF (1970) Sobre o efeito da temperatura na ocorrência de fixação noturna de CO_2 em orquídeas e bromélias. An Acad Bras Ciênc 42:843–849

Croat TB (1991) A revision of *Anthurium* sect. *pachyneurium* (Araceae). Ann Mo Bot Gard 78:539–855

Cuatrecasas J (1968) Páramo vegetation and its life forms. In: Troll C (ed) Geoecology of the mountainous regions of tropical America. Colloquium geographicum 9. Ferdinand Dümmler, Bonn, pp 163–186

Damuth JE, Fairbridge RW (1970) Equatorial Atlantic deep-sea arkosic sands and ice-age aridity in tropical south America. Geol Soc Am Bull 86:198–206

Dean W (1995) With broadax and firebrand. The destruction of the Brazilian Atlantic forest. University of California Press, Berkeley

de Grande DA, Lopes EA (1981) Plantas da restinga da Ilha do Cardoso (São Paulo, Brasil). Hoehnea 9:1–22

Eiten G (1982) Brazilian "savannas". In: Huntley BJ, Wallzer BH (eds) Ecology of tropical savannas. Ecological studies, vol 42. Springer, Berlin Heidelberg New York, pp 25–29

Fleig M (1990) Liquens saxícolas, cortícolas e terrícolas do Morro Santana. Rio Grande do Sul. II. Espécies e novas ocorrências. Pesqui Bot 41:33–50

Fonseca GAB de (1985) The vanishing Brazilian Atlantic forest. Biol Conserv 34:17–34

Frahm J-P (1991) Dicranaceae: Campylopodioideae, Paraleucobryoideae. Flora Neotropica. Monograph No. 54. New York Botanical Garden, New York

Giulietti AM, Menezes NL, Pirani JR, Meguro M, Wanderley MGL (1987) Flora da Serra do Cipó: caraterização e lista das espécies. Bol Bot/USP 9:1–151

Golte W (1993) *Araucaria*. Verbreitung und Standortansprüche einer Coniferengattung in vergleichender Sicht. Franz Steiner, Stuttgart

Grandi TSM, Siqueira JC, Paula JA (1988) Levantamento florístico da flora fanerogâmico dos campos rupestres da Serra da Piedade, Caeté, Minas Gerais. Pesq Bot 39:89–104

Guedes-Bruni RR, Lima HC de (1996) Serranias do estado do Rio de Janeiro – O conhecimento florístico atual e as implicações para a conservação da diversidade na Mata Atlântica. Eugeniana 22:9–22

Hoehne FC (1953) Flora Brasilica, vol XII. Orchideas. Instituto de Botânica. São Paulo

Hueck K (1966) Die Wälder Südamerikas. Ökologie, Zusammensetzung und wirtschaftliche Bedeutung. Gustav Fischer, Stuttgart

Ibisch PL, Rauer G, Rudolph D, Barthlott W (1995) Floristic, biogeographical, and vegetational aspects of Pre-Cambrian rock outcrops (inselbergs) in eastern Bolivia. Flora 190:299–314

IEF [Fundação Instituto Estadual de Florestas (RJ)]. 1994. Mapa da Reserva da Biosfera da Mata Atlântica. Estado do Rio de Janeiro, Rio de Janeiro

King LC (1967) The morphology of the earth: a study and synthesis of world scenery. Oliver and Boyd, London

Leme EMC, Marigo LC (1993) Bromélias na natureza. Marigo Communicações Visual, Rio de Janeiro

Leoni LS (1997) Catálogo preliminar das fanerógamas ocorrentes no Parque Nacional do Caparaó - MG. Pabstia 8:1–28

Lima HC de, Guedes-Bruni RR (1994) Reserva ecológia de Macaé de Cima. Nova Friburgo. Rio de Janeiro. Aspectos floristicos das especies vasculares, vol 1. Jardim Botânico do Rio de Janeiro

Lima HC de, Guedes-Bruni RR (1997a) Serra de Macaé de Cima: diversidade florística e conservação em Mata Atlântica. Jardim Botânico do Rio de Janeiro

Lima HC de, Guedes-Bruni RR (1997b) Diversidade de plantas vasculares na Reserva Ecológica de Macaé de Cima. In: Lima HC de, Guedes-Bruni RR (eds) Serra de Macaé de Cima: diversidade florística e conservação em Mata Atlântica. Jardim Botânico do Rio de Janeiro, pp 29–39

Lima HC de, Guedes-Bruni RR (1997 c) Plantas arbóreas da Reserva Ecológica de Macaé de Cima. In: Lima HC de, Guedes-Bruni RR (eds) Serra de Macaé de Cima: diversidade florística e conservação em Mata Atlântica. Jardim Botânico do Rio de Janeiro, pp 53–64

Luteyn JL (1989) Speciation and diversity of Ericaceae in Neotropical montane vegetation. In: Holm-Nielsen LB, Nielsen IC, Balslev H (eds) Tropical forests. Botanical dynamics, speciation and diversity. Academic Press, London, pp 297–310

Lynch JD (1979) The amphibians of the lowland tropical forest. In: Düllman WE (ed) The South American herpetofauna: Its origin, evolution, and dispersal. Mus Nat Hist Monogr 7, University of Kansas, Lawrence, pp 189–215

Mabberley DJ (1996) The plant book. A portable dictionary of the higher plants. Cambridge University Press, Cambridge

Major J (1988) Endemism: a botanical perspective. In: Myers AA, Giller PS (eds) Analytical biogeography. Chapman & Hall, London, pp 117–146

Martinelli G (1997) Biologia reprodutiva de bromeliaceae na Reserva Ecológia de Macaé de Cima. In: Lima HC de, Guedes-Bruni RR (eds) Serra de Macaé de Cima: diversidade floristica e conservação em Mata Atlântica. Jardim Botânico do Rio de Janeiro, pp 213–250

Martinelli G, Bandeira J (1989) Campos de altitude. Editora Index, Rio de Janeiro

Martinelli G, Vaz AMSF (1988) Padrões fitogeográficos em Bromeliaceae dos campos de altitude da floresta pluvial tropical costeira, no Estado do Rio de Janeiro. Rodriguésia 64/66:3–10

Mickel J (1962) An annotated list of the *Anemia* collection in the Herbário Barbosa Rodrigues. Sellowia 14:47–50

Miranda FELF, Oliveira RR (1983) Orquídeas rupícolas do Morro do Pão de Açúcar, Rio de Janeiro. Atas Soc Bot Brasil (RJ) 1:99–105

Modenesi MC (1988) Significado dos depósitos correlativos quaternários em Campos do Jordão-São Paulo: Implicações paleoclimáticas e paleoecológicas. Bol Inst Geol 7:1–155

Moreira AAN (1977) Relevo. Geografia do Brasil. Região Nordeste. IBGE, Rio de Janeiro, pp 1–46

Mori SA, Boom BM, Prance GT (1981) Distribution patterns and conservation of eastern Brazilian coastal forest tree species. Brittonia 33:233–245

Nimer E (1977) Clima. Geografia do Brasil. Região Sudeste. IBGE, Rio de Janeiro, pp 51–89

Oliveira-Filho AT, Machado JNM (1993) Composição florística de uma floresta semidecídua montana, na Serra do São José, Tiradentes, Minas Gerais. Acta Bot Bras 7:71–88

Öllgard B (1993) Lycopodiaceae – diversity in neotropical montane forests. In: Balslev H (ed) Neotropical montane forests. Biodiversity and conservation. AAU Reports 31. Aarhus University Press, Aarhus, pp 13

Osmond CB, Austin MP, Berry JA, Billings WD, Boyer JS, Dacey JWH, Nobel PS, Smith SD, Winner WE (1987) Stress physiology and the distribution of plants. Bioscience 37:38–48

Phillips DL (1982) Life forms of granite outcrop plants. Am Midl Nat 107:206–208

Pirani JR, Giulietti AM, Mello-Silva R, Meguro M (1994) Checklist and patterns of geographic distribution of the vegetation of Serra do Ambrósio, Minas Gerais, Brazil. Rev Bras Bot 17:133–147

Por FD (1992) Sooretama: the Atlantic rain forest of Brazil. SPB Academic Publishing, Den Haag

Porembski S (1996) Notes on the vegetation of inselbergs in Malawi. Flora 191:1–8

Porembski S, Barthlott W (1993) Ökogeographische Differenzierung und Diversität der Vegetation von Inselbergen in der Elfenbeinküste. Animal-plant interactions in tropical environments. Annu Meet German Society for Tropical Ecology, Results. Bonn, 1992, pp 149–158

Porembski S, Barthlott W, Dörrstock S, Biedinger N (1994) Vegetation of rock outcrops in Guinea: granite inselbergs, sandstone table mountains and ferricretes – remarks on species numbers and endemism. Flora 189:315–326

Porembski S, Szarzynski J, Mund J-P, Barthlott W (1996) Biodiversity and vegetation of small-sized inselbergs in a West-African rain forest (Taï, Ivory Coast). J Biogeo 23:47–55

Porembski S, Martinelli G, Ohlemüller R, Barthlott W (1998) Diversity and ecology of saxicolous vegetation mats on inselbergs in the Brazilian Atlantic rainforest. Divers Distrib 4:107–119

Rambo B (1956) A flora fanerogâmica dos Aparados Riograndenses. Sellowia 7:235–298

Ratisbona LR (1976) The climate of Brazil. In: Schwerdtfeger W (ed) Climates of Central and South America. World Survey of Climatology, vol 12. Elsevier, Amsterdam, pp 219–288

Rauh W (1978) Die Wuchs- und Lebensformen der tropischen Hochgebirgsregionen und der Subantarktis, ein Vergleich. In: Troll C, Lauer W (eds) Geoökologische Beziehungen zwischen der Südhalbkugel und den Tropengebirgen. Franz Steiner, Wiesbaden, pp 62–71

Raunkiaer C (1934) The life-forms of plants and statistical plant geography. Oxford University Press, London

Riccomini C (1989) O rift continental do sudeste brasileiro. Doctoral Thesis. Universidade de São Paulo

Richards PW (1996) The tropical rainforest. Cambridge University Press, Cambridge

Rizzini CT (1954) Flora Organensis. Arq Jard Bot Rio d J 13:116–246

Rizzini CT (1979) Tratado de fitogeografia do Brasil. Editora Universidade de São Paulo

Rundel PW, Smith AP, Meinzer FC (eds) (1994) Tropical alpine environments. Plant form and function. Cambridge University Press, Cambridge

Ruschi A (1950) Fitogeografia do Estado do Espírito Santo. Bol Mus Mello Leitão 1:1–353

Safford HD (1999) Brazilian páramos I. An introduction to the physical environment and vegetation of the campos de altitude. J Biogeogr 26:693–712

Schobbenhaus C, Campos DA, Derze GR, Asmus H (1984) Geologia do Brasil. Ministério das Minas e Energia, Brasília

Segadas-Vianna F, Dau L (1965) Ecology of the Itatiaia Range, southeastern Brazil. II. Climate and altitudinal climatic zonation. Arquiv Mus Nacional LIII:31–53

Smith LB, Ayensu ES (1976) A revision of American Velloziaceae. Smithsonian Contrib Bot 30:1–172

Smith LB, Downs RJ (1974) Pitcairnioideae (Bromeliaceae). Flora Neotropica, Monograph Number 14. New York Botanical Garden, New York

Thomas MF (1974) Tropical geomorphology. John Wiley, New York

Trompette R (1994) Geology of Western Gondwana (2000–500 Ma). Pan-African-Brasiliano aggregation of South America and Africa. AA Balkema, Rotterdam

Tryon RM (1972) Endemic areas and geographic speciation in tropical American ferns. Biotropica 4:121–131

Ururahy JCC, Collares JER, Santos MM, Barreto RAA (1983) Vegetação. Projeto Radambrasil. Folhas SF, 23/24, Rio de Janeiro/Vitória. Levantamento de Recursos Naturais 32. Ministério das Minas e Energia, Rio de Janeiro, pp 553–623

Valeton I, Beissner H, Carvalho A (1991) The Tertiary bauxite belt on tectonic uplift areas in the Serra da Mantiqueira, South-East Brazil. Contributions to Sedimentology Nr. 17. E Schweizerbart'sche Verlagsbuchhandlung, Stuttgart

Vareschi V (1992) Ecología de la vegetación tropical. Sociedad Venezolana de Ciencias Naturales, Caracas

Zomlefer WB (1994) Guide to flowering plant families. The University of North Carolina Press, Chapel Hill

10.9 Floristics of Australian Granitoid Inselberg Vegetation

S.D. Hopper

10.9.1 Introduction

Perhaps the most famous Australian landscape icon is the inselberg Uluru (Ayer's Rock), a 350 m tall arkose sandstone monolith emergent from the red sandplain of central Australia (Breeden 1994). Although rarely as big or striking in their insularity, inselbergs of granitoid rocks are found in local regions across Australia, being especially common and conspicuous in the flat terrain of Western Australia (Withers and Hopper 1997).

Constituting among the few reliable catchments of freshwater in a largely arid continent, granitoid inselbergs played a vital role in aboriginal and colonial European life. Today, this role continues in some areas, but granitoid inselbergs also are increasing in importance for nature conservation and tourism (e.g., Girraween and Bald Rock National Parks, McDonald et al. 1995; Wave Rock, Twidale and Bourne 1998). Granitoid inselbergs are found from the tropical north of Australia through the vast arid zone to the temperate southern margins of the nation. Consequently, inselberg vegetation is diverse and especially noteworthy in southwestern Australia for its high floristic endemism (Hopper et al. 1997).

This chapter provides a brief overview of Australian granitoid inselberg vegetation. Floristic knowledge of these inselbergs has accumulated over more than two centuries since naturalists accompanying maritime explorers first set foot on the granitic mainland and adjacent islands of southern Australia (e.g., Brown 1810). However, systematic study of the regional floristics of inselbergs is very much a recent research focus, and much remains to be investigated. This work is, therefore, more of a progress report than a definitive synthesis. Useful and more detailed reference works for Australian inselberg research are found in the recent granite outcrops symposium proceedings published by the Royal Society of Western Australia (Withers and Hopper 1997), and in the book Rock of Ages (Bayly 1999).

Ecological Studies, Vol. 146
S. Porembski and W. Barthlott (eds.) Inselbergs

10.9.2 Geography and Geology

Granitoid bedrock of varying age underlies vast areas of Australia. The continent comprises three stabilized crustal regions (cratons), western, central, and eastern, of decreasing age. Thus, most of eastern Australia has rocks 600 Ma or younger, substantial parts of central Australia date further back up to 2500 Ma ago, while even older rocks dominate cratons in the west. Indeed, Western Australia contains the oldest known fragments of the earth's crust – minute crystals of zircon from Jack Hills-Mt. Narryer's granite rocks that began to solidify 4300 Ma ago, just 300 Ma after the origin of the planet itself. The granitoid gneiss containing these crystals is among the oldest rocks known, formed 3600 Ma ago (Myers 1997). The youngest Australian granite rocks are found on the east coast, e.g., in northeast Queensland, aged 230–310 million years (Willmott and Stephenson 1989).

Inselbergs occur in all areas of Australian granitoid bedrock, and display the range of forms (bornhardts, nubbins, castle koppies, etc.) found on other continents (Campbell 1997). They are emergent from landscapes that have had varying environmental histories, ranging from recent glaciation and tectonic uplift in the Australian Alps and northeast Tasmania, to prolonged tectonic stability and the absence of glaciation since the Permian in southwestern Australia.

10.9.3 Flora, Climate and Vegetation

The Australian flora differs dramatically at the species and generic level from that elsewhere, due to the Gondwan origins of significant components and a long period of insular evolution in the Tertiary (Barlow 1981). An estimated 25 000 species of vascular plants (10 % of the global total) occur in Australia, with at least 85 % endemism. At the level of plant family, Australia shows very little endemism, primarily because the evolution of most families predated the fragmentation of Gondwana. Cosmopolitan weeds introduced from other continents number in excess of 3000 species in Australia (an estimated 15 % of the flora – Humphries et al. 1991).

From the earliest analysis of the composition of the native Australian vascular flora by Hooker (1860), three key floristic elements have been identified:

1. The bulk of species (15 000+) are endemic evergreen sclerophyllous plants of forests, woodlands, mallee (i.e., lignotuberous multistemmed

eucalypts), shrublands, sedgelands, and grasslands. The continent is especially noteworthy in the dominance of two woody but evergreen genera – *Eucalyptus* (800+ species) and *Acacia* (1000+ species). Spinifex hummock grasses of the genus *Triodia* are prominent over vast desert areas. Species-rich Gondwanan families in Australia include the Proteaceae (*Banksia*, *Grevillea*, etc.), Myrtaceae (*Eucalyptus*, *Melaleuca* etc.), Epacridaceae (southern heaths), and Restionaceae (southern rushes). Occupying about 5% of the continent, the southwest of Western Australia with mediterranean climate is especially rich in this major Gondwanan element of the Australian flora. It has an estimated 8000 species, 75% endemic to the region itself (Hopper 1979, 1992). In contrast, the extensive central Australian desert region occupying one third of the continent has only 2000 species (Jessop 1981).

2. An ancient rainforest element that covered only 1% of the Australian landmass at European colonization. Rainforests occur in scattered sites from Tasmania to north Queensland and westwards across the tropical Northern Territory to the Kimberley region of Western Australia, and contain upwards of 2000 species, many endemic to Australia (Tracey 1981; Webb et al. 1984). The rainforests are living Gondwan museums, i.e., fragmented and depleted relicts of vegetation that covered much of the continent prior to the onset of late-Tertiary aridity when Australia drifted north away from Antarctica. These rainforest patches differ significantly in composition, with three major floristic groups recognized – cool-wet temperate rainforests of Tasmania, Victoria, and New South Wales, hot-wet subtropical-tropical rainforests from near Sydney north to Queensland and the wettest parts of the Northern Territory, and hot-dry semideciduous or deciduous rainforests and vine thickets extending from the Kimberley across to north Queensland and south into semiarid New South Wales.
3. A cosmopolitan element of up to 3000 species occupying coastal habitats, saltlands, wetlands, and alpine or mountainous areas. Typically, endemism is lower in this component of the flora.

Two thirds of Australia is arid (Barker and Greenslade 1982), with less than 300 mm annual rainfall, including the Pilbara region of the northwest and the southern Nullarbor Plain where the desert meets the coast. Arid Australia is unusually well vegetated, with the deserts dominated by spinifex (*Triodia*) hummock grasslands and scattered low shrubs, mallees, or small trees. Extensive red sand dune systems are interspersed among low rocky ranges and hills throughout much of the arid zone. Large salt lakes occupy desert basins and support fringing halophytic vegetation of samphires (chenopods) and other salt-tolerant plants. The arid zone flora is relatively recent in origin, with few endemics (Barlow 1981).

Surrounding the arid zone on three sides are broad semiarid belts in which woodlands, mallee, and sclerophyll shrublands are prevalent (Groves 1994). Again, salt lakes with uncoordinated drainage are common in broad valley floors, particularly in the west and south of the semi-arid zone, but the east is occupied by the bulk of Australia's largest river system, the Murray-Darling. The temperate semiarid zone is rich in Australian endemics, particularly the southwest with its kwongan shrublands (Pate and Beard 1984), and the mallee regions of Western Australia, South Australia, Victoria, and New South Wales (Noble and Bradstock 1989; Noble et al. 1991).

The wetter parts of Australia occupy less than a fifth of the continent, south and eastwards of the eastern highlands, across the tropical north, and in a small isolated region of the southwest. Vegetation is complex, matching the topography and geology of these areas. The wettest places with deep fertile soils in eastern and northern Australia are occupied by rainforests, while shallow soils in the highest rainfall areas have stunted woodlands and heaths. Wet sclerophyll forests occupy fire-prone sites, and contain immense hardwoods such as mountain ash (*Eucalyptus regnans*) in the southeast and karri (*E. diversicolor*) in the southwest. Mountains support alpine vegetation and stunted sclerophyll communities. Less fertile soils on rock outcrops throughout the wetter parts of Australia also have sclerophyll shrublands and stunted woodlands. Freshwater lakes and streams are strongly seasonal, as are coastal estuaries and embayments into which the latter discharge (McComb and Lake 1990).

Coastal floras include mangroves and the usual cosmopolitan plants of dunes and strand (e.g., Cribb and Cribb 1985), while adjacent Australian marine environments are noteworthy in the north for their coral communities of the Great Barrier Reef on the east coast and such places as Ningaloo Reef on the west coast. Temperate marine algal floras are among the richest in the world (Womersley 1981, 1990).

10.9.4 Inselberg Vegetation – Life-Forms, Adaptations, Systematics

Habitats on Australian inselbergs include bare rock, cryptogamic crusts, gnammas (rock pools or weather pits, seasonal and permanent, Fig. 10.9.1), soil-filled crevices, caves/tafoni/shade of boulders or exfoliated

Fig. 10.9.1. On Western Australian inselbergs gnammas comprise a highly specialized flora including *Isoetes*, *Glossostigma*, and *Myriophyllum*. (Photograph N. Biedinger)

slabs, herbfields on shallow well-drained soil (Fig. 10.9.2), herbfields on shallow seasonally waterlogged soil (ephemeral flush vegetation), shrublands and woodlands on deep well-drained soil, shrublands, woodlands and forests on deep seasonally waterlogged soil, and permanent springs. Occasionally, significant watercourses and fringing salt lakes will be found adjacent to Australian inselbergs.

Granite outcrop habitats have elicited remarkable parallel evolution in many plants and animals (Mares 1997; Porembski et al. 1997; Wyatt 1997). The combination of high solar radiation, rapid runoff of rainfall, and shallow soils on a rocky substrate provide microhabitats of accentuated seasonal and diurnal stresses. Conditions may vary over a few meters from cool permanently moist shaded caves with water seepage to drought-afflicted shallow soils and rock surfaces fully exposed to all the elements.

Few organisms can tolerate the harshest of these rock-surface habitats, which tend to be occupied by cryptogamic crusts of cosmopolitan cyanobacteria, lichens, and mosses such as *Grimmia laevigata*. Australian outcrops conform to this trend, but have not been especially well studied. Given the climatic diversity across the nation where inselbergs are found,

Fig. 10.9.2. Depressions filled with shallow soil are characterized by a species-rich community of mainly annual plants. (Photograph N. Biedinger)

cryptogamic diversity is likely to vary. Pigott and Sage (1997) listed 33 lichen species from 11 families on a rock in semiarid Western Australia. On southern Australian outcrops experiencing high rainfall, moss cushions become prominent, with the luxuriant brown carpets of *Breutelia affinis* and the verdant green *Campylopus bicolor* and *C. australis* noteworthy.

Noticeably absent from Australian inselbergs are monocotyledonous mats on bare rock, formed by species of Cyperaceae and Velloziaceae on tropical inselbergs in Africa and Madagascar, and by species of Bromeliaceae and Velloziaceae in South America (Porembski et al. 1997). Also absent in Australia are mats of the fern ally *Selaginella* as seen in East Africa, Brazil, and the Appalachian granite outcrops of the USA (Porembski et al. 1997; Wyatt 1997).

Coping with seasonal or unpredictable drought is, undoubtedly, the most significant survival strategy faced by granite outcrop plants. Among Australian herbaceous perennials on outcrops, pincushion lilies (*Borya* spp., Boryaceae, Fig. 10.9.3) are abundant and remarkable in their capacity as resurrection plants (Gaff 1981) to withstand dessication to less than 5 % of normal leaf moisture content, turning orange in the process, and

Fig. 10.9.3. Several species of *Borya* (pincushion lilies) occur as resurrection plants on mainly Western Australian outcrops. (Photograph N. Biedinger)

rehydrating to normal green leaves within a day of rainfall. There are at least seven species of *Borya* found on Australian inselbergs. These plants do not extend to arid-zone outcrops. Other resurrection plants commonly seen on Australian granites include the rock ferns *Cheilanthes* spp. and *Pleurosorus rutifolius*.

Persisting underground as a tuber during drought is a common strategy among Australian granite outcrop herbs, especially lilioids (e.g., *Wurmbea* spp., Colchicaceae), orchids (*Caladenia, Thelymitra, Pterostylis, Diuris, Prasophyllum*), and sundews (*Drosera*), as well as the unusual fern ally *Phylloglossum* (Pate and Dixon 1982). On northern Australian outcrops, tuberous Cyperaceae tend to occupy this niche (e.g., *Cyperus bulbosa*).

Carnivorous plants are also a feature of Australian tuberous inselberg plants. Southwestern Australian inselbergs have at least 15 species of *Drosera* ranging in form from rosetted (e.g., *D. bulbosa*) to scrambling climbers up to 1 m tall (e.g., *D. macrantha*). There is a single annual species in this genus on inselbergs (*D. glanduligera*). Ephemeral flush herb fields also commonly have species of *Utricularia* (Lentibulariaceae), while *Byblis* may be found on tropical inselbergs together with species such as *Drosera indica*.

Succulence is not prominent in the Australian granite outcrop perennial floras, but has evolved in a few taxonomically unrelated species, e.g., *Spiculaea ciliata* (Orchidaceae), *Carpobrotus* spp. (Aizoaceae), baobab (*Adansonia gibbosa*, Bombacaceae), *Sarcostemma australe* (Asclepiadaceae), and *Hoya* spp. (Asclepiadaceae). Outcrops that abut salt lakes have various perennial succulents in such genera as *Halosarcia* (Chenopodiaceae), *Tetragonia* (Aizoaceae), and *Zygophyllum* (Zygophyllaceae). Some perennial shrubs of south coast granites have unusually thick leaves (e.g., *Hakea clavata*, Proteaceae; *Anthocercis viscosa*, Solanaceae), and at least one has tuberous roots (*Calothamnus tuberosus*, Myrtaceae). In the tropics some normally epiphytic genera become epilithic to take advantage of the extra light on inselbergs. Species of orchid genera such as *Dendrobium* feature here.

The graminoid habit has evolved in several unrelated Australian groups, with common perennials forming tussocks including diverse Cyperaceae (especially *Lepidosperma* spp.), blind grass (*Stypandra* spp., Phormiaceae), the sedge-like grass *Spartochloa scirpoidea* (Poaceae), and lemon grass *Cymbopogon* spp. (Poaceae). Desert and subtropical outcrops are dominated by uniquely Australian hummock grasses with inrolled pungent leaves (*Triodia* spp., Poaceae).

Drought avoidance through an annual life history is a major feature of Australian granite herbs in families such as the Asteraceae, Stylidiaceae, Poaceae, Goodeniaceae, Amaranthaceae, and Centrolepidaceae. Most of these annuals have relatively small inconspicuous flowers suggestive of inbreeding (Ornduff 1987), but some are brightly colored and adundant. Most outcrops in southern Western Australia are bedecked with colorful swards of annual triggerplants (*Stylidium*, Stylidiaceae) and everlastings (e.g., *Rhodanthe, Waitzia*, Asteraceae) that enliven the herbfields. Succulence is a feature of some annuals, especially those that occupy the dry spectrum of shallow soils (e.g., *Calandrinia* spp., Portulacaceae; *Crassula* spp., Crassulaceae, Fig. 10.9.4).

Annuals are also found commonly in seasonally water-logged shallow soils or ephemeral flush communities (Pignatti and Pignatti 1994). These annuals include species of Centrolepidaceae (*Centrolepis* and *Aphelia*), Apiaceae (*Hydrocotyle*), Juncaginaceae (*Triglochin*), Cyperaceae (*Schoenus, Isolepis, Cyperus*), Asteraceae (*Quinetia, Millotia, Rutidosis, Siloxerus, Toxanthes, Hyalosperma, Podotheca* etc.), and Stylidiaceae (*Stylidium, Levenhookia*).

Deeper inselberg soils enable survivorship of woody perennials, of which the tallest on Australian inselbergs outside rainforests include eucalypts (*Eucalyptus* and *Corymbia*, Myrtaceae), wattles (*Acacia*, Mimosaceae), she-oaks (*Allocasuarina*, Casuarinaceae), and rock figs (*Ficus*,

Fig. 10.9.4. Succulent annuals, like *Calandrinia* spp. and *Crassula* spp., are typical colonizers of shallow soils. (Photograph N. Biedinger)

Moraceae). In Western Australia, species richness of inselberg woody perennials is greatest in the Myrtaceae (at least 162 taxa, including 44 *Eucalyptus*, 26 *Melaleuca*, and 20 *Verticordia*), Mimosaceae (82+ taxa, all *Acacia*), Papilionaceae (66+ taxa) and Proteaceae (54+ taxa). The inselberg woody perennials are depauperate in Epacridaceae, which in other habitats is among Australia's largest plant families. Sclerophyllous leaves characterize the vast majority of woody perennials found on Australian granite outcrops outside rainforests.

A noteworthy feature of these woody perennials in southwestern Australia is the high proportion of bird-pollinated species (Hopper 1981) in families such as Myrtaceae (e.g., *Eucalyptus, Calothamnus, Melaleuca, Verticordia*), Proteaceae (*Grevillea, Hakea, Banksia*), Myoporaceae (*Eremophila*), Fabaceae (*Kennedia, Crotalaria*), Loranthaceae (*Amyema, Lysiana*), and Sterculiaceae (*Brachychiton*). Also, southwestern outcrops have an unusually high number of woody perennials that are obligate seeders – plants that are killed by fire and recruit only from seed. For example, 77 % of the perennials in a granite community at Chiddarcooping Nature Reserve northeast of Merredin in Western Australia were obligate seeders (S.D. Hopper et al., unpubl.).

Gnammas on Australian inselbergs (Fig. 10.9.1) have a unique flora comprising annual species of quillworts (*Isoetes*, Isoetaceae), mudmats (*Glossostigma*, Scrophulariaceae), millfoils (*Myriophyllum*, Haloragaceae) and others, including the introduced South African annual *Crassula natans* (Crassulaceae).

Ornduff (1987) and Ohlemüller (1997) found that introduced weed species, mainly annuals, comprised 23.7 and 17.0% of the granite herb field floras they sampled respectively in Western Australia. These figures are high compared with the ca. 9% that weeds represent of the Western Australian flora as a whole (Green 1985; Keighery 1995). They are even more significant when considered with granite outcrop herbfield floras elsewhere on earth, which have far fewer invasive weeds (Porembski et al. 1997; Wyatt 1997).

Weed growth is most pronounced in full sun on inselbergs, especially where soil has been disturbed and enriched by rabbit dung or agricultural activity. In these situations, annual grasses such as *Briza maxima*, *Avena fatua*, and *Ehrharta longifolia* often dominate and replace native annuals throughout much of southwestern Australia. Weeds are rare only where a dense shrub layer or low forest of native woody perennials persists. It is clear that a persistent seed rain of weeds occurs over vast regions of Australia, as weeds appear in open areas on outcrops where little or no disturbance is evident and native plants dominate.

Hopper et al. (1997) have attributed the high invasibility of disturbed Western Australian plant communities to the absence of major glacial soil stripping as an evolutionary force acting on the flora. Native species are unable to compete against weeds from habitats where soil disturbance is a regular perturbation. Further experimental study of weeds in granite herbfields could test this hypothesis, and assist attempts at restoration of invaded outcrops.

10.9.5 Biogeography, Endemism, and Speciation

Inselberg vegetation in Australia reflects the Gondwan origin of the continent as a whole. Endemism is high at the species level, reflecting the long period of insular evolution in the Tertiary and Quaternary, but some genera and almost all families are shared with other Gondwan land masses, especially South America, New Zealand, Africa, and Madagascar. Tropical inselbergs also share species and genera with New Guinea and Indonesia, not unexpectedly given the relatively recent collision of the Australian and Southeast Asian continental plates.

As documented for other continents, Australian inselberg floras grade from showing little discontinuity in species from the surrounding landscape matrix in the vast arid zone to a highly discordant aggregation of species rich in local endemics in high-rainfall temperate forests and, to a lesser extent, in tropical forests (Webb 1972; McDonald et al. 1995; Hopper et al. 1997; Hunter and Clarke 1998).

Species richness patterns mirror those for the flora at large, being greatest on southwest Australian outcrops, regularly exceeding 100 taxa on rocks 0.5 ha or larger, and sometimes 200+ taxa on larger inselbergs (Hopper et al. 1997). These outcrops appear to be the richest recorded on earth. Reasons include a stable terrestrial landscape that has not been glaciated since the Permian, and an insular environment that has experienced repeated climatic change through the late Tertiary and Quaternary conducive to plant speciation (Hopper 1979, 1992).

In contrast, inselbergs of a similar size, climate, and surrounding forest matrix in Tasmania have fewer species and lower endemism, reflecting the generalist postglacial flora of the island as a whole. Inselbergs in Victoria and on the New England batholith are intermediate in species richness between those of Tasmania and of southwestern Australia (Webb 1972; McDonald et al. 1995; Hunter and Clarke 1998).

Western Australian granite outcrops display plant biogeographic patterns that mirror that of the whole flora (Hopper 1979, 1992; Hopper et al. 1996) – species richness and endemism are pronounced in the transitional rainfall zone (wheat belt) of the southwest, and attenuate as rainfall decreases through the pastoral country to the deserts, Pilbara and Kimberley. For example, the total number of vascular plant taxa recorded by the author and colleagues on a sample of rocks ranged from 142 (Point Matthew, near Augusta) and 201 (Mt. Frankland) in the highest rainfall forests, 192 (Mt. Ney) and 187 (Yilliminning Rock near Narrogin, Pigott and Sage, 1997) in the transitional rainfall zone, to 85 (Daggar Hills, near Yalgoo) in the pastoral zone, and 90 (Moolyella Rocks, east of Marble Bar) and 80 (Spear Hill, west of Marble Bar) in the arid Pilbara.

There are no angiosperm species shared between northern (Kimberley, Pilbara) outcrops and those in the southwest. Moreover, the northern outcrop floras are virtually identical with that from the matrix of surrounding terrain, save for outcrop specialists like rock figs and rock ferns. Walters and Wyatt (1982) similarly recorded low endemism and little discontinuity between vascular plants on granite outcrops and adjacent landforms of the arid Central Mineral Region of Texas. In contrast, southwestern and adjacent pastoral zone rocks have higher levels of local endemism, especially in the high-rainfall forest region, where the outcrops present the most striking difference in habitat to the surrounding vegetation matrix (e.g., Brooker and Margules 1996; Wardell Johnson and Williams 1996).

Biogeographical relationships of outcrop floras across the southwest are under ongoing study by the author. As a precursor, Hopper and Brown (in Hopper et al. 1997) documented the distribution of 126 orchid taxa on 41 outcrops ranging from the highest rainfall forests through the transitional zone wheatbelt to the arid zone. Each rock outcrop was treated as a site in a classification of the orchid data.

The study highlighted a number of significant trends. A primary division occurred between the 15 rocks found in the forested High Rainfall Zone (>800 mm a^{-1}) and the rest ranging from the Transitional Rainfall Zone of the wheat belt into the Arid Zone. Subsequent divisions established as much difference among forest rocks as among the wheat belt/arid rocks, even though the forest rocks were confined to a much smaller area. Moreover, remarkably, closely adjacent rocks were widely separated in the classification, indicating significant differences in their orchids [e.g., 10.3 km rock and 10.9 km rock of Ornduff's (1987), separated by just 600 m of jarrah forest on Albany Highway]. Conversely, rocks separated geographically often had similar orchid floras (e.g., Boyagin Rock and Pingaring Rock on the western and eastern sides of the south-central wheat belt, respectively).

These patterns suggest significant barriers to orchid dispersal, particularly between forest rocks, and high levels of local extinction and stochastic events underlying the presence of orchids on individual rocks in the southwest and adjacent arid zone. There have been dynamic climatic fluctuations across the southwest for several million years as Australia drifted northwards and arid conditions overtook much of central Australia (Hopper 1979; Hopper et al. 1996). The diversity of microhabitats on granite outcrops provided refuge for plants adapted to both dry or wet conditions as the surrounding matrix waxed and waned climatically (Marchant 1973). Survivorship in small populations on granite refuges undoubtedly was a matter of chance in the face of such repeated climatic turmoil.

Interestingly, the above conditions of small disjunct outcrop populations undergoing recurrent stresses is predicted as ideal for genetic divergence and speciation (Grant 1981). Is this prediction borne out by studies of WA outcrop plants? Western Australian granite outcrops have 141 recorded orchid taxa. Of these, 22 (16%) are more or less endemic. The endemics have geographical ranges from widespread on outcrops throughout the southwest (e.g., *Spiculaea ciliata*) to highly restricted to a few adjacent outcrops less than 10 km apart (e.g., *Caladenia caesarea* ssp. *maritima*, *Thelymitra dedmaniarum*).

In terms of evolutionary origins, these endemics display at least three patterns: (1) relictual, with no obvious close relatives and therefore likely to have been on granites for a long period of time (e.g., *Spiculaea ciliata*, a monotypic genus); (2) derived by speciation from allopatric congeners of habitats other than granite (e.g., *Caladenia granitora*, from coastal granites east of Albany, sister species to *C. infundibularis* of western high-rainfall forests and coastal heaths; *C. hoffmanii* ssp. *graniticola*, of east-central wheat-belt outcrops, sister to *C. hoffmanii* ssp. *hoffmanii* of lateritic loams well to the northwest in the Northampton region); and (3) derived by speciation from allopatric congeners of other granite outcrops (e.g., *C. exstans*, from outcrops east of Esperance, sister to *C. integra* of western wheat-belt outcrops). Thus, the orchid data do indeed support the hypothesis that conditions on southwest granite outcrops have facilitated genetic divergence and speciation. Genetic studies of a few other granite taxa lend further support, e.g., the herb *Isotoma petraea* (James 1965, 1970, 1982, 1984, 1992; James et al. 1983, 1989, 1990; Bussell and James 1997), and eucalypts endemic to granite (Hopper and Burgman 1983; Sampson et al. 1988). Clearly, more work along these lines is needed.

A start has also been made in the Wet Tropics World Heritage Area of north Queensland by Dr. Julia Playford (pers. comm.). She has examined genetic divergence among mountain-top species of *Rhododendron, Agapetes, Austromyrtus*, and *Cryptocaria* on rocky outcrops (mainly granite) and found a recurring division between populations on mountains to the south and north of Cairns.

There are a large number of granite endemics in southwestern Australia, especially among the perennials that dominate the woody vegetation and herbfields. Hopper et al. (1997) have shown that 16% of orchids on outcrops are endemic. For eucalypts, around 24% are endemic (S.D. Hopper, unpubl.). The level of endemism for the whole Australian inselberg flora is difficult to determine without more penetrating research, but there is no doubt that southwestern Australia has higher levels than any other system documented (e.g., Walters and Wyatt 1982; Porembski et al. 1995; this Vol.).

The refugial opportunities offered by Australian granite inselbergs are also evident in species that have highly disjunct outliers well removed from the main geographical distribution. Examples from southwest Australian inselberg eucalypts include populations on wet outcrop sites in much lower rainfall areas than the main species' stand (e.g., the Jilakin Rock stand of jarrah *Eucalyptus marginata*, the Twine Rock stand of *E. wandoo*, and the Kuendar stand of *E. rudis*). Conversely, arid-adapted species penetrate high rainfall areas on dry north-facing slopes of granites (e.g., populations of *Eucalyptus drummondii* west of Margaret River).

10.9.6 Biodiversity of Australian Inselbergs

In conclusion, Australian inselbergs vary significantly in the relationship between regional and local floristic diversity. Relatively recently occupied climatic regions such as postglacial Tasmania and the vast Australian arid center display species-poor inselberg vegetation with very little local endemism and low beta diversity.

In contrast, inselbergs of southwestern Australia have enjoyed similar climatic conditions for many millions of years, and display high alpha and beta floristic diversity. Even common dominants may appear and disappear across just a few kilometers between adjacent outcrops [e.g., the large inselbergs Kokerbin Rock, Mt. Caroline, and Mt. Stirling in the central wheat belt within 10 km of each other either have (Kokerbin Rock, Mt. Caroline) or do not have (Mt. Stirling) the 3–5 m tall dominant shrub *Calycopeplus ephedroides* (Euphorbiaceae)].

Ornduff (1987) and Ohlemüller (1997) found significant variation in floristic diversity of herbfields in southwestern Australia, with species richness and alpha diversity increasing from high rainfall to semiarid (transitional rainfall) areas, and then declining in the arid zone. However, gamma diversity between outcrops was remarkably higher in the high rainfall region, perhaps reflecting the barrier to migration posed by tall eucalypt forest surrounding inselbergs. While some interesting statistical correlations such as these are emerging, experimental studies are required to gain an incisive understanding of the causes of floristic biodiversity patterns on Australian inselbergs.

Acknowledgements. Many colleagues have helped with fieldwork, discussion and ideas since my first encounter with Australian inselberg vegetation in 1972. Too numerous to name, I thank them all. John Hunter kindly made available an unpublished manuscript relevant to this chapter. Robert Wyatt and Robert Ornduff were supportive hosts during my tenure as a Fulbright Senior Scholar and Miller Visiting Research Professor in the USA in 1990, and greatly facilitated development of an international perspective on granite outcrop plants. Stefan Porembski (Rostock) has recently broadened this international perspective, and been an extremely patient editor as well. I am grateful for the support of these colleagues.

References

Barker WR, Greenslade PJM (eds) (1982) Evolution of the flora and fauna of arid Australia. Peacock Publications, Adelaide

Barlow BA (1981) The Australian flora: its origin and evolution. Flora of Australia 1:25–75

Bayly IAE (1999) Rocks of ages. University of Western Australia Press, Nedlands
Breeden S (1994) Uluru looking after Uluru-Kata Tjuta – the Anangu Way. Simon and Schuster, East Roseville, New South Wales
Brooker MG, Margules CR (1996) The relative conservation value of remnant patches of native vegetation in the wheatbelt of Western Australia: I. Plant diversity. Pac Conserv Biol 2:268–278
Brown R (1810) Prodromus Florae Novae Hollandiae et Insulae Van-Diemen exhibens characteres plantarum quas annis 1802–1805. Taylor, London
Burgman MA (1987) An analysis of the distribution of plants on granite outcrops in southern Western Australia using Mantel tests. Vegetatio 71:79–86
Bussell JD, James SH (1997) Rocks as museums of evolutionary processes. In: Withers PC, Hopper SD (eds) Granite outcrops symposium. J R Soc West Aust 80:221–229
Campbell EM (1997) Granite landforms. In: Withers PC, Hopper SD (eds) Granite outcrops symposium. J R Soc West Aust 80:101–112
Cribb AB, Cribb JW (1985) Plant life of the Great Barrier Reef and adjacent shores. University of Queensland Press, St Lucia
Gaff DF (1981) The biology of resurrection plants. In: Pate JS, McComb AJ (eds) The biology of Australian plants. The University of Western Australia Press, Nedlands, pp 114–146
Grant V (1981) Plant speciation, 2nd edn. Colombia University Press, New York
Green JW (1985) Census of the vascular plants of Western Australia. Western Australia Herbarium, Department of Agriculture, Perth
Groves RH (ed) (1994) Australian vegetation. 2nd edn. Cambridge University Press, Cambridge
Hooker JD (1860) Introductory essay: Botany of the Antarctic voyage of H.M. Discovery ships Erebus and Terror in the years 1839–1843, III. Flora Tasmaniae, Reeve, London
Hopper SD (1979) Biogeographical aspects of speciation in the south west Australian flora. Annu Rev Ecol Syst 10:399–422
Hopper SD (1981) Honeyeaters and their winter food plants on granite rocks in the central wheatbelt of Western Australia. Aust Wildl Res 8:187–197
Hopper SD (1992) Patterns of diversity at the population and species levels in south-west Australian mediterranean ecosystems. In: Hobbs RJ (ed) Biodiversity of Mediterranean ecosystems in Australia. Surrey Beatty, Sydney, pp 27–46
Hopper SD, Burgman MA (1983) Cladistic and phenetic analyses of phylogenetic relationships among populations of *Eucalyptus caesia*. Aust J Bot 31:161–172
Hopper SD, Harvey MS, Chappill JA, Main AR, Main BY (1996) The Western Australian biota as Gondwanan heritage – a review. In: Hopper SD, Chappill JA, Harvey MS, George AS (eds) Gondwanan heritage: past, present and future of the Western Australian biota. Surrey Beatty, Chipping Norton, New South Wales, pp 1–46
Hopper SD, Brown AP, Marchant NG (1997) Plants of Western Australian granite outcrops. In: Withers PC, Hopper SD (eds) Granite outcrops symposium. J R Soc West Aust 80:141–158
Humphries SE et al. (1991) Plant invasions: the incidence of environmental weeds in Australia. Kowari 2. Australian National Parks and Wildlife Service, Canberra
Hunter JT, Clarke PJ (1998) The vegetation of granitic outcrop communities on the New England batholith of eastern Australia. Cunninghamia 5:547–618
James SH (1965) Complex hybridity in *Isotoma petraea* I. The occurrence of interchange heterozygosity, autogamy and a balanced lethal system. Heredity 20:341–53
James SH (1970) Complex hybridity in *Isotoma petraea* II. Components and operation of a possible evolutionary mechanism. Heredity 25:53–77

James SH (1982) The relevance of genetic systems in *Isotoma petraea* to conservation practice. In: Groves RH, Ride WDL (eds) Species at risk: research in Australia. Academy of Science, Canberra, pp 63–71

James SH (1984) The pursuit of hybridity and population divergence in *Isotoma petraea*. In: Grant WF (ed) Plant biosystematics. Academic Press, Toronto, pp 169–177

James SH (1992) Inbreeding, self-fertilization, lethal genes and genomic coalescence. Heredity 68:449–456

James SH, Wylie AP, Johnson MS, Carstairs SA, Simpson GA (1983) Complex hybridity in *Isotoma petraea* V. Allozyme variation and the pursuit of hybridity. Heredity 51:653–63

James SH, Sampson JF, Playford J (1989) Complex hybridity in *Isotoma petraea* VII. Assembly of the genetic system in the O6 Pigeon Rock population. Heredity 64:289–295

James SH, Playford J, Sampson JF (1990) Complex hybridity in *Isotoma petraea* VIII. Variation for seed-aborting lethal genes in the O6 Pigeon Rock population. Heredity 66:173–180

Jessop J (ed) (1981) Flora of Central Australia. Reed Books, Sydney

Keighery GJ (1995) How many weeds? In: Burke G (ed) Invasive weeds and regenerating ecosystems in Western Australia. Institute of Science and Technology Policy, Murdoch University, Perth, pp 8–12

Marchant NG (1973) Species diversity in the south-western flora. J R Soc West Aust 56:23–30

Mares MA (1997) The geobiological interface: granite outcrops as a selective force in mammalian evolution. In: Withers PC, Hopper SD (eds) Granite outcrops symposium. J R Soc West Aust 80:131–139

McComb AJ, Lake PS (1990) Australian wetlands. Angus & Robertson, North Ryde, New South Wales

McDonald W, Gravatt C, Grimshaw P, Williams J (1995) The flora of Girraween and Bald Rock National Parks. Queensland Department of Environment and Heritage, Brisbane

Moran GF, Hopper SD (1983) Genetic diversity and the insular population structure of the rare granite rock species, *Eucalyptus caesia* Benth. Aust J Bot 31:161–172

Myers JS (1997) Geology of granite. In: Withers PC, Hopper SD (eds) Granite outcrops symposium. J R Soc West Aust 80:87–100

Newbey KR, Hnatiuk RJ (1985) Vegetation and flora. In: Dell J et al. (eds) The biological survey of the eastern goldfields. Part 3. Jackson-Kalgoorlie study area. Rec West Aust Mus Suppl 23:11–38

Newbey KR, Keighery GJ, Hall NJ (1995) Vegetation and flora. In: Keighery GJ, McKenzie NL, Hall NJ (eds) The biological survey of the eastern goldfields. Part 11. Boorabbin-Southern Cross study area. Rec West Aust Mus Suppl 49:17–30

Noble JC, Bradstock RA (eds) (1989) Mediterranean landscapes in Australia. Mallee ecosystems and their management. CSIRO, East Melbourne

Noble JC, Joss PJ, Jones GK (eds) (1991) The Mallee lands: a conservation perspective. CSIRO, East Melbourne

Ohlemüller R (1997) Biodiversity patterns of plant communities in shallow depressions on Western Australian granite outcrops (inselbergs). MSc Thesis, University of Bonn

Ornduff R (1986) Comparative fecundity and population composition of heterostylous and non-heterostylous species of *Villarsia* (Menyanthaceae) in Western Australia. Am J Bot 73:282–286

Ornduff R (1987) Islands on islands: plant life on the granite outcrops of Western Australia. University of Hawaii, Harold L. Lyon Arboretum Lecture No. Fifteen. University of Hawaii Press, Honolulu

Ornduff R (1996) An unusual floral monomorphism in *Villarsia* (Menyanthaceae) and its proposed origin from distyly. In: Hopper SD, Chappill JA, Harvey MS, George AS (eds) Gondwanan heritage: past, present and future of the Western Australian biota. Surrey Beatty, Chipping Norton, New South Wales, pp 212–222

Pate JS, Beard JS (1984) Kwongan plant life of the sandplain. University of Western Australia Press, Nedlands

Pate JS, Dixon KW (1982) Tuberous, cormous and bulbous plants. The University of Western Australia Press, Nedlands

Pignatti E, Pignatti S (1994) Centrolepidi-Hydrocotyletea alatae, a new class of ephemeral communities in Western Australia. J Veg Sci 5:55–62

Pigott PJ, Sage LW (1997) Remnant vegetation, priority flora and weed invasions at Yilliminning Rock, Narrogin, Western Australia. In: Withers PC, Hopper SD (eds) Granite outcrops symposium. J R Soc West Aust 80:201–208

Porembski S, Brown G, Barthlott W (1995) An inverted latitudinal gradient of plant diversity in shallow depressions on Ivorian inselbergs. Vegetatio 117:151–163

Porembski S, Seine R, Barthlott W (1997) Inselberg vegetation and the biodiversity of granite outcrops. In: Withers PC, Hopper SD (eds) Granite outcrops symposium. J R Soc West Aust 80:193–199

Sampson JF, Hopper SD, James SH (1988) Genetic diversity and the conservation of *Eucalyptus crucis* Maiden. Aust J Bot 36:447–460

Tracey JG (1981) Australia's rain forests: where are the rare plants and how do we keep them? In: Synge H (ed) The biological aspects of rare plant conservation. John Wiley, London, pp 165–178

Twidale CR, Bourne JA (1998) Multistage landform development, with particular reference to a cratonic bornhardt. Geogr Ann 80A:79–94

Walters TW, Wyatt R (1982) The vascular flora of granite outcrops in the Central Mineral Region of Texas. Bull Torrey Bot Club 109:344–364

Wardell Johnson G, Williams M (1996) A floristic survey of the Tingle Mosaic, south-western Australia: applications in land use planning and management. J R Soc West Aust 79:249–276

Webb LJ, Tracey JG, Williams WT (1984) A floristic framework of Australian rainforests. Aust J Ecol 9:169–198

Webb R (1972) Vegetation of granite outcrops with special reference to Wilson's Promontory. MSc Thesis, University of Melbourne

Willmott WF, Stephenson PJ (1989) Rocks and landscapes of the Cairns district. Queensland Department of Mines, Brisbane

Withers PC, Hopper SD (eds) (1997) Granite outcrops symposium. J R Soc West Aust 80:87–237

Womersley HBS (1981) Aspects of the distribution and biology of Australian marine macro-algae. In: Pate JS, McComb AJ (eds) The biology of Australian plants. University of Western Australia Press, Perth, pp 294–306

Womersley HBS (1990) Biogeography of Australasian marine macroalgae. In: Clayton MN, King RJ (eds) Biology of marine plants. Longman, Cheshire, pp 367–381

Wyatt R (1997) Reproductive ecology of granite outcrop plants from the south-eastern United States. In: Withers PC, Hopper SD (eds) Granite outcrops symposium. J R Soc West Aust 80:123–129

10.10 Flora and Vegetation of Granite Outcrops in the Southeastern United States

R. Wyatt and J.R. Allison

10.10.1 Geography and Geology

10.10.1.1 Geographical Distribution

Much of North America is underlain by granite rocks, which are a major component of the ancient crystalline blocks or shields that form the nucleus of the continent (Twidale 1982). Within the United States, there appear to be three major areas in which granitic inselbergs occur: (1) western states (especially California), (2) south-central states (Texas, Oklahoma, and Arkansas), and (3) southeastern states (Georgia, Alabama, North Carolina, South Carolina, and Virginia). These regions correspond rather closely to a zone of orogenic granitic rocks, chiefly of magmatic origin, mapped by Twidale (1982).

The flora and vegetation of inselbergs in the western and south-central United States appear to have been studied much less than those in the southeast. Rundel (1975) studied primary succession on granite outcrops in the southern Sierra Nevada of California and listed a number of lichens, mosses, ferns, and seed plants that are characteristic of, or limited to, specific habitats. Floristic surveys of the Wichita Mountains Wildlife Refuge, which includes granitic exposures, provide information on the characteristic plants and their role in succession (Eskew 1938; Diehl 1953; Buck 1964; Crockett 1964). Best-studied of the inselbergs in the south-central United States are the extensive, large outcrops in the Central Mineral Region of Texas. Our knowledge of their vascular flora and vegetation has been summarized by Walters and Wyatt (1982).

The granite outcrops of the southeastern United States have been more thoroughly investigated floristically and ecologically than those of any other comparable region. Within the southeast, granitic rocks are exposed in both the Piedmont and Appalachian Mountain physiographic provinces. Granites and gneisses are exposed as domes or cliffs in the mountains. These commonly begin at the top as gentle slopes with patches of vege-

Ecological Studies, Vol. 146
S. Porembski and W. Barthlott (eds.) Inselbergs

tation or mat communities that then curve more steeply to end as vertical cliffs near their base (Quarterman et al. 1993). Succession on these mountain slopes has been described by Oosting and Anderson (1937), and the ecology of several rare species from these communities has been studied by Johnson (1995). The vascular plants that are characteristic of these mountain outcrops, however, are distinctly different from those of the Piedmont granite outcrops.

In the southeastern United States, the Piedmont physiographic province slopes gradually eastward from the Blue Ridge Escarpment, which marks the boundary of the Appalachian Mountains, to the Fall Line, which marks the boundary of the Atlantic Coastal Plain (Fenneman 1938). Typical Piedmont topography consists of a series of gently rolling hills interrupted periodically by deeper, steeper valleys of large rivers (McVaugh 1943). For the most part, these rivers cut directly across the Piedmont and Coastal Plain to empty into the Atlantic Ocean or the Gulf of Mexico. Most of the Piedmont is covered with mixed mesophytic forest dominated by oaks, hickories, and pines (Braun 1950). Dotted throughout the region, however, are occasional expanses of bare rock, which harbor a distinctively different flora and vegetation. The zone in which typical Piedmont granite outcrops occur ranges from east-central Alabama, through Georgia and the Carolinas (McVaugh 1943), to south-central Virginia (Harvill 1976). In Georgia, where outcrops are largest and most numerous, Wharton (1978) has published a map (based on data provided by S. Pickering) of most of the exposures that exceed 1ha in area.

10.10.1.2 Geological Age and Weathering

Geologists estimate that most of the granitic rocks that outcrop in the Piedmont of the southeastern United States are approximately 300–350 Ma old (Watson 1910; Whitney et al. 1976). Molten granite (magma) was intruded into preexisting country rock at a depth of about 16 km below the surface (Atkins and Joyce 1980). Over millions of years, erosion removed thousands of meters of overlying rock, exposing the more resistant granitic plutons (Atkins and Griffin 1977). The present level of the Piedmont is believed to have come about as the result of a general uplift in the Tertiary, following base-leveling of a Cretaceous peneplain (McVaugh 1943).

Oosting and Anderson (1939) speculated that granite outcrops in eastern North Carolina were of recent origin, having arisen from accelerated erosion following fires and other human disturbances. McVaugh (1943) argued, however, that "the date of the original exposure of the granitic surfaces is not easy to fix, but it is probably very remote." He and others (e.g., Burbanck and Platt 1964; Murdy 1968; Wyatt and Fowler 1977) pointed to the apparent antiquity of the endemic flora as evidence for the long-continued presence of granitic exposures in the Piedmont. Individual outcrops may not be this ancient; for example, Atkins and Joyce (1980) estimated that Stone Mountain, the largest inselberg in the southeast, became exposed only about 15 million years ago. Nevertheless, it is likely that the characteristic habitats and organisms of exposed rock have existed somewhere in the Piedmont for a considerably longer period of time (Wyatt and Fowler 1977). The organisms have managed to disperse from one rock outcrop island to another as the weathering cycle of the Piedmont has covered some exposures with soil and uncovered new ones (Burbanck and Platt 1964).

Granite outcrops in the southeast vary tremendously in size, shape, and position in the landscape (Fig. 10.10.2, 2–5). Some consist of small, flat-lying exposures (flatrocks) only a few square meters in area, whereas Stone Mountain is a steeply sloping inselberg that covers about 237 ha and rises approximately 230 m above the surrounding countryside (Atkins and Joyce 1980). The rock units themselves are often very heterogeneous in mineral composition and/or texture as well (Watson 1910; Campbell 1921), yet this seems to have little impact on the nature of the vegetation. Often, neighboring outcrops that would appear to have developed on the same rock unit actually occur on distinctly different igneous intrusions. For example, Stone Mountain, Panola Mountain, and Arabia Mountain all occur within about 10 km of each other, southeast of Atlanta, Georgia. All three inselbergs support many of the characteristic species of granite outcrops. Nevertheless, each represents a distinct and separate rock type: Stone Mountain Granite, Panola Granite, and Arabia Mountain Gneiss, respectively (Atkins and Joyce 1980).

The differences in mineral composition and/or texture of various rock units may have their greatest significance for the vegetation in determining the resulting patterns of chemical and physical weathering that produce characteristic rock outcrop habitats for plants. The rock surface is commonly subject to exfoliation (Baker 1945, 1956), which can create crevices and talus piles (Fig.10.10.2, 2). Uneven weathering of the rock surface results in shallow depressions commonly called weathering pits (Atkins and Griffin 1977) or solution pits (Quarterman et al. 1993). It is in these depressions that the most distinctive granite outcrop endemics occur. Those rock-rimmed pits that are deep enough to hold water between convective rainstorms harbor a number of rare aquatic plants (Fig.10.10.2, 3). Other depressions, which have accumulated a thin layer of soil and which lack a complete rim to retain water, support many unusual annuals (Fig. 10.10.2, 4). Other distinctive habitats on the outcrops include deeper accumulations of soil over bare granite, seepage areas created by the slow discharge of water from patches of forest on the rock, and woods margins (Wyatt and Fowler 1977).

10.10.2 Climate

10.10.2.1 Temperature and Precipitation

The southeastern United States is characterized as having a humid temperate climate, where mean annual temperatures are approximately 15–18 °C, with the coldest months averaging 5–10 °C and the warmest, 22–27 °C (Fig.10.10.2, 1). Annual precipitation typically exceeds 100 cm, with little seasonality, as the average for every month generally is greater than 6–8 cm (Fig. 10.10.1).

10.10.2.2 Microenvironments on Granite Outcrops

Within this region of generally mild and mesic conditions, granite outcrops create distinctly different, extreme microenvironments. Temperatures on the outcrops typically are much higher than in the surrounding forest because of high incident radiation, absorption of heat by the rock, and low evapotranspiration (Wyatt 1997). Temperatures in excess of 50 °C are common at the rock surface during summer months. Moreover, the shallow, mineral soil overlying impervious rock and the sparse vegetation cover lead to extraordinarily high runoff. It has been estimated that more than 95 % of the annual precipitation in outcrop communities is lost as direct runoff versus only 10 %, on average, for the Piedmont of Georgia (Duke and Crossley 1975). These conditions make the granite outcrops "islands of desert embedded in a sea of mesic deciduous forest" (Wyatt 1997).

In addition to providing "microenvironmental deserts" (Phillips 1982), however, the outcrops also provide "microenvironmental oases". The deeper rock-rimmed pools that retain water between rainstorms afford aquatic habitats for highly specialized plants. Moreover, the slow, steady discharge of water from islands of forest on the rock surface creates a constantly wet seepage area in which a number of wetland species flourish.

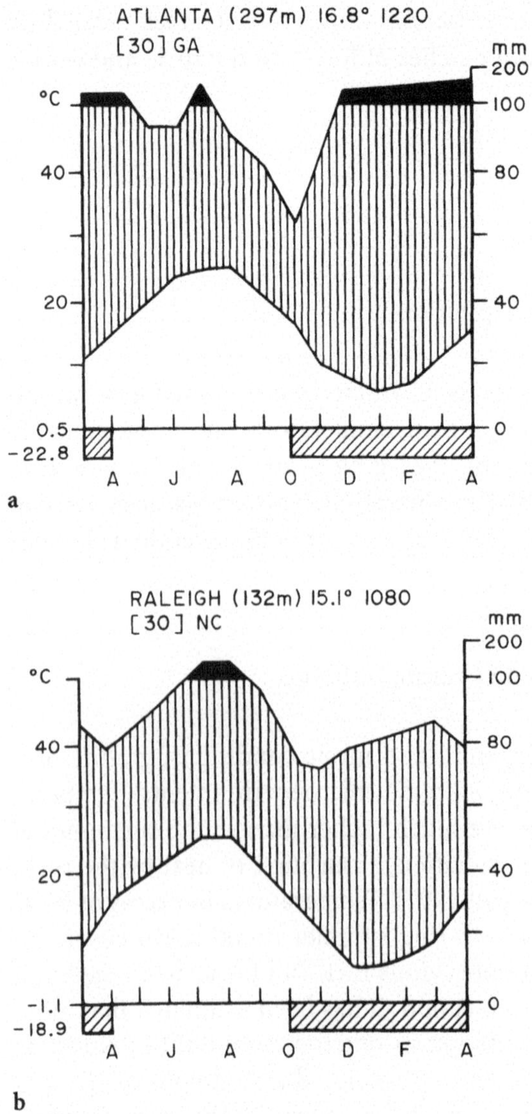

Fig. 10.10.1a,b. Climate diagrams for Atlanta, Georgia, and Raleigh, North Carolina. These diagrams were constructed following the guidelines of Walter and Lieth (1967) and Walter (1979). Data were obtained from the US Weather Bureau as compiled by Visher (1954)

2
3
4
5
6
7
8
9
10

10.10.3 Zonal Vegetation

10.10.3.1 Primary Forest

The presettlement vegetation of the Piedmont was an oak-hickory-pine forest (Braun 1935, 1950; Oosting 1942; Küchler 1964). The canopy was dominated by several species of oaks, including *Quercus alba*, *Q. falcata*, *Q. rubra*, *Q. velutina*, and *Q. coccinea*. On somewhat drier upland sites, *Q. stellata* and *Q. marilandica* were often abundant; on wetter bottomland sites, *Q. nigra* and *Q. phellos* were common. Occasionally, the dominant successional pines, *Pinus taeda* and *P. echinata*, might persist. Other important canopy hardwoods included hickories, especially *Carya tomentosa* and *C. glabra*, *Ulmus alata*, *Liriodendron tulipifera*, *Sassafras albidum*,

---

Fig. 10.10.2, 2–10. Photographs illustrating some of the characteristic granite outcrop habitats and vegetation described in the text. **2** Granite inselberg in Wilkes County, North Carolina. Such steeply sloping exposures belie the name flat rocks, which is commonly applied to granite outcrops in the southeastern United States. **3** This large, rock-rimmed pool at the summit of Heggie's Rock in eastern Georgia harbors the rare and endangered mat-forming quillwort, *Isoetes tegetiformans*. Note the zonal vegetation of oak-hickory-pine forest in the background. **4** A small, flat granite outcrop in eastern Alabama. Vegetation mats develop as soil accumulates in depressions weathered in exposed granite. The smaller, shallower depression *on the left* contains *Diamorpha smallii* (Crassulaceae), a red-bodied succulent; the larger, deeper depression *on the right* contains fruticose lichens, haircap moss (*Polytrichum commune*), and *Andropogon virginicus* (Poaceae). **5** An individual of the dominant successional tree in the surrounding forest, *Pinus taeda*, has invaded a deep soil island on this outcrop in eastern North Carolina. In the *left foreground* is an island dominated by lichens, *Senecio tomentosus* (Asteraceae), and *Andropogon virginicus*; in the *right foreground*, *Diamorpha smallii*. **6** The bare rock surface is covered by a foliose lichen (*Xanthoparmelia conspersa*) and the moss *Grimmia laevigata*, in mats of which *Diamorpha smallii* grows in abundance. **7** Characteristic of the many winter annuals that occur on granite outcrops, *Sedum pusillum* (Crassulaceae) germinates in the fall and overwinters as frost-resistant rosettes. It grows in mats of the moss *Hedwigia ciliata*, typically in the shade of red cedar (*Juniperus virginiana*) trees. **8** Vegetation mats over granite develop concentric zones, with earlier colonists displaced to shallower soil at the edges. In this annual-perennial herb community, we see, progressing centripetally, *Diamorpha smallii*, *Arenaria uniflora* (Caryophyllaceae), *Senecio tomentosus*, and *Andropogon virginicus*. **9** *Phacelia maculata* (Hydrophyllaceae) is another winter annual that is restricted to granite outcrops in the Piedmont. Here, it is growing on an outcrop east of Atlanta, Georgia. **10** A number of species on the Piedmont outcrops are disjunct hundreds of kilometers from their main ranges. *Amsonia ludoviciana* grows here on an outcrop near Walnut Grove, Georgia, whereas the next nearest localities are on the Gulf Coastal Plain of Louisiana and (historically) adjacent Mississippi.

Liquidambar styraciflua, and *Diospyros virginiana*. Among the tree species of the understory were *Cornus florida*, *Cercis canadensis*, *Ilex opaca*, *Oxydendrum arboreum*, *Nyssa sylvatica*, and *Acer rubrum*. The shrub layer included species such as *Euonymus americana*, *Aesculus sylvatica*, *Vaccinium stramineum*, and *Viburnum* spp. Characteristic woody vines included *Vitis rotundifolia*, *Gelsemium sempervirens*, *Bignonia capreolata*, *Toxicodendron radicans*, and *Smilax* spp. Common in the herbaceous layer were *Senecio anonymus*, *Penstemon australis*, *Chasmanthium laxum*, *Ruellia carolinensis*, *Hexastylis arifolia*, *Viola* spp., *Panicum* spp., and *Carex* spp. Oosting (1942) has provided a detailed description of Piedmont plant communities, based principally on studies in North Carolina; Wharton (1978) has done the same for Georgia.

10.10.3.2 Secondary Forest

Oak-hickory forest formerly covered more than 75% of the Piedmont uplands (Wharton 1978), but today only isolated fragments remain, mostly in areas where rocky terrain or steep local topography has constrained agricultural use. Nearly all of the Piedmont was cleared of primary forest and put into agricultural production by 1850; the landscape today represents a mosaic of patches in various stages of second growth following disturbance. In fact, the Piedmont of North Carolina (Keever 1950; Peet and Christensen 1980) and Georgia (Golley 1962; Odum 1971) has served as a model for studies of old-field succession. Abandoned fields dominated by crabgrass (*Digitaria sanguinalis*) in the first year support horseweed (*Conyza canadensis*) and *Aster pilosus*, which are replaced about the third year by broomsedge (*Andropogon virginicus*). Pines (*Pinus taeda* or *P. echinata*) invade and become dominant during years 25–100, after which the zonal vegetation (oak-hickory forest) gradually holds sway. Typically, therefore, one sees in the Piedmont of the southeast a complex mixture of various stages in the recovery of the oak-hickory forest.

10.10.4 Inselberg Vegetation

10.10.4.1 Terrestrial/Xeric Habitats

Granite outcrops stand out in stark contrast to the surrounding matrix of old-field areas and oak-hickory-pine forest. They are populated by a unique, highly specialized group of plant species that differ strikingly in aspect from adjacent areas that support zonal vegetation. Perhaps the most obvious differences are seen in the terrestrial plants that occur in xeric areas associated with the bare rock.

One immediate difference that strikes an observer is the paucity of woody plants. Whereas oaks, hickories, and pines dominate the zonal vegetation, annual and perennial herbs with adaptations to avoid or resist drought reign supreme on granite outcrops. Among the few characteristic trees is Georgia oak (*Quercus georgiana*), which appears to hybridize on many of the outcrops with its close congener, water oak (*Q. nigra*). Pines often occur in association with granite outcrops, the local successional dominant (*P. taeda* or *P. echinata*) eventually invading the deepest accumulations of soil over the rock (Fig.10.10.2, 5). Another local name for granitic flatrocks is cedar rocks, reflecting the common occurrence of red cedar (*Juniperus virginiana*).

A few shrubs and woody vines, though by no means limited to granite outcrops, may grow especially well in the thin woods at the margins of outcrops. Fringe tree (*Chionanthus virginicus*), for example, seems to flower much more prolifically in such settings than in the dense shade of forests, as do *Ptelea trifoliata*, *Aesculus sylvatica*, and *Amelanchier arborea*. Among the woody vines are such spectacular plants as *Gelsemium sempervirens*, *Bignonia capreolata*, and *Smilax smallii*.

It is the herbaceous plants, however, that give the granite outcrops their unique, distinctive character. Burbanck and Platt (1964) devised a scheme for classifiying granite outcrop communities into four categories, based on maximum depth of the soil over the rock. The "diamorpha community", with soil depths of 2–9 cm, consists of a single vascular plant: *Diamorpha smallii* (Fig. 10.10.2, 6). This diminutive winter annual is able to invade and dominate because of its ability to tolerate low moisture levels in the soil and because of freedom from competition in this stressful microenvironment (Wiggs and Platt 1962; Sharitz and McCormick 1973; Quarterman et al. 1993). Occasionally, a few cryptogams occur with *D. smallii* in these areas, including lichens (*Cladonia leporina*, *C. caroliniana*) and mosses (*Campylopus tallulensis*, *Grimmia laevigata*).

Lichen-annual herb communities occupy depressions over granite where soil depths reach 7–15 cm (Fig. 10.10.2, 7). In the spring these areas are dominated by annuals such as *Arenaria uniflora*, *Nuttallanthus canadensis*, *Rumex hastatulus*, and *Agrostis elliottiana*. In the summer, *Bulbostylis capillaris*, *Croton willdenowii*, and *Hypericum gentianoides* become conspicuous. The fall aspect in these communities is dominated by the Confederate daisy, *Viguiera porteri*. More constant, but rarely dominating these areas, are lichens, principally several species of *Cladonia* and *Cladina*.

Annual-perennial herb communities occur on soils 14–41 cm deep. An important soil builder in these areas is the moss *Polytrichum commune* (Quarterman et al. 1993). On thinner soil, *Viguiera porteri* and the succulent perennials *Talinum mengesii* or *T. teretifolium* may dominate.

Toward the center of the depression, on deeper soils, perennial herbs such as *Senecio tomentosus*, *Schoenolirion croceum*, *Nothoscordum bivalve*, and *Andropogon virginicus* hold sway. An interesting aspect of development of the vegetation in these soil islands over bare rock is the displacement of earlier colonizing species to the periphery, where the soil is most shallow. This produces concentric rings of vegetation, with the perennials at the center and *Diamorpha* limited to the shallow soil on the periphery (Fig. 10.10.2, 8).

Burbanck and Platt's (1964) herb-shrub communities, with soil depths of 40–50 cm, contain many of the same species as the annual-perennial herb community with the addition of some woody species (Fig. 10.10.2, 5). With respect to a single soil island over granite, this often means just one small tree (often *Pinus taeda* or *Juniperus virginiana*). Other woody species might include *Gelsemium sempervirens*, *Smilax smallii*, *Vaccinium arboreum*, or *Rhus copallina*.

Many authors have interpreted these four major communities as seral stages in primary succession on granite (e.g., Oosting and Anderson 1939; Keever et al. 1951; Burbanck and Platt 1964; Shure and Ragsdale 1977). Unfortunately, there have been few long-term studies of individual soil islands. Burbanck and Phillips (1983) reported on a 22-year study of 34 soil island communities in the Georgia Piedmont. They concluded that the preponderance of evidence supported the view that changes were occurring in soil depth that were sufficient to drive transitions in plant communities, at least between the annual-perennial herb stages (Quarterman et al. 1993). Similar studies initiated in North Carolina by Oosting and Anderson, however, revealed little change in the plant communities and uncovered at least a few examples in which the proposed successional process regressed (L.E. Anderson, pers. comm.)! In any event, it appears that the rate of change in these plant communities is extremely slow, and solid documentation of the processes involved will be exceedingly difficult to obtain.

In addition to the supposed successional processes in depressions, it has been proposed that characteristic replacements of species occur on the bare rock itself. These were originally supposed to involve crustose lichens, such as *Verrucaria* spp., followed by the moss *Grimmia laevigata*, then *Diamorpha smallii*, and finally the other outcrop communities described above (Fig. 10.10.2, 6). Oosting and Anderson (1939) and Keever et al. (1951) cast doubt on this hypothesis, arguing that soil accumulation by crustose, and even foliose, lichens (such as *Xanthoparmelia conspersa*) is minimal, and mosses are really the first soil-building colonists on rock surfaces. It is certainly true that *Diamorpha smallii* can grow abundantly in mats of *Grimmia laevigata* on granite outcrops, but it is not clear that this commonly leads to a transition to later successional stages. Another common moss on the granite outcrop is *Hedwigia ciliata*, within the mats of which the rare endemic *Sedum pusillum* grows (Fig. 10.10.2, 7), typically in the shade of *Juniperus virginiana* (Wyatt 1983). Such areas are also common habitat for the stem succulent *Opuntia* spp.

Seepage areas on the outcrops afford habitat for a few unusual wetland species, including a number of taxa that are normally confined to the Coastal Plain, such as the peat moss *Sphagnum cyclophyllum*, *Lindernia monticola*, and *Utricularia* spp. Rock crevices and rubble piles at the bases of outcrops support a number of ferns. Among these are *Cheilanthes lanosa* (rarely *C. tomentosa*), and *Woodsia obtusa*. A number of rare, disjunct species have also been discovered on isolated outcrops (see below).

10.10.4.2 Aquatic Habitats

Some of the most unique, and certainly the most endangered, species on granite outcrops are those aquatic taxa that occur in the rock-rimmed pools. Perhaps the rarest of these is the quillwort *Isoetes tegetiformans* (Fig. 10.10.2, 3). Originally known from a single pool on one outcrop (Rury 1978), its known range has been extended only to a handful of nearby sites in Georgia (Allison 1993). Similarly, *I. melanospora* occurred on fewer than 20 outcrops in Georgia and South Carolina, from half of which it has already been extirpated (Allison 1993). The slightly more widespread endemic *Amphianthus pusillus* is also recognized as a federally protected (threatened) species, which has suffered significant habitat loss in this century (Allison 1993).

10.10.5 Life-Forms and Adaptations

10.10.5.1 Morphology and Physiology

The extremely hot, dry conditions in certain microenvironments on granite rocks has led to a number of morphological, physiological, and life-history adaptations of the characteristic plants. Some perennial herbs tolerate drought by storing water in succulent stems or leaves (e.g.,

Opuntia humifusa, Talinum teretifolium). A number of annuals also have succulent tissues (e.g., *Portulaca smallii, Diamorpha smallii, Sedum pusillum*). Many of the dominant species of outcrops differ from their congeners in being significantly more hairy (e.g., *Senecio tomentosus, Croton willdenowii, Tradescantia hirsuticaulis*). It seems likely that a dense covering of white hairs acts to increase the reflectivity of the plant's leaves and thereby decrease the heat load. Moreover, the tomentum may help to reduce evapotranspiration. The conspicuous silvery-white hairpoints of the dominant moss on bare granite surfaces, *Grimmia laevigata*, may similarly work to increase albedo and reduce water loss.

Some authors have speculated that outcrop plants further enhance their water-use efficiency by employing alternative pathways of carbon fixation (C_4 and CAM). Baskin and Baskin (1988) reviewed this literature, however, and concluded that nearly all of the endemic plants use the C_3 pathway. The exceptions were *Portulaca smallii* and *Cyperus granitophilus*, which are C_4 plants, and *Isoetes tegetiformans*, which uses CAM (Keeley 1982). Among the nonendemic, but characteristic, species of the outcrops, *Opuntia humifusa* and *Agave virginica* are reported to be CAM plants (Martin et al. 1982) and *Andropogon virginicus* uses C_4 (Williams 1969).

10.10.5.2 Life-Forms

Perhaps the most common solution to the stresses posed by microenvironmental deserts on granite outcrops is drought avoidance. Poikilohydric mosses and lichens are natural resurrection plants that can photosynthesize soon after rehydration. Some ferns, like *Pleopeltis polypodioides*, which is common on the limbs of red cedar trees on the outcrops, can also accomplish this feat. The quillworts that occupy rock-rimmed pools, *Isoetes melanospora* and *I. tegetiformans*, are able to go dormant when the pools dry late in the spring, then resume growth whenever rainstorms refill the pools, regardless of season (Allison 1993).

Most of the endemic flowering plants on the outcrops have evolved a winter annual life history. Seeds germinate after late September or early October rains to produce rosettes of frost-resistant leaves. When temperatures begin to rise in late March or early April, the plants bolt and flower profusely. By early May, when the shallow soil depressions have become bone-dry, the winter annuals have already matured their crop of seeds, which typically require an after-ripening period of 4–5 months. Thus, they are able to survive the extremely hot, dry summer months as dormant populations of drought-resistant seeds. In addition to *Diamorpha smallii*, *Sedum pusillum*, and *Amphianthus pusillus*, there are many other annuals

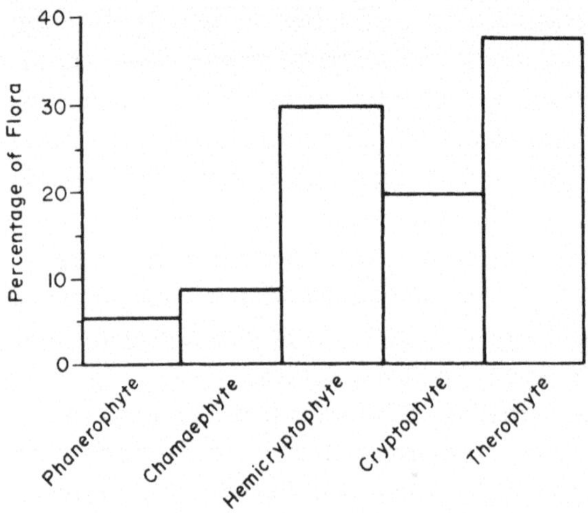

Fig. 10.10.3. Life-form spectrum for granite outcrop vegetation in the southeastern United States. The life-form classification used is that of Raunkiaer (1934)

on the outcrops, some of which, like *Viguiera porteri*, are annual members of groups that are typically perennial. *Arenaria glabra*, which has been treated as merely a variety of the arctic-alpine perennial *A. groenlandica* (e.g., Radford et al. 1964), differs chiefly in its annual life history, whose evolution presumably enabled the species to invade granite outcrops (Wyatt 1986).

Taken as a whole, the vascular flora of the granite outcrops in the southeastern United States presents a life-form spectrum that contrasts starkly with that of the Piedmont flora at large (Fig. 10.10.3). The most striking differences are the increased percentage of therophytes and the decreased percentage of phanerophytes. These same trends were noted previously by Walters and Wyatt (1982) and by Phillips (1982). The life-form spectrum for granite outcrops is, in fact, very similar to those published for deserts (e.g., Death Valley, California: Raunkiaer 1934).

10.10.6 Systematic and Ecological Characteristics

Deciding upon a definitive list of granite outcrop plants is necessarily arbitrary, as difficult decisions must be made regarding species that occur in marginal habitats, weedy species, occasional species, etc. McVaugh (1943) recognized "about 200" species "that may be definitely identified with the flat rocks" in the southeast. Phillips (1982) distinguished only 155 for his survey of the same region, whereas Walters and Wyatt (1982) expanded the list to include a total of 302. For our purposes here, we have identified a total of 118 species that we view as constituting the characteristic assemblage of species common on granite outcrops in the southeastern United States (Appendix). Our list is much shorter than that of previous authors because we have tried to be extremely careful to include only the elements of the outcrop flora that we believe are distinctive. Thus, such common components of the vegetation as *Juniperus virginiana*, *Pinus taeda*, and *Andropogon virginicus*, which were included in lists by earlier researchers, are excluded from our list of characteristic plants.

The 118 taxa on our list include eight strict endemics, for which no natural occurrences are known except on granite outcrops of the Piedmont. A few of these, such as *Phacelia dubia* var. *georgiana* and *Portulaca smalii* have weedy tendencies that permit them to spread to human-maintained openings, such as mowed highway shoulders. The 13 near-endemics have the great majority (>75 %) of their reported occurrences on granite outcrops in the southeast. There are 13 taxa that we characterize as half-endemics. The majority of reported occurrences of these taxa are on granite outcrops, but they are suspected to be nearly as abundant on other substrates, mostly outside of the Piedmont. We recognize 54 provincial endemics, taxa for which the great majority of occurrences are *not* on granite outcrops, but for which >95 % of the known occurrences in the Piedmont are on granite outcrops. The six disjunct endemics are plants that (1) have their center of distribution or abundance outside the Piedmont, (2) are known to occur in the Piedmont only on granite outcrops, and (3) are more than 100 km disjunct from their occurrences in other provinces. Our category of incidentals includes 15 taxa with >95 % of their global occurrences outside the Piedmont, but whose few known Piedmont occurrences are entirely or prinipally on granite outcrops. We also include here seven additional taxa that were not counted in the official totals, because they have been collected from fewer than five outcrops. Some of these are freak disjuncts, known only from a single outcrop. Finally, we did include nine weeds. There are native plants found today in a variety of early successional habitats, in and beyond the Piedmont, usually associated with human disturbance (e.g., old fields, gardens) or artificial maintenance of successional community stages (e.g., mowed areas, such as road shoulders or cemeteries). At least some of these are believed to have been confined to areas such as granite outcrops prior to widespread disturbance of the landscape following European settlement.

The family spectrum based on our list of characteristic plants shows a very uneven distribution (Fig. 10.10.4). Approximately 27 % of the characteristic flora belong to only three families: Cyperaceae, 17 (14.4 %), Asteraceae, 9 (7.6 %), and Scrophulariaceae, 6 (5.1 %). At the other extreme, 20 families and 62 genera are represented by a single species. In broad outline, these results are very similar to the patterns reported for the granite outcrop flora of Texas (Walters and Wyatt 1982).

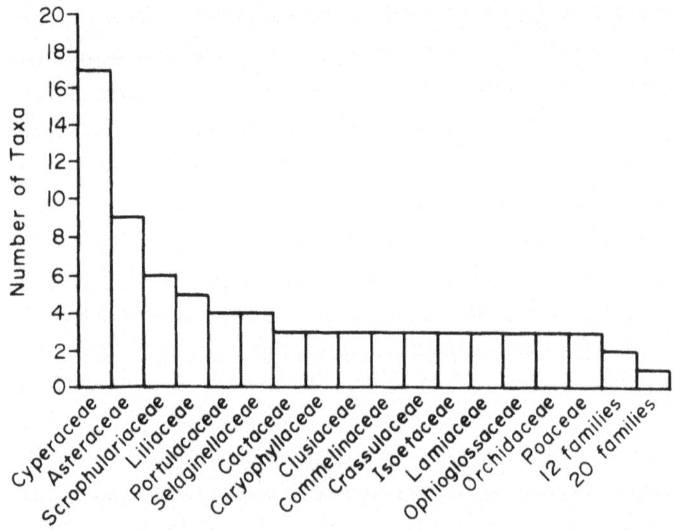

Fig. 10.10.4. Family spectrum for granite outcrop flora in the southeastern United States

It cannot be overemphasized that the distribution of the characteristic plants across the archipelago of granite outcrops in the southeast is tremendously uneven. As might be expected of an "island" distribution, the flora is not uniform across the wide range of outcrops. Many species are highly localized in their distribution, and others are very sporadic in occurrence, despite ranges that might span the entire zone of occurrence of outcrops in the Piedmont. Our account of the vegetation here focuses principally on outcrops east of Atlanta, where the characteristic habitats and plants are "best" developed.

10.10.7 Biogeography and Endemism/Speciation

10.10.7.1 Level of Endemism

As noted previously, it is somewhat arbitrary to determine which species merit inclusion on a definitive list of granite outcrop plants, and this decision will obviously influence the final calculations of the percentage of endemism in the flora. Our admittedly very restricted list includes only eight species and infraspecific taxa (7 %) that are entirely restricted to granite outcrops (strict endemics). In addition to these are another 26 % that represent taxa which find their optimum conditions on granite outcrops, but which may range more widely (near-endemics, half-endemics, disjunct endemics, and weeds). As discussed below, some of these may represent currently weedy species that were originally restricted to granite rocks.

In any case, the level of endemism on granite outcrops in the southeast is highly unusual for the Piedmont. Other than those species limited to outcrops, there are few truly distinctive elements in the Piedmont flora. Most of the taxa occurring there have a wider range centered on either the Appalachian Mountains or the Atlantic Coastal Plain. The Piedmont flora also includes a large number of weedy species, many of which have been introduced from Eurasia. These taxa are rarely present in large numbers in the most distinctive, open habitats of granite outcrops. Some ecotonal environments have been invaded by introduced pest species, such as *Ligustrum sinense.*

The comparatively high number of endemic taxa on the outcrops appears to reflect their antiquity and isolation. Apparently, the outcrop pool habitat has persisted long enough for a monotypic genus like *Amphianthus*, with its highly specialized morphology and life history, to evolve. Among others, McVaugh (1943) and Murdy (1968) have argued that *Amphianthus* has no close relatives in the Scrophulariaceae. It remains to be seen, however, what its relationships might be to certain genera, such as *Glossostigma* from Australia and certain *Lindernia* from Africa, which occupy similar niches on granite rocks in those parts of the world. If molecular genetic evidence does not suggest a link between these taxa, it will indeed be a remarkable example of convergent evolution.

Another example that supports the idea of an ancient origin and long stability of the outcrop habitats is *Isoetes tegetiformans.* This quillwort is remarkable in several respects, including its unbranched roots, distichous phyllotaxy even at maturity, and ability to spread clonally and form mats (Rury 1978). It is unclear if the monotypic genus *Diamorpha* is truly another example of an isolated lineage, as its relationship to several species of *Sedum* seems clear (see below). Walters and Wyatt (1982) argued that the comparatively low percentage of endemism on the granite rocks of Texas (about 1 %) reflects their relative lack of distinctness from the surrounding arid countryside. Among others, Carlquist (1974) has argued that an "island" pattern of distribution may foster plant speciation. Certainly, the granite outcrops of the Piedmont, occurring as an archipelago within a sea of mesic oak-hickory-pine forest, would seem to fit this model.

10.10.7.2 Origins/Affinities of the Flora

McVaugh (1943) pointed out the strong affinities of the granite outcrop flora with elements of the Madro-Tertiary Geoflora, as defined by Axelrod

(1958), of the southwestern United States and northern Mexico. The presence of these xerophytes on granite outcrops in the southeast was discussed by Braun (1955). She suggested that, during a period of warmer and drier climates in the Late Tertiary, plants from the West extended their ranges eastward into the Piedmont. The Ozark Mountains served as a land bridge to facilitate this migration. As available moisture increased and temperatures decreased in the Pleistocene, those xerophytic elements remaining in the southeast took refuge on granite outcrops. As evidence of this migration from the west, McVaugh (1943) cited *Viguiera porteri*, the only eastern representative of a group otherwise restricted to western North America. Despite evidence suggesting that *V. porteri* actually has closer affinities with *Helianthus* (Schilling and Heiser 1981; Schilling and Jansen 1989), it remains true, nonetheless, that its close congeners occur in the desert southwest. A number of other taxa characteristic of granite outcrops in the southeast belong to genera that are most diverse and abundant in the southwest, such as *Yucca*, *Opuntia*, *Talinum*, and *Portulaca* (Walters and Wyatt 1982).

McVaugh (1943) argued that "the basic flora of the Blue Ridge, on the other hand, is of quite different derivation from that of the Piedmont." There are, however, a few elements of the granite outcrop flora that appear to have been derived from the Arcto-Tertiary geoflora of the northern United States. Mention has already been made of *Arenaria glabra*, a rock outcrop endemic that may have evolved an annual life history, diverged from its close relative *A. groenlandica*, and invaded the granite outcrops. Wyatt (1986) speculated that this may have occurred during the Pleistocene, when the range of the arctic-alpine ancestral species may have been forced southward and to lower elevations in the southern Appalachians. A similar origin for the outcrop endemic *Phacelia dubia* var. *georgiana* seems likely (Murdy 1968), although its relationship to another endemic, *P. maculata* (Fig. 10.10.2, 9) and to other varieties remains unresolved (Levy et al. 1996). It is possible that further studies will reveal closer affinities of some other endemic species in such large, complex genera as *Cyperus*, *Scirpus*, and *Panicum* to taxa with more northern distributions.

Another distribution pattern to which McVaugh (1943) drew attention involved taxa which occurred discontinuously across a wide range, including granite outcrops in the southeast. Among these are the fern *Pilularia americana* and *Portulaca umbraticola* ssp. *coronata*. There are also a number of species, especially ferns with wind-dispersed spores, that have been reported as disjuncts from their main ranges further west or south (Fig. 10.10.2, 10). For example, Wagner (1965) reported *Pellaea wrightiana* from a single outcrop in North Carolina. More recently, *Astrolepis sinuata*, a characteristic fern of granite outcrops in central Texas, has been discovered in Meriwether County, Georgia (J.R. Allison, pers. obs.). A few rare taxa characteristic of limestone outcrops in central Tennessee and of prairie communities in the Midwest have recently been discovered on a few isolated outcrops in the Piedmont of Georgia and South Carolina. It seems likely that all of these represent relatively recent long-distance dispersal events.

As mentioned previously, the weedy species on granite outcrops seem to represent a very distinct subset of the Piedmont weed flora. As McVaugh (1943) noted, "the aggressive old-field and forest species are excluded." This observation led Wyatt and Fowler (1977) to propose that the few weedy species common on granite outcrops may, indeed, have once been endemics restricted to outcrop habitats (e.g., *Andropogon virginicus*). These particular taxa have been noted even on highly isolated rock outcrops by early explorers, such as Harper (1939), who observed on a remote granite outcrop in eastern Alabama "a surprising number of weeds, though the place was not close to any house or field." These outcrop weeds may have possessed a number of traits that enabled them to prosper in newly opened habitats, such as old fields, following the arrival of humans. Marks (1983) has advanced a similar argument with respect to successional species in the northeast.

Wyatt (1997) has taken this view further, suggesting that granite outcrops may have been the original evolutionary springboard for such present-day widespread Coastal Plain plants as *Rumex hastatulus* and *Nuttallanthus canadensis*. These preeminent weeds of Coastal Plain agricultural fields may have perfected their ability to grow in open, disturbed areas of poor, sandy soil while existing primarily on granite outcrops. When similar habitats became available on the Atlantic and Gulf Coastal Plains, they may have undergone range expansion. In fact, it would appear that their distributions and abundances have increased even since the time of McVaugh (1943). Other species that might fit this scenario are *Nothoscordum bivalve*, *Krigia virginica*, *Trifolium carolinianum*, *Triodanis perfoliata*, *Houstonia pusilla*, *Valerianella radiata*, and *Agrostis elliottiana*.

10.10.7.3 Modes of Speciation

At least two of the most characteristic species endemic to granite outcrops in the southeastern United States appear to have originated via allopolyploidy. Using isozymes as genetic markers, Murdy and Carter (1985) showed that *Talinum teretifolium* is an allotetraploid. This species combines one genome of diploid *T. mengesii*, from outcrops in Georgia, with one genome of diploid *T. parviflorum*, from the south-central United States and a small area of Chilton and Coosa Counties, Alabama (Carter and Murdy 1985). Despite conflicting evidence from some loci and a disquietingly high number of null alleles, they were able to reject unequivocally the alternative hypothesis that *T. calycinum* was the second progenitor. Definitive proof is still lacking, but it seems clear that *Diamorpha smallii* also is an allotetraploid ($n=9$), one of whose diploid progenitors is *Sedum pusillum* ($n=4$), also endemic to outcrops in the southeast. As Baldwin (1940) suggested, the other diploid progenitor is likely to be an annual species of *Sedum*, such as *S. nuttallianum* of the south-central United States, with $n=5$. Alternatively, *D. smallii* could be related to *Parvisedum*, a small genus endemic to rock outcrops in California and northern Mexico with $n=9$ (Clausen 1975). Murdy (1968) has also suggested that the outcrop endemic *Cyperus granitophilus* may have originated via autopolyploidy from the more widespread *C. aristatus*.

For allopolyploid speciation, hybridization is a necessary prerequisite. It does appear, from limited evidence, that granite outcrops are hot spots for interspecific hybridization. To some extent, this may be due to their offering a wide range of habitats, some of which mimic those of other physiographic provinces. For example, the Coastal Plain *Senecio tomentosus* extends its range into the Piedmont on the outcrops, where it comes into contact with *S. anonymus*. Chapman and Jones (1971) documented hybridization between these species, showing reduced pollen viability of the F_1 hybrids. Other taxa known to hybridize in association with granite outcrops include *Quercus* (*Q. georgiana* × *nigra*: R. Wyatt and J.R. Allison, pers. obs.), *Aesculus* (*A. pavia* × *sylvatica*: dePamphilis and Wyatt 1989, 1990) and *Isoetes* (*I. piedmontana* × *tegetiformans*: Van De Genachte 1996; Barr and Musselman 1997; *I. melanospora* × *piedmontana*: Matthews and Murdy 1969).

It also seems quite likely that the geographical isolation of populations on granite outcrops has stimulated allopatric speciation. This process is believed to be the mode of origin of the outcrop endemics *Phacelia dubia* var. *georgiana* and *P. maculata* (Murdy 1966, 1968; Levy 1991a). Subsequent studies of these taxa have suggested how reproductive barriers can arise as populations diverge and strongly hint that new incipient varieties are still being formed (Levy 1991b; Levy et al. 1996). Two other outcrop species that Murdy (1968) advanced as examples of ecogeographical speciation are *Portulaca smallii* and *Rhynchospora globularis* var. *saxicola*. This interpretation, however, may be backwards, as the Coastal Plain habitats occupied by their progenitors (presumably, *P. pilosa* and *R. globularis* var. *globularis*) are probably younger than the Piedmont outcrops. Hence, the situation may be analogous to the derivation of Coastal Plain weeds from species originally restricted to Piedmont granite outcrops. These problems cry out for more careful, detailed study.

10.10.8 Appendix

Strict Endemics	Family
Amphianthus pusillus	Scrophulariaceae
Aster avitus	Asteraceae
Isoetes melanospora	Isoetaceae
Isoetes tegetiformans	Isoetaceae
Juncus georgianus	Juncaceae
Phacelia dubia var. *georgiana*	Hydrophyllaceae
Phacelia maculata	Hydrophyllaceae
Sedum pusillum	Crassulaceae
Near-endemics	
Arenaria uniflora	Caryophyllaceae
Coreopsis grandiflora var. *saxicola*	Asteraceae
Cyperus granitophilus	Cyperaceae
Diamorpha smallii	Crassulaceae
Fimbristylis brevivaginata	Cyperaceae
Isoetes piedmontana	Isoetaceae
Lindernia monticola	Scrophulariaceae
Oenothera fruticosa var. *subglobosa*	Onagraceae
Portulaca smallii	Portulacaceae
Portulaca umbraticola ssp. *coronata*	Portulacaceae
Quercus georgiana	Fagaceae
Rhynchospora globularis var. *saxicola*	Cyperaceae
Viguiera porteri	Asteraceae
Half-endemics	
Allium cuthbertii	Alliaceae
Anemone berlandieri	Ranunculaceae
Croton willdenowii	Euphorbiaceae
Gratiola gracilis	Scrophulariaceae
Liatris microcephala	Asteraceae
Minuartia glabra	Caryophyllaceae
Pycnanthemum curvipes	Lamiaceae
Schoenolirion croceum	Hyacinthaceae
Scirpus koilolepis	Juncaceae
Senecio tomentosus	Asteraceae
Talinum mengesii	Portulacaceae
Talinum teretifolium	Portulacaceae
Tradescantia hirsuticaulis	Commelinaceae
"Provincial" endemics	
Aletris farinosa	Nartheciaceae
Anagallis minima	Primulaceae
Arabis missouriensis	Brassicaceae

Bigelowia nuttallii	Asteraceae
Botrychium lunarioides	Ophioglossaceae
Bulbostylis capillaris	Cyperaceae
Callisia rosea	Commelinaceae
Cheilanthes lanosa	Adiantaceae
Cyperus aristatus	Cyperaceae
Cyperus haspan	Cyperaceae
Delphinium carolinianum	Ranunculaceae
Eleocharis microcarpa	Cyperaceae
Euphorbia commutata	Euphorbiaceae
Fimbristylis annua	Cyperaceae
Forestiera ligustrina	Oleaceae
Fuirena squarrosa	Cyperaceae
Helianthus longifolius	Asteraceae
Hypericum denticulatum var. *acutifolium*	Hypericaceae
Hypericum gymnanthum	Hypericaceae
Hypericum lloydii	Hypericaceae
Ipomopsis rubra	Polemoniaceae
Juncus secundus	Juncaceae
Lepuropetalon spathulatum	Saxifragaceae
Manfreda virginica	Agavaceae
Mecardonia acuminata	Scrophulariaceae
Nolina georgiana	Nolinaceae
Oenothera linifolia	Onagraceae
Ophioglossum crotalophoroides	Ophioglossaceae
Opuntia humifusa	Cactaceae
Opuntia humifusa × pusilla	Cactaceae
Opuntia pusilla	Cactaceae
Pilularia americana	Marsileaceae
Polygala curtissii	Polygalaceae
Quercus margaretta	Fagaceae
Rhexia mariana	Melastomataceae
Rhynchospora globularis	Cyperaceae
Rhynchospora rariflora	Cyperaceae
Sabatia quadrangula	Gentianaceae
Scleria ciliata	Cyperaceae
Scleria pauciflora	Cyperaceae
Scutellaria parvula	Lamiaceae
Selaginella rupestris	Selaginellaceae
Seymeria cassioides	Scrophulariaceae
Sideroxylon lanuginosum	Sapotaceae
Silene caroliniana var. *caroliniana*	Caryophyllaceae
Solidago gracillima	Asteraceae
Spiranthes cernua	Orchidaceae
Spiranthes praecox	Orchidaceae
Spiranthes vernalis	Orchidaceae
Steinchisma hians	Poaceae
Tradescantia ohiensis	Commelinaceae
Utricularia cornuta	Lentibulariaceae
Utricularia subulata	Lentibulariaceae
Xyris jupicai	Xyridaceae

Disjunct endemics	
Allium speculae	Alliaceae
Amsonia ludoviciana	Apocynaceae
Danthonia compressa	Poaceae
Draba aprica	Brassicaceae
Eriocaulon koernickianum	Eriocaulaceae
Selaginella tortipila	Selaginellaceae
Incidentals (counted)	
Burmannia capitata	Burmanniaceae
Crassula aquatica	Crassulaceae
Cuscuta harperi	Cuscutaceae
Drosera brevifolia	Droseraceae
Drosera capillaris	Droseraceae
Elatine brachysperma	Elatinaceae
Eleocharis baldwinii	Cyperaceae
Fimbristylis puberula	Cyperaceae
Ophioglossum nudicaule	Ophioglossaceae
Osmanthus americana	Oleaceae
Polygonella americana	Polygonaceae
Scleria reticularis	Cyperaceae
Scutellaria multiglandulosa	Lamiaceae
Selaginella arenicola ssp. *arenicola*	Selaginellaceae
Selaginella arenicola ssp. *riddellii*	Selaginellaceae
Incidentals (not counted)	
Astrolepis sinuata	Polypodiaceae
Ophioglossum engelmannii	Ophioglossaceae
Pellaea wrightiana	Adiantaceae
Ptilimnium nodosum	Apiaceae
Ribes curvatum	Grossulariaceae
Saxifraga texana	Saxifragaceae
Viburnum rafinesquianum	Caprifoliaceae
Weeds	
Agrostis elliottiana	Poaceae
Houstonia pusilla	Rubiaceae
Krigia virginica	Asteraceae
Nothoscordum bivalve	Alliaceae
Nuttallanthus canadensis	Scrophulariaceae
Rumex hastatulus	Polygonaceae
Trifolium carolinianum	Fabaceae
Triodanis perfoliata	Campanulaceae
Valerianella radiata	Valerianaceae

References

Allison JR (1993) Technical draft recovery plan for three granite outcrop species. US Fish and Wildlife Service, Jackson, Mississippi, pp 1–41

Atkins RL, Griffin MM (1977) Geologic guide to Panola Mountain State Park: Rock outcrop trail. Georgia Dept Nat Res, Atlanta, Georgia, pp 1–12

Atkins RL, Joyce LG (1980) Geologic guide to Stone Mountain Park. Georgia Dept Nat Res, Atlanta, Georgia, pp 1–29

Axelrod I (1958) Evolution of the Madro-Tertiary geoflora. Bot Rev 24:433–509

Baker WB (1945) Studies of the flora of the granite outcrops of Georgia. Emory Univ Q 1:162–171

Baker WB (1956) Some interesting plants on granite outcrops of Georgia. Georgia Mineral Newsl 9:10–19

Baldwin JT (1940) Cytophyletic analysis of certain annual and biennial Crassulaceae. Madroño 5:184–192

Barr JR (1997), Musselman LJ (1997) New hybrids in the genus *Isoetes* (quillworts) in the southeastern United States. ASB Bull 44:119–120.

Baskin JM, Baskin CC (1988) Endemism in rock outcrop plant communities of unglaciated eastern United States: an evaluation of the roles of the edaphic, genetic and light factors. J Biogeogr 15:829–840

Braun EL (1935) The undifferentiated deciduous forest climax and the association-segregate. Ecology 16:514–519

Braun EL (1950) Deciduous forests of eastern North America. Blakiston, Philadelphia

Braun EL (1955) The phytogeography of unglaciated eastern United States and its interpretations. Bot Rev 21:297–375

Buck P (1964) Relationships of the woody vegetation of the Wichita Mountains Wildlife Refuge to geological formations and soil types. Ecology 45:336–344

Burbanck MP, Phillips DL (1983) Evidence for plant succession on granite outcrops of the Georgia Piedmont. Am Midl Nat 109:94–104

Burbanck MP, Platt RB (1964) Granite outcrop communities of the Piedmont Plateau in Georgia. Ecology 45:292–306

Campbell EG (1921) Some aspects of Stone Mountain and its vegetation. Proc Indiana Acad Sci 31:91–100

Carlquist S (1974) Island biology. Columbia Univ Press, New York

Carter MEB, Murdy WH (1985) Systematics of *Talinum parviflorum* Nutt. and the origin of *T. teretifolium* Pursh (Portulacaceae). Rhodora 87:191–130

Chapman GC, Jones SB (1971) Hybridization between *Senecio smallii* and *S. tomentosus* (Compositae) on the granite flatrocks of the southeastern United States. Brittonia 23:209–216

Clausen RT (1975) *Sedum* of North America north of the Mexican Plateau. Cornell Univ Press, Ithaca, New York

Crockett JJ (1964) Influence of soils and parent materials on grasslands of the Wichita Mountains Wildlife Refuge, Oklahoma. Ecology 45:326–335

dePamphilis CW, Wyatt R (1989) Hybridization and introgression in buckeyes (*Aesculus*: Hippocastanaceae): a review of the evidence and a hypothesis to explain long-distance gene flow. Syst Bot 14:593–611

dePamphilis CW, Wyatt R (1990) Electrophoretic confirmation of interspecific hybridization in *Aesculus* (Hippocastanaceae) and the genetic structure of a broad hybrid zone. Evolution 44:1295–1317

Diehl SG (1953) The vegetation of the Wichita Mountains Wildlife Refuge. MSc Thesis, Oklahoma State Univ, Stillwater

Duke KM, Crossley DA (1975) Population energetics and ecology of the rock grasshopper, *Trimerotropis saxatilis*. Ecology 56:1106– 1107

Eskew DT (1938) The flowering plants of the Wichita Mountains Wildlife Refuge. Am Midl Nat 20:695–703

Fenneman NM (1938) Physiography of the eastern United States. McGraw-Hill, New York

Golley FB (1965) Structure and function of an old-field broomsedge community. Ecol Monogr 35:113–131
Harper RM (1939) Granite outcrop vegetation in Alabama. Torreya 39:153–159
Harvill AM (1976) Flat-rock endemics in Gray's Manual range. Rhodora 78:145–147
Johnson BR (1995) The ecology and restoration of a high montane plant community. PhD Thesis, University of Georgia, Athens
Keeley JE (1982) Distribution of diurnal acid metabolism in the genus *Isoetes*. Am J Bot 69:254–257
Keever C (1950) Causes of succession on old fields of the Piedmont, North Carolina. Ecol Monogr 20:230–250
Keever C, Oosting HJ, Anderson LE (1951) Plant succession on Rocky Face Mountain, Alexander County, North Carolina. Bull Torrey Bot Club 78:401–421
Küchler AW (1964) Potential natural vegetation of the conterminous United States. Am Geogr Soc Spec Publ 36
Levy F (1991a) Morphological differentiation in *Phacelia dubia* and *P. maculata*. Rhodora 93:11–25
Levy F (1991b) A genetic analysis of reproductive barriers in *Phacelia dubia*. Heredity 67:331–345
Levy F, Antonovics J, Boynton JE, Gillham NW (1996) A population genetic analysis of chloroplast DNA in *Phacelia*. Heredity 76:143–155
Marks PL (1983) On the origin of the field plants of the northeastern United States. Am Nat 122:210–228
Martin CE, Lubbers AE, Teeri JA (1982) Variability in crassulacean acid metabolism: a survey of North Carolina succulent species. Bot Gaz 143:491–497
Matthews JM, Murdy WH (1969) A study of *Isoetes* common to the granite outcrops of the southeastern Piedmont, United States. Bot Gaz 130:53–61
McVaugh R (1943) The vegetation of the granite flatrocks of the southeastern United States. Ecol Monogr 7:404–443
Murdy WH (1966) The systematics of *Phacelia maculata* and *P. dubia* var. *georgiana*, both endemic to granite outcrop communities. Am J Bot 53:1028–1036
Murdy WH (1968) Plant speciation associated with granite outcrop communities of the southeastern Piedmont. Rhodora 70:394–407
Murdy WH, Carter MEB (1985) Electrophoretic study of the allopolyploidal origin of *Talinum teretifolium* and the specific status of *T. appalachianum* (Portulacaceae). Am J Bot 72:1590–1597
Odum EP (1971) Fundamentals of ecology. Saunders, Philadelphia
Oosting HJ (1942) An ecological analysis of the plant communities of Piedmont, North Carolina. Am Midl Nat 28:1–126
Oosting HJ, Anderson LE (1937) The vegetation of a barefaced cliff in western North Carolina. Ecology 18:280–292
Oosting HJ, Anderson LE (1939) Plant succession on granite rock in North Carolina. Bot Gaz 100:750–768
Peet RK, Christensen NL (1980) Succession: a population process. Vegetatio 43:131–140
Phillips DL (1982) Life-forms of granite outcrop plants. Am Midl Nat 107:206–208
Quarterman E, Burbanck MP, Shure DJ (1993) Rock outcrop communities: limestone, sandstone, and granite. In: Martin WH, Boyce SG, Echternacht AC (eds) Biodiversity of the southeastern United States: Upland terrestrial communities. Wiley, New York, pp 35–86
Radford AE, Ahles HE, Bell CR (1964) Manual of the vascular flora of the Carolinas. University of North Carolina Press, Chapel Hill
Raunkiaer C (1934) The life-forms of plants and statistical plant geography. Oxford University Press, London

Rundel PW (1975) Primary succession on granite outcrops in the montane southern Sierra Nevada, USA. Madroño 23:209–220

Rury PM (1978) A new and unique, mat-forming Merlin's grass (*Isoetes*) from Georgia. Am Fern J 68:99–108

Schilling EE, Heiser CB (1981) An infrageneric classification of *Helianthus* (Compositae). Taxon 30:393–403

Schilling EE, Jansen RK (1989) Restriction fragment analysis of chloroplast DNA and the systematics of *Viguiera* and related genera (Asteraceae: Heliantheae). Am J Bot 76: 1769–1778

Sharitz RR, McCormick JF (1973) Population dynamics of two competing annual plant species. Ecology 54:723–740

Shure DJ, Ragsdale HL (1977) Patterns of primary succession on granite outcrop surfaces. Ecology 58:993–1006

Twidale CR (1982) Granite landforms. Elsevier, Amsterdam

Van De Genachte EE (1996) Conservation genetics of the granite outcrop quillworts *Isoetes melanospora* and *Isoetes tegetiformans*. MSc Thesis, University of Georgia, Athens

Visher SS (1954) Climatic atlas of the United States. Harvard University Press, Cambridge, Massachusetts

Wagner WH (1965) *Pellaea wrightiana* in North Carolina and the question of its origin. J Elisha Mitchell Sci Soc 81:95–103

Walter H, Lieth H (1967) Klimadiagramm-Weltatlas. Gustav Fischer, Jena

Walter H (1979) Vegetation und Klimazonen. Ulmer, Stuttgart

Walters TW, Wyatt R (1982) The vascular flora of granite outcrops in the Central Mineral Region of Texas. Bull Torrey Bot Club 109:344–364

Watson TL (1910) Granites of the southeastern Atlantic states. US Geol Survey Bull 426

Wharton GH (1978) The natural environments of Georgia. Georgia Dept Nat Res, Atlanta

Whitney JA, Jones LM, Walker RL (1976) Age and origin of the Stone Mountain Granite, Lithonia District, Georgia. Geol Soc Am Bull 87: 1067–1077

Wiggs DN, Platt RB (1962) Ecology of *Diamorpha cymosa*. Ecology 43:654–670

Williams JE (1969) Photosynthesis in seven old-field plants and the contributions of each to total community biomass. Bull Georgia Acad Sci 27:1–12

Wyatt R (1983) Reproductive biology of the granite outcrop endemic *Sedum pusillum* (Crassulaceae). Syst Bot 8:24–28

Wyatt R (1986) Ecology and evolution of self-pollination in *Arenaria uniflora* (Caryophyllaceae). J Ecol 74:403–418

Wyatt R (1997) Reproductive ecology of granite outcrop plants from the southeastern United States. In: Withers PC, Hopper SD (eds) Granite outcrops symposium. J R Soc West Aust 80:123–129

Wyatt R, Fowler NL (1977) The vascular flora and vegetation of the North Carolina granite outcrops. Bull Torrey Bot Club 104:245–253

11 Phytogeography

R. SEINE, S. POREMBSKI, and U. BECKER

11.1 Introduction

Inselbergs occur worldwide, mainly in the tropics and subtropics. Their scattered distribution, uniform structure, and limited number of habitats render inselbergs ideal ecosystems for comparative studies. The regional reports on the vegetation of inselbergs in Chapter 10 (this Vol.), have presented a wealth of information that will be discussed in a more general understanding here.

11.2 Systematic Composition of Inselberg Florulae

The vegetation of inselbergs in Africa, America, and Australia shows a differing systematic composition for each region. This is reflected in the ten most speciose vascular plant families of individual areas presented in Table 11.1.

It has been argued before that the systematic composition of inselberg florulae is not a random subset of the regional flora (Ornduff 1987; Gröger 1995; Seine 1996). Among the most speciose families, three groups can be distinguished: first, the world's largest vascular plant families (Asteraceae, Orchidaceae, Fabaceae, Rubiaceae, Euphorbiaceae), which are more or less omnipresent in inselberg florulae; second, plant families which are very speciose regionally, such as Bromeliaceae, Melastomataceae, Centrolepidaceae, and Stylidiaceae; third, relatively small families with preadaptations or speciation on inselbergs (poikilohydric Adiantaceae and Scrophulariaceae, succulent Portulacaceae, Aloaceae).

Of course, more than one criterion can be adequate for individual plant families, e.g., Euphorbiaceae are a large family and include speciating, succulent species in the genus *Euphorbia*.

Restriction of systematic groups to inselbergs and similar habitats is rarely encountered, however, Boryaceae, Myrothamnaceae, and Vellozia-

Ecological Studies, Vol. 146
S. Porembski and W. Barthlott (eds.) Inselbergs

ceae among vascular plants and Peltulaceae among lichens are almost exclusively found on inselbergs. Certain genera have high species numbers on inselbergs. In Africa, *Aloe* (approx. 40 spp.), *Cyperus* (28 spp.), *Eriocaulon* (15 spp.), *Euphorbia* (14 spp.), *Indigofera* (17 spp.), *Lindernia* (13 spp.), *Tephrosia* (13 spp.), and *Utricularia* (15 spp.) are among the most speciose genera on inselbergs.

Only few vascular plant species on inselbergs have a distribution that covers several floristic kingdoms. Seine (1996) found merely five species that occur naturally on inselbergs in both Venezuela and Zimbabwe (*Cyperus esculentus*, *Jacquemontia tamnifolia*, *Polycarpaea corymbosa*, *Utricularia subulata*, *Waltheria indica*). Therefore, floristic similarity between different regions should be measured on higher taxonomic level. Floristic composition on the level of vascular plant genera has been used to calculate the Sørensen similarity index. Values for spermatophytes and pteridophytes have been calculated separately; the results are presented in Fig. 11.1. The inventories used are from Biedinger (pers. comm.), Ornduff (1987), Gröger (1995), Porembski (unpubl. data), Seine (1996), and Wyatt and Allison (Chap. 10.10, this Vol.).

In seed plants, floristic similarity between different regions obviously decreases with distance. De Granville (1979) and Reitsma et al. (1992) obtained basically the same results from comparisons of inselberg vegetation on a smaller geographic scale. The impact of distance on floristic change is certainly modulated by regional paleoclimate and connected regional vegetation history. This modulation may explain why Reitsma et al. (1992) found punctuations of the general trend. The two mainland countries studied within the Paleotropics, Côte d'Ivoire, and Zimbabwe, are floristically very close to each other. Comparisons between floristic kingdoms generally result in low figures for the Sørensen index. It is interesting to note that floristically, inselberg vegetation of the USA and Venezuela are less similar to each other than both are in comparison to African countries.

Pteridophyte flora of all included regions is much more homogenous; Sørensen indices are higher than those obtained for seed plants. There is surprisingly high correspondence between inselberg vegetation in the USA and Western Australia (genera in common: *Cheilanthes*, *Isoetes*, *Ophioglossum*), which even exceeds the similarity between Côte d'Ivoire and Zimbabwe. This result should, however, not be overestimated as the overall number of genera included in the calculation is quite small.

Although our knowledge of cyanobacteria, lichens, and bryophytes on inselbergs is only scant, a tendency emerges from the data. Cyanobacterial genera on inselbergs in Africa and America show a high degree of congruence (Büdel et al., this Vol.) and it might be inferred that this is also true for Asia and Australia. Many lichen genera are also present on more than

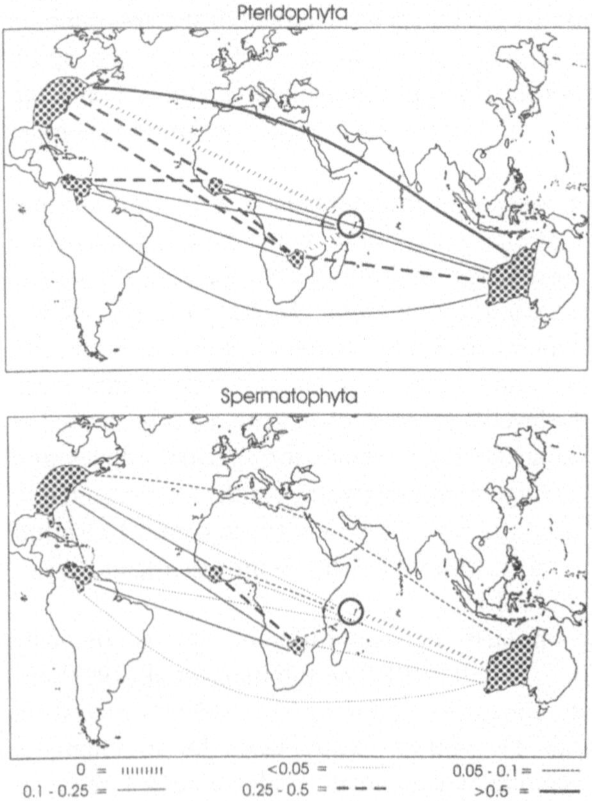

Fig. 11.1. Floristic similarities of inselberg florulae. Similarity is expressed as Sørensen index calculated for genera in common between the regions connected by the lines

one continent (Büdel et al., this Vol.; U. Becker, unpubl.) and even cosmopolitic. Frahm (Chap. 6, this Vol.) reports that the majority of bryophytes from inselbergs are widespread, cosmopolitan, or pantropic species. Obviously, phylogenetically old groups (cyanobacteria, lichens, pteridophytes, ferns) have a more homogenous distribution than spermatophyta. This may be attributed to the antiquity of both inselbergs and groups of organism which had sufficient time to colonize them.

11.3 General Tendencies in Distribution Patterns

Inselbergs are scattered like islands on the land surface of the continents. As their distribution is discontinuous, assemblages of highly localized

Table 11.1. The ten most speciose vascular plant families in inselberg vegetation according to region and rank (number of species). Species numbers of families not belonging to these are omitted (–). Data used: Porembski, Chap. 15, this Vol. (Côte d'Ivoire), Seine 1996 (Zimbabwe), Fischer and Theisen, Chap. 10.4, this Vol. (Madagascar), Biedinger and Fleischmann, Chap. 10.5, this Vol. (Seychelles), Gröger 1995 (Venezuela), Wyatt and Allison, Chap. 10.10, this Vol. (USA), Ornduff 1987 (Western Australia)

	Côte d'Ivoire	Zimbabwe	Madagascar	Seychelles[a]	Venezuela	USA	Western Australia
Fabaceae s.l.	1 (75)	1 (58)	10 (19)	7 (3)	4 (35)	–	9 (4)
Poaceae	2 (72)	2 (53)	2 (36)	2 (4)	6 (31)	1 (9)	2 (16)
Cyperaceae	3 (68)	3 (40)	5 (26)	–	1 (40)	3 (7)	9 (4)
Scrophulariaceae	4 (23)	5 (22)	8 (20)	–	–	6 (3)	8 (6)
Rubiaceae	5 (22)	7 (18)	–	2 (4)	2 (40)	10 (2)	–
Orchidaceae	6 (15)	–	3 (33)	1 (5)	5 (33)	–	–
Commelinaceae	7 (14)	–	–	–	–	4 (3)	–
Lentibulariaceae	8 (14)	–	–	–	–	–	–
Malvaceae	9 (14)	–	–	–	–	–	–
Euphorbiaceae	10 (13)	6 (18)	6 (22)	2 (4)	9 (15)	–	–
Asteraceae	–	4 (33)	1 (79)	–	–	2 (7)	1 (35)
Lamiaceae	–	8 (14)	–	–	–	–	–
Adiantaceae	–	9 (12)	–	–	–	–	–
Acanthaceae	–	10 (10)	–	–	–	–	–
Asclepiadaceae	–	–	4 (31)	–		–	–
Aloaceae	–	–	7 (21)	–		–	–
Gentianaceae	–	–	8 (20)	–		–	–
Apocynaceae	–	–	–	2 (4)	8 (18)	–	–
Myrtaceae	–	–	–	7 (3)	10 (14)	–	–
Arecaceae	–	–	–	2 (4)	–	–	–
Melastomataceae	–	–	–	–	3 (36)	–	–
Bromeliaceae	–	–	–	–	7 (20)	–	–
Liliaceae	–	–	–	–	–	5 (3)	–
Hypericaceae	–	–	–	–	–	7 (2)	–

Table 11.1 (*continued*)

	Côte d'Ivoire	Zimbabwe	Madagascar	Seychelles[a]	Venezuela	USA	Western Australia
Portulacaceae	–	–	–	–	–	8 (2)	–
Rosaceae	–	–	–	–	–	9 (2)	–
Stylidiaceae	–	–	–	–	–	–	3 (11)
Centrolepidaceae	–	–	–	–	–	–	4 (10)
Anthericaceae	–	–	–	–	–	–	5 (8)
Apiaceae	–	–	–	–	–	–	5 (8)
Droseraceae	–	–	–	–	–	–	7 (7)

[a] Only eight families have been included for the Seychelles as species numbers are low

species might be expected to dominate inselberg vegetation. The reports from Africa (Côte d'Ivoire, Porembski, this Vol.; Zimbabwe, Seine and Becker, Chap. 10.2, this Vol.), Madagascar (Fischer and Theisen, Chap. 10.4, this Vol.) and South America (Bolivia, Ibisch et al. 1995; Guianas, Raghoenandan, Chap. 10.7, this Vol.; Venezuela, Gröger 1995) point rather in the opposite direction. The relative number of widespread species is larger than in the regional flora.

In Africa, for example, the Guineo-Congolian element is commonly of little importance while widespread savanna elements (Sudano-Zambezian, Zambesi-Masaian, and Sudano-Zambesi-Masaian) are well represented. The prevalence of savanna elements is observed even in the Taï-rainforest of the Côte d'Ivoire (Porembski et al. 1995).

The most striking example is from inselbergs in Venezuelan Guayana, where Gröger (1995) reports 74% of species having a distribution that exceeds Guyana (neotropical, South American, Caribbean), while Maguire (1970) estimates some 75% of the regional flora to be restricted to Guayana.

Evidence, however, is not completely unambiguous; in the remote and long-isolated islands of the Seychelles (Braithwaite 1984) a different situation has been found. Fleischmann et al. (1996) found some 78% of the indigenous plants on inselbergs to be endemic. The general figure for the granitic islands is approximately 18% of endemics (Renvoize 1979). The high degree of endemism found in inselberg vegetation may be a consequence of the long isolation of the islands since the breaking-up of Gondwana.

Although widespread species are over-represented, characteristic elements of the regional flora are important in inselberg vegetation. Classification (with MULVA-software) of inselbergs in Zimbabwe grouped inselbergs in the Afromontane region apart from inselbergs of the lower areas which belong to the Zambezian floristic region (Seine 1996). These inselbergs show drastic floristic changes over a distance of a few kilometers. Further research in other boundary zones between floristic regions is needed to confirm these results.

There is some evidence for habitat-specific tendencies in distribution patterns. In West and Southeast Africa, constituents of herbaceous vegetation of soil-filled depressions, ephemeral flush, and monocotyledonous mats are commonly widespread species. On the other hand, species recorded in rock pools often show a more regional distribution (Seine 1996; Barthlott and Porembski 1998; Porembski, Chap. 15, this Vol.).

11.4 Endemism

The proportion of endemic species in inselberg vegetation exhibits no common tendency. While it is higher than the regional average on inselbergs in Zimbabwe (Seine et al., 1998), the Georgia Piedmont (Wyatt and Allison this volume), and the Seychelles (Fleischmann et al. 1996) the figures are below average in Venezuelan Guayana (Gröger 1995), the Guianas (Raghoenandan, Chap. 10.7, this Vol.) and Madagascar (Fischer and Theisen, Chap. 10.4, this Vol.). This discrepancy may be due to regional paleoclimate and vegetation history. Further research will be needed to elucidate why the number of endemics does not follow a uniform trend.

The highest proportions of endemics in inselberg florulae are reached in the Seychelles with 78 % (Fleischmann et al. 1996), Australia with approximately 70 % (R. Ohlemüller et al., in prep.), and Madagascar with 72 % (Fischer and Theisen, Chap. 10.4, this Vol.).

The data on the occurrence of paleo- and neoendemics on inselbergs is too patchy for global comparative study. In some cases we can infer the status of endemics from the systematic position of the species. Systematically isolated taxa, e.g., *Encephalartos* (Africa), *Isoetes* (Africa, America, Australia), and *Medusagyne* (Seychelles) can be regarded as paleoendemics on this base. The inverse reasoning can be used for endemics from modern, not isolated taxa such as *Portulaca* (Zimbabwe), *Cordia* (Venezuela), or *Malaxis* (Seychelles).

Several endemic species are known from only a single (*Lindernia syncerus*) or few (*Aloe tauri*, *Lindernia yauendensis*, *Worsleya rayneri*) inselberg localities.

11.5 Influence of Regional Diversity

Inselbergs have been studied in regions of different vascular plant diversity. In spite of the regional diversity, the number of species recorded from inselbergs in the tropics seemed to be relatively constant.

When the area is included into the comparison, however, the species number of inselberg florulae diverge largely from each other (Fig. 11.2). Data used for the species/area plot are from Ornduff (1987), Gröger (1995), Seine (1996), Fischer and Theisen (Chap. 10.4, this Vol.), S. Porembski (unpubl. data), Wyatt and Allison (Chap. 10.10, this Vol.). The slope of a hypothetic straight line through each point gives a value indicating the species richness per area. The magnitude of the values gener-

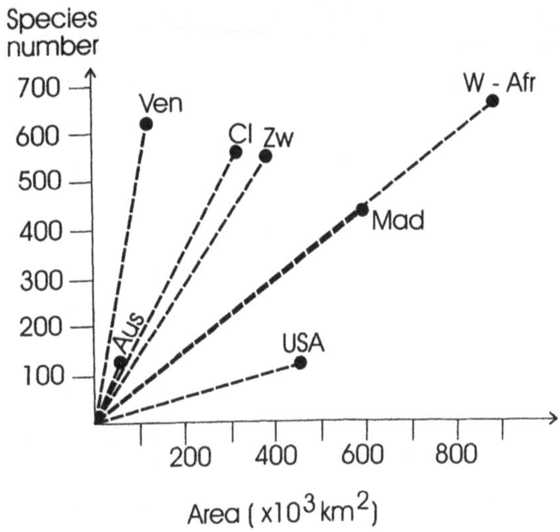

Fig. 11.2. Species/area plot of regional inselberg florulae

ally follows the observed trend of phytodiversity as mapped by Barthlott et al. (1996).

Note the difference between the slope for Côte d'Ivoire and West Africa (including Senegal, Côte d'Ivoire, Ghana and Benin). The inselberg vegetation of West Africa is well studied and homogenous over the whole area (Porembski, Chap. 15, this Vol.). The species number on inselbergs in all of West Africa is not much higher than that recorded for the Côte d'Ivoire, resulting in a much lower slope for the entire area. Unfortunately, the inventories in southeast Africa are not as complete, but the published reports (Porembski 1996; Seine 1996) indicate that the number of species on inselbergs, e.g., in Zimbabwe and Malawi, is much higher than in Zimbabwe alone. This would lead to a higher slope for southeast African inselbergs, as expected from the generally higher phytodiversity of the region.

These results may partly be explained with an influence of the regional species richness upon the beta-diversity found in inselberg vegetation. While in relatively low-diverse West Africa inselberg vegetation remains almost unchanged over vast distances, more differences are recorded in medium-diverse Southeast Africa, and tremendous variation in species composition is found even between adjacent inselbergs in Atlantic Brazil.

The most diverse inselberg flora recorded so far is that of Venezuelan Guayana followed by that of Western Australia while the least diverse was found in the USA.

11.6 Influence of Paleogeography and Paleoclimate

Inselbergs are very old landscape elements. Büdel (1978) estimates an age of 10 to 100 Ma for an inselberg of 100 m height. It can be deduced that inselbergs were most probably available for colonization in the Tertiary.

Inselbergs may have acted as refugia in times of climatic change for both xerophilic and hydrophilic species. This is quite surprising at first sight, but the results presented in the preceding section support this hypothesis. As xeric islands, they support species that are usually found in savannas even on inselbergs in tropical rainforest (Porembski et al. 1995). On the other hand, Giess (1971) and Jürgens and Burke (Chap. 10.3, this Vol.) found relatively hygrophilic species at the foot of inselbergs in the dry parts of Namibia. Tree species that usually grow in mesic savannas take advantage of the runoff water from inselbergs collected either in deep soil-filled depressions or at the foot of the slopes. Seine (1996) reported several species otherwise found at stream margins from inselbergs in Zimbabwe.

During the Pleistocene conversions of rainforest to savanna and back to rainforests some hygrophilic forest species might have persisted at the relatively moist foot of an inselberg. They could then have extended their distribution under pluvial conditions.

As rainforests shrank in dry periods, savannas and other xerophilic formations extended their range of distribution. The species that migrated with the formations may then have remained on inselbergs as the climate became moister again. Examples have been reported from Africa and North America. In West Africa *Aloe buettneri* and *A. schweinfurthii* occur on inselbergs while the genus *Aloe* has its center in South and East Africa. Walters and Wyatt (1982) report species of *Yucca*, *Opuntia*, *Talinum*, and *Portulaca* as outliers on granite outcrops in the southeastern USA while these genera are most diverse in the southwestern USA.

Inselbergs harbor persistent ancient elements such as *Encephalartos ituriensis* and *E. septentrionalis* in the Kongo, *E. equatorialis* in Uganda (Hurter et al. 1996), *E. munchii* (Osborne 1993b), and *E. concinnus* in Zimbabwe (Osborne 1993a). Another taxonomically isolated group of plants well represented on inselbergs throughout the world is the genus *Isoetes*, with two spp. in the USA (Wyatt and Allison, Chap. 10.10, this Vol.), two spp. in Africa (S. Porembski unpubl.; Seine 1996) and two spp. in Australia (Ornduff 1987). Their presence bears living proof of the long-term existence of inselbergs and their typical habitats.

Angiosperms were already well developed when the southern supercontinent Gondwana disintegrated in the Cretaceous. In disjunct families such as the Velloziaceae, a distribution pattern that may stem from the

Gondwana period is found. The Velloziaceae have centers of diversity in South America, Southeast Africa, and Madagascar. A disjunction on a smaller geographic scale is found in *Microdracoides squamosus*, a Cyperaceae which occurs in Cameroon to Nigeria, and Sierra Leone to Guinea. This disjunction was probably caused by Pleistocene climatic fluctuations which fragmented a previously continuous area. However, not all disjunctions that have been observed should be interpreted in terms of paleogeography; e.g., the presence of the American *Utricularia juncea* on inselbergs in the Côte d'Ivoire rather seems to be a result of relatively recent long-distance dispersal (Dörrstock et al. 1996) as it belongs to the "modern" Lentibulariaceae. The same is true for Malagasy populations of *Rhipsalis baccifera* which have developed as true lithophytes on inselbergs in Madagascar (Barthlott 1983).

Vicariating taxa are also well represented in inselberg vegetation. The dominant mat-forming species on African inselbergs are the Cyperaceae *Afrotrilepis pilosa* and *Coleochloa setifera*, which occur in West Africa and East and Southeast Africa, respectively. Within the Velloziaceae, *Xerophyta* and *Vellozia* vicariate in Africa, Madagascar, and America.

11.7 Speciation

Inselbergs occur as a mosaic of islands in a homogenous surrounding. Distances between individual inselbergs or inselberg groups vary largely. Microclimatic conditions on inselbergs differ significantly from the surrounding (Szarzynski, Chap. 3, this Vol.). Therefore, speciation through geographic isolation (allopatric) or ecological isolation (sympatric) should be possible on inselbergs. Wyatt and Allison (Chap. 10.10, this Vol.) give a survey of hybrid origin of species found on inselbergs in the Piedmont of the USA.

Species pairs with one species restricted to inselbergs and the second, closely related species found in the surrounding landscape have been reported by Gröger (1995), Seine (1996), and Wyatt and Allison (Chap. 10.10, this Vol.). These species are obviously ecologically different nowadays but whether they evolved allopatrically or sympatrically can only be speculated. For one of these species (*Arenaria glabra*), however, Wyatt (1986) proposed an ecological isolation through a change in life history.

Speciation in geographic isolation may be the mechanism underlying the formation of the *Encephalartos munchii* group occurring on inselbergs on the Zimbabwe-Mozambique border (Osborne 1993b). Several species have been recorded to occur on inselbergs in close vicinity to each other. While most *Encephalartos* species on inselbergs are relictual endemics, relatively recent speciation is involved in the *E. munchii* group. Wyatt and Allison (Chap. 10.10, this Vol.) cite a number of examples of allopatric speciation on inselbergs in the USA.

S. Porembski (unpubl.) noted the presence of morphologically different forms of *Cyanotis lanata* and *Afrotrilepis pilosa* on different inselbergs in West Africa, Seine (unpubl.) found similar morphological variation between stands of *Coleochloa setifera* and *Lindernia pulchella* in Zimbabwe. It might be speculated that morphological differences indicate the beginning of a divergent evolution of isolated populations. This tendency is also observed in the taxonomically notorious genus *Xerophyta* which exibits an astonishing amount of morphological plasticity already over short distances.

11.8 Exotics and Weeds

Exotics are plant species introduced to a country where they do not occur naturally. Unfortunately, weeds are not easily defined (Crawley 1997). We will here consider weeds to be exotics which invade natural vegetation in their new surrounding. Weeds are usually species with an ability to colonize early stages of succession and with a short, often annual, life cycle. The relation between inselbergs and weeds and exotics is twofold: inselbergs may be a source for invasive species but they may also be an ecosystem susceptible to weed and exotics invasion.

11.8.1 Inselbergs as a Source of Exotics and Weeds

Many species recorded as indigenous vegetation on inselbergs have a potential of invading early successional or disturbed habitats. If they are cultivated as exotics for economic or horticultural purposes, they may escape cultivation and become naturalized or obnoxious weeds.

Cyanotis lanata is a typical species of inselbergs in Africa. In the Côte d'Ivoire, *Cyanotis lanata* occurs along tarred roadsides, where it may form extensive, monospecific colonies. The species is otherwise almost exclusively found on inselbergs and on lateritic crusts (Porembski et al. 1994). Other species, such as *Bryophyllum tubiflorum* from inselbergs in Madagascar and other Malagasy Crassulaceae, are widespread exotics in the tropics and have been imported for horticultural purposes. *Melinis*

Fig. 11.3. *Furcraea foetida* growing on an inselberg on the island of Mahé, Seychelles. (Photograph N. Biedinger)

repens, a common grass of inselbergs in southern Africa, is today widely naturalized in the tropics and USA (Hitchcock 1950; Zizka 1989). *Bryophyllum tubiflorum*, native to inselbergs in Madagascar, has been recorded from inselbergs in Zimbabwe (Seine 1996) and Brazil (Porembski et al. 1998). Wyatt (1997) suggested that a number of weeds in the USA originated from inselbergs in the Piedmont.

A serious problem for conservationists is *Furcraea foetida*, which has been planted throughout the tropics for its fibers that were used for the production of textiles. The species is highly invasive. It has been recorded on inselbergs in the Seychelles (Fig. 11.3) and Madagascar (Biedinger and Fleischmann, Chap. 10.5, this Vol.; W. Barthlott, pers. comm.). Raghoenandan (Chap. 10.7, this Vol.) reports *Furcraea foetida* to occur naturally on inselbergs in the Guianas and considers it a typical plant there.

Dematiaceous fungi, naturally occurring on rocks, might be called weeds as they are even found growing on marble buildings, where they may cause considerable damage through biopitting (Sterflinger and Krumbein 1997).

The remarkable success of inselberg species as weeds may partly be explained by the harsh environmental conditions they face in their natural habitats. Species that endure these conditions and also possess a high potential to colonize or recolonize habitats after disturbance have excellent preadaptations to invade disturbed ground in other ecosystems. Coupled with well-spreading propagules, such inselberg species would be perfect invaders.

11.8.2 Alien Invasion on Inselbergs

Inselbergs are only reluctantly colonized by alien species. Only few weeds or exotics have been recorded on inselbergs in Africa (Porembski, Chap. 15, this Vol.; Jürgens and Burke, Chap. 10.3, this Vol.; Seine 1996), Madagascar (Fischer and Theisen, Chap. 10.4, this Vol.), the USA (Wyatt and Allison, Chap. 10.10, this Vol.), South America (Gröger, Chap. 10.6, this Vol.; Raghoenandan, Chap. 10.7, this Vol.). They were mostly recorded on only few inselbergs, usually near human settlements, and did not replace natural vegetation on a large scale. On a single inselberg in the Côte d'Ivoire, Mt. Brafouédi, *Ananas comosus* has excluded *Afrotrilepis pilosa* and now forms mats. This is most probably due to continuous supply of plantlets from discarded tips of fruits. Other relatively common exotics or weeds on inselbergs are *Catharanthus roseus*, *Lantana camara*, *Canna* spp., *Opuntia* spp., and several Crassulaceae.

Contrastingly, alien species account for approximately 25 and 40% of the inselberg flora in Western Australia (Ornduff 1987) and the Seychelles (Biedinger and Fleischmann, Chap. 10.5, this Vol.). Although species numbers of exotics are that high, regeneration of indigenous plants is not impeded in the Seychelles (Fleischmann et al. 1996) and nothing similar is mentioned for Western Australia.

Most probably, the adverse abiotic conditions on inselbergs are too forbidding for most exotics. Even highly disturbed inselbergs near towns usually maintain a vegetation cover of indigenous plants.

Acknowledgements. We would like to thank the Deutsche Forschungsgemeinschaft (DFG, grant no. Ba 605/2) for funding research on inselbergs. Scholarships granted to R. Seine and U. Becker by the Studienstiftung des deutschen Volkes and Graduiertenförderung des Landes Nordrhein-Westfalen, respectively, are gratefully acknowledged. The authorities of Côte d'Ivoire, Malawi, and Zimbabwe kindly granted permission to conduct research. The friendly support of R. B. Drummond (Harare), T. Müller (Harare), N. Nobanda (Harare), and J. Seyani (Zomba) is gratefully acknowledged. H. Geithmann (Bonn) assisted in the preparation of Fig. 11.1 and 11.2. N. Biedinger (Rostock) kindly granted permission to use Fig. 11.3.

References

Barthlott W (1983) Biogeography and evolution in Neo- and Palaeotropical Rhipsalinae (Cactaceae). Sonderber Naturwiss Ver Hamb 7:241–248

Barthlott W, Porembski S (1998) Diversity and phytogeographical affinities of inselberg vegetation in tropical Africa and Madagascar. In: Huxley CR, Lock JM, Cutler DF (eds) Chorology, taxonomy and ecology of the floras of Africa and Madagascar. Royal Botanic Gardens, Kew, pp 119–129

Barthlott W, Lauer W, Placke A (1996) Global distribution of species diversity in vascular plants. Erdkunde 50:317–327

Braithwaite CJR (1984) Geology of the Seychelles. In: Stoddart DR (ed) Biogeography and ecology of the Seychelles Islands. Dr W Junk, The Hague, pp 17–33

Büdel J (1978) Das Inselberg-Rumpfflächenrelief der heutigen Tropen und das Schicksal seiner fossilen Altformen in anderen Klimazonen. Z Geomorphol Suppl 31: 79–110

Crawley MJ (1997) Plant ecology. 2nd ed. Blackwell, Oxford

De Granville JJ (1979) Forest flora and xeric flora refuges in French Guyana during the late Pleistocene and Holocene. Communication, 5th Int Symp Assoc Trop Biol Caracas, 1978, p 45

Dörrstock S, Seine R, Porembski S, Barthlott W (1996) First record of the American *Utricularia juncea* (Lentibulariaceae) for tropical Africa. Kew Bull 51:579–583

Fleischmann K, Porembski S, Biedinger N, Barthlott W (1996) Inselbergs in the sea: vegetation of granite outcrops on the islands of Mahé, Praslin and Silhouette (Seychelles). Bull Geobot Inst ETH 62:61–74

Giess W (1971) A preliminary vegetation map of South West Africa. Dinteria 4:5–114

Gröger A (1995) Die Vegetation der Granitinselberge Südvenezuelas. Thesis, Bonn

Hitchcock AE (1950) Manual of the grasses of the United States. United States Government Printing Office, Washington

Hurter J, Glen H, Claasen I (1996) *Encephalartos equatorialis* Hurter. Encephalartos 44:4–9

Ibisch PL, Rauer G, Rudolph D, Barthlott W (1995) Floristic, biogeographical, and vegetational aspects of Pre-Cambrian rock outcrops (inselbergs) in eastern Bolivia. Flora 190:299–314

Maguire B (1970) On the flora of the Guayana Highland. Biotropica 2:85–100

Ornduff R (1987) Islands on islands: plant life on the granite outcrops of Western Australia. Harold L. Lynn Arboretum Lecture No. 15, University of Hawaii Press, Honolulu

Osborne R (1993a) *Encephalartos concinnus* R.A. Dyer & Verdoorn. Encephalartos 34:4–11

Osborne R (1993b) *Encephalartos munchii* R.A. Dyer & Verdoorn. Encephalartos 35:4–9

Porembski S (1996) Notes on the vegetation of inselbergs in Malawi. Flora 191:1–8

Porembski S, Barthlott W, Dörrstock S, Biedinger N (1994) Vegetation of rock outcrops in Guinea: granite inselbergs, sandstone table mountains and ferricretes – remarks on species numbers and endemism. Flora 189:315–326

Porembski S, Brown G, Barthlott W (1995) An inverted latitudinal gradient of plant diversity in shallow depressions on Ivorian Inselbergs. Vegetatio 117:151–163

Porembski S, Martinelli G, Ohlemüller R, Barthlott W (1998) Diversity and ecology of saxicolous vegetation mats on inselbergs in the Brazilian Atlantic rainforest. Divers Distrib 4:107–119

Reitsma JM, Louis AM, Floret JJ (1992) Flore et végétation des inselbergs et dalles rocheuses; première étude au Gabon. Bull Mus Natl Hist Nat B Adansonia 14:73–97

Renvoize SA (1979) The origins of Indian Ocean island floras. In: Bramwell D (ed) Plants and islands. Academic Press, London, pp 107–129

Seine R (1996) Vegetation von Inselbergen in Zimbabwe. Archiv naturwissenschaftlicher Dissertationen, vol. 2. Martina Galunder, Wiehl

Seine R, Becker U, Porembski S, Follmann G, Barthlott W (1998) Vegetation of inselbergs in Zimbabwe. Edinb J Bot 55:267–293

Sterflinger K, Krumbein WE (1997) Dematiacous fungi as a major agent for biopitting on Mediterranean marbles and limestones. Geomicrobiol J 14:219–230

Walters TW, Wyatt R (1982) The vascular flora of granite outcrops in the Central Mineral Region of Texas. Bull Torrey Bot Club 109:344–364

Wyatt R (1986) Ecology and evolution of self-pollination in *Arenaria uniflora* (Caryophyllaceae). J Ecol 74:403–418

Wyatt R (1997) Reproductive ecology of granite outcrop plants from the southeastern United States. In: Withers PC, Hopper SD (eds) Granite outcrops synposium. J R Soc West Aust 80:123–129

Zizka G (1989) *Melinis* Beauv. In: Exell AW, Wild H (eds for Flora Zambesiaca Committee) Flora Zambesiaca, vol 10. Crown Agents, London

12 Factors Controlling Species Richness of Inselbergs

S. Porembski, R. Seine, and W. Barthlott

12.1 Introduction

Understanding patterns of biodiversity is one of the major topics of current ecological research. There is still much debate about the decisive factors regulating the species richness of plant communities. Attempts to explain biodiversity patterns, and moreover, to address the question of the maintenance of local species richness focus on certain critically discussed concepts, such as long-term equilibrium versus nonequilibrium conditions or metapopulation dynamics (for an overview see Tilman and Pacala 1993). Many problems concerning the ecological significance of diversity (e.g., does diversity promote community stability?) are still far from resolved (Schulze and Mooney 1993). This is mostly due to our lack of understanding of the fundamental attributes of both species and ecosystems.

For a long time, oceanic islands have attracted much biological interest and have been of great importance for much outstanding research. A classical example is given by Darwin's studies on the Galápagos finches, which were influential for his theory of natural selection (Darwin 1845). Unique biological phenomena, like the presence of peculiar woody plants (belonging mainly to herbaceous families) on islands have been described in detail (Carlquist 1974), and islands have provided the basis for the formulation of theoretical fundaments, such as *The Theory of Island Biogeography* by MacArthur and Wilson (1967). Biodiversity on islands is still a very vivid and promising field of research. Vitousek et al. (1995) provide an overview of recent developments in this field.

Inselbergs are characterized by harsh environmental conditions (see Szarzynski, Chap. 3, this Vol.) and bear a very characteristic vegetation. These rock outcrops therefore constitute biological islands, despite the fact that there is a certain degree of floristic exchange between inselbergs and

Ecological Studies, Vol. 146
S. Porembski and W. Barthlott (eds.) Inselbergs

adjacent vegetation types. Over the past years an increasing number of detailed floristic and vegetational studies covering a large geographical area have been published which provide insights into the composition of different habitats on inselbergs. In contrast, a biodiversity approach that considers, for example, small-scale spatial and temporal distribution patterns and population dynamics has been largely neglected hitherto, with only a few papers (e.g., Sharitz and McCormick 1973; Collins et al. 1989; Houle and Phillips 1989; Houle 1990; Porembski et al. 1995) devoted to these topics.

There are a number of reasons which may render inselbergs highly suitable models for addressing different aspects of biological diversity.

1. Granitic and gneissic inselbergs occur in both tropical and temperate regions thereby extending throughout nearly all climatic and vegetational zones. In being geologically more or less homogeneous, this allows for comparative studies between uniform objects distributed along steep ecological gradients.
2. Inselbergs cover a broad spectrum of sizes, from several square kilometers to only a few square meters. Large inselbergs may attain an absolute height of more than 600 m above the surroundings, small outcrops rise only a few centimeters. Contrary to oceanic islands, which lose their characteristic vegetation below a certain minimum size (small island effect) even small inselbergs (i.e., only a few m^2 in size) retain their typical vegetation. Their relatively small size, the low degree of habitat complexity, and a comparatively low number of species and small populations are methodologically advantageous for biodiversity studies.
3. Due to their small agricultural potential and their protected state because of religious reasons (see Seine, Chap. 14, this Vol.) inselbergs belong to the least disturbed ecosystems, which is in stark contrast to the situation of most oceanic islands.

The intention of this chapter is to provide insights into common patterns of diversity on inselbergs and to identify possible factors responsible for structuring their plant communities. Due to our long-term experience in the Côte d'Ivoire, the focus will be on this country. However, this account aims to give a survey of general mechanisms which maintain diversity in inselberg plant communities. For general introductions into methods of biodiversity research and definitions we refer to Gaston (1996), Magurran (1988), and Whittaker (1972).

12.2 Species Richness of Inselbergs: the Relationship Between Local and Regional Diversity

There is still some controversy concerning the important determinants of local diversity. For recent accounts on this topic see, e.g., Ricklefs (1987), Ricklefs and Schluter (1993), Westoby (1993), and Caley and Schluter (1997). According to the latter authors, "a range of relationships between local and regional species richness is possible, from those in which local species richness is dependent on regional species richness to those in which it is not". A methodological problem for investigating the relationship between local and regional diversity is the choice of appropriate objects. The search for pertinent patterns of biodiversity has led to very different approaches on various systematic, taxonomic, and ecological levels. It has been suggested that the subjectivity of these approaches should be reduced by selecting indicator taxa (Pearson 1995). Another objective approach, however, could be the choice of indicator ecosystems. Of particular interest is the examination of sites that are more or less uniform in their abiotic characteristics over broad geographic ranges. Due to their relatively homogeneous geology (at least from the viewpoint of plants), granitic and gneissic inselbergs appear promising candidates.

As mentioned above, indicator ecosystems can help to understand the relationship between local and regional diversity. In the following, this relationship is examined by comparing species richness of both complete inselbergs and individual habitats (e.g., monocotyledonous mats) in different geographical regions. Assessing the role of regional diversity for local species richness of inselbergs is possible by direct comparison between rock outcrops located in clearly distinguished regions. In order to carry out meaningful comparisons, inselbergs of similar size and geomorphology were chosen from Côte d'Ivoire, Zimbabwe, and Venezuela. Profound field experience enabled the proper selection of suitable sites which are representative for a whole region. There are large regional differences both in species richness and the number of endemics (Table 12.1).

Data in Table 12.1 clearly show the inselbergs in Venezuela to be richer in species number and endemics (both Venezuela and Zimbabwe) in comparison to West African rock outcrops. Several factors might be responsible for this significant differentiation. For example, it may be due to the longer continuous development of rock outcrop habitats (i.e., no drastic climatic changes) and the generally larger number of available azonal localities (such as the Precambrian crystalline chains in East Africa, like Usambara and Uluguru, or white sand savannas and sandstone outcrops of the Guayana shield).

Table 12.1. Species richness of vascular plants and number of endemics of equally sized inselbergs in Côte d'Ivoire, Zimbabwe, and Venezuela

	Species number	No. of endemics
Côte d'Ivoire	97	0
Zimbabwe	94	5
Venezuela	170	38

In respect of individual habitats, monocotyledonous mats, which occur like carpets on exposed rocky slopes, are one of the most conspicuous inselberg communities throughout the tropics. The sedge *Afrotrilepis pilosa* is the dominant mat-forming species on West African inselbergs, where it is the main constituent of an extremely species-poor community. It is apparent that competitive displacement and the low degree of disturbance prevent higher diversity in the almost monospecific mats. *Coleochloa*-mat communities on East African and Madagascan inselbergs are only slightly richer in species. The comparison with mat communities on inselbergs in other tropical regions reveals drastic differences in species richness. For instance, the Brazilian rock outcrops situated in the Mata Atlântica forest support a floristically much more diverse (i.e., higher alpha diversity) mat community that is characterized by a large percentage of endemics (Porembski et al. 1998).

Mat communities on inselbergs in the Mata Atlântica in eastern Brazil are characterized by a high degree of species richness, particularly when compared to their West African counterparts. A major underlying cause of this discrepancy is the extraordinary high regional diversity of the Mata Atlântica in general. It can therefore be assumed that regional processes may have had and still have a considerable impact on mat communities in this part of the world. This is, for instance, demonstrated by the rich lithophytic vegetation on East Brazilian inselbergs, which is in marked contrast to the situation in West Africa. Most typical lithophytic components (e.g., Bromeliaceae and Orchidaceae) of rock outcrops in the Mata Atlântica are closely related to epiphytic species occurring in the vicinity.

Further important factors contributing to the high degree of species richness on eastern Brazilian rock outcrops are the long evolutionary history largely uninterrupted by "catastrophic" events, as well as the large number of available azonal localities. It is conceivable that this factor has resulted in the evolution of a large number of species, that are adapted to withstand the harsh environmental conditions on rock outcrops and in the canopy. The difference in diversity between mat communities on West

African and East Brazilian inselbergs appears, therefore, to be a consequence of the size of the available species pool.

The mat vegetation of the Brazilian inselbergs is also characterized by a relatively high degree of spatial floristic variability (i.e., beta diversity). This fact probably indicates the importance of random colonization events within mat communities. It can be speculated that their unpredictable floristic composition is due to chance immigration (i.e., randomness of colonization as suggested by "lottery" models of community organization, Yodzis 1986) out of a rich pool of mat-forming species which are possibly very similar in their habitat preferences. It therefore seems apparent that the above-mentioned differences in mat diversity are indicative of the fact that local species richness is determined by regional diversity on the basis that if the species pool of the surrounding area is larger, there will, theoretically at least, be a greater number of taxa able to colonize inselberg-specific habitats.

Numerous examples confirm the existence of a latitudinal gradient in diversity with a decline in species richness from the equator towards the polar regions (Fischer 1961). How does tropical inselberg plant diversity compare with temperate zone inselberg plant diversity? If we examine gamma diversities (i.e., the number of species in a whole region) of selected inselberg habitats, it becomes clear that diversity of inselberg plants does not show a clear latitudinal trend. For example, a total of 72 species have been recorded in shallow depressions on Ivorian inselbergs compared to 134 species in Western Australia (Ohlemüller 1997). Our own preliminary data indicate that there is large regional variation in gamma diversity of individual habitats, with no specific region being the most diverse for all types of habitats. The reasons that may explain why the almost ubiquitous latitudinal gradient is not clearly reflected by the inselberg plant communities are open to speculation. Independent of latitude (excluding regions of extreme cold or dryness), regional plant diversity of inselbergs is promoted by the following factors (in descending order of importance):

1. The presence of a large number of inselberg-like sites (e.g., rock outcrops, canopy) increases species richness because of lower regional extinction rates due to the wider spread of populations.
2. Environmental fluctuation prevents a state of competitive equilibrium being reached. For example, moderate perturbations of growth encourage diversity by reducing the competitive ability of potential dominants, therefore facilitating the coexistence of inferior competitors.
3. Environmental stability over long periods of time (e.g., without climatic extremes such as glaciation, aridity) can contribute to a rise in

diversity because continuous environmental conditions allow evolutionary processes to play a greater role, often expressed as an increased degree of endemism.
4. Inselberg geomorphology is a determinant of diversity since parameters like steepness of slopes will decide whether certain communities are able to develop or not.

It is also important to mention a further component of plant diversity, i.e., the genetic divergence of species occurring in geographically isolated populations on rock outcrops. There is a considerable amount of literature available concerning the effects of ecological isolation and ecotypic differentiation of rock outcrop plants (for survey see Baskin and Baskin 1988). Prominent in this respect are the detailed studies on the evolutionary ecology of plants restricted to serpentine biota in California (Kruckeberg 1984), which have delivered considerable insight into speciation processes within discontinuously distributed taxa (e.g., *Streptanthus*, Mayer et al. 1994). For granite outcrop plants, several studies (e.g., Moran and Hopper 1983; Murdy and Brown Carter 1985; Sampson et al. 1988; Wyatt et al. 1992) have revealed minor genetic differences within isolated populations, but high genetic diversity between isolated populations. For example, this pattern is found in *Eucalyptus caesia*, a rare tree which has a disjunct distribution on inselbergs in southwestern Australia. This is also the case with *Arenaria uniflora* und *Talinum mengesii*, granite outcrop endemics in southeastern USA. High levels of population divergence and low levels of within-population genetic diversity are the consequence of several factors, such as genetic drift and self-pollination. A large degree of infraspecific population divergence could be demonstrated for the granite outcrop species *Isotoma petraea* in Western Australia. In selfing populations of this species, translocation heterozygosity has evolved in association with inbreeding and the presence of recessive lethals (James 1965; James et al. 1983). In contrast to these examples, however, the genetic structure of the Commelinaceae *Tradescantia hirsuticaulis* (found primarily on rock outcrops in southeastern USA) displays the opposite tendency by exhibiting high levels of genetic variation within populations and a low level of genetic differentiation among populations (Godt and Hamrick 1993). In this case, the low level of divergence between populations is difficult to explain, since there are no pollinators and no obvious means of long-distance dispersal that serve to increase gene flow among noncontiguous populations.

12.3 Species Richness, Size, and Isolation of Inselbergs

12.3.1 Species Richness of Inselbergs and Size

The number of vascular plant species on inselbergs is a function of their size, with large inselbergs harboring more species than smaller ones (Fig. 12.1). This has been observed for various tropical and temperate regions. This general pattern of diversity is not surprising, and has often been reported from many biota (see survey in Rosenzweig 1995). In the Côte d'Ivoire, nearly 100 inselbergs ranging in size from 200 m^2 to 7 km^2 were sampled (excluding forest-type vegetation). The lowest number of species recorded per outcrop was 3, the most species-rich outcrop contained 204 species (own unpubl. data). The rise in species richness with increasing area is mainly due to the fact that large inselbergs contain more habitat types, for instance mat communities and ephemeral flush vegetation. Moreover, the size of populations increases with inselberg size, thus reducing the risk of local extinction and contributing to higher species richness. Inselberg size not only influences the number of plant species, but also the relative abundance of life-forms. Figure 12.2 shows the life-form spectra (according to Raunkiaer 1934) of three inselbergs situated in the savanna zone which vary considerably in size. The smallest outcrop (500 m^2) is dominated by therophytes (75% of species), followed by hemicryptophytes and cryptophytes, whereas chamaephytes and phanero-

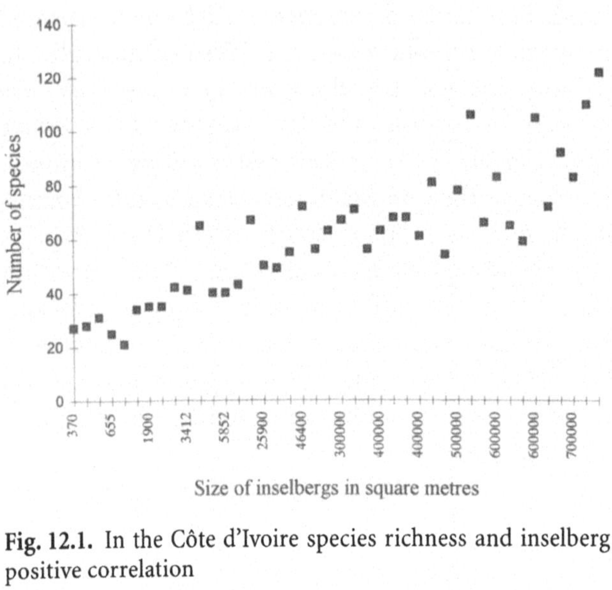

Fig. 12.1. In the Côte d'Ivoire species richness and inselberg size display a significant positive correlation

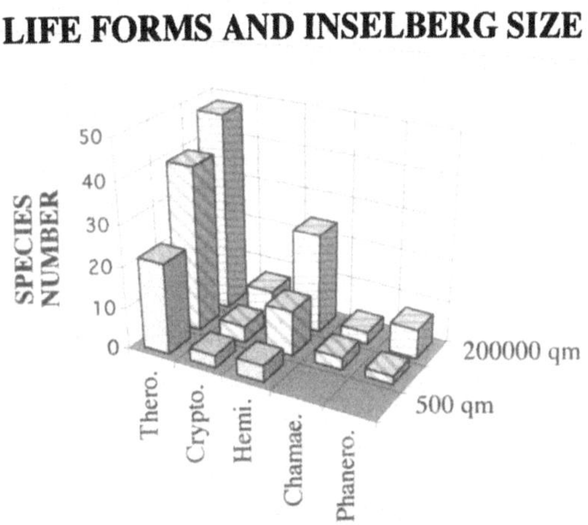

Fig. 12.2. Inselberg size and life-form spectra (data obtained in the Côte d'Ivoire)

phytes are absent. On larger outcrops (15 000 m² or 200 000 m²) the importance of therophytes diminishes (60 and 45 %, respectively) and species number of all other life-forms (including chamaephytes and phanerophytes) increases. These size-specific life-form spectra probably reflect the relative importance of disturbance in relation to inselberg area. The impact of unpredictable climatic fluctuations (i.e., amount and distribution of rainfall) is particularly high on small rock outcrops and therefore promotes the extraordinary high percentage of annuals. Many annuals on Ivorian inselbergs are highly successful pioneers of sites characterised by short-term disturbance. Perennial plants prefer the more stable growth conditions typical of larger inselbergs. For example, the chamaephyte *Afrotrilepis pilosa* is usually not found on outcrops less than 1 ha in size, and for the successful establishment of trees such as *Hildegardia barteri* and *Hymenodictyon floribundum*, inselbergs larger than 5 ha are needed. For inselbergs in the Côte d'Ivoire, therefore, a species-specific size classification can be made (Table 12.2).

However, these species-specific size requirements are modified by several factors including frequency and degree of isolation of rock out-

Table 12.2. Life-form-specific size classification of inselbergs in the Côte d'Ivoire

Inselberg type	Percentage therophytes	Minimal inselberg size (m²)
Lichen inselberg	–	ca. 1
Therophyte inselberg	>80 %	ca. 50
Perennial herb inselberg	ca. 60 %	ca. 10 000
Phanerophyte inselberg	ca. 45 %	ca. 50 000

crops within a particular area. For instance, *Afrotrilepis pilosa* was recorded on rock outcrops less than 1 ha in size which were situated in close proximity (ca. 300 m) to larger inselbergs. It can be assumed that the presence of *Afrotrilepis pilosa* on small rock patches where reproductive success is low but local extinction rates are high is the result of a source-sink relationship, with the larger inselbergs acting as diaspore donors.

The comparison between equal-sized Ivorian inselbergs located in the savanna zone reveals that there is only minor variation in species richness. Qualitatively, however, large differences between individual outcrops can be found. Table 12.3 lists similarity coefficients (as a measure of beta diversity) for a number of inselbergs studied. On the face of it, the similarity coefficients are relatively low, which indicates high beta diversity. This would underline the importance of stochastic factors in regulating species composition. A closer inspection, however, reveals that stochasticity is mainly due to "generalists" (i.e., weed-like species occurring in a broad range of localities) which randomly colonize vacant sites where nonequilibrium conditions prevail. In contrast, the inselberg vegetation contains a deterministic fraction, i.e., species whose presence can be predicted with almost certainty. Although they may not predominate in terms of species number, these "specialists" such as *Afrotrilepis pilosa* and *Cyanotis lanata*, form the most important components of the inselberg vegetation in respect of certain parameters, including phytomass (i.e., physiognomic appearance) and constancy. It can be concluded, therefore, that most habitats on West African inselbergs as well as those in other regions are in equilibrium since they are dominated by relatively few, often abundant specialists. At the same time, however, there is a strong small-scale nonequilibrium component indicated by the presence of a high proportion of rare generalists.

The dichotomy between relatively few but abundant taxa and a large number of rare species has been demonstrated for inselbergs in north-

Table 12.3. Similarity coefficients (after Sørensen) for inselbergs in the savanna zone of the Côte d'Ivoire. Approximate size of inselbergs ranged between 80–100 ha

	No. 1	No. 2	No. 3	No. 4	No. 5	No. 6	No. 7
No. 1		0.30	0.28	0.31	0.29	0.26	0.31
No. 2			0.28	0.27	0.28	0.28	0.32
No. 3				0.29	0.29	0.27	0.30
No. 4					0.30	0.29	0.30
No. 5						0.31	0.30
No. 6							0.29

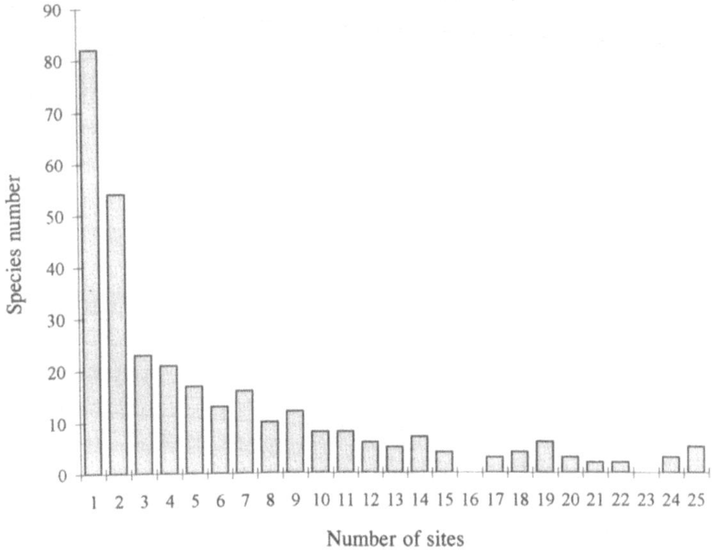

Fig. 12.3. Number of records of vascular plant species found on 25 inselbergs in the Comoé National Park (NE Côte d'Ivoire). Only five species were present on all inselbergs

eastern Côte d'Ivoire (Porembski and Brown 1995). Figure 12.3 shows that more than 75 % of all species occurred on fewer than 5 (out of 25) outcrops, with 37 % of species restricted to a single inselberg. Species present on 90 % of the rock outcrops studied were mainly specialists. Similar species-abundance distributions have been frequently observed elsewhere (Gaston 1994).

If only rock outcrop specialists are considered, the similarity in species composition between individual inselbergs in the northeast of the Côte d'Ivoire seems to be influenced by their size. Figure 12.4 shows that variation in species composition is greatest between small rock outcrops, whereas large outcrops are more similar to each other. It can be assumed that small inselbergs are more frequently hit by catastrophic events (e.g., prolonged drought) which result in increased rates of local extinction. Recolonization is primary initiated by generalists which have a relatively large species pool. This process is therefore largely unpredictable and consequently leads to greater variation in species composition. Such similarity patterns have also been reported for tree populations on islands (Barnes 1991).

Preliminary studies on the vegetation of Brazilian inselbergs have revealed a sharp contrast to the situation in West Africa concerning the low beta diversity of rock outcrop specialists. Even over short distances, it is

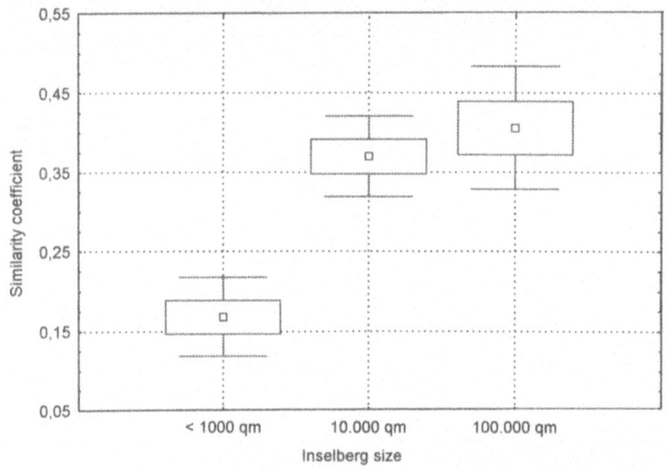

Fig. 12.4. Similarity in species composition between inselbergs of different sizes

not possible to predict which species might be present on an individual inselberg (Porembski et al. 1998). In a study of granite outcrops in the Mojave Desert with homogeneous abiotic characteristics, Cody (1978) observed considerable differences not only in species composition, but also in the niche position of the individual species. According to this author, such divergences could be the result of island-like extinction processes.

12.3.2 Species Richness of Inselbergs and Spatial Isolation

Despite a lack of detailed distribution data, it is obvious that inselbergs are not regularly distributed over the continental crystalline shields. There is a large amount of regional variation in inselberg frequency causing varying degrees of isolation. From the viewpoint of a plant species the degree of isolation is modified primarily by two factors. First, its dependence on inselbergs as growth sites (a close affinity to inselbergs enhances isolation) and second, by its dispersal ability (low ability favors isolation).

In the Côte d'Ivoire the distance between individual inselbergs varies considerably. Using aerial photographs and topographic maps, it is possible to generate a frequency distribution map of them (Fig. 12.5). This clearly reveals that inselbergs are not evenly spaced throughout the country. They are concentrated in the northern and central areas, declining markedly towards the south. The degree of spatial isolation of plant communities on inselbergs situated in the southern parts of the Côte

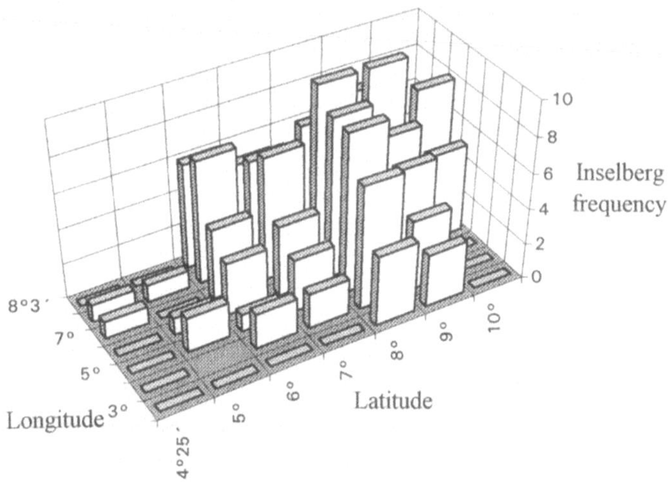

Fig. 12.5. Frequency distribution of inselbergs in the Côte d'Ivoire. They are particularly well represented in central and northern parts of the country

d'Ivoire is further emphasized by the low floristic affinity between rock outcrops and the surrounding rainforests. As shown below, there is a pronounced gradient of diversity of plant communities on Ivorian inselbergs from the savanna region in the north to the rainforests in the south of the country, with the latter being the least diverse. It can be assumed that, apart from other factors such as biotic interactions, isolation is a major factor in determining rates of immigration and extinction, and thus has to be considered as a basic determinant of species richness for inselberg plant communities in general.

12.4 Species Richness: Diversity Patterns on Inselbergs

Inselbergs support isolated patches of different habitats which are readily distinguished physiognomically (see Porembski et al., Chap. 4, this Vol.). According to own observations on inselbergs in the Côte d'Ivoire, there are large habitat-specific differences in both alpha (i.e., species number and evenness) and beta diversity (i.e., the degree of change in species diversity along a transect or between habitats, Magurran 1988) when comparing the major habitat types. Table 12.4 presents such data for *Afrotrilepis pilosa* mats, ephemeral flush communities, and seasonal rock pools. The mat vegetation extends over large areas of rock, whereas ephemeral flush communities usually cover only a few square meters or occur as narrow fringes

Table 12.4. Relationship between disturbance rate, beta diversity and species richness for selected habitats on inselbergs in the Côte d'Ivoire

Habitat	Species number	Disturbance rate	Beta diversity
Afrotrilepis pilosa mats	Low	Low	Low
Ephemeral flush communities	High	Intermediate	Intermediate
Seasonal rock pools	Low	High	High

surrounding mats. Rock pool size is frequently less than 1 m^2. The mats, which are dominated by the highly competitive Cyperaceae *Afrotrilepis pilosa* throughout West Africa, are almost monospecific and floristically uniform (Porembski et al. 1996). Similar results have been reported from East Africa (i.e., Zimbabwe, Malawi). Despite the small size of ephemeral flush communities, they represent the most species-rich type of vegetation on West African inselbergs. Each year, during the rainy season, the component species of such communities must reestablish themselves. This provides ample opportunities for stochastic colonization and establishment. Seasonal fluctuations of this sort prevent the occurrence of competitive exclusion, allowing more species to coexist and therefore promote diversity. In contrast, the species-poor mats dominated by *Afrotrilepis pilosa* can be regarded as having reached a state of equilibrium, leaving only few opportunities for the establishment of less competitive species. Due to its adaptations, *Afrotrilepis pilosa* is only stressed by the harsh abiotic conditions on inselbergs but not seriously disturbed. Seasonal rock pools on Ivorian inselbergs are characterized by low species numbers but high beta diversity. This can probably be attributed to the frequency and intensity of disturbing climatic events.

The narrow ephemeral flush belts fringing the mats represent an ecotone between the more-or-less static cryptogamic crust on rocks and the mat community. Ecotones are characterized by a sudden change in ecological structure, and it is well known that they form relatively species-rich transitional zones between different plant communities (edge effect, for survey see Forman 1995).

However, these are patterns for particular habitat types on Ivorian inselbergs and cannot be rigorously applied to other biogeographical regions. For instance, our studies in Brazil indicate that species richness of certain habitats is strongly determined by local and regional availability of species (species pool hypothesis, see Taylor et al. 1990) and rock outcrop geomorphology. As is shown in more detail below, the mat communities on inselbergs in the Brazilian Atlantic rainforest are by far more species-rich due to a large regional species pool. In contrast, Brazilian ephemeral flush com-

Table 12.5. Diversity (average Shannon index) of inselberg habitats in different geographic regions. Data based on Porembski et al. (1995, 1996, 1998), Dörrstock (1994), Seine (1996), Ohlemüller (1997)

Habitat	Côte d'Ivoire	Zimbabwe	Brazil	Australia
Monocotyledonous mats	0.3	0.9	1.3	–
Ephemeral flush communities	2.6	1.9	0.8	–
Shallow depressions	2.1	1.2	0.6	2.6

munities are poor in species compared to the situation in West Africa. This is probably a consequence of the extreme steepness of many inselbergs, which does not allow this vegetation type to develop to any extent. An overview of the diversity of certain inselberg habitats in distinct geographical regions is given in Table 12.5.

12.5 Species Richness of Inselbergs: Coexistence of Equilibrium and Nonequilibrium Communities

12.5.1 Temporal Species Turnover

Analyzing the mechanisms that regulate the composition of plant communities is one of the major tasks of plant ecology. A vast array of empirical and conceptual studies (for surveys see Crawley 1986; Tilman 1988) exist which indicate that communities are either in equilibrium (i.e., regulated by biotic interactions) or in nonequilibrium (i.e., regulated by chance and abiotic disturbances). The importance of these different concepts can be tested by measuring seasonal variation in the composition and abundance of species within particular communities. However, only few studies have dealt with seasonal variation in plant communities on inselbergs. The extensive research of Houle (1990), Houle and Phillips (1989), Isichei and Longe (1984), and Burbanck and Phillips (1983) in this field was mainly concerned with successional processes. In this context, shallow depressions have been the subject of several studies. It was demonstrated that soil depth increases with time, resulting in a sequence of different successional stages (Burbanck and Platt 1964; Shure and Ragsdale 1977).

In our own studies, which have been conducted on inselbergs in the Côte d'Ivoire since 1990, emphasis was placed on the analysis of temporal changes in species number and diversity within different plant communities. Both short-term (i.e., over the course of a single rainy season) and long-term (i.e., over a period of 10 years) vegetation dynamics were documented.

As an example to illustrate short-term dynamics, seasonal vegetation development within shallow depressions, *Afrotrilepis pilosa* mats, and ephemeral flush vegetation on granite inselbergs in the Comoé National Park (NE Côte d'Ivoire) will be described for the rainy season 1991 (see Porembski and Barthlott 1997). These habitats are clearly differentiated in respect of species richness. The study area (situated at 8°5'N and 3°1'W) has a seasonal climate with a dry season from October/November to March/April, and a rainy period from April/May to October. In 1991, total annual rainfall was 1017 mm. During April and May, almost 400 mm of rain fell, followed by several relative dry months and a secondary maximum of precipitation in October. Precipitation in March and April caused germination in all three habitats and numerous seedlings emerged. In both the shallow depression and *Afrotrilepis pilosa* mat, the first flowering individuals belonged to the tiny Scrophulariaceae *Lindernia exilis* (30 May). Within the ephemeral flush community, *Utricularia subulata* was the first species to flower (15 June). Fruiting specimens were first found in the shallow depression (the grass *Microchloa indica*, 15 June). Between the individual communities significant differences could be detected in number of species as well as in species diversity (Fig. 12.6).

Ephemeral flush vegetation was the most speciose community, followed by the shallow depression and the *Afrotrilepis pilosa* mat. Whereas species number and diversity (due to changes in percentage cover, the latter was more variable than the number of species) of the *Afrotrilepis pilosa* mat did not change within nearly 5 months (i.e., from 20 May to 15 October 1991), a considerable decrease was recorded in the other communities. Between 15 June and 25 August (in August, the longest uninterrupted period without rain was recorded: 12 days) the ephemeral flush community lost three therophytes (i.e., *Drosera indica*, *Rotala stagnina*, *Xyris straminea*), as did the shallow depression (i.e., *Lindernia exilis*, *Merremia pinnata*, *Mollugo nudicaulis*) before seed maturity. From 5 September to 15 October the number of species in all three habitats remained constant. Between 15 October and 10 November a sharp decrease in number of species and in percentage cover occurred within the ephemeral flush community as well as in the shallow depression due to the onset of the dry season. These results indicate a close relationship between rainfall distribution and the seasonal dynamics of species number and diversity. The maxima in species number and diversity were already reached after 4 weeks, whereas percentage cover values increased more slowly (except for the *Afrotrilepis pilosa* mat). Dry weather in August and September caused a considerable

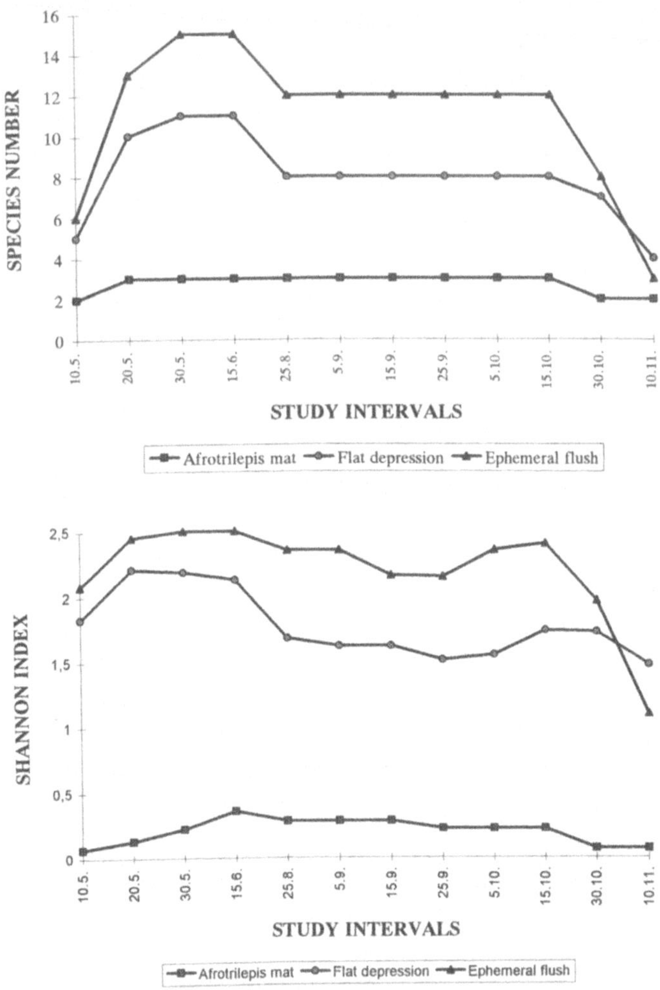

Fig. 12.6. Seasonal dynamics of plant communities on inselbergs in the Comoé National Park (NE Côte d'Ivoire). *Above* Species number in selected habitats during the rainy season 1991; *below* Species diversity during the rainy season 1991. The *x-axis* indicates days of field monitoring. (Porembski and Barthlott 1997)

decline in species number and diversity in the shallow depression and in the ephemeral flush vegetation, resulting in the local extinction of 27 and 20 %, respectively, of the species during this period before seed maturity was reached. In contrast, the *Afrotrilepis pilosa* mat was barely influenced by the rainfall regime because of the preponderance of the poikilohydric *Afrotrilepis pilosa*, which outcompetes less competitive therophytic species. Ephemeral flush vegetation and shallow depression were dominated by annuals and were both much richer in species than the *Afrotrilepis pilosa* mat. Disturbance in the form of extreme climatic events (i.e., highly variable annual rainfall regimes, but also large year-to-year fluctuations) obviously exerts a considerable stochastic influence on inselberg plant communities characterized by large percentages of r-strategists. This may cause local mortality or even extinction of the often small populations, thus leading to nonequilibrium conditions. This effect was observed in both ephemeral flush vegetation and in the shallow depression, where a high proportion of the species inventory disappeared without having produced mature seeds following a period of prolonged drought. Preliminary data from germination experiments indicate the existence of seedbanks in certain Ivorian inselberg communities. Persistent seedbanks of annuals on inselbergs could be an important precondition for the long-term maintenance of their populations. In addition, it can be expected that extinction of a local population on a small spatial scale (i.e., on an individual inselberg) can be compensated for by recruitment from persistent populations in close vicinity, i.e., via migration from a neighboring rock outcrop.

Of general concern is the fact that we have frequently observed a considerable asynchrony in the phenology of plant populations on inselbergs located in seasonally wet regions in West and East Africa. In our opinion, the highly localized distribution of rainfall events is the driving force behind this spatiotemporal asynchrony of local populations. This phenomenon is displayed most prominently by inselbergs affected by drought and therefore with populations threatened by extinction when occurring side by side with moist inselbergs, supporting flourishing local populations.

It can therefore be concluded that the impact of unforseeable climatic disturbances does not allow the establishment of a community equilibrium resulting in competitive exclusion of low competitive r-strategists by highly competitive k-strategists in certain habitat types on inselbergs. Moreover, climatic year-to-year fluctuations may constantly alter the competitive performance of each species, which consequently influences the direction of competitive exclusion (Shmida and Ellner 1984). In habitats such as shallow depressions and ephemeral flush communities, environmental fluctuations guarantee the coexistence of a larger set of species, as was suggested by Hutchinson (1959) and Grubb (1977).

It seems likely that small-scale spatial dynamics in nonequilibrium plant communities on rock outcrops do not follow certain rules of community organization, but are driven by stochastic colonization events. Probably the most important factor for the long-term maintenance of species-rich nonequilibrium communities on inselbergs is the availability of similar habitats in the vicinity which serve as potential sources for recolonization following local extinction.

It can be assumed that nonequilibrium communities on inselbergs harbor species which persist in metapopulations (i.e., their local populations are linked by dispersal, Gilpin and Hanski 1991; Hanski and Simberloff 1997). A large proportion of them are high-risk species (in the sense of Rosenzweig 1995), which disappear locally from a certain site but are able to recolonize readily if suitable environmental conditions prevail. This interpretation may explain why nonequilibrium communities on inselbergs are more speciose than equilibrium communities which are dominated by a few highly competitive species, and provides further support for the assumption of Tilman (1994). According to him, there is an interspecific trade-off between the competitive ability of a plant species and its ability to disperse. This implies that inferior competitors are superior colonists.

As a tool for understanding the mechanisms of vegetation change on inselbergs, permanent plots comprising different habitat types were established in the Côte d'Ivoire. The data reported here focus on seasonally water-filled rock pools and *Afrotrilepis pilosa* mats. They were regularly monitored within the frame of a long-term study that was initiated in 1990. Rock pools represent old habitats which may have persisted in their present geomorphological state for hundreds or even thousands of years. Based on the monitoring of nearly 200 rock pools on different outcrops, information on the frequency of local extinction and rates of recolonization has become available. Using presence/absence data, species turnover was calculated for each individual rock pool. The results show a significant positive correlation between this parameter and species richness (Fig. 12.7), indicating a negative correlation between diversity and stability for rock pools (Krieger 1997). This relationship is the consequence of an increasing proportion of therophytes with rising species numbers in rock pools. The observations reveal a high rate of local extinction and subsequent recolonization (Fig. 12.8). Climatic stochasticity is the main driving force behind these dynamics, whereas biotic interactions seem to be less significant. Species composition of rock pools cannot usually be predicted for the following year. Combined with unforseeable local extinction, random colonization of vacant sites may be responsible for the high variation in species composition between rock pools. Therefore one has to assume that the ability for rapid colonization is of crucial importance for the persistence of many species occurring in rock pools. The high number of generalist species (i.e., not restricted to rock pools but occurring in a broad range of habitats) indicates that competitive ability and resistance to environmental stresses are not as important as good dispersability.

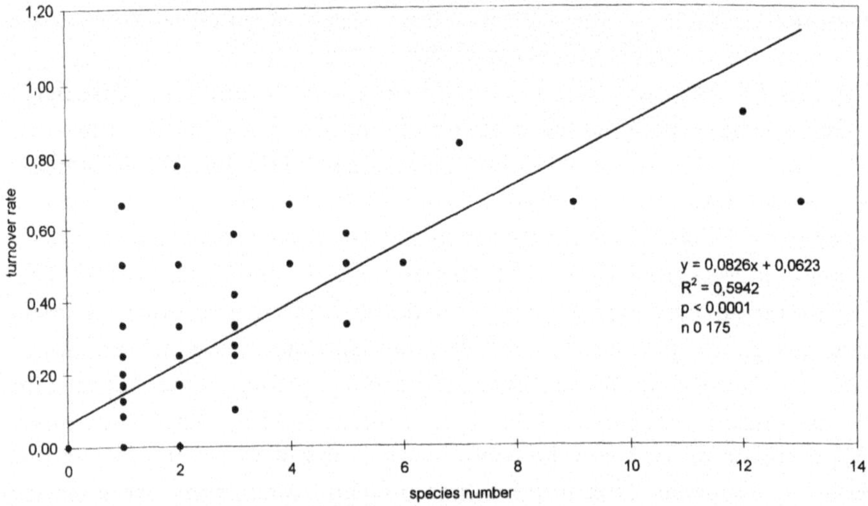

Fig. 12.7. Relationship between species richness and turnover in rock pools on inselbergs in the Côte d'Ivoire for the period 1991–1996. (After Krieger 1997)

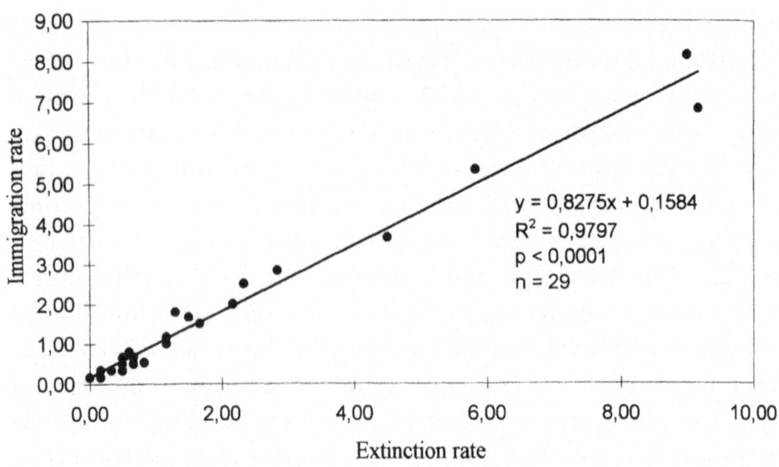

Fig. 12.8. Local extinction and colonizations in rock pools on inselbergs in the Côte d'Ivoire. (After Krieger 1997)

Most of the species found in rock pools are weak competitors that persist within these relatively small habitat fragments due to their high dispersal capabilities allowing them to migrate between appropriate sites. Intuitively, one may consider plant populations in rock pools on inselbergs to represent metapopulations. However, continued long-term observations are needed in order to decide whether the metapopulation concept can really be applied.

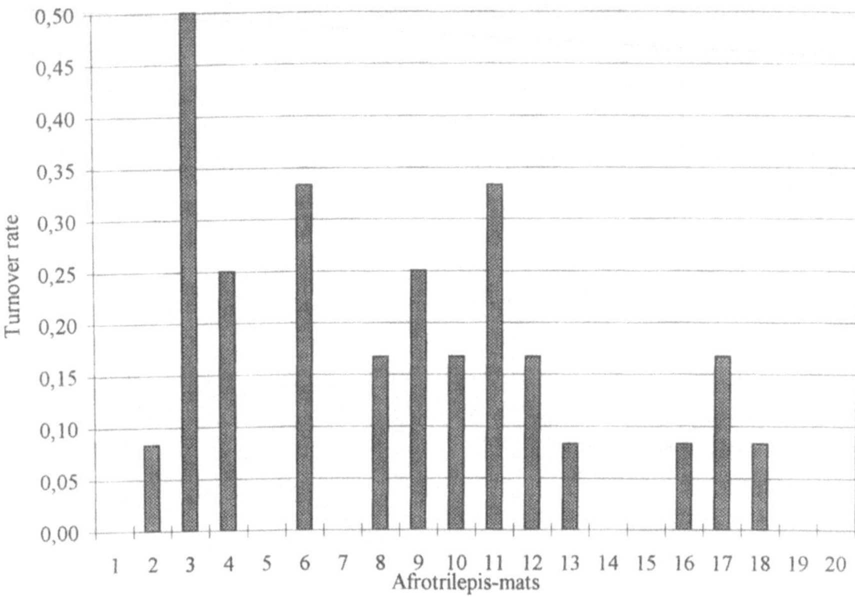

Fig. 12.9. Species turnover in *Afrotrilepis pilosa* mats on inselbergs in the Côte d'Ivoire. (own unpubl. data)

Mats of *Afrotrilepis pilosa* are characterized by very modest numbers of species. The measured turnover rates of these mats are much lower than those obtained for rock pools (Fig. 12.9). An explanation for this observation could be the fact that the mat community is influenced mainly by deterministic processes, i.e., competitive displacement of inferior competitors by *Afrotrilepis pilosa* which prevents higher general diversity and also decreases turnover rates.

Apart from permanent plots that were used to document vegetation dynamics of certain habitat types, complete species inventories of small-

Table 12.6. Inselberg size (in m^2), extinction rate, immigration rate, absolute and relative species turnover on rock outcrops in the Comoé National Park (NE Côte d'Ivoire), 1990–1996. (After Krieger 1997)

Size (m^2)	Species number	Extinction rate	Immigration rate	Turnover (abs.)	Turnover (rel.)
510	28	1.5	1.3	1.4	3.3
655	49	1.3	1.0	1.2	6.3
1001	58	0.7	0.7	0.7	5.3
3920	76	2.5	1.3	1.9	2.9

sized inselbergs (i.e., covering an area up to 4000 m^2) were recorded on a regular basis. As can be seen from Table 12.6, there is a positive correlation between turnover rate and species richness. The forces causing enhanced turnover with increasing species richness on the rock outcrops may be quite similar to those most relevant to rock pools, primarily environmental stochasticity.

12.5.2 Spatial Species Turnover

Species richness and diversity of plant communities changes along environmental gradients. Several hypotheses (for surveys, see Rosenzweig and Abramsky 1993; Tilman and Pacala 1993) indicate that plant diversity is a unimodal function of resource availability, with diversity reaching a maximum at intermediate resource availability.

There are several regions where inselbergs occur along conspicuous environmental gradients, which allows them to be used as models to address the issue of spatial species turnover. For this purpose, we have deliberately chosen inselbergs in the tropics and also in the temperate zone in order to detect possible differences concerning the determinants of biodiversity patterns. We tested the spatial distribution of plant diversity of inselbergs along an environmental gradient of annual precipitation in West Africa (Côte d'Ivoire) and Western Australia.

In the Côte d'Ivoire, data were compiled for several inselberg habitats along a latitudinal gradient from the savanna zone in the north to the rainforests in the south of the country. Figure 12.10 shows the diversity

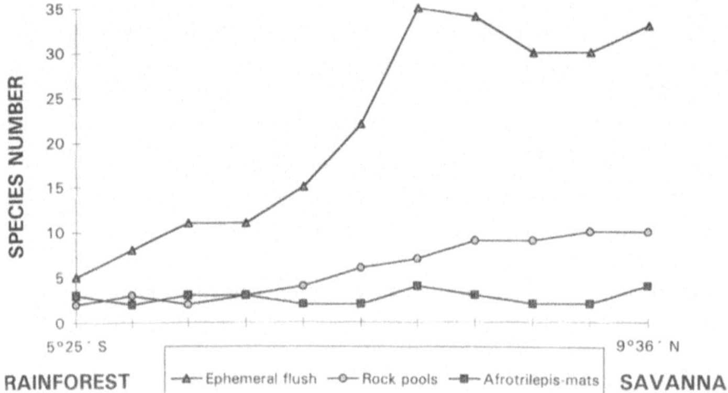

Fig. 12.10. Species diversity of inselberg plant communities along a savanna-rainforest gradient in the Côte d'Ivoire. (own unpubl. data)

indices of individual sample plots for shallow depressions, ephemeral flush vegetation, and *Afrotrilepis pilosa* mats. It is evident that there is an inverted (compared to the general pattern of phytodiversity) latitudinal gradient for shallow depressions and ephemeral flush vegetation, whereas the diversity of *Afrotrilepis pilosa* mats remains nearly constant (Porembski et al. 1995, 1996).

In order to explain this inverted gradient of diversity, two important aspects should be taken into consideration. The first assumption is that the vegetation of both shallow depressions and ephemeral flush communities is in a state of nonequilibrium due to stochastic environmental disturbance. In the savanna zone of the Côte d'Ivoire, prolonged droughts are a frequent occurrence, even during the rainy season. This prevents a state of equilibrium being attained with highly competitive species dominating (a situation, however, which can be observed on inselbergs in the rainforest zone) and enables the maintenance of species rich nonequilibrium communities. In contrast, the low diversity of *Afrotrilepis pilosa* mats along the entire latitudinal gradient indicates that this community is in equilibrium. A lesser degree of isolation may be a second factor that is responsible for the increased species richness on savanna vs. rainforest inselbergs. In the latter zone, inselbergs and similar azonal localities are less frequent. This could result in enhanced extinction rates and consequently lower species richness (cf. sect. 12.3.2).

A similar pattern of diversity was revealed for plant communities in shallow depressions on granite outcrops along a climatic gradient encompassing arid woodland (*Acacia-Casuarina* thicket), mallee (today mainly converted into farmland), and warm temperate Karri forest (with *Eucalyptus diversicolor* dominating) in Western Australia (Ohlemüller 1997). Species diversity in shallow depressions increases from relatively humid Karri forest towards mallee and arid woodland where on the driest outcrops diversity is decreasing again (Fig. 12.11). Plant diversity in shallow depressions therefore corresponds closely with the general phytodiversity pattern in southwestern Australia, where the Transitional Rainfall Zone shows the highest species richness (Hopper 1992). This pattern may reflect the consequences of the "intermediate disturbance hypothesis" (Connell 1978) since the Transitional Rainfall Zone has been subjected to recurrent and unpredictable environmental perturbations in the late Tertiary and Quaternary, and has led to a pronounced evolution and persistence of taxa.

Several abiotic and biotic factors in shallow depressions on Western Australian inselbergs were investigated and their possible impact on community structure tested by using multivariate methods. The increase in diversity towards the arid zone is probably due to the reduced competitive ability of certain species (most of all *Borya* spp.), together with en-

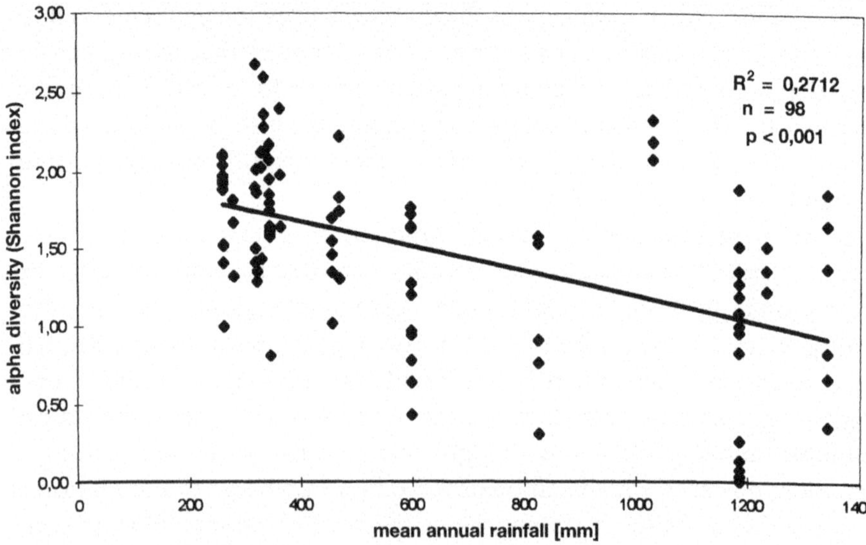

Fig. 12.11. Species diversity in shallow depressions on rock outcrops in southwestern Australia along an environmental gradient. (after Ohlemüller 1997)

hanced stochastic disturbance and the presence of a larger species pool of xero-tolerant species in the more arid regions. The above-mentioned examples thus illustrate that plant diversity of tropical as well as temperate inselbergs is largely influenced by the same determinants.

Shallow depressions not only exhibit changes in species richness on large geographical scales but, as demonstrated above, they also provide an opportunity to study spatial differentiation patterns on very small scales, for example within a single shallow depression. The controlling factors which determine the floristic structure of tropical inselberg communities have not been clearly identified to date. However, studies concerning this topic have been conducted on granite inselbergs in temperate regions (in southeastern USA by, e.g., Burbanck and Platt 1964; Sharitz and McCormick 1973; Collins et al. 1989; Houle and Phillips 1989). In these cases it was demonstrated that both biotic (i.e., competition) and abiotic factors are important regulators of community structure. Within shallow depressions, soil depth and moisture increase from the periphery towards the center. On very shallow soils (<9 cm) where environmental conditions are particularly harsh, the annual Crassulaceae *Diamorpha smallii* dominates due to its tolerance of low moisture levels and the absence of competitors (Wiggs and Platt 1962). On deeper soils, the influence of biotic factors becomes more pronounced and perennial plant cover increases. A similar vegetation zonation of shallow depressions with annuals dominating the peripheral zone can also be observed on tropical inselbergs (e.g., in the Côte d'Ivoire). In addition, there is a correlation between soil depth and species richness, with diversity peaking at an intermediate soil depth. As pointed out by Harper (1967), plant species react to stressful abiotic and biotic conditions by reducing their size (stunting) or by reducing their numbers (thinning). Certain species occurring in shallow depressions on Ivorian inselbergs exhibit both stunting and thinning, as is the case with the Scrophulariaceae *Lindernia exilis* (Fig. 12.12).

Another way to characterize spatial diversity patterns is to consider small-scale variations in species richness and composition as found between habitat patches on a single inselberg, or between rock outcrops in close proximity. In this respect, beta diversity is a useful measure of expressing uniformity or variability of inselberg vegetation. Taking monocotyledonous mats on inselbergs in Brazilian Atlantic rainforest as an example, it can be demonstrated that the extent of spatial species turnover within rock outcrop habitats can be remarkably high, despite the fact that the environmental conditions are extremely stressful (Porembski et al. 1998). Concerning the degree of floristic similarity between individual mats (beta diversity), there are great intra- and interinselberg differences. Mats on the inselberg Pedra Alegre showed the lowest similarity in floristic composition (high beta diversity), while mats on Serra dos Aimores were characterized by a high degree of similarity (low beta diversity, Table 12.7).

Fig. 12.12. Reduction in size (stunting) of *Lindernia exilis* (Scrophulariaceae) along a gradient in soil depth in a shallow depression from the periphery (*left*) towards the center (*right*)

Table 12.7. Beta diversity (expressed by using the Sørensen index as similarity coefficient) between mat communities on inselbergs in Brazilian Atlantic rainforest. (Porembski et al. 1998)

	Morro do Cuca	Parati Mirim	Serra dos Aimores	Nova Venezia	Pancas	Pedra Alegre
Morro do Cuca	1	0.105	0.118	0.211	0.222	0.174
Parati Mirim		1	0.125	0.222	0.118	0.273
Serra dos Aimores			1	0.375	0.133	0.400
Nova Venezia				1	0.471	0.636
Pancas					1	0.476
Pedra Alegre						1

High beta diversity of the Pedra Alegre mat community may indicate a considerably higher importance of local stochastic influences on colonization processes within this habitat compared to Serra dos Aimores. Moreover, there is a large amount of interinselberg variation (i.e., high beta diversity) between the mat communities of the rock outcrops studied.

In this respect, Morro do Cuca is the most distinct rock outcrop (Sørensen max. 0.222). The mat community of Pedra Alegre showed the closest links to other sites. For instance, over 60% of the species were found at Nova Venezia, situated close by. The high beta diversity of mat communities on Brazilian rock outcrops is in marked contrast to the situation of *Afrotrilepis pilosa* mats on West African inselbergs (Porembski et al. 1996). Similar large small-scale differences in species composition among habitat islands have also been observed for rock pools (own unpubl. data from the Côte d'Ivoire) and shallow depressions on inselbergs in the Côte d'Ivoire (own unpubl. data) and the United States (Collins et al. 1989). The latter authors interpret the compositional differences among individual shallow depressions as a result of both variations in habitat quality as well as stochastic variation in dispersal and establishment.

12.6 Conclusions

Inselbergs represent clearly delimited ecosystems that provide the opportunity to address questions related to the controlling mechanisms of biological diversity. In particular, they allow the examination of factors influencing the species richness within naturally fragmented communities. Overall, it becomes evident that inselbergs comprise a variety of habitats that vary greatly in regard to the spatiotemporal dynamics of their species composition. For example, species-poor equilibrium assemblages, such as *Afrotrilepis pilosa* mats, which show a large degree of biotic interactions, occur side by side with species-rich nonequilibrium communities (e.g., ephemeral flush vegetation) where stochastic environmental disturbance leads to the coexistence of species. In regard to species numbers, the nonequilibrium component, that mainly consists of vagrants, dominates over inselberg specialists. However, species belonging to the latter category constitute the bulk of the biomass. As already pointed out, plant diversity of inselberg communities varies along environmental gradients. For inselbergs in tropical and temperate regions, diversity peaks under circumstances where migration between isolated patches is promoted (mainly in nonforest surroundings or in the presence of other azonal sites) and where nonequilibrium conditions (such as in seasonally dry regions) prevent competitive exclusion by a few highly competitive species. High regional plant diversity is another factor promoting a high local species richness of inselberg communities. This was demonstrated above by comparing West African with Brazilian mat communities.

The relationship between the stability and diversity of ecosystems is still being rigorously debated (see Pimm 1993). From the viewpoint of inselbergs, it seems apparent that species richness is positively correlated with turnover of species caused by local extinction and recolonization. However, a major precondition for the maintenance of a certain level of species richness in inselberg communities, where rates of extinction can be very high, is the degree to which populations are linked to those of similar habitats via gene flow.

There are numerous attempts to assess the possible effects of global change (such as effects on climate, atmospheric composition, land use) on species diversity and on the functioning of ecosystems. Concerning the implications of global change, the main emphasis has been placed on the relationship between the physical climate system and the dynamics of the vegetation. A major factor which will possibly lead to changes in the distribution of certain vegetation belts is the greenhouse gas, CO_2, whose increasing atmospheric concentration is expected to have profound effects on terrestrial ecosystems (Bazzaz et al. 1996). Predicting the implications of global change in regard to functional aspects of the inselberg ecosystem is hardly possible since we are still at the beginning of our understanding about functional species groups and key organisms in this system. Despite our lack of knowledge, a tentative prediction can be made for a relatively "simple" type of competitive interactions on tropical inselbergs, namely between nitrogen-fixing cyanobacteria and cyanobacterial lichens. These organisms are constituents of the biofilm on free exposed rocks, with cyanobacteria prevailing under humid conditions and lichens dominating in drier climates. Changes in precipitation amount and distribution will possibly result in changing dominance patterns between these biofilm components, which could have positive or negative feedback on productivity and nutrient supply in the inselberg vegetation and the surroundings, which are influenced by runoff water.

However, not only changes in climate and atmospheric composition will influence the vegetation of inselbergs, but also changes in the landscape structure (e.g., fragmentation, expansion of ecotones) could have serious consequences for the diversity and species composition of this ecosystem. Apart from quarrying and some leisure activities which have caused local damage to the vegetation of some inselbergs, rock outcrops have not been subjected to any significant agricultural pressures in the past. In many regions, therefore, they represent the last remains of untouched ecosystems, and form habitat islands within landscapes dominated by agroforestry. Despite data indicating a certain degree of resistance to invasive alien plants (Fleischmann 1997 reporting on the situation of granitic outcrops in the Seychelles), it is to be expected that the fragmentation of contiguous vegetation types will result in the loss of distinct island-like attributes of inselbergs. This effect is already to be seen, for example, in the Atlantic rainforest of Brazil, where exotic weeds successfully use roads and other ruderal sites as a way to gain access to inselbergs which otherwise harbor a largely endemic flora. The exact consequences of these plant invasions are unclear, but one has to fear drastic losses amongst the indigenous vegetation. This is demonstrated by the aggressive behavior of the neophytic grasses *Melinis repens* and *M. minutiflora*, which have become dominant in certain habitats on East Brazilian inselbergs. However, there is no one-way relationship between inselbergs and invading weeds. Rock outcrops also act as "evolutionary springboard" (Wyatt 1997) for opportunistic species, some of which are becoming frequent in open, disturbed localities. Originally restricted to inselbergs, they have managed to invade open sites via man-made transport routes, as is the case of the Piedmont region in southeastern USA (e.g., *Rumex hastatulus*, *Nutallanthus canadensis*) or the Mojave Desert. In the latter case, Cody (1978) presumes that the traits of the Asteraceae *Chrysothamnus teretifolius*, that are of adaptive value on island-like, small granite outcrops characterized by high population turnover rates, favor the colonization of similar unstable or patchy man-made sites. In summary, further increasing landscape fragmentation will lead to an enhanced migration of species to and from inselbergs with serious consequences (e.g., higher extinction rates) for this unique ecosystem. As has been demonstrated for other ecosystems (e.g., Kricher 1973), fragmentation and ecotonal expansion will result in increased species richness, which is attributable mainly to the invasion of widespread ruderal species within formerly isolated habitat patches.

Acknowledgements. Financial support by the Deutsche Forschungsgemeinschaft (grant no. Ba 605/4-3, SPP Mechanismen der Aufrechterhaltung tropischer Diversität) is gratefully acknowledged. For valuable discussions we like to thank G. Brown (Rostock), A. Krieger (Rostock), and R. Ohlemüller (Dunedin).

References

Barnes WJ (1991) Tree populations on the islands of the lower Chippewa River in Wisconsin. Bull Torrey Bot Club 118:424–431

Baskin JM, Baskin C (1988) Endemism in rock outcrop plant communities of unglaciated eastern United States: an evaluation of the roles of the edaphic, genetic and light factors. J Biogeogr 15:829–840

Bazzaz FA, Bassow SL, Berntson GM, Thomas SC (1996) Elevated CO_2 and terrestrial vegetation: implications for and beyond the global carbon budget. In: Walker B, Steffen W (eds) Global change and terrestrial ecosystems. Cambridge University Press, Cambridge, pp 43–76

Burbanck MP, Phillips DL (1983) Evidence of plant succession on granite outcrops of the Georgia Piedmont. Am Nat 109:94–104

Burbanck MP, Platt RB (1964) Granite outcrop communities of the Piedmont Plateau in Georgia. Ecology 45:292–306

Caley MJ, Schluter D (1997) The relationship between local and regional diversity. Ecology 78: 70–80

Carlquist S (1974) Island biology. Columbia Univ Press, New York

Cody ML (1978) Distribution ecology of *Haplopappus* and *Chrysothamnus* in the Mojave Desert. I. Niche position and niche shifts on north-facing granitic slopes. Am J Bot 65:1107–1116

Collins SL, Mitchell GS, Klahr SC (1989) Vegetation-environment relationships in a rock outcrop community in southern Oklahoma. Am Midl Nat 122:339–348

Connell JH (1978) Diversity in tropical rain forests and coral reefs. Science 199:1302–1310

Crawley MJ (1986) The structure of plant communities. In: Crawley MJ (ed) Plant ecology. Blackwell, Oxford, pp 1–50

Darwin C (1845) The voyage of the Beagle. Everyman's Library, JM Dent London

Dörrstock S (1994) Vegetation der Hangmoore auf Inselbergen in der Côte d'Ivoire. MSc Thesis, University of Bonn

Fischer AG (1961) Latitudinal variations in organic diversity. Am Sci 49:50–74

Fleischmann K (1997) Invasion of alien woody plants on the islands of Mahé and Silhouette, Seychelles. J Veg Sc 8:5–12

Forman RTT (1995) Land mosaic. The ecology of landscapes and regions. Cambridge University Press, Cambridge

Gaston KJ (1994) Rarity. Chapman and Hall, London

Gaston KJ (1996) Biodiversity. A biology of numbers and difference. Blackwell, Oxford

Gilpin ME, Hanski I (1991) Metapopulation dynamics: empirical and theoretical investigations. Cambridge University Press, Cambridge

Godt MJW, Hamrick JL (1993) Genetic diversity and population structure in *Tradescantia hirsuticaulis* (Commelinaceae). Am J Bot 80:959–966

Grubb PJ (1977) The maintenance of species richness in plant communities: the importance of the regeneration niche. Biol Rev 52:107–145

Hanski I, Simberloff D (1997) The metapopulation approach. Its history, conceptual domain, and application to conservation. In: Hanski I, Gilpin ME (eds) Metapopulation biology: ecology, genetics, and evolution. Academic Press, San Diego, pp 5–26

Harper JL (1967) A Darwinian approach to plant ecology. J Appl Ecol 4:267–290

Hopper SD (1992) Patterns of plant diversity at the population and species levels in south-west Australian mediterranean ecosystems. In: Hobbs RJ (ed) Biodiversity of mediterranean ecosystems in Australia. Surrey Beatty, Chipping, Norton, pp 27–46

Houle G (1990) Species-area relationship during primary succession in granite outcrop plant communities. Am J Bot 77:1433–1439

Houle G, Phillips DL (1989) Seed availability and biotic interactions in granite outcrop plant communities. Ecology 70:1307–1316

Hutchinson GE (1959) Homage to Santa Rosalia, or why are there so many kinds of animals? Am Nat 93:145–159

Isichei AO, Longe PA (1984) Seasonal succession in a small isolated rock dome plant community in western Nigeria. Oikos 43:17–22

James SH (1965) Complex hybridity in *Isotoma petraea*. I. The occurrence of interchange heterozygosity, autogamy and a balanced lethal system. Heredity 20:341–353

James SH, Wylie AP, Johnson MS, Carstairs SA, Simpson GA (1983) Complex hybridity in *Isotoma petraea*. V. Allozyme variation and the pursuit of hybridity. Heredity 51:653–663

Kricher JC (1973) Summer bird species diversity in relation to secondary succession on the New Jersey Piedmont. Am Midl Nat 89:121–137

Krieger A (1997) Vegetationsdynamik und Species Turnover in saisonalen Felsgewässern auf Inselbergen der Côte d'Ivoire (Westafrika). MSc Thesis, University of Bonn

Kruckeberg AR (1984) California serpentines: flora, vegetation, geology, soils, and management problems. University California Publ Bot 78:1–180

MacArthur RH, Wilson EO (1967) The theory of island biogeography. Princeton University Press, Princeton

Magurran AE (1988) Ecological diversity and its measurement. Chapman and Hall, London

Mayer MS, Soltis PS, Soltis DE (1994) The evolution of the *Streptanthus glandulosus* complex (Cruciferae): genetic divergence and gene flow in serpentine endemics. Am J Bot 81:1288–1299

Moran GF, Hopper SD (1983) Genetic diversity and the insular population structure of the rare granite rock species *Eucalyptus caesia* Benth. Aust J Bot 31:161–172

Murdy WH, Brown Carter ME (1985) Electrophoretic study of the allopolyploidal origin of *Talinum teretifolium* and the specific status of *T. appalachianum* (Portulaceae). Am J Bot 72:1590–1597

Ohlemüller R (1997) Biodiversity patterns of plant communities in shallow depressions on Western Australian granite outcrops (inselbergs). MSc Thesis, University of Bonn

Pearson DL (1995) Selecting indicator taxa for the quantitative assessment of biodiversity. In: Hawksworth DL (ed) Biodiversity. Measurement and estimation. Chapman and Hall, London, pp 75–79

Pimm SL (1993) Biodiversity and the balance of nature. In: Schulze E-D, Mooney HA (eds) Biodiversity and ecosystem function. Springer, Berlin Heidelberg New York, pp 347–359

Porembski S, Barthlott W (1997) Seasonal dynamics of plant diversity on inselbergs in the Ivory Coast (West Africa). Bot Acta 110:466–472
Porembski S, Brown G (1995) The vegetation of inselbergs in the Comoé National Park (Ivory Coast). Candollea 50:351–365
Porembski S, Brown G, Barthlott W (1995) An inverted latitudinal gradient of plant diversity in shallow depressions on Ivorian inselbergs. Vegetatio 117:151–163
Porembski S, Brown G, Barthlott W (1996) A species-poor tropical sedge community: *Afrotrilepis pilosa* mats on inselbergs in West Africa. Nord J Bot 16:239–245
Porembski S, Martinelli G, Ohlemüller R, Barthlott W (1998) Diversity and ecology of saxicolous vegetation mats on inselbergs in the Brazilian Atlantic rainforest. Divers Distrib 4:107–119
Raunkiaer C (1934) The life-forms of plants and statistical plant geography. Oxford University Press, London
Ricklefs RE (1987) Community diversity: relative roles of local and regional processes. Science 235:167–171
Ricklefs RE, Schluter D (eds) (1993) Species diversity in ecological communities. University of Chicago Press, Chicago
Rosenzweig ML (1995) Species diversity in space and time. Cambridge University Press, Cambridge
Rosenzweig ML, Abramsky Z (1993) How are diversity and productivity related? In: Ricklefs RE, Schluter D (eds) Species diversity in ecological communities. University of Chicago Press, Chicago, pp 52–65
Sampson JF, Hopper SD, James SH (1988) Genetic diversity and the conservation of *Eucalyptus crucis* Maiden. Aust J Bot 36:447–460
Schulze E-D, Mooney HA (1993) Biodiversity and ecosystem function. Springer, Berlin Heidelberg New York
Seine R (1996) Vegetation von Inselbergen in Zimbabwe. Archiv naturwissenschaftlicher Dissertationen, vol 2. Martina Galunder, Wiehl
Sharitz RR, McCormick JF (1973) Population dynamics of two competing annual plant species. Ecology 54:723–740
Shmida A, Ellner S (1984) Coexistence of plant species with similar niches. Vegetatio 58:29–55
Shure DJ, Ragsdale HL (1977) Patterns of primary succession on granite outcrop surfaces. Ecology 58:993–1006
Taylor DR, Aarssen LW, Loehle C (1990) On the relationship between r/K selection and environmental carrying capacity: a new habitat templet for plant life-history strategies. Oikos 58:239–250
Tilman D (1988) Plant strategies and the dynamics and function of plant communities. Princeton University Press, Princeton
Tilman D (1994) Competition and biodiversity in spatially structured habitats. Ecology 75:2–16
Tilman D, Pacala S (1993) The maintenance of species richness in plant communities. In: Ricklefs RE, Schluter D (eds) Species diversity in ecological communities. University of Chicago Press, Chicago, pp 13–25
Vitousek PM, Loope LL, Adsersen H (1995) Islands. Biological diversity and ecosystem function. Springer, Berlin Heidelberg New York
Westoby M (1993) Biodiversity in Australia compared with other continents. In: Ricklefs RE, Schluter D (eds) Species diversity in ecological communities. University of Chicago Press, Chicago, pp 170–177
Whittaker RH (1972) Evolution and measurement of species diversity. Taxon 21:213–251
Wiggs DN, Platt RB (1962) Ecology of *Diamorpha cymosa*. Ecology 43:654–670

Wyatt R (1997) Reproductive ecology of granite outcrop plants from the southeastern United States. J R Soc West Aust 80:123–129

Wyatt R, Evans EA, Sorenson JC (1992) The evolution of self-pollination in granite outcrop species of *Arenaria* (Caryophyllaceae). VI. Electrophoretically detectable genetic variation. Syst Bot 17:201–209

Yodzis P (1986) Competition, mortality and community structure. In: Diamond J, Case TJ (eds) Community ecology. Harper and Row, New York, pp 480–491

13 The Fauna of Inselbergs

M.A. Mares and R.H. Seine

13.1 Introduction

Other chapters in this volume have made clear that inselbergs are of enormous interest and importance in themselves as structural components of the environment, as well as through their effects on vegetation from providing substrates on which bacteria, lichens, and mosses develop, to providing special microhabitats that can permit the existence of forests in association with the rock habitat within an otherwise barren landscape.

Many of the attributes that make inselbergs important to plants also make them important to animals. First, inselbergs are quite old, with some formations dating back more than 10 Ma (Büdel 1978). This means that the fauna will have had a long period of time over which taxa could have evolved special attributes that permit them to inhabit the inselberg rock habitat. Moreover, inselbergs often occur as habitat islands within larger areas that have a very different substrate and physiognomic structure. Examples of this abound, but the tree-covered kopjes that occur within the arid Namib Desert, or the forested serrotes found within the semiarid Caatinga scrubland of northeastern Brazil are good examples of how different from the surrounding area the inselberg habitat can be. Additionally, as will be discussed below, the three-dimensional structure of the rock itself is important to many animals.

Zoologists have not dedicated a great deal of attention to delineating how inselbergs function as evolutionary and ecological forces affecting speciation and diversification. However, Withers (1979), Mares and Lacher (1987), and Mares (1997) considered a number of ways in which the physical characteristics of inselbergs can influence the evolution and persistence of mammals in a region. Among these are the direct effects of the inselbergs themselves as structural and climatic components of an area, as well as the indirect effects of the rock habitat on vegetation, which then becomes a major niche element for mammals. These authors also considered how inselbergs function as isolated and specialized habitats, thus

Ecological Studies, Vol. 146
S. Porembski and W. Barthlott (eds.) Inselbergs

influencing speciation in various groups of mammals, as well as contributing to regional species diversity.

In this chapter, we review these factors, as well as other qualities of the rocky inselberg habitat, that have pronounced effects on the overall fauna of an area, as well as on the adaptive diversity and species richness within particular taxonomic groups when the inselberg habitat is considered at a global scale.

13.2 How Inselbergs Influence Faunal Development

Inselbergs vary greatly in size, shape, and overall structural complexity, but each of these attributes is of some importance to the animals that specialize on the rock habitat. Additionally, inselbergs influence those species that regularly inhabit the rocks, although they are not rock specialists, as well as those species that utilize the inselberg only on an occasional basis. Moreover, the effects of inselbergs on elements of the fauna may be different for different groups of animals. Thus, an isolated inselberg in Namibia may be little more than a feeding station, watering area, or shady refuge for a large, 200 kg gemsbok, while functioning at the same time as a complete and isolated island for a rock-specialized lizard.

An inselberg is a major structural component of the niche space of organisms inhabiting the rocks. Rocks offer a hard substrate that requires special adaptions of the feet or body surface if an animal is to specialize on the rocks as a major part of its adaptive space. Thus, in the inselbergs of the Caatinga of Brazil, for example, a bat (family Molossidae, or free-tail bats) has become specialized for inhabiting the narrow spaces created by the exfoliating granite of the inselbergs. This species, *Molossops mattogrossensis*, has developed a dorsoventrally flattened head and body for entering the narrow cracks, as well as tubercular projections on its forearms that permit it to develop traction against the rock surface when it is wedged in the cracks (Willig and Jones 1985). Remarkably, in rocky areas of the arid Namib Desert of Africa, and in Ethiopia, Kenya, and the Sudan, other bats in the same family have also become specialized as inhabitants of the cracks in exfoliating granite. *Platymops setiger*, the species in eastern Africa, has developed the same unusual characteristics as its Brazilian analog, a flattened, dorsoventrally compressed head and body, and tubercular projections on its forearms (Nowak 1991). In the southern desert, *Sauromys petrophilus* shows the same adaptations of morphology, ecology, and behavior, differing mainly in the fact that it has not developed the tubercles on the forearm (Skinner and Smithers 1990). Examples of similar convergent equivalents developing in association with inselbergs are not uncommon, nor are they limited to bats.

Mares and Lacher (1987), for example, examined how medium-sized rodents and rodent-like mammals throughout the world have specialized for life on inselbergs, describing many of the adaptations of these rock specialists. They found that the suites of adaptations associated with life in the rocks included morphological specializations (specialized feet, teeth, tail, eyes, snout), ecological specializations (similar diets, habitat selection, nest-site similarities, similar reproductive ecology), and behavioral characteristics. These similarities even extended to the development of a harem-based social system, which has developed in response to the fact that the rock habitat is both a critical resource, and one that can be defended by a single male. This permits the males to accrue females, that must have access to the rocks to raise their young. This trait, which is uncommon in small mammals, has appeared repeatedly in mammals that inhabit inselbergs.

The influence on the fauna of inselbergs is not limited to mammals, however. In the inselbergs of the Brazilian Caatinga there is a rock-specialized lizard, family Tropiduridae, *Tropidurus semitaeniatus*, that is dorsoventrally flattened and has roughened scales that help wedge it into the cracks so that it cannot be removed easily from the rocks by a predator. This species has marked cryptic coloration and matches the rock substrate perfectly (Vitt 1981). The effect of inselbergs as habitats in which specialization and speciation can take place is not limited to rock outcrops occurring in arid areas, as the above examples might indicate. Within the vast Amazon rainforest of Brazil, several other species of *Tropidurus* have also evolved morphological, reproductive, and behavioral specializations like those of the inselberg specialists of the xeric Caatinga (Vitt 1993). In some of the Amazonian populations of *Tropidurus hispidus*, it was shown through morphoecological analysis and DNA sequencing that populations that colonize the rock habitat can diverge fairly rapidly from nonrock-dwelling conspecifics in the development of traits that are associated with life on an inselberg (Vitt et al. 1997). In the Namib Desert of southern Africa, there is a lizard in a different familiy (Cordylidae), *Platysaurus intermedius*, that shows adaptations that are very similar to the Brazilian rock-specialized lizards, including dorsoventral flattening of the head and body, cryptic coloration, and the habit of using exfoliating granite cracks to wedge the body and escape predation (Broadly 1978).

In the southern United States, there is a frog, *Syrrophus marnockii* (family Leptodactylidae), that inhabits the rock outcrops of the Edward's Plateau in south-central Texas. The frog uses crevices in the rock as refugia from predators and is characterized by a flattened head and body (Conant and Collins 1991).

The influence of the morphology of inselbergs on the suites of adaptive traits of the various species discussed above illustrates that the rock habitat, per se, is an important evolutionary force. The rocks provide structural niche components, such as nest sites, areas of shade, sites of water accumulation, high points from which sentinels can observe the approach of predators, and microsites, such as exfoliating granite, or larger cracks in the rock, that are used as predator escape refugia. Organisms, whether vertebrates (mammalian, avian, reptilian, amphibian), or invertebrates, have all developed species that have specialized on the unusual structure of the inselberg, or have at least become adapted to use inselbergs occasionally.

The influence that inselbergs have on plants also makes the rock outcrop a special habitat component for animals. Due to their three-dimensional structure and rock substrate, inselbergs clearly offer microclimates that differ from the surrounding habitat. The rocks may offer quiet pools of water and relatively lush green vegetation within areas of more pronounced aridity (e.g., as in the inselbergs of Namib Desert). Conversely, the rocky islands may offer sites of relative aridity compared to the surrounding rainforest (e.g., the rock outcrops lying within the Amazon rain forest may support plants such as cacti that are adapted to much more xeric conditions than the rain forest). Thus, those species possessing the evolutionarily option to exploit the rock resource may tend to become either more adapted to aridity or more adapted to mesic conditions than their founding populations. For those organisms that colonize and exploit the inselberg habitat, the pronounced influence of inselbergs on vegetation and microclimate translates directly into increased opportunities for specialization (with concomitant reduction in competitive interactions with ancestral populations).

Since inselbergs permit specialization and increased habitat complexity within an area, they also are a major biodiversity component of a region. Inselbergs permit the existence of rock specialists like those listed above, but also function as shelters, hunting locales, foraging sites, and other parts of the niche of animals that permit casual occupation of the inselberg to occur. Thus, for example, an inselberg in Namibia supports such rock specialists as a rock lizard, a flat-headed bat, a pygmy rock mouse (*Petromyscus*), a small rock elephant shrew (*Elephantulus*), a dassie rat (*Petromus*), a mid-sized hyrax (*Procavia*), or a large-bodied klipspringer (*Oreotragus*). These species not only show the unusual adaptations characeristic of rock specialists, but they would not exist in the region were it not for the inselbergs. In addition, the inselbergs are frequented by a large number of animals that use the inselberg as a part of their range (e.g., leopards, *Panthera* hunt on inselbergs and take shelter in the rocks, as does the deadly black mamba snake, *Dendroaspis*). Hawks and owls frequent the inselbergs as places of food and shelter, as do gemsbok (*Oryx*), hyenas (*Crocuta*), and other mammals.

Although the list of species that are either specialized for inselbergs, or that inhabit a region in part because inselbergs provide special microhabitats and food resources differs from region to region (Appendix), the influence of the rocky isolated inselbergs on patterns of adaptation, evolution, coexistence, and regional biodiversity is profound. Inselbergs, on a global scale, have made a significant contribution to speciation events in animals. Moreover, since many of the species that occur on inselbergs are limited to small, isolated populations that cannot exist elsewhere, except on other inselbergs, isolated granitic outcrops merit special attention from conservationists.

13.3 Appendix: Animal Species Closely Associated with Inselbergs

The following paragraph provides a list of animal species known to be closely associated with inselbergs. It was compiled from data published in (1) Branch (1994), (2) Haltenorth and Diller (1977), (3) Maclean (1993), (4) Mares and Lacher (1987), (5) Quarterman et al. (1993), (6) Rödel (1996), (7) Rödel (1997), (8) R. Seine (pers. comm.), (9) Withers and Edward (1997) and it is by no means complete. Listed are the species (or higher taxon) names, a description of the type of association with inselbergs, usually direct quotes from the original text, and the region from which this type of association is known.

Arachnida, Scorpiones: *Rhopalurus rochai*, E South America, "occur only in the rockpiles of the Caatinga …" (4); *R. laticaudata*, SE America, "occur only in the rockpiles of the Caatinga …" (4); **Araneae:** *Teyl luculentus*, W Australia, "… all Teyl are restricted to granite outcrops or granite-related habitats …" (9); **Pseudoscorpiones:** *Synsphyronus* spp., Australia, "… may be restricted to granite outcrops …" (9).

Insecta, Embioptera: *Notoligotoma* spp., W Australia, "… may be restricted, in part, to granite outcrops …" (9); **Diptera:** *Archaeochlus* spp., SW Australia, "… are restricted to granite outcrops …" (9).

Amphibia, Urodela: *Notophthalmus viridescens*, N America, "… is found in permanent pools …" (5); *Plethodon dorsalis*, N America, "… well adapted to cedar glades since it's larval stages do not require an aquatic habitat for development ..." (5); **Anura:** *Acris crepitans*, N America, "... generally associated with temporary pools or streams within glades ..." (5); *Pseudacris triseriata*, N America, "... generally associated with temporary pools or streams within glades ..." (5); *Hyla versicolor*, N America, "... generally associated with temporary pools or streams within glades ..." (5); *Silurana tropicalis*, W Africa, "single adults in rock pools ..." (6); *Bufo maculatus*, W Africa, "... spawns in rock pools ..." (6); *Hemisus marmoratus*, W Africa, "... larvae develop in rock pools ..." (6); *Ptychadena maccarthyensis*, W Africa, "... spawns in rock pools ..." (6); *Hoplobatrachus occipitalis*, W Africa, "... adults present in rock pools; spawning and development of larvae in rock pools ..." (6); *Leptopelis viridis*, W Africa, "... development of larvae in savanna rock pools ..." (6).

Reptilia, Chelonia: *Homopus boulengeri*, W Africa, "... shelters on rocky outcrops ..." (1); *Pelomedusa subrufa olivacea*, W Africa, "... young *P. subrufa olivacea*, but no adults were found in small rock-pools on an isolated inselberg ..." (7); **Squamata:** *Platynotus semitaeniatus*, E South America, "occur only in the rockpiles of the Caatinga ..." (4); *Telescopus beetzii*, W Africa, "... lives on rock outcrops ..." (1); *Aspidelaps lubricus*, W Africa, "... is fond of rocky outcrops..." (1); *Mabuya quinquetaeniata*, W Africa, "... active, rock-living-species; runs around on exposed granite domes and other hard rock faces ..." (1); *M. lacertiformis*, W Africa, "... mainly found on hard rock outcrops ..." (1); *M. laevis*, W Africa, "... forage (...) in large rock cracks of granite outcrops ..." (1); *M. striata*, W Africa, "... also live on rock outcrops ..." (1); *M. sulcata*, W Africa, "... these rock-living skinks can be seen chasing over rock outcrops ..." (1); *M. variegata*, W Africa, "... active during the day on rocky outcrops ..." (1); *Lacerta australis*, W Africa, "... on rugged sandstone outcrops ..." (1); *Cordylosaurus subtesselatus*, W Africa, "... forages among succulent vegetation on small rock outcrops ..." (1); *Gerrhosaurus major*, W Africa, "... lives in cracks in small, well-vegetated rock outcrops ..." (1); *G. validus*, W Africa, "... rock-living (...), prefer the upper slopes of large granite koppies ..." (1); *Cordylus cataphractus*, W Africa, "... lives in large cracks on low rock outcrops ..." (1); *C. polyzonus*, W Africa, "... living in sun-split rocks of small rock outcrops ..." (1); *C. rhodesianus*, W Africa, "... lives (...) in rock cracks on rocky outcrops ..." (1); *C. vittifer*, W Africa, "... lives in cracks in small rock outcrops ..." (1); *C. warreni*, W Africa, "... Montane, well-wooded rocky outcrops ..." (1), *Platysaurus imperator*, W Africa, "... lives on massive boulders on top of gneiss hills ..." (1); *P. intermedius*, W Africa, "... lives (...) on smooth outcrops of granite, gneiss or sandstone ..." (1); *P. relictus*, W Africa, "... lives on sandstone outcrops ..." (1); *Pseudocordylus melanotus*, W Africa, "... Rock outcrops on mountain plateaus and rolling grassland ..." (1); *Agama atra*, W Africa, "... usually lives on rock outcrops ..." (1); *A. kirkii*, W Africa, "... lives on granite and paragneiss outcrops ..." (1); *A. planiceps*, W Africa, "... They live on rock outcrops ..." (1); *Ctenophorus ornatus*, SW Australia, "... restricted to rock outcrops ..." (9); *C. yinnietharra*, SW Australia, "... restricted to rock outcrops ..." (9); *C. rufescens*, SW Australia, "... restricted to rock outcrops ..." (9); *C. caudicinctus*, SW Australia, "... found more widely on the other rock forms (...), as well as granite outcrops ..." (9); *C. reticulatus*, SW Australia, "... found more widely on the other rock forms (...), as well as granite outcrops ..." (9); *Gehyra montium*, W Australia, "... apparently restricted to granite outcrops ..." (9); *Afroedura africana*, W Africa, "... on the shaded, overhanging surfaces of large granite boulders ..." (1); *A. amatolica*, W Africa, "... on a granite outcrop ..." (1); *A. transvaalica*, W Africa, "... beneath rock flakes in granite and sandstone outcrops ..." (1); *Homopholis wahlbergii*, W Africa, "... Rock fissures, particularly on overgrown koppies ..." (1); *Lygodactylus methueni*, W Africa, "... forage on (...) rock outcrops ..." (1); *Pachydactylus bicolor*, W Africa, "... rock-living, favouring thin cracks in small, shattered rock outcrops ..." (1); *P. bibronii*, W Africa, "... live on rock outcrops ..." (1); *P. laevigatus*, W Africa, "... restricted to rock outcrops ..." (1); *P. oculatus*, W Africa, "... is found only on rock outcrops..." (1); *P. namaquensis*, W Africa, "... live in rock cracks, usually on large outcrops ..." (1); *P. tigrinus*, W Africa, "... rock-living, these geckos inhabit narrow crevices in granite and sandstone outcrops ..." (1); *P. tsodiloensis*, W Africa, "... inhabit quartzite and dolomitic limestone outcrops ..." (1); *Phelsuma ocellata*, W Africa, "... running and jumping between boulders on rocky hillsides and outcrops ...." (1); *Rhoptropus biporosus*, W Africa, "... active on low rock outcrops and boulders, preferring flat surfaces..." (1).

Aves, Caprimulgiformes: *Caprimulgus tristigma*, W Africa, "... Habitat: Wooded and bushy rocky hills, kopjes, outcrops ..." (3).

Mammalia, Marsupialia: *Pseudocheirus dahli*, N & NW Australia, "... found in rock outcrops in savannas ..." (4); *Sminthopsis longicaudata*, W Australia, "... found in rocky habitats, including but certainly not exclusive to granite outcrops ..." (9); *Peradorcas concinna*, N Australia, "... occur in rocky outcrops ..." (4); *Petrogale* spp. (7), Australia, "... inhabit rocky ridges and boulder piles ..." (4); **Chiroptera:** *Molossops mattogrossensis*, E South America, "occur only in the rockpiles of the Caatinga ..." (4); *Chiroptera* spp., W Africa, "... roost in caves and under exfoliating rock on rock outcrops ..." (8); **Hyracoidea:** *Heterohyrax brucei*,

E Africa, "... highly specialized for the rocky kopje habitat ..." (4); *Procavia johnstoni*, Africa, Sinai, Palestine, "... occupies kopjes and other rocky outcrops ..." (4); *Kerodon rupestris*, E South America, "... the only endemic mammal occuring in the Caatinga, (...) inhabits boulder piles and rocky outcrops ..." (4); *Chinchilla* spp., W South America, "... live in and among rock crevices and rocky outcrops ..." (4); *Ctenodactylidae* spp., N Africa, "... inhabit rock outcrops ..." (4); *Zyzomys argurus*, W Australia, "... always are associated with rocky outcrops, particularly sandstones ..." (9); *Z. woodwardi*, W Australia, "... always are associated with rocky outcrops, particularly sandstones ..." (9); **Artiodactyla:** *Oreotragus oreotragus*, W Africa, "... live on rock outcrops ..." (2).

References

Branch B (1994) Field guide to snakes and other reptiles of southern Africa. Struik, Cape Town

Broadly DG (1978) A revision of the genus *Platysaurus* A. Smith (Sauria: Cordylidae). Occas Pap Natl Mus Monum Rhod Ser B 6:129–185

Büdel J (1978) Das Inselberg-Rumpfflächenrelief der heutigen Tropen und das Schicksal seiner fossilen Altformen in anderen Klimazonen. Z Geomorphol Suppl 31:79–110

Conant R, Collins JT (1991) A field guide to amphibians and reptiles: eastern and central North America. Houghton Mifflin, Boston

Haltenorth T, Diller H (1977) Säugetiere Afrikas und Madagaskars. BLV, München

Maclean GL (1993) Robert's birds of Southern Africa. John Voelcker Bird Book Fund, Cape Town

Mares MA (1997) The geobiological interface: granitic outcrops as a selective force in mammalian evolution. J R Soc West Aust 80:131–140

Mares MA, Lacher TE Jr (1987) Ecological, morphological and behavioral convergence in rock-dwelling mammals. Curr Mammal 1:307–348

Nowak RM (1991) Walker's mammals of the world, vol 1. Johns Hopkins University Press, Baltimore

Quarterman E, Burbanck MP, Shure DJ (1993) Rock outcrop communities: Limestone, sandstone, and granite. In: Martin WH, Boyce SG, Echternacht AC (eds) Biodiversity of the Southeastern United States. Upland Communities. Wiley, New York, pp 35–86

Rödel M-O (1996) Amphibien der westafrikanischen Savanne. Edition Chimaira, Frankfurt

Rödel M-O (1997) Lebensräume und Entwicklung junger Starrbrust-Pelomedusen *Pelomedusa subrufa olivacea* (Schweigger, 1812) im Comoé-Nationalpark, Elfenbeinküste (Testudines: Pelomedusinae). Herpetozoa 10:23–33

Skinner JD, Smithers RHN (1990) The mammals of the Southern African subregion. University of Pretoria, Pretoria

Vitt LJ (1981) Lizard reproduction: habitat specificity and constraints on relative clutch mass. Am Nat 117:506–514

Vitt LJ (1993) Ecology of isolated open-formation *Tropidurus* (Reptilia, Tropiduridae) in Amazonian lowland rain forest. Can J Zool 71:2370–2390

Vitt LJ, Caldwell JP, Zani PA, Titus TA (1997) The role of habitat shift in the evolution of lizard morphology: evidence from tropical *Tropidurus*. Proc Natl Acad Sci USA 94:3828–3832

Willig MR, Jones JK Jr (1985) *Neoplatymops mattogrossensis*. Mammal Species 244:1–3

Withers PC (1979) Ecology of a small mammal community on a rocky outcrop in the Namib Desert. Madoqua Ser II 12:229–246

Withers PC, Edward DH (1997) Terrestrial fauna of granite outcrops in Western Australia. In: Withers PC, Hopper SD (eds) Granite outcrops symposium. J R Soc West Aust 50:159–166

14 Human Dimensions and Conservation

R. Seine

14.1 Introduction

With the human population rising steadily, virtually no part of the world's land surface remains unaffected by man's activities. If the diversity of both life and habitats on earth is to be maintained for future generations, it is of crucial importance to first understand the biological role of the habitats and human activities connected with them. The former has been dealt with in the foregoing chapters; this chapter provides a short outline of the human dimension and some proposals for the conservation of inselberg vegetation.

A thorough description of the ethnological importance of inselbergs is far beyond the scope of the biologist. The following report will, therefore, necessarily be fragmentary in this respect. On the other hand, field observations reported here may prove useful for future investigators into this truly fascinating field.

14.2 Cultural Aspects

The image of certain inselbergs is commonplace today: almost every tourist knows the Sugar Loaf of Rio de Janeiro or Ayers Rock in Australia. Like mountains in general, inselbergs have probably always been attractive and interesting places for man. This assumption is also supported by the fact that many inselbergs bear a name. The names are often descriptive, such as Sugar Loaf, La Tortuga (the tortoise) or Domboshava (brown rock).

We know that inselbergs have played an important role in human life for a long time: they were used as shelters by the San people (bushmen) in southern Africa (Summers 1960) as early as 30 000 B.P. Ever since, inselbergs have been used by human beings for everyday life and for outstanding religious or profane purposes. The examples presented here represent a broad range of human interaction with inselbergs.

Ecological Studies, Vol. 146
S. Porembski and W. Barthlott (eds.) Inselbergs

The dry microclimate (Szarzynski, Chap. 3, this Vol.) so characteristic for inselbergs quickens the process of drying crops (Fig. 14.1) and cloth. This simple use of inselbergs has been observed all over Africa. It was probably widespread in early cultures.

In West and Central Africa, oval grinding holes of uniform shape have been recorded on many shield inselbergs. Grinding was effected by placing the grain on the inselberg and moving a flat stone over the grain. Over long periods of time, holes were shaped through abrasion. The widespread occurrence of these holes illustrates the importance of inselbergs as village grinding places. This traditional method of producing flour is still practiced in certain parts of Africa (Vande Weghe 1990).

In cleared landscapes, inselbergs often still bear some tree cover, which is then strenuously exploited for firewood. Herbaceous plants are harvested for their medicinal properties or as fodder. Inselbergs with gentle slopes are even used as pastures.

Fig. 14.1. Drying of corn (*Zea mays*) using inselberg microclimate. Near Niangbo, Côte d'Ivoire

The water catchment and storage capacity of inselbergs has been of importance to nomadic cultures. Australian Aborigines used rock pools and rock ponds as cisternae (locally called gnamma) which they covered with slabs of stone to reduce evaporation and to keep the water clean. Some ponds were probably enlarged and excavated to contain more water. Early European settlement in Western Australia was likewise dependent on water storage on inselbergs (Laing and Hauck 1997). In the dry parts of Western Australia, the runoff water is still collected for cattle and, as in the town of Hyden, for public water supply. In Zimbabwe, ephemeral flush vegetation is used as watering place for cattle (Fig. 14.2). Within towns, water reservoirs are often built on top of the hills to convert the elevation into water pressure.

The natural weathering of inselbergs leads to the formation of more or less plane sheets of rock which have been used for construction (Rudd 1968; Bindon 1997). Today, inselbergs are quarried to supply building material (Fig. 14.3).

Hill caves were used as shelter and for housing during times of war in Nigeria by the Yoruba (Afolabi 1966). The elevated position clearly is useful as an outlook controlling the surrounding and offers strategic advantages. Australian Aborigines used inselbergs as landmarks on their journeys

Fig. 14.2. A ring of stones opens a well for cattle in ephemeral flush vegetation. Domboshava, Zimbabwe

Fig. 14.3. Inselberg used to quarry granite. Near Rusape, Zimbabwe

Fig. 14.4. Rock paintings in Nswatugi Cave, Rhodes Matopos National Park, Zimbabwe

(Ornduff 1987; Bindon 1997). In Uganda and the Sudan, inselbergs were important meeting points (Murray 1981). Rock paintings have been found on inselbergs in Australia, Asia, South America, and Africa. Their occurrence in Africa runs from Kenya to South Africa and Namibia. A stylistic gradient can be recognized from north to south (Summers 1960). A wealth of several hundreds of rock paintings is still preserved in Zimbabwe (Summers 1960). Among the most important of these are the San paintings in the Nswatugi Cave (Fig. 14.4). Namibian rock paintings are currently the focus of detailed studies (Pager 1995; Lenssen-Erz 1997). In Australia, Roberts et al. (1997) could date rock paintings to an age of 17 500 years BP using mud-wasp nests. In Venezuela, pictures have been carved into the rock rather than painted onto it (Lüttge 1997).

During the zenith of their power in the 13th to 15th century, the Zimbabwean Shona built a multitude of representative palaces (Fig. 14.5) and shrines on the top of inselbergs. The most important center of power was Great Zimbabwe, with as many as 100 000 inhabitants (Garlake 1982).

Fig. 14.5. Ruins of Naletale. Near Gweru, Zimbabwe

The famous Ndebele king Mzilikazi was buried sitting upright under a cliff on an inselberg overlooking his realm (Thomas 1996). This demonstration of power was repeated by Cecil John Rhodes (Fig. 14.6), the colonizer of Southern Africa, who asked to be buried on a higher hill within sight of the grave of the Ndebele king Lobengula (Thomas 1996). Subsequently, the hill was developed into a national cemetry of honor.

The practical use of inselbergs seems to have been the base for religious worshipping of hills and hill goddesses in the Yoruba culture (Afolabi 1966). In Zimbabwe, the palace of Great Zimbabwe remained a religious center well into the 19th century (Murray 1981). Inselbergs are also sacred, e.g., in the Côte d'Ivoire (montagne fétiche), with the East African Kikuyu and Shona, and the Venezuelan Piaroa (Mbiti 1974; Gröger 1995). Reitsma et al. (1992) reports that inselbergs in Gabon are believed to be the abode of ghosts. The Zimbabwean Mambo use caves for the rainmaker's ceremony (Seine 1996). In Australia, inselbergs have an important position in indigenous mythology (Bindon 1997).

A number of Indian shrines are associated with inselbergs. Hindu temples are built on the hills in Trichinopoly and Elura, the latter built between the 5th and 10th century. The Jaina worship at a shrine on Mt. Abu in Rajputan. Most impressive are 7th century temples in Mahabalipur,

Fig. 14.6. "View of the World", Rhodes Matopos National Park, Zimbabwe. C.J. Rhodes was entombed between the boulders on top of the inselberg according to his will

which have artistically been chiseled from granite outcrops (Glasenknapp 1928). In Lalibela (Ethiopia) a Christian church in the form of a cross has been shaped into an inselberg (Murray 1981).

14.3 Human Impact

Inselberg vegetation has been reported to be quite tolerant of human influence and remains stable under moderate use (Porembski and Barthlott 1992; Porembski et al. 1994; Gröger 1995; Fleischmann et al. 1996; Seine 1996; Seine et al. 1998). Inselbergs are therefore often the last natural habitats remaining in densely populated areas (Porembski and Barthlott 1992; Porembski et al. 1997) and deserve some conservation attention for that reason. In general, their vegetation is not yet under severe threat.

The vegetation may, however, be considerably damaged by human activities: burning of inselberg vegetation for mere amusement is practized in Côte d'Ivoire (S. Porembski pers. comm.) and Gabon (Reitsma 1992). Fires set in the surrounding vegetation increase the frequency of fires on inselbergs and thus change their vegetation (Gröger 1995). The influence of fire is especially critical, as a number of species on inselbergs seem to be extremely sensitive to fire. Logging of woody species for firewood, as recorded in Zimbabwe, stunts or destroys the woody vegetation of inselbergs. Overgrazing may also lead to changes in the vegetation (Seine 1996). Resulting erosion of soil has been observed in Zimbabwe (Fig. 14.7). Likewise, tourism may cause erosion if not channeled to avoid sensitive vegetation. The construction of cable cars (as on the Sugar Loaf, Rio de Janeiro) and increased rock climbing (as in the USA, S. Porembski, pers. comm.) may lead to a huge amount of tourists which, in turn, might degrade inselberg vegetation.

Small-scale collecting of natural products such as traditional medicine and dry matter for the production of brooms, etc. usually does not harm the vegetation (Seine 1996). Although quarrying ultimately destroys complete inselbergs, it is practized in few cases and therefore poses no serious problems. Drying of crops and installation of water tanks and antennae commonly inflict small-scale changes or destruction of vegetation but have no dramatic effect upon vegetation.

Near settlements the danger of exotic species being introduced is considerably higher than otherwise (Seine et al., Chap. 11, this Vol.). Ornamentals from gardens have often been found to invade inselberg vegetation in the vicinity.

Fig. 14.7. Erosion of soil after overgrazing. Near Domboshava, Zimbabwe

14.4 Genetic Resources

Inselberg vegetation worldwide harbors a wealth of species. Only few of these are currently used on a more than regional scale. Examples of economically relevant plants are *Furcraea foetida*, planted for fiber production, and horticulturally interesting species of *Bryophyllum*, *Melocactus*, and *Kalanchoe*.

In general, plants used in local alimentation or traditional medicine have some economic importance (Pearce and Moran 1994). Today, medicinal plants provide approximately 75% of drugs in traditional medicine (Principe 1991) and perhaps 25% in industrialized nations (Pearce and Moran 1994). A number of pharmaceutical companies have started screening programs, testing medical properties of plant species (Reid et al. 1993), which may lead to an increasing economic importance of relevant species. So far, little is known about the number of medicinal plants that grow on inselbergs.

A comparison of plant species reported for inselbergs in Zimbabwe (Seine 1996) and a list of traditional medical plants (Gelfand et al. 1985) shows that 80 out of 549 plants from inselbergs are used by the local population (see Appendix). A thorough test of these species for active components has not yet been performed.

Alimentation with wild species helps to lessen the risk of starvation in times of drought (Cunningham 1995). Some 30 species from inselbergs in Zimbabwe have edible parts (see Appendix) and contribute to a varied diet of the local population (Tredgold 1986). Among these is the poikilohydric *Myrothamnus flabellifolia*, whose leaves are used for tea.

Chemical components of lichens from inselbergs might gain future commercial importance, e.g., Peltulaceae and Lichinaceae have been reported to contain highly effective UV light-blocking substances (Garcia-Pichel and Castenholz 1993; Büdel et al., Chap. 5, this Vol.).

Inselbergs are centers of diversity for poikilohydric vascular plants (Seine et al. 1995). These species are able to endure complete desiccation without major damage to the plant's tissue. Several hours to a few days after rehydration they take up normal photosynthesis. Bartels et al. (1990) reported that dehydration stimulates biosynthesis of components believed to protect the cells from damage caused by desiccation. The crucial step for desiccation tolerance being dehydration, not rehydration (Bernacchia et al. 1996).

Poikilohydric vascular plants could have significant future relevance. If poikilohydrous habit could, by traditional breeding or modern biotechnology, be transferred to crops such as rice or wheat, this would be an important step towards reliable food production in the semiarid tropics (Gaff and Ellis 1974; Gaff and Latz 1978).

14.5 Proposals for Conservation and Sustainable Use

Without doubt, inselbergs deserve at least some protection because of their natural beauty and the fascination they exert upon man. Their contribution to the regional flora and their wealth of endemic species, often restricted to this ecosystem, are further good reasons for the conservation of their indigenous, natural vegetation.

As pointed out above, inselbergs are often the last refuge for species that have suffered extinction in the surrounding landscape. A group of inselbergs or even a large individual inselberg might offer just enough room for plant species to persist. The relatively low frequency of fire on inselbergs is another asset to vegetation, especially for trees and fire-sensitive species.

The ideal structure for conservation would be a group of inselbergs plus fringing forests and a stretch of land between the hills. It would comprise several populations of inselberg species and their natural surrounding, catching the offrunning water, and storing it for some time. Logging and burning should be stopped within the territory, while extraction of a limited amount of natural products will most probably have no negative consequences.

As inselbergs generally are of little agricultural interest, concepts to protect them might be realized without confronting local economic interests. In the contrary, local communities might profit from the conservation of inselbergs and their vegetation: income could be generated through sustainable use of genetic resources and labor connected with collecting and screening. Capacity building and technology transfer could be further benefits for local communities and national economies (Nader and Mateo 1998). Where appropriate, visiting tourists could also contribute to meet the costs of conservation.

The touristic potential of inselbergs is not to be neglected: they are prime attractions in Australia [Mt. Uluru (Ayers Rock)], Zimbabwe (Great Zimbabwe, Matopos), and Brazil (Sugar Loaf). In this context, it is important to note that, while the sites should be open to the public, local taboos must be heeded by visitors. Where a high number of visitors is expected, these should preferably be guided and walk on the bare slopes to avoid soil erosion (Fig. 14.8).

Awareness of the natural and cultural heritage among a country's residents is essential to motivate communities to conserve important sites. A dual charge system as operated in Zimbabwe is a helpful instrument to attract local residents: a reasonable and affordable admission fee is asked from residents in local currency, while tourists from abroad are charged a higher amount of money payable either in foreign currency or in local money. At least part of the admission fees should then be spent on conservation efforts within the area or, if possible, for further projects.

14.6 Appendix

The species listed below occur on inselbergs in Zimbabwe and are used either medicinally (m) or for alimentation (a). The information on traditional use are from Gelfand et al. (1985) and Tredgold (1986). Occurrence of species on inselbergs in Zimbabwe has been reported by Seine (1996).

Fig. 14.8. Tourists ascending Uluru (Ayers Rock), Australia, by the only route open to visitors

Pteridophyta

Adiantaceae: *Pellaea calomelanos* (m). **Selaginellaceae:** *Selaginella imbricata* (m), *S. dregei* (m).

Angiospermae

Acanthaceae: *Hypoestes forskaolii* (m). **Aloaceae:** *Aloe chabaudii* (m), *A. excelsa* (m). **Amaryllidaceae:** *Boophone disticha* (m). **Anacardiaceae:** *Lannea stuhlmannii* (a), *Rhus leptodictya* (m), *Sclerocarya birrea* (a, m). **Apiaceae:** *Heteromorpha trifoliata* (m), *Steganotaenia araliacea* (m). **Apocynaceae:** *Carissa edulis* (a, m), *Diplorhynchus condylecarpon* (m), *Holarrhena pubescens* (m). **Asclepiadaceae:** *Ceropegia distincta* (a), *C. stenantha* (m), *Sarcostemma viminale* (m). **Asparagaceae:** *Asparagus virgatus* (m). **Asteraceae:** *Acanthospermum australe* (m), *Aspilia pluriseta* (m), *Bidens biternata* (m), *Galinsoga parviflora* (a, introduced), *Schkuhria pinnata* (m), *Tagetes minuta* (m, introduced). **Bombacaceae:** *Adansonia digitata* (a, m). **Burseraceae:** *Commiphora marlothii* (a), *C. mollis* (a, m), *C. mossambicensis* (m). **Caesalpiniaceae:** *Bauhinia petersiana* (m), *Brachy-*

stegia spiciformis (m), *Julbernardia globiflora* (m), *Peltophorum africanum* (m), *Tylosema fassoglensis* (a, m). **Campanulaceae:** *Wahlenbergia undulata* (a). **Capparidaceae:** *Cleome monophylla* (a). **Celastraceae:** *Maytenus senegalensis* (m). **Chrysobalanaceae:** *Parinari curatellifolia* (a, m). **Clusiaceae:** *Garcinia buchananii* (m), *G. livingstonei* (a). **Colchicaceae:** *Gloriosa superba* (m). **Combretaceae:** *Combretum apiculatum* (m), *C. molle* (m), *C. zeyheri* (m), *Terminalia brachystemma* (m), *T. sericea* (m), *T. stenostachya* (m). **Commelinaceae:** *Commelina africana* (a). **Convolvulaceae:** *Evolvulus alsinoides* (m), *Ipomoea verbascoides* (m), *Merremia tridentata* (m). **Cyperaceae:** *Cyperus angolensis* (m), *C. esculentus* (a). **Dioscoreaceae:** *Dioscorea cochleari-apiculata* (m). **Ebenaceae:** *Diospyros lycioides* (a, m). **Eriospermaceae:** *Eriospermum abyssinicum* (m). **Euphorbiaceae:** *Bridelia mollis* (a), *Croton gratissimus* (m), *Euphorbia ingens* (m), *E. matabelensis* (m), *E. schinzii* (m), *Pseudolachnostylis maprouneifolia* (m), *Uapaca kirkiana* (a). **Fabaceae:** *Erythrina abyssinica* (m), *Indigofera astragalina* (m), *Mundulea sericea* (m), *Pterocarpus angolensis* (m), *Zornia glochidiata* (m). **Lamiaceae:** *Hemizygia bracteosa* (m), *Leucas martinicensis* (m), *Pycnostachys urticifolia* (m). **Liliaceae:** *Dipcadi viride* (a). **Loganiaceae:** *Buddleja salviifolia* (a). **Malvaceae:** *Azanza garckeana* (a, m). **Melianthaceae:** *Bersama abyssinica* (m). **Mimosaceae:** *Elephantorrhiza burkei* (m). **Moraceae:** *Ficus glumosa* (m), *F. thonningii* (a, m). **Myrothamnaceae:** *Myrothamnus flabellifolia* (a, m). **Ochnaceae:** *Ochna pulchra* (m). **Olacaceae:** *Ximenia caffra* (m). **Orchidaceae:** *Ansellia africana* (m). **Oxalidaceae:** *Biophytum petersianum* (m). **Pedaliaceae:** *Ceratotheca triloba* (a, m). **Poaceae:** *Cynodon dactylon* (m), *Hyparrhenia filipendula* (m), *Pogonathria squarrosa* (m). **Portulacaceae:** *Anacampseros rhodesica* (m). **Proteaceae:** *Faurea saligna* (m). **Rhamnaceae:** *Ziziphus mucronata* (a, m). **Rubiaceae:** *Hymenodictyon floribundum* (m), *Oldenlandia herbacea* (m), *Vangueria infausta* (a, m). **Sapindaceae:** *Allophyllus alnifolius* (m). **Sapotaceae:** *Bequaertiodendron magalismontanum* (a, m), *Mimusops zeyheri* (a). **Scrophulariaceae:** *Craterostigma plantagineum* (m). **Simaroubaceae:** *Kirkia acuminata* (m). **Solanaceae:** *Solanum delagoense* (m), *S. incanum* (m). **Sterculiaceae:** *Waltheria indica* (m). **Tiliaceae:** *Grewia bicolor* (a), *G. flavescens* (a, m). **Velloziaceae:** *Xerophyta equisetoides* (m). **Verbenaceae:** *Clerodendrum glabrum* (m), *C. ternatum* (m), *Lantana camara* (a, m), *Vitex payos* (a, m). **Vitaceae:** *Ampelocissus africana* (a), *A. obtusata* (a, m), *Cyphostemma buchananii* (a, m), *Rhoicissus revoilii* (a, m).

Acknowledgements. The Deutsche Forschungsgemeinschaft (DFG, grant Ba 605/2) kindly funded research in Zimbabwe. The Studienstiftung des deutschen Volkes is thanked for a scholarship grant. Through fruitful discussions, U. Becker, N. Biedinger, and S. Porembski contributed to this chapter.

References

Afolabi OGJ (1966) Yoruba culture. University of Ife and University of London Press, London

Bartels D, Schneider K, Terstappen G, Piatkowski D, Salamini F (1990) Molecular cloning and abscisic acid-modulated genes which are induced during desiccation of the resurrection plant *Craterostigma plantagineum*. Planta 187:27–34

Bernacchia G, Salamini F, Bartels D (1996) Molecular characterization of the rehydration process in the resurrection plant *Craterostigma plantagineum*. Plant Physiol 111:1043–1050

Bindon P (1997) Aboriginal people and granite domes. In: Withers PC, Hopper SD (eds) Granite outcrops symposium. J R Soc West Aust 80:173–180

Cunningham AB (1995) Wild plant use and resource management. In: Bennun LA, Aman RA, Crafter SA (eds) Conservation of biodiversity in Africa. National Museum of Kenya, Nairobi

Fleischmann K, Porembski S, Biedinger N, Barthlott W (1996) Inselbergs in the sea: vegetation of granite outcrops on the islands of Mahé, Praslin and Silhouette (Seychelles). Bull Geobot Inst ETH 62:61–74

Gaff DF, Ellis RP (1974) Southern African grasses with foliage that revives after dehydration. Bothalia 11:305–308

Gaff DF, Latz PK (1978) The occurrence of resurrection plants in the Australian flora. Aust J Bot 26: 485-492

Garcia-Pichel F, Castenholz RW (1993) Occurrence of UV-absorbing, mycosporine-like compounds among cyanobacterial isolates and an estimate of their screening capacity. Appl Environ Microbiol 59:163–169

Garlake P (1982) Life at Great Zimbabwe. Mambo, Gweru

Gelfand M, Mavi S, Drummond RB, Ndemera B (1985) The traditional medical practitioner in Zimbabwe. Mambo, Gweru

Glasenknapp H von (1928) Heilige Stätten Indiens. Georg Müller, München

Gröger A (1995) Die Vegetation der Granitinselberge Südvenezuelas. PhD thesis, Bonn

Laing IAF, Hauck EJ (1997) Water harvesting from granite outcrops in Western Australia. In: Withers PC, Hopper SD (eds) Granite outcrop symposium. J R Soc West Aust 80: 173–180

Lenssen-Erz T (1997) Metaphers of intactness of environment in Namibian rock paintings. In: Faulstich P (ed) Rock art as visual ecology. IRAC Proceedings, vol. 1. American Rock Art Research Association, Tucson, pp 43–54

Lüttge U (1997) Physiological ecology of tropical plants. Springer, Berlin, Heidelberg, New York

Mbiti JS (1974) Afrikanische Religion und Weltanschauung. De Gruyter, Berlin

Murray J (1981) Weltatlas der alten Kulturen. Christian, München

Nader W, Mateo N (1998) Biodiversity – resource for new products, development and self-confidence. In: Barthlott W, Winiger M (eds) Biodiversity. A challenge for development research and policy. Springer, Berlin Heidelberg New York

Ornduff R (1987) Islands on islands. Harold L. Lyon Arboretum Lecture 15, University of Hawaii Press, Honolulu
Pager H (1995) The rock paintings of the upper Brandberg III, southern Gorges. Heinrich Barth-Institut, Köln
Pearce D, Moran D (1994) The economic value of biodiversity. Earthscan, London
Porembski S, Barthlott W (1992) Struktur und Diversität der Vegetation westafrikanischer Inselberge. Geobot Kollq 8:69–80
Porembski S, Barthlott W, Dörrstock S, Biedinger N (1994) Vegetation of rock outcrops in Guinea: granite inselbergs, sandstone table mountains and ferricretes - remarks on species numbers and endemism. Flora 189:315–326
Porembski S, Fischer E, Biedinger N (1997) Vegetation of inselbergs, quarzitic outcrops and ferricretes in Rwanda and eastern Zaire (Kivu) Bull Jard Bot Nat Belg 66:81–99
Principe P (1991) Monetizing the pharmacological benefits of plants. US Environmental Protection Agency, Washington DC
Reid WV, Laird SA, Meyer CA, Gamez R, Sittenfeld A, Janzen DH, Gollin MA, Juma C (1993) Biodiversity prospecting: using genetic resources for sustainable development. World Resources Institute, Washington DC
Reitsma JM, Louis AM, Floret JJ (1992) Flore et végétation des inselbergs et dalles rocheuses: première étude au Gabon. Bull Mus Nat Hist Nat, B, Adansonia 14:73–97
Roberts R, Walsh G, Murray A, Olley J, Jones R, Morwood M, Tuniz C, Lawson E, Macphall M, Bowdery D, Naumann I (1997) Luminescence dating of rock art and past environments using mud-wasp nests in northern Australia. Nature 387:696–699
Rudd S (1968) Lekkerwater ruins, Tsindi Hill, Theydon, Rhodesia. Rhod Sci Assoc Proc Trans 52:38–50
Seine R (1996) Vegetation von Inselbergen in Zimbabwe. Archiv naturwissenschaftlicher Dissertationen, vol 2. Martina Galunder, Wiehl
Seine R, Fischer E, Barthlott W (1995) Notes on the Scrophulariaceae of Zimbabwean inselbergs, with the description of *Lindernia syncerus* spec. nov. Feddes Repert 106:7–12
Seine R, Becker U, Porembski S, Follmann G, Barthlott W (1998) Vegetation of inselbergs in Zimbabwe. Edinburgh J Bot 55:267–293
Summers R (1960) Environment and culture in Southern Rhodesia. Proc Am Philos Soc 104:266–292
Thomas A (1996) Rhodes - the race for Africa. African Publishing Group, Borrowdale, Harare
Tredgold MH (1986) Food plants of Zimbabwe. Mambo, Gweru
Vande Weghe JP (1990) Akagera. WWF, Brussels

15 Biodiversity of Terrestrial Habitat Islands – the Inselberg Evidence

S. Porembski

15.1 Introduction

The detailed consequences of human impacts on the biodiversity and functioning of ecosystems are still not understood. There exists, however, much evidence today that the rapid conversion of landscapes leads to the fragmentation of populations and subsequently in many cases to the loss of genetic diversity. The implications of fragmentation processes for the stability of ecosystems are likewise a matter of debate (see contributions in Schulze and Mooney 1993; and Mooney et al. 1996). The study of oceanic islands and terrestrial habitat islands has already enriched our knowledge considerably concerning the understanding of ecological phenomena related to isolation and fragmentation.

The present volume deals with inselbergs as models for biodiversity research and aims to provide a synthesis of both abiotic and biotic data obtained from this ecosystem. We have deliberately chosen a broad approach to cover as much integral components of the ecosystem inselberg as possible. Since inselbergs have been largely neglected in the past by biologists, it has been essential to provide some basic information on their abiotic characteristics, habitat structure, and floristic inventory. In the following, a concise survey is given of our knowledge concerning the ecosystem structure (including considerations on adaptive traits) of inselbergs and the factors which influence their spatial and temporal dynamics in regard to species richness.

15.2 Inselbergs – an Old and Environmentally Harsh Ecosystem

The preoccupation with the genesis and geomorphology of inselbergs has a long tradition (Chap. 2, this Vol.; Bremer and Sander), which is certainly

Ecological Studies, Vol. 146
S. Porembski and W. Barthlott (eds.) Inselbergs

due to their remarkable physiognomy. Granitic or gneissic inselbergs are old and stable land forms which frequently are millions of years old. In certain regions they are considered to represent paleoforms that developed under different climatic conditions in the history. Despite their widespread occurrence on the crystalline continental shields, granitic and gneissic inselbergs show only relatively small differences in their parent rock. In particular, dome-shaped inselbergs mainly consist of vast and apparently uniform rocky slopes. However, of some importance for the structuring of inselberg habitats are microforms (e.g., shallow depressions) which are due to weathering and which are characterized by the accumulation of relatively nutrient-poor substrate. Attaining a maximum size of a few meters in width and a depth between 2 and 50 cm, they form important microsites for the establishment of vascular plants. Mostly the substrate consists of fine-textured soil material, flakes, or grus which originated from the parent rock. Otherwise fine-textured soils are largely absent from inselbergs, rendering them edaphically dry azonal sites.

There are detailed data available concerning the microclimatic conditions on inselbergs. Their environmental harshness has been demonstrated for inselbergs situated in different biomes (Szarzynski, Chap. 3, this Vol.; Phillips 1982). The size of individual inselbergs ranges from a few m^2 to several km^2. Already small-sized inselbergs show the typical microclimatic attributes rendering them dry azonal sites. Of overriding importance are microclimatic parameters affecting water availability and heat load. Since rainfall is largely lost to runoff due to the lack of water-storing soil, inselbergs are edaphically dry. As a consequence of strong solar radiation at the rock surface causing high temperatures (> 60 °C) and evaporation rates, inselbergs form "microenvironmental deserts" even when situated in a humid surrounding, thus emphasizing their island-like character. The physical attributes of the rock surface influence the radiation and heat balance of inselbergs. For example, the high amount of absorbed energy causes a strong heating of the rock and parallel warming of the air near the rock. Rising masses of warm air are frequently to be observed on sun-exposed rocky slopes. According to Oke (1987), bubbles of heated air rise at the hot inselberg surface with increasing velocity. The air deficit near the inselberg is compensated for by air volumina from the surroundings. The convective uplift of air could be important for the dispersal of diaspores. In particular on inselbergs situated in rainforest, this transport mechanism could play a crucial role for anemochorous plants.

15.3 The Importance of Cryptogams on Inselbergs

The seemingly bare rock surface of inselbergs is more or less completely covered by a crust consisting of cryptogams. Most prominent are cyanobacteria and lichens, which are responsible for the frequently dark coloration of inselbergs (Büdel at al., Chap. 5, this Vol.). Additionally, dematiaceous fungi (frequently belonging to the genus *Lichenothelia*) have been found on inselbergs in both tropical and temperate regions. Rock-inhabiting cyanobacteria, lichens, and fungi are poikilohydric and are effectively protected against UV damage by the possession of UV-absorbing substances (Büdel et al. 1997). In contrast to vascular plant species, on inselbergs most components of cryptogamic crusts (in particular cyanobacteria) have fairly large distributional areas with a considerable percentage being pantropics or even cosmopolitics. However, future molecular analyses could possibly reveal a more heterogenous picture of morphologically uniform and widespread cryptogams.

Large distributional areas are likewise typical for bryophytes on inselbergs. Frahm (Chap. 6, this Vol.) shows that cosmopolitic, pantropical, and tropical African species represent the majority of bryophytes on African inselbergs. This fact is the more surprising since bryophytes on tropical inselbergs are, apart from a few exceptions, sterile. Spore production was observed only in the hepatics (*Riccia*) and in the moss *Archidium ohioense*, which probably are epizoochorous. All other mosses on tropical inselbergs lack not only sporophytes but also vegetative propagation. The lack of any means of dispersal in nearly all inselberg bryophytes is in contrast to their wide ranges. It can be speculated whether they are trapped in sterile condition on inselbergs, which could be changed under different climatic conditions.

15.4 Adaptive Traits Towards Life on Inselbergs

The specific environmental constraints on inselbergs are reflected by certain ecophysiological and morphological/anatomical adaptations (Kluge and Brulfert, Chap. 9, Biedinger et al., Chap. 8, this Vol.). Most important stressors to plants on inselbergs are short supply of water, high irradiance and temperature, and low availability of nutrients. One of the most spectacular adaptations to scarcity of water is desiccation tolerance, i.e., poikilohydry (Gaff 1977; Lüttge 1997). Inselbergs form a center of diversity for desiccation-tolerant vascular plants (resurrection plants) which evolved independently in different families on inselbergs in various geographic regions. According to Gaff (1980), poikilohydry is mainly a protoplasmatic property. During de- and rehydration the plants undergo drastic structural, physiological, and biochemical alterations (Hartung et al. 1998). In the dehydrated state the concerned plants may survive rainless periods for months.

Very remarkable in morphology are poikilohydric monocotyledonous plants on inselbergs. These may attain a height of up to 4 m and possess woody-fibrous pseudostems with leaves crowded at the top. The pseudostems consist of persistent leaf bases and adventitious roots. The adventitious roots of tree-like Cyperaceae and Velloziaceae possess a one- to multi-layered velamen radicum (Porembski and Barthlott 1995), which is important for immediate water uptake. On inselbergs, many Velloziaceae and Cyperaceae form clonal colonies which have an estimated age of several hundred years. Clonal growth of mat-forming species provides substantial advantages (e.g., long-term persistence and rapid occupation of suitable sites). Pachycaulous and caudiciformous plants as well as succulents represent another group of plants on inselbergs which is well adapted to cope with edaphic and climatic dryness. In certain regions therophytes are of considerable importance among the vegetation of inselbergs. They dominate in particular in seasonally dry areas, where they survive the dry season as seeds in a dormant state stored in a seedbank.

In regard to pollination ecology, inselbergs have only few highly specific traits to offer. According to Wyatt (1997), a large percentage of granite outcrop species in the southeastern USA are outcrossers which are self-incompatible or have evolved self-fertilization. Wyatt (1981) could also demonstrate the rare case of ant pollination for *Diamorpha smallii*, a tiny Crassulaceae with flowers nearly lying on the ground. Another unusual case are several species of the genus *Anthospermum* (on inselbergs in East Africa, Madagascar) which is unique among the Rubiaceae in being wind-pollinated.

Animals which are specialized for inhabiting inselbergs have developed unusual adaptations to cope with harsh environmental conditions (Mares and Seine, Chap. 13, this Vol.). For example, bats (family Molossidae) became specialized in both Brazil and tropical Africa for inhabiting cracks in exfoliating granite. They are characterized by dorsoventrally flattened head and body for entering narrow cracks. Moreover, these bats possess tubercular projections on their forearms which can be used to develop traction against the rock when wedged in the cracks (Willig and Jones 1985; Skinner and Smithers 1990; Nowak 1991). Medium-sized rodents and rodent-like mammals have developed many adaptations for life on inselbergs (Mares and Lacher 1987). Specializations include morphological (feet, teeth, tail, eyes, snout), ecological (e.g., similar diets, habitat selection) and behavioral traits. Rock-specialized lizards (family Tropiduridae) show similar adaptations (e.g., dorsoventral flattening of head and body, cryptic coloration). Inselbergs are frequented not only by animals which are particularly adapted to this ecosystem but are regularly visited by other animals that use them as part of their range for hunting and shelter.

15.5 Phytogeographical Aspects

Inselbergs occur over a vast geographic range and show great qualitative and quantitative differences in the composition of their vegetation (see the various contributions in Chap. 10 and Seine et al., Chap. 11, this Vol.). Families such as Asteraceae, Fabaceae, and Rubiaceae form ubiquitous elements on inselbergs throughout whereas other families show strong regional preferences (e.g., Bromeliaceae). Floristic similarity in regard to flowering plants between different inselbergs depends on distance and on the regional vegetation history. In general there is little floristic similarity between widely separated regions, i.e., only a few species are common to both African and South American inselbergs. This observation is in strong contrast to the cryptogams which frequently occur on inselbergs throughout the tropics. For certain regions (i.e., tropical Africa) habitat-specific tendencies in distribution patterns exist. For example, species in shallow depressions are usually widespread, whereas those occurring in rock pools show a more regional distribution.

Concerning the number of locally endemic plant species there are drastic regional contrasts. Particularly rich in endemics are inselbergs in the Seychelles, Australia, and Madagascar, thus reflecting the generally high degree of endemism there. A conclusive statement about the phytogeographical affinities of the vegetation of inselbergs is, however, complicated by the lack of information from large parts of Brazil and the Indian subcontinent.

15.6 Species Richness of Inselberg Plant Communities

Between both individual inselbergs and habitats may exist drastic differences in diversity (Porembski et al., Chap. 12, this Vol.). Plant species richness of inselbergs is strongly influenced by regional phytodiversity and the vegetation structure in the surroundings. For example, in regard to species richness, inselbergs in tropical Africa can be classified as either relatively poor in species and endemics (e.g., Côte d'Ivoire) or as comparatively rich (e.g., Zimbabwe). These contrasting diversity patterns in the vegetation of inselbergs are probably a consequence of several factors: historical events (e.g., a migration of major vegetation belts), stochastic factors (i.e., short-term climatic disturbances), and the degree of isolation. In general, species richness of inselbergs seems to be promoted by the following circumstances:

- presence of a large number of inselbergs or ecologically similar sites (rock outcrops in general, white sand savannas);
- a certain degree of short-term environmental fluctuation which prevents competitive equilibrium;
- long-term environmental stability provides opportunities for differentiations between isolated populations.

The isolated habitats on inselbergs show large differences in both alpha and beta diversity. Interestingly, on West African inselbergs the ephemeral flush community, which is characterized by its small extent and large percentage of tiny annuals, is the most species-rich community. The narrow ephemeral flush belts which occur around larger carpet-like mats form an ecotone between open rock and the mat community. Ecotones show a sudden change in ecological structure, and are frequently relatively species-rich transitional zones (edge effect). The ephemeral flush community is characterized by stochastic colonization and establishment, which promotes a higher diversity. In contrast, the species-poor mats have reached a state of equilibrium in which only a few species can coexist.

Species rich habitats on inselbergs are characterized by nonequilibrium plant communities. Stochastic local extinction and colonization events occur frequently, as could be shown by regular monitoring of permanent plots (Porembski et al., Chap. 12, this Vol.). An important precondition for the long-term maintenance of species-rich nonequilibrium communities on inselbergs is the presence of similar habitats in the vicinity. In this way, local populations are linked by dispersal and persist despite local extinctions in metapopulations. Nonequilibrium communities on inselbergs harbor a large proportion of species which occasionally undergo local extinctions but are good dispersers able to recolonize lost sites rapidly. This implies relatively high temporal turnover rates in certain inselberg habitats.

A conspicuous change in the diversity of inselberg plant communities is detectable along environmental gradients in both tropical and temperate regions. A transect along a humid-arid precipitation gradient revealed that the species richness on inselbergs is relatively low in wet zones but increases towards seasonally dry areas (Porembski et al. 1995). This pattern of diversity with a maximum in seasonally dry zones could be due to stochastic environmental disturbances (e.g., unpredictable droughts). The latter prevent a state of equilibrium from being attained under more arid conditions and allow for the maintenance of species-rich non-equilibrium communities. Moreover, the number of favorable azonal sites suitable for colonization by inselberg species likewise increases along a gradient from humid towards seasonally dry zones. Thus, the higher number of localities available for the establishment of inselberg species may also contribute to the higher diversity of inselbergs in drier regions.

References

Büdel B, Karsten U, Garcia Pichel F (1997) Ultraviolet-absorbing scytonemin and mycosporine-like amino acid derivatives in exposed, rock-inhabiting cyanobacterial lichens. Oecologia 112: 165–172

Gaff DF (1977) Desiccation tolerant vascular plants of southern Africa. Oecologia 31: 95–109

Gaff DF (1980) Protoplasmic tolerance of extreme water stress. In: Turner NC, Kramer PJ (eds) Adaptation of plants to water and high temperature stress. John Wiley, New York, pp 207–231

Hartung W, Schiller P, Dietz K-J (1998) Physiology of poikilohydric plants. Prog Bot 59: 299–322

Lüttge U (1997) Physiological ecology of tropical plants. Springer, Berlin Heidelberg New York

Mares MA, Lacher TE Jr (1987) Ecological, morphological and behavioral convergence in rock-dwelling mammals. Curr Mammal 1: 307–348

Mooney HA, Cushman JH, Medina E, Sala OE, Schulze E-D (eds) (1996) Functional roles of biodiversity. A Global Perspective. SCOPE 55. John Wiley, Chichester

Nowak RM (1991) Walker's mammals of the world, vol 1. Johns Hopkins Univ Press, Baltimore

Oke TR (1987) Boundary layer climates. Methuen, London

Phillips DL (1982) Life-forms of granite outcrop plants. Am Midl Nat 107: 206–208

Porembski S, Barthlott W (1995) On the occurrence of a velamen radicum in Cyperaceae and Velloziaceae. Nord J Bot 15: 625–629

Porembski S, Brown G, Barthlott W (1995) An inverted latitudinal gradient of plant diversity in shallow depressions on Ivorian inselbergs. Vegetatio 117: 151–163

Schulze E-D, Mooney HA (1993) Biodiversity and ecosystem function. Ecological Studies 99. Springer, Berlin Heidelberg New York

Skinner JD, Smithers RHN (1990) The mammals of the Southern African subregion. University of Pretoria, Pretoria

Willig MR, Jones JK Jr (1985) *Neoplatymops mattogrossensis*. Mammalian Species 244: 1–3

Wyatt R (1981) Ant pollination of the granite outcrop endemic *Diamorpha smallii* (Crassulaceae). Am J Bot 68: 1212–1217

Wyatt R (1997) Reproductive ecology of granite outcrop plants from the southeastern United States. In: Withers PC, Hopper SD (eds) Granite outcrops symposium. J R Soc West Aust 80: 123–129

Subject Index

abscisic acid 146
algae 69, 72–74, 219
adaptations 101, 110, 123, 130, 150, 159, 283, 300, 327, 359, 419, 509
  anatomical 99
  ecophysiological 143
  morphological 99
  photosynthetic 166
  reproductive 117, 301
annuals 130, 283, 360–361, 400, 420, 467
arkose 18
Australia 78, 97, 391, 436, 495
Ayers Rock 13, 18–20, 391, 494
azonal sites 201, 204

biodeterioration 76, 85
biodiversity 1, 286, 404, 451–453, 507
biogeography 232, 380, 400
biomembranes 158
Brandberg 245
Brazil 339, 442
Brazilian shield 309
bryophytes 91

caatinga 352, 382
campos de altitude 350, 359, 375
campos rupestres 380, 382, 384
carnivorous plants 61, 66, 110, 131, 160, 191, 269, 284, 397
chamaephytes 117, 299
cliffs 3, 409–410
climate
  evaporation 41
  evapotranspiration 145
  microclimate 37–47, 508
  temperature 41
climate change 179, 198, 293, 307, 336, 344, 380, 402, 443
clonal populations 193
coexistence 202
competition 97, 169, 252, 361, 417, 473
competitive equilibrium 455
competitive exclusion 463, 467, 476
competitive interactions 195
conservation 255, 259, 287, 391, 493, 501
convection 42–43
Corcovado 340
Côte d'Ivoire 38, 46, 75, 86, 92, 134, 179, 442, 453, 457
cryptogamic crusts 66, 182, 295, 353, 394, 509
cyanobacteria 51, 59, 69, 87, 182, 187, 190, 217, 219, 246, 263, 279, 295, 323
cytorrhysis 146–147

Dahomey Gap 205
degradation 256
desiccation tolerance 75, 123, 145, 149
deterministic processes 102
disjunctions 205, 233, 307, 309, 402, 456
dispersal 43, 109, 137, 230, 286, 301–302, 362, 402, 456, 469, 509
  anemochorous 102, 137, 230, 301, 362
  barochorous 230
  endozoochorous 102, 137, 362
  epizoochorous 102, 137
  long-distance 100–101, 136, 426, 444
  ornithochorous 230, 301
disturbance 138, 140, 202, 385, 400, 422, 447, 454, 458, 464–473, 476, 513
diversity 4, 97, 139, 169, 335, 356, 404, 442, 465
  alpha 274, 384, 404
  beta 274, 280, 376, 384, 404, 442, 455, 459–463, 474
  gamma 384, 404, 455
  genetic 456
  local 4, 453
  regional 4, 441, 453–455
dolerite ridges 245

double planation 8
drought 46

ecotone 65, 129, 190, 477
edge effect 512
endemics 102, 206, 247, 253–254, 267, 271, 279, 286, 304–307, 334–336, 347, 380, 383–384, 392, 400–403, 419–427, 440–445, 453–456, 511
environmental conditions 37
environmental fluctuations 5, 202
epiphytes 63, 65, 103, 110, 113, 194–195, 299, 325
exotics 447
extinction 203, 206, 307, 457–462, 467–468, 476–477

fauna 484
ferns 56– 57, 437
ferricretes 66, 178, 190
fire 119, 124, 193, 198, 228, 385, 499
flatrocks 411, 417
floristic affinities 96, 113, 307–309
floristic composition 103, 200
fog 242, 249–253
Fouta Djalon 205
fragmentation 1–2, 143, 204, 477
French Guiana 315
fungi 75–76, 86
  dematiaceous 69, 183, 446

generalists 375, 459
genetic divergence 403, 456
geology 343
geomorphology 507
geophytes 130, 299
germination 130, 196
gnamma 16, 58, 394, 400, 495
Gondwana 213, 287, 392, 400, 440, 444
gradient 471–473
grinding holes 186–187
Guayana region 134, 304–305
Guayana shield 291, 316
Guineo-Congolian elements 204, 440
Guyana 315

habitat types 50
  boulders 52
  clefts 56, 221
  crevices 55, 220, 296–297
  drainage channels 53, 82, 182–183, 218, 246, 263, 281, 295, 353
  ephemeral flush vegetation 61, 97, 182, 190, 201, 263, 395, 457
  monocotyledonous mats 62–63, 182, 192, 195, 247, 265–267, 297, 326, 357–358, 454
  rock pools 57, 72, 82–83, 97–98, 113, 123, 149, 182, 187, 204, 222, 246, 281, 297, 394, 440, 468, 475, 495
  rock surface 51, 182
  shallow soil-filled depressions 59, 182, 355–356
  wet flush vegetation 54, 86, 131, 219
Harmattan 42, 179
hemicryptophytes 118, 299
homoiochlorophyllous plants 123
human influence 499
hybridization 427

Idanre Hills 186
inbreeding 133, 398
indicator taxa 453
inselbergs 1
  age 13, 443
  archepelagoes 250–251
  bornhardts 8, 61, 177, 186
  crystalline 239
  definition 7
  koppjes 11, 177
  landscapes 7, 343
  massivs 245
  monadnock 8
  plateau 239
  shield 8, 177
  whaleback 8, 320
inselberg specialists 109–113, 117, 188, 190, 192, 196, 201, 204, 251, 459
inselgebirge 11
irradiation 44, 158
island 1, 5, 49, 251, 277, 412, 418, 423, 437, 443, 451
  biogeography 1
  hopping 307
  oceanic 118, 136, 452
isolation 284, 427, 444, 458, 461, 472, 512
isotope ratio 157

Jos Plateau 180, 186

lichens 182, 217, 230, 246, 263, 323, 353
  chlorophytic 53, 79, 84
  cyanobacterial 51, 53, 76, 79, 84, 86
life-forms 110, 117, 152, 196–197, 228, 299, 356–362, 421, 457–458
lithosols 25
liverworts 59, 188

Madagascar 161, 233, 263, 445
mammals 483
  rodent-like 485
mass effect 228
Mata Atlântica 347–350, 385, 454
microforms 23–24, 32
microorganisms 79
migration 232, 425
Minas Gerais 339
mineral content 28–29
minimum size 203, 452
monocots
  arborescent 119
montagne fétiche 498
mosses 54, 56, 59, 188
mycosporine-like amino acid compunds (MAAs) 87

Namaqualand 255
Namibia 233, 497
Namib desert 237
neophyte 195, 202, 207
nitrogen 85
nonequilibrium communities 464, 468, 476
nutrients 34, 76, 85, 160

opferkessel 53
oriçanga 16, 23, 27–31, 57
overheating 158

Pão de Açúcar 340, 363, 383
phenology 135
phenotypic plasticity 121, 169
photosynthesis 148, 161
  C_3/CAM intermediate plants 156
  C_3 photosynthesis 163–164, 169
  C_4 photosynthesis 152–153
  C_4 plants 163, 420
  carbon gain 84
  CO_2 fixation 152, 157
  crassulean acid metabolism (CAM) 127, 151–157, 163–164, 168, 420
  gas exchange 156
  light saturation 85
phytogeographic composition 101
phytogeography 334, 511
Piedmont 409, 422, 441, 446, 477
pioneers 62, 195, 315, 325–326, 458
poikilochlorophyllous plants 123–124
poikilohydry 62, 75, 101, 109–115, 123–125, 145–148, 158, 189, 192–197, 219–220, 226, 229, 246, 269, 297, 327, 359, 384, 420, 435, 501, 509
pollination 109, 133, 136, 230, 362, 510
  anemophily 134, 198, 230, 362
  ant-pollination 133
  chiropterophily 134, 198, 230
  entomophily 134, 230
  ornithophily 134, 198, 230
protozoa 160
pseudocarst 291
pseudostems 62–63, 118, 193, 197–198, 510

quarrying 496, 499

refugia 98, 285
rehydration 146–149
resurrection plants 122, 145, 229, 269, 396–397
restinga 352, 380, 382, 384
Rio de Janeiro 8, 339
rock
  age 392, 410
  color 49, 79, 263, 323
  engravings 245
  paintings 496–497
roots 130, 148, 197
Roraima 328

Sahara 183, 186
sclerophylly 129
scytonemin 87
serpentine 456
Serra da Mantiqueira 340
Serra do Itatiaìa 343
Serra do Mar 340
Serra dos Órgãos 340
Seychelles 75, 92, 233, 440, 446
Shea Rock 323
similarity 97, 200, 436, 459, 474
soil 59, 241, 357
  chemistry 355

depth 190, 418, 474
pH 30
types 25–27, 241
solution pits 411
source-sink dynamics 380
speciation 233, 272, 402, 424–427, 435, 444–445
species
diversity 93
pool 455, 460–463, 473
richness 114, 250–251, 284, 286, 401, 441, 451, 462, 511
turnover 188, 203, 376, 468–474, 477
Spitzkoppe 241
Sri Lanka 13
stability 139, 455, 476, 512
stochastic processes 102
stomata 119, 150
Stone Mountain 411
stress 144
temperature stress 158
succession 195, 296, 326, 356, 410, 418–419, 464
succulence 126–127, 150, 228, 300, 359, 384, 398, 420
succulents 186, 204, 228, 233, 247, 255, 269, 322
Sudano-Zambezian elements 101, 204
Suriname 315
systematic composition, inselberg vegetation 105–107, 199, 216, 268–271, 302–304, 328, 362–374, 421–423, 435

Taï National Park 38
temperature resistance 128
tepuis 3, 291
therophytes 58, 63, 110, 117, 188, 196, 204, 225, 299, 361, 421, 457
tintenstriche 52, 69
tourism 256, 391
Tumac Humac Mountains 291, 315, 326, 328

ultraviolet
absorbing substances 52
radiation 76, 87
sunscreens 87
Uluru 391
United States 409, 444

vegetation dynamics 465
vegetative propagation 62, 119, 131, 192
velamen radicum 63, 115, 120, 148, 510
Venezuela 295, 305, 436, 453, 497
Voltzberg 320, 322

water catchment 495
Wave Rock 391
weathering 15, 32–34, 86, 344, 411
weeds 5, 138, 184–185, 202, 264, 285, 335, 377, 380, 400, 422, 426, 445–447, 477

xanthophylls 159

Zimbabwe 83, 86, 92, 216, 436, 442, 453, 495, 502
zonation 245

Species Index

Acanthella sprucei 296, 301
Acarospora 72, 78, 82, 217
Actiniopteris 229
 radiata 185, 264
 semiflabellata 185
Adenium 228
 obesum 221
Aechmea 110
Aeollanthus 229
 densiflorus 221
 pubescens 60, 137
Aeschynomene lateritia 185
Afrotrilepis
 jaegeri 195
 pilosa XIII, XIV, XIX, 62, 66, 98, 101, 113, 121, 124, 139–140, 148, 192–194, 197–198, 203, 444, 454, 458, 463–465, 470–476
Agave virginica 420
Aloe 184, 197, 205, 225, 229, 269
 buettneri 184, 443
 cameronii 227
 capitata 159, 163
 chabaudii 227
 inyangensis 232
 lateritica 219
 pearsonii 255
 schweinfurthii 184, 443
Amphianthus 113, 424
 pusillus 419
Amsonia ludoviciana 415
Anacampseros rhodesica 220, 229
Ananas
 ananassoides 158
 comosus 138, 184–185, 195, 207, 286, 289
Andropogon virginicus 415, 422, 426
Anemia 109, 323, 335
Angraecum 128, 269
 sesquipedale 159, 165
 sororium 134, 162, 164
Antherotoma naudinii 227
Anthocercis viscosa 398
Anthocerotae 91
Anthospermum 134, 510
 whyteanum 221
Anthurium XVIII, 359
 bonplandii 300
Aphelia 398
 brizula XVII
Aponogeton monostachyus 222
Archidium ohioense 98, 100, 509
Arenaria
 glabra 444
 uniflora 133, 415, 456
Arthraerua leubnitziae 245
Asplenium 102, 186
 stuhlmannii 185, 194, 202
Avena fatua 400
Axonopus
 compressus 202
 ramosus 325
Ayensua 122

Baccharis 360
Barbacenia seubertiana 356
Bombax costatum 185
Borya XVII, 121, 396–397, 472
 sphaerocephala XIX
Brachymenium 99
 acuminatum 93
 exile 93, 98, 183, 281
Brachystelma 205
Breutelia affinis 396
Briza maxima 400
Bryatae 91
Bryum
 arachnoideum 93, 98–99, 183, 218, 223
 argenteum 99

Buellia 78, 217
Buiningia brevifolia XVIII
Bulbophyllum 161, 184, 269
Bulbostylis 302, 323
  coleotricha 185, 220
  leucostachya XIX, 296, 301
Burmannia
  dasyantha 308
  madagascariensis XIII, 191, 204
  sanariapoana 308

Calandrinia 127, 130, 398–399
  granulifera 133
Caloplaca 72, 78, 82–83, 217
Calothamnus tuberosus 398
Calothrix parietina 73
Calycopeplus ephedroides 404
Campylopus
  australis 396
  bicolor 396
  brevirameus 287
  tallulensis 417
Caralluma 221, 228
  prigonium 225
Catapyrenium lacinulatum 78
Catharanthus roseus 447
Centrolepis 398
Cephalocarpus 122
Ceropegia 269
Chaetomium 75
Chamaecrista 110, 302
  mimosoides 224
Chamaegigas intrepidus 123, 148–149, 241, 246
Cheilanthes 102, 221
Chlorophytum 136
Chroococcidiopsis 72–73, 85–86, 182, 295
Cinnamomum verum 285
Cipura rupicola 305
Cissus quadrangularis 184, 186, 219
Cladonia
  corallifera 78
  didyma 78
  leporina 417
Cladosporium 75
Clappertonia polyandra 190
Cleome elegantissima 241, 246
Clidemia 302
Clusia 127, 156, 325
Cnidoscolus 320, 335
  urens 325
Coleocephalocereus XI, XVIII, VI, 111, 134
  buxbaumianus 128
Coleochloa 148, 454
  abyssinica 195, 205
  setifera XV, 62, 101, 113, 121, 162, 226, 266, 444
Commiphora leptophloeos 309
Comolia nummularioides 305
Coniosporium 75
Cordia 441
Crassula 127, 398–399
  natans 400
  schlechteri 225
Craterostigma 123, 205
  plantagineum 223
Cyanotis arachnoidea 39, 46, 184, 195
  lanata XIV, 60, 127, 130, 137, 188–189, 202–203, 208, 221, 445
  nodiflorum 264
Cynanchum meyeri 252
Cyperus 110, 187, 197, 201, 220
  bulbosa 397
  granitophilus 153, 420
  submicrolepis 58
Cyrtocarpa velutiniflora 308
Cyrtopodium 320
  punctatum 297

Decagonocarpus oppositifolius 298
Dermatiscum 217
  thunbergii 72, 78, 83–84
Deuterocohnia 62
Diamorpha smallii XII, 133, 415, 473, 510
Dimelaena 72
  oreina 78, 83
Dissotis 111, 190, 197, 201, 226
  rotundifolia 208
Djaloniella ypsilostyla 206
Dolichos 190
Dopatrium
  junceum 222
  longidens XIV, 59, 187
  senegalense 200
Doryopteris 335
  collina 323
Dracaena reflexa XVI
Drosera 160, 191, 267
  indica XIII, 61, 227, 465

madagascariensis 267
Dyckia 111

Echinopsis hammerschmidii 64
Elephantulus 486
Encephalartos 102, 205, 441, 443
Encholirium XVIII, 63
Encyclia granitica 325
Eremastrella crystallifera 78
Eremothamnus marlothianus 253
Ernestia cordifolia 305–306
Eriocaulon 191, 222
Eryngium histrix 64
Erythroxylum 302
Esterhazya splendida 356
Eucalyptus
caesia 456
drummondii 403
marginata 403
Eugenia 302
Euphorbia 137, 198, 228
milii 264, 272
letestui XIII
unispina XIII, 186
virosa 250
Everardia 122
Exormotheca 223

Fernseea itatiaiae 354
Ficus abutilifolia 186
Fimbristylis dichotoma 185, 204, 281
Furcraea 320
foetida 138, 264, 286, 289, 324, 446, 500
gigantea XV

Garckea 99
Gelsemium sempervirens 417
Genlisea 131, 160, 190, 192, 205, 227
barthlottii 206
margaretae 267, 269
sanariapoana 301
stapfii 202
Glaziophyton mirabile 384
Gloeocapsa 73, 323
sabulosa 280
sanguinea 182, 295
Glossostigma XVII, 113, 395, 400
Grimmia laevigata 97, 395, 415

Habenaria procera 204
Hakea clavata 398
Haematococcus 72
Haematostaphis barteri 186
Hedwigia ciliata 415
Helicteres heptandra 309
Hemionitis rufa 325
Heppia lutosa 78, 82
Heterodermia 78
Hildegardia barteri 186, 195
Holarrhena floribunda 190
Hydrocotyle 398
Hymenodictyon floribundum 43, 137, 190, 195, 221
Hyophila involuta 93

Impatiens baroni 264
Indigofera 110, 197, 201, 220, 225
Irlbachia alata 134, 326
Isoetes XVII, 109, 113, 156, 187, 197, 222, 395, 400, 441, 443
melanospora 419–420
nigritiana 130, 132
tegetiformans XII, 415, 419–420, 424
Isotoma petraea 133, 456

Jensenobotrya lossowiana 251

Kalanchoe 167–169, 228, 269, 500
brasiliensis 359
campanulata 164
daigremontiana 158
lanceolata 225
wildii 233
Klattia 122
Klebsormidium 72
Kleinia longiflora 254

Lecanora 78
Lemna 58
Lepraria 78
Levenhookia 398
Lichenothelia 72
intermedia 75
Lindernia 59, 111, 113, 271
andringitrae 269
conferta 222–223
exilis 43, 130, 188, 196, 465, 474
monroi 222–223
parviflora 267
pulchella 225, 445

syncerus 441
yaundensis 188, 197, 441
Lophoschoenus hornei XVI, 120, 281–282
Ludwigia 58

Malaxis 441
Mandevilla 131, 302, 335, 363
caurensis 306
lancifolia 296
Manihot brachyloba 326
Marchantiatae 91
Marsilea polycarpa 200
Medusagyne oppositifolia 279, 285, 441
Melinis
minutiflora 355, 477
repens 477
Melocactus 111, 137, 320, 323, 359, 500
mazelianus 295
neryi 324, 334
smithii 334
Miconia 302
Microchloa indica 189, 224
Microdracoides squamosus XIX, 195, 197–198, 205–206, 444
Microloma penicillatum 252
Millotia 398
Mimosa 302
Molossops mattogrossensis 484
Moringa ovalifolia 247
Mougeotia 72
Murdannia semiteres 222
Myriophyllum XVII, 395, 400
Myrothamnus 123, 137
flabellifolia XV, 134, 149, 226, 247, 501
moschata XV, 134, 148, 266

Navia 297
Nematostylis anthophylla 164
Nemum spadiceum 191
Nepenthes pervillei XVI, 284
Neurotheca loeselioides 205, 335
Nivenia 122
Nostoc 73, 86
Notothylas javanica 191

Oedogonium 72
Ophioglossum 109, 191, 227
costatum 130, 189, 197
Opuntia 233
humifusa 420
Orectanthe 122
Oreonesion testui 202
Oreotragus 486
Orthophytum XI, XVIII, 62, 110

Pachypodium 272
densiflorum XV, 266, 273
Paepalanthus 122
Pandanus XVI
balfourii 281
multispicatus 281
seychellarum 281
Panicum
griffonii 191
laxum 202
maximum 355
tenellum 191
Parmelia 217
Parmotrema cristiferum 78
Pelargonium
cortusifolium 253
mirabile 252
Pellaea 109, 186
calomelanos 221
doniana 194
wrightiana 425
Peltula 39, 51, 217–218, 263, 323
bolanderi 78
boletiformis 82
clavata 78–79, 82
congregata 82, 182
cylindrica 82
euploca 72, 78, 82, 84
impressa 82
lingulata 54, 79, 82, 182–183, 187
obscurans 78, 82
patellata 78, 82
placodizans 72, 82, 86–87
santessonii 82
tortuosa 72, 76, 78, 82, 85, 182, 295
umbilicata 72, 78, 82, 84, 182
zahlbruckneri 78, 82
Petromus 486
Petromyscus 486
Phacelia
dubia var. georgiana 422
maculata 415
Pilosocereus glaucescens XVIII
Pinus taeda XII, 417, 422
Pitcairnia 62, 110, 297, 302
armata 296

bulbosa 306
feliciana 205
geyskesii XI, 309, 335
pruinosa 158
Platymops setiger 484
Pleurosorus rutifolius 397
Polygala 110, 227
adenophora 326
Polystachya 121
johnstonii 227
microbambusa XIX, 194
rosea 136
Polytrichum commune 415
Porphyrosiphon 86
Portulaca 302, 441
pygmaea 296, 305
rhodesica 224
sedifolia 295
smallii 153, 420, 422
Prepusa 360, 383
Prescottia plantaginea 355
Procavia 486
Pseudobombax croizatii 296, 300
Pseudolaelia vellozicola XVIII
Pyrenopsis 82
Pyxine 78

Quercus georgiana 417
Quinetia 398

Raphionacme 124
Rhacocarpus purpurascens 99, 218
Rhipsalis baccifera 228, 283, 444
ssp. horrida XV, 151, 272
Rhynchospora 110, 302
Riccia 93, 98–100, 188, 191, 223, 509
Rinodina 78, 217
Rotala 187
wildii 222

Sansevieria 197
Sarcostemma viminale XVI, 221, 283
Sauromys petrophilus 484
Schizothrix 52, 73, 86, 323
Schlumbergera 359, 383
Schwenkia americana 335
Scleria 267
Scolecobasidium 75
Scytonema 51–52, 73, 323, 353
crassum 72
ocellatum 73
Sedum pusillum 415
Selaginella XI, XII, 59, 64, 186, 222–223, 297, 302
dregei 224
njam-njamensis 195
sellowii 64
tenerrima 186
Senecio
phonolithicus 255
tomentosus 415
Sinningia 111, 134
Solenostemon graniticola 127
Spiculea ciliata 402
Spirodela polyrrhiza 188
Sporobolus
festivus 185
pectinellus 191
Stapelianthus decaryii XV
Starria zimbabweensis 72
Stereospermum acuminatissimum 43
Stigonema 51, 73, 323, 353
mamillosum 183
Streptanthus 456
Streptocarpus 123, 264
Stylidium 122, 398
Stylosanthes guianensis 325

Tabebuia 302
Talbotia 148
Talinum
mengesii 418, 456
paniculata 359
teretifolium 418, 426
Tephrosia 197, 220, 225
mossiensis 56
Thrasya granitica 328
Tibouchina 204, 363
Tillandsia 363
kurt-horstii XVIII
Tolypothrix byssoidea 73
Toninia bumamma 78
Tradescantia hirsuticaulis 420, 456
Triglochin 398
Trilepis 62, 335, 356
ihotzkiana 66, 357
kanukuensis 326
Tripogon 148
minimus 101, 189, 197
Tropidurus semitaeniatus 485
Trypethelium aeneum 78
Tylecodon schaeferanus 255

Usnea 72, 78, 83, 182, 217, 263
pulvinata 218
welwitschiana 218
Utricularia XIII, 54, 110–111, 136, 188, 190, 198, 219, 227, 263, 358
juncea 202, 205, 444
nelumbifolia 131
oliverana 295
stellaris 223
subulata 183, 267, 296, 465

Vanilla phalaenopsis 283
Vellozia XI, XVIII, 110, 444
candida XVIII
tubiflora 301, 335
Viguiera porteri 417
Virectaria 190
Vriesea 62, 127, 363

Worsleya rayneri XI, 383–384, 441

Xanthoparmelia 78, 82, 217
Xerophyta 110, 121, 148, 205, 226, 229, 444–445
dasylirioides XV, 161–163, 267
humilis 223
pinifolia 120, 266
scabrida 147–148
schnitzleinia 195, 205
splendens XV, XIX, 119
viscosa 247
Xyris 55, 110, 302
straminea 183, 191, 465

Zygogonium 72
Zygopetalum mackayi 355
Zygophyllum stapffii 245

Ecological Studies

Volumes published since 1994

Volume 106
Minimum Animal Populations (1994)
H. Remmert (Ed.)

Volume 107
The Role of Fire in Mediterranean-Type Ecosystems (1994)
J.M. Moreno and W.C. Oechel (Eds.)

Volume 108
Ecology and Biogeography of Mediterranean Ecosystems in Chile, California and Australia (1995)
M.T.K. Arroyo, P.H. Zedler, and M.D. Fox (Eds.)

Volume 109
Mediterranean-Type Ecosystems. The Function of Biodiversity (1995)
G.W. Davis and D.M. Richardson (Eds.)

Volume 110
Tropical Montane Cloud Forests (1995)
L.S. Hamilton, J.O. Juvik, and F.N. Scatena (Eds.)

Volume 111
Peatland Forestry. Ecology and Principles (1995)
E. Paavilainen and J. Päivänen

Volume 112
Tropical Forests: Management and Ecology (1995)
A.E. Lugo and C. Lowe (Eds.)

Volume 113
Arctic and Alpine Biodiversity. Patterns, Causes and Ecosystem Consequences (1995)
F.S. Chapin III and C. Körner (Eds.)

Volume 114
Crassulacean Acid Metabolism. Biochemistry, Ecophysiology and Evolution (1996)
K. Winter and J.A.C. Smith (Eds.)

Volume 115
Islands. Biological Diversity and Ecosystem Function (1995)
P.M. Vitousek, L.L. Loope, and H. Adsersen (Eds.)

Volume 116
High Latitude Rainforests and Associated Ecosystems of the West Coast of the Americas: Climate, Hydrology, Ecology and Conservation (1996)
R.G. Lawford,, P. Alaback, and E. Fuentes (Eds.)

Volume 117
Global Change and Mediterranean-Type Ecosystems (1995)
J. Moreno and W.C. Oechel (Eds.)

Volume 118
Impact of Air Pollutants on Southern Pine Forests (1996)
S. Fox and R.A. Mickler (Eds.)

Volume 119
Freshwater Ecosystems of Alaska. Ecological Syntheses (1997)
A.M. Milner and M.W. Oswood (Eds.)

Volume 120
Landscape Function and Disturbance in Arctic Tundra (1996)
J.F. Reynolds and J.D. Tenhunen (Eds.)

Volume 121
Biodiversity and Savanna Ecosystem Processes. A Global Perspective (1996)
O.T. Solbrig, E. Medina, and J.F. Silva (Eds.)

Volume 122
Biodiversity and Ecosystem Processes in Tropical Forests (1996)
G.H. Orians, R. Dirzo, and J.H. Cushman (Eds.)

Volume 123
Marine Benthic Vegetation. Recent Changes and the Effects of Eutrophication (1996)
W. Schramm and P.H. Nienhuis (Eds.)

Volume 124
Global Change and Arctic Terrestrial Ecosystems (1997)
W.C. Oechel et al. (Eds.)

Volume 125
Ecology and Conservation of Great Plains Vertebrates (1997)
F.L. Knopf and F.B. Samson (Eds.)

Volume 126
The Central Amazon Floodplain: Ecology of a Pulsing System (1997)
W.J. Junk (Ed.)

Volume 127
Forest Decline and Ozone: A Comparison of Controlled Chamber and Field Experiments (1997)
H. Sandermann, A.R. Wellburn, and R.L. Heath (Eds.)

Volume 128
The Productivity and Sustainability of Southern Forest Ecosystems in a Changing Environment (1998)
R.A. Mickler and S. Fox (Eds.)

Volume 129
Pelagic Nutrient Cycles: Herbivores as Sources and Sinks (1997)
T. Andersen

Volume 130
Vertical Food Web Interactions: Evolutionary Patterns and Driving Forces (1997)
K. Dettner, G. Bauer, and W. Völkl (Eds.)

Volume 131
The Structuring Role of Submerged Macrophytes in Lakes (1998)
E. Jeppesen et al. (Eds.)

Volume 132
Vegetation of the Tropical Pacific Islands (1998)
D. Mueller-Dombois and F.R. Fosberg

Volume 133
Aquatic Humic Substances: Ecology and Biogeochemistry (1998)
D.O. Hessen and L.J. Tranvik (Eds.)

Volume 134
Oxidant Air Pollution Impacts in the Montane Forests of Southern California (1999)
P.R. Miller and J.R. McBride (Eds.)

Volume 135
Predation in Vertebrate Communities: The Białowieża Primeval Forest as a Case Study (1998)
B. Jędrzejewska and W. Jędrzejewski

Volume 136
Landscape Disturbance and Biodiversity in Mediterranean-Type Ecosystems (1998)
P.W. Rundel, G. Montenegro, and F.M. Jaksic (Eds.)

Volume 137
Ecology of Mediterranean Evergreen Oak Forests (1999)
F. Rodà et al. (Eds.)

Volume 138
Fire, Climate Change and Carbon Cycling in the North American Boreal Forest (2000)
E.S. Kasischke and B. Stocks (Eds.)

Volume 139
Responses of Northern U.S. Forests to Environmental Change (2000)
R. Mickler, R.A. Birdsey, and J. Hom (Eds.)

Volume 140
Rainforest Ecosystems of East Kalimantan: El Niño, Drought, Fire and Human Impacts (2000)
E. Guhardja et al. (Eds.)

Volume 141
Activity Patterns in Small Mammals: An Ecological Approach (2000)
S. Halle and N.C. Stenseth (Eds.)

Volume 142
Carbon and Nitrogen Cycling in European Forest Ecosystems (2000)
E.-D. Schulze (Ed.)

Volume 143
Global Climate Change and Human Impacts on Forest Ecosystems: Postglacial Development, Present Situation and Future Trends in Central Europe (2001)
J. Puhe and B. Ulrich

Volume 144
Coastal Marine Ecosystems of Latin America (2001)
U. Seeliger and B. Kjerfve (Eds.)

Volume 145
Ecology and Evolution of the Freshwater Mussels Unionoida (2001)
G. Bauer and K. Wächtler (Eds.)

Volume 146
Inselbergs: Biotic Diversity of Isolated Rock Outcrops in Tropical and Temperate Regions (2000)
S. Porembski and W. Barthlott (Eds.)

The manufacturer's authorised representative in the EU is Springer Nature Customer Service Centre GmbH, Europaplatz 3, 69115 Heidelberg, Germany. If you have any concerns regarding our products, please contact ProductSafety@springernature.com

Printed and bound by CPI Group (UK) Ltd, Croydon, CR0 4YY
12/08/2026
02194532-0007